1st International Conference Science Physics and Education (ICSPE 2021)

Journal of Physics: Conference Series Volume 2165

Mataram City, Indonesia
10 – 11 September 2021

ISBN: 978-1-7138-5136-3
ISSN: 1742-6588

Printed from e-media with permission by:

Curran Associates, Inc.
57 Morehouse Lane
Red Hook, NY 12571

Some format issues inherent in the e-media version may also appear in this print version.

Institute of Physics
Dirac House, Temple Back
Bristol BS1 6BE UK

Phone: 44 1 17 929 7481
Fax: 44 1 17 920 0979

techtracking@iop.org

Additional copies of this publication are available from:

Curran Associates, Inc.
57 Morehouse Lane
Red Hook, NY 12571 USA
Phone: 845-758-0400
Fax: 845-758-2633
Email: curran@proceedings.com
Web: www.proceedings.com

TABLE OF CONTENTS

Preface

Assalamu'alaikum, Wr. Wb.

The International Conference Science Physics and Education (ICSPE) is a conference organized by Physics Education Program Study, Faculty of Teacher Training and Education, University of Mataram. The 1st ICSPE held by virtual conference from Lombok, which is one of the Beautiful Island in Indonesia, on 10-11 September 2021.

International Conference Science Physics and Education (ICSPE) has offered a platform for bringing together students, postdocs, academics, and industrial experts to exchange their ideas and contributing an integrative approach to research in theoretical and applied physics. The program of the conference consisted of oral invited lectures of the leading experts on selected topics. The international seminar with the theme " Innovation and Creation on Education and Physics Research in Covid-19 Era" is expected to provide great benefits for developing physics research in the future's challenging 4.0 era. It is aimed at promoting, developing, and disseminating interdisciplinary research from many different fields of physics.

Every submitted paper was reviewed using a single-blind-reviewed process. Note that their authors' articles presented in these Proceedings have been considerably modified after discussion during presentations or in the review process. Furthermore, let me express my appreciation and gratitude to the committee and the entire International Conference committee (ICSPE), who have prepared this conference. ICSPE 2021 was attended by 12 Keynote Speakers and 72 participants representing various institutions or higher education institutions.

Finally, I would like to thank all presenters and participants in the 1st ICSPE 2021 with the hope that it will provide great motivation for us, especially those who are always involved in theoretical and applied physics research, physics education research, and learning and the application of physics in our lives.

Wassalamu'alaikum, Wr. Wb.

Thank you,
The 1st[h] ICSPE 2021 Chairman
Assoc. Prof. Drs. Aris Doyan, M.Si., Ph.D.

Journal of Physics: Conference Series 2165 (2022) 011001 doi:10.1088/1742-6596/2165/1/011001

Brief of Event

The 1st International Conference Science Physics and Education (ICSPE) 2021 was held by virtual conference from Lombok, one of the most beautiful islands in Indonesia, on 10-11 September 2021. The conference was held virtually due to the COVID-19 pandemic. At that time, the Indonesian government issued a policy of restricting travel and gathering activities. Therefore, according to the committee's agreement, the center for hosting the conference from Mataram City, Lombok Island.

Conference held in plenary for two days, which each keynote speaker talk in 20 minutes, and for the parallel session, each presenter talk for 10 minutes. The conference held virtually using the Zoom meeting platform. The presenter attended the conference from their place in around South East Asia and East Asia. The total participants is 115, consisting of keynote speaker, presenter, and participant.

The first day of ICSPE 2021 took place starting at 8.00 am with an opening by the Vice-Rector of University of Mataram, followed by a photo session by the committee, who attended in person and with speakers, participants, and invitations who were present virtually. The opening of the ICSPE 2021 was still a vivacious event, with all the limited to meet.

(a)

(b)

(c)

Figure 1. a) Remarks by the Chairman of the committee, Assoc. Prof. Drs. Aris Doyan, M.Si., Ph.D. b) Opening by the Vice-Rector of University of Mataram, Agusdin, SE., MBA., DBA. c) The situation of the opening of the ICSPE 2021

The event continued with 6 Keynote Speakers presentations continued with a parallel session in the form of presentations by each presenter divided into 10 virtual meeting rooms.

The second day's conference began with presentations by 6 Keynote Speakers continued with a parallel session and closing by the Vice-Dean of Teacher Training and Education Faculty, University of Mataram, Dr. Gunawan, M.Pd. The 1st ICSPE 2021 was successfully held with 74 articles. However, 72 articles were accepted to be continued after the reviews to be published with a percentage of 97%. The selection and review process is carried out through an initial screening by the editor, including a similarity check and grammar check, followed by peer reviews by at least 2 reviewers for each article. Revisions by the author and final editing made published articles have good quality.

Overall, there are no technical difficulties in delivering the conference. However, there are fewer social aspects in the virtual conference. Commonly, out of the conference schedule (offline conference), the participant can talk to each other, e.g., dinner sessions. Because of the main schedule limit for discussion, most of the topics conference can discuss in the dinner session. But the virtual conference did not. Hopefully, all challenges in the pandemic period will end soon so that life can return to normal and the next conference can be carried out offline as usual.

Journal of Physics: Conference Series **2165** (2022) 011001 doi:10.1088/1742-6596/2165/1/011001

The 1st ICSPE 2021 Lombok, Indonesia

ORGANIZING COMMITTEE

Advisory Board

Prof. HongJoo Kim (Kyungpook National University, Republic of Korea)
Prof. Dr. Eng. Kuwat Triyana (Universitas Gadjah Mada, Indonesia)
Prof. Jakrapong Kaewkhao, Ph.D (Nakhon Pathom Rajabhat University, Thailand)
Prof. Dr. Drs. Edi Istiyono, M.Si. (Yogyakarta State University, Indonesia)
Assoc. Prof. Muhammad Mustafa Awang Kechik, Ph.D (Universiti Putra Malaysia, Malaysia)
Christine Pueblo Abo, Ph.D (University of San Carlos, Philippines)
Prof. Dr. Agus Suyatna, M.Si (Universitas Lampung, Indonesia)
Prof. Nur Islami, Ph.D (Universitas Riau, Indonesia)
Assoc. Prof. Dra. Susilawati, M.Si., Ph.D.(University of Mataram, Indonesia)
Assoc. Prof. Dr. Joni Rokhmat, M.Si.(University of Mataram, Indonesia)
Nazarudin, S.Si., M.Si., Ph.D. (University of Jambi, Indonesia)
Dr. Gunawan, M.Pd.(University of Mataram, Indonesia)
Prof. Aripin, M.Si., Ph.D. (University of Siliwangi, Indonesia)
Prof. Didik Santoso, M.Si., Ph.D. (University of Brawijaya, Indonesia)
Prof. Adi Susilo, M.Si., Ph.D. (University of Brawijaya, Indonesia)
Assoc. Prof. Drs. Aris Doyan, M.Si., Ph.D (University of Mataram, Indonesia)
Assoc. Prof. Dr. Edy Supriyanto, M.Si. (University of Jember, Indonesia)
Dr. Wan Yusmawati Wan Yussof (UNPM, Malaysia)
Assoc. Prof. Dr. Ida Kaniawati, M.Si. (Universitas Pendidikan Indonesia, Indonesia)
Assoc. Prof. Dr. Hamdi, M.Si. (Universitas Negeri Padang, Indonesia)
Assoc. Prof. Dr. Matheus Souisa, M.Si. (University of Patimura, Indonesia)
Assoc. Prof. Syuhendri, M.Pd., Ph.D. (University of Sriwijaya, Indonesia)
Prof. Dr. Baba Musta (Univesiti Malaysia Sabah, Malaysia)
Assoc. Prof. Drs. Maison, M.Si., Ph.D (University of Jambi, Indonesia)
Assoc. Prof. Syahrial Ayub, S.Pd., M.Si (University of Mataram, Indonesia)
Byon Nugraha Jayawiguna, M.Si. (Universitas Indonesia, Indonesia)

Chairman

Assoc. Prof. Aris Doyan, Ph.D. (University of Mataram, Indonesia)

Steering Committee

Prof. Dr. H. A. Wahab Jufri, M.Sc.(University of Mataram, Indonesia)
Sukardi, M.Pd. (University of Mataram, Indonesia)
Dr. Gito Hadiprayitno, S.Pd., M.Si. (University of Mataram, Indonesia)
Dr. Gunawan, M.Pd(University of Mataram, Indonesia)
Dr. Harjono, M.Pd. (University of Mataram, Indonesia)
Dr. M. Makhrus, M.Pd. (University of Mataram, Indonesia)

Financial Manager

Ni Nyoman Sri Putu Verawati, M.Pd. (**University of Mataram, Indonesia**)

Journal of Physics: Conference Series **2165** (2022) 011001 doi:10.1088/1742-6596/2165/1/011001

Committee

Wahyudi, M.Si (University of Mataram, Indonesia)
Ahmad Busyairi, S.Pd., M.Pd (University of Mataram, Indonesia)
Ni Nyoman Sri Putu Verawati, S.Pd., M.Pd (University of Mataram, Indonesia)
Sutrio, M.Si. (University of Mataram, Indonesia)
Hairunisyah Sahidu, M.Pd (University of Mataram, Indonesia)
Syahrial A, M.Si (University of Mataram, Indonesia)
Muhammad Taufik, M.Si (University of Mataram, Indonesia)
Muhammad Zuhdi, M.Si (University of Mataram, Indonesia)
I Wayan Gunada, M.Si (University of Mataram, Indonesia)
Fathoroni, M.Pd. (University of Mataram, Indonesia)
Hikmawati, S.Pd., M.Pd. (University of Mataram, Indonesia)
Jannathin Ardhuda, M.Sc. (University of Mataram, Indonesia)
Fuad Tamami, S.Pd. (University of Mataram, Indonesia)

IT Management

I Gusti Ngurah Yudi Handayana, M.Sc.
Syamsul Hakim, M.Pd.
Lalu Muliyadi, M.Pd.
Haris Munandar M.Pd.

Contact for Declaration
Assoc. Prof. Drs. Aris Doyan, Ph.D
Email: aris_doyan@unram.ac.id
University of Mataram, Indonesia

Journal of Physics: Conference Series **2165** (2022) 011002 doi:10.1088/1742-6596/2165/1/011002

Peer review declaration

The 1th International Conference Science Physics and Education (ICSPE 2021)

All papers published in this volume of Journal of Physics: Conference Series have been peer reviewed through processes administered by the Editors. Reviews were conducted by expert referees to the professional and scientific standards expected of a proceedings journal published by IOP Publishing.

- **Type of peer review:**

  Single-blind Review

- **Review Criteria:**

  All submitted papers received by the Conference Committee will be to determine whether they are properly prepared and meet the scope of conference. Paper that do not fit the meet the standards of the conference will be rejected before peer-review. Once a paper passes the initial checks, it will be continued to similarity check (turnitin) and assigned to at least two independent experts for peer-review by single-blind review. Peer review comments are sent to the author for revision or accepted/decline the article with reviewer's name is confidental. All the papers go through the peer-review process and receive at least two reviews by three main criteria (Technical Criteria, Quality Criteria, and Presentation Criteria); then the decision which will be one of the following:

  Accepted:

  The paper is accepted and ready to publish without any flawed

  Accept after Minor Revisions:

  The paper is in principle accepted after revision based on the reviewer's comments. Authors are given one week for minor revisions.

  Reconsider after Major Revisions:

  The acceptance of the paper would depend on the revisions. The author needs to provide a point by point response or provide a rebuttal if some of the reviewer's comments cannot be revised. Authors will be asked to resubmit the revised paper within a suitable time frame, and the revised version will be returned to the reviewer for further comments.

  Reject:

  The paper has serious flaws, and/or makes no original significant contribution. No offer of re-submission to the journal is provided.

Journal of Physics: Conference Series **2165** (2022) 011002 doi:10.1088/1742-6596/2165/1/011002

- **Conference submission management system:**
  Submission conference using the website that is https://icspe.unram.ac.id/registration/
- **Number of submissions received:** 74
- **Number of submissions sent for review:** 72
- **Number of submissions accepted:** 53
- **Acceptance Rate (Number of Submissions Accepted / Number of Submissions Received X 100):**72%
- **Average number of reviews per paper**: 2 reviews per paper
- **Total number of reviewers involved:** 24 reviewers
- **Any additional info on review process (eg Plagiarism check system):** All submissions were checked for plagiarism with Turnitin and Grammar checking by Grammarly
- **Contact person for queries (Full name, affiliation, institutional email address)**
  Assoc. Prof. Drs. Aris Doyan, M.Si., Ph.D
  University of Mataram
  aris_doyan@unram.ac.id

Development of Timor local red peanut with multi gamma irradiation technique for obtaining several superior mutant cultivars that tolerant of drought stress, disease pests, and high production

B Pasangka[1], Refli[2], A Wahid[3], I G Pasangka[4]

[1,3] Physics Department, Faculty of Sciences and Engineering Nusa Cendana University Indonesia
[2] Biology Department, Faculty of Sciences and Engineering Nusa Cendana University Indonesia
[4] Mathematics Department, Faculty of Sciences and Engineering Nusa Cendana University Indonesia

E-mail: b.pasangka@staf.undana.ac.id, refliundana@gmail.com, awundana@gmail.com, irvandi.p@staf.undana.ac.id

Abstract. The main problem studied in this research is the development of local Timor red peanuts using multigamma irradiation. This study aims to develop local Timo red peanut varieties through breeding using standard multi-gamma irradiation and careful selection to obtain superior mutant cultivars. Research methods include: irradiation, observation, sampling, selection, purification, comparative, and interpretation. The results obtained: The development of local Timor red peanuts varieties through breeding with multigamma irradiation techniques and careful selection resulted in superior mutant cultivars that could adapt to several abiotic and biotic conditions and significantly increased production. The use of multigamma irradiation in the development of local Timor red peanuts resulted in three varieties of selected high yielding mutant varieties with relatively similar production and significantly increased compared to the parent variety. The average total percentage increase in production for the three varieties of selected high yielding mutant varieties resulting from multigamma irradiation was 49.16% with an average total dry seed moisture content of 11.07%.

1. Introduction

Peanut (Arachis *hypogaea* L) is the second most important legumes after soybeans cultivated by farmer groups for the fulfillment of food needs, the food industry, people's food diversification, and nutritional fulfillment because peanuts have various properties such as: anti-cancer, anti-bacterial, antioxidant, improve function red blood cells, swelling, increased appetite, back pain, and so on [1].

Peanut cultivation in Indonesia is generally grown on dry land which is dominated by Podzolic Red Yellow and Latosol content with a soil slope of less than 8%, and can also be planted in Alluvial and *Regosol* type of rice fields [2]. The province of East Nusa Tenggara has great opportunities for the development and cultivation of peanuts because it is dominated by dry land with an available area of 18,396 ha and an average productivity of 1.12 tons/ha of peanuts [3,4,5]. The average productivity of peanuts according to the research results of Balitkabi [6] is 2.61 tons/ha and BPS NTT [3] is 1.12 t/ha

Journal of Physics: Conference Series **2165** (2022) 012001 doi:10.1088/1742-6596/2165/1/012001

much lower. The average production of peanuts in Indonesia from 2014 to 2018 has relatively decreased, namely 638,896 tons (2014), 605,449 tons (2015), 570,470 tons (2016), 495,447 tons (2017), and 512,198 tons (2018) (BPS, 2018). Especially in East Nusa Tenggara, peanut productivity is relatively low. The average yield of peanut productivity per hectare at the national level is around 1.29 tons/ha [2]. BPS research results show that the national average productivity of peanuts is 1.3 tons/ha, so that in that year Indonesia had to import 130,000 tons of peanuts to meet domestic needs [7]. The average productivity specifically for local red peanuts at the farmer level has only reached 0.7 tons/ha.

The low productivity of peanuts in each region is caused by various different factors, but in general, biotic and abiotic factors that affect growth, plant development, and peanut production include [2]: 1) poor drainage, 2) excess water at the beginning of growth in paddy fields in the rainy season and drought in the dry season in the late growth phase, 3) Lack of major nutrients such as N, P, K, and Ca, 4) Competition with weeds in the vegetative growth phase, weeding is rarely done or late, 5) Shallow tillage so that root growth and pod development is not optimal, 6) The seeds used have low growth ability (not superior), and 7) Disease attacks such as: bacterial wilt disease and fungal wilt, rust and leaf spot, Peanut Stripe Virus (*PStV*) stripe virus, and attacks by rats, whitefly, caterpillars, pod borer, and nematodes.

The main problem studied in this study is the improvement of local Timor red peanut seed cultivars using multi-gamma irradiation to obtain superior cultivars that are able to adapt to drought stress, pests and diseases, and high production. The advantage of this method is that superior varieties can be obtained in a relatively short time and several variations of varieties so as to facilitate careful selection to obtain superior local red bean seed cultivars [8]. The conventional method of crossing up to F7 or F9 with a long enough time to get superior cultivars.

This study aims to: 1) develop local Timorese red bean varieties through breeding using standard multi-gamma irradiation and careful selection to obtain superior mutant cultivars, 2) improve human resource capabilities through the application of nuclear technology in the field of food security..

2. Methods

2.1. Research Location and Time
Sampling of local Timor red peanuts is at the cultivation centers in Atambua, Kefa, Malaka, and Soe, Timor Island. Sample irradiation was carried out at the Mini Nuclear Laboratory of the Physics Study Program, Faculty of Science and Engineering, Nusa Cendana University. The planting sites are centered on the pilot lands of East Penfui Kupang NTT, Tinoring Tanah Toraja South Sulawesi, and Mamuju West Sulawesi.

2.2. Research Instruments
The main research instruments include: 1) Standard multi-gamma radiation sources: sample irradiation. 2) Radiation dose counter: measures the multi-gamma radiation dose used. 3) Water content analysis tool, 5) Tractor: land management. 6) Digital scales: weigh the mass of the beans. 7) Other supporting tools: crowbar, hoe, bucket, *paralon* pipe, pest sprayer, shovel, saw, hammer, sickle, etc.

2.3. Research Method
The main methods applied in the research are irradiation and careful selection. Other supporting methods include: observation/survey, sampling, selection, purification, comparison, and interpretation. The advantage of this method is that superior mutant cultivars are obtained in a relatively short time and several variations of superior mutant cultivars are obtained, making it easy to select as desired.

Journal of Physics: Conference Series **2165** (2022) 012001 doi:10.1088/1742-6596/2165/1/012001

2.4. *Research Design*

This research was carried out with the following procedures: 1) Conducted observations/surveys for the selection of local Timor red peanut seeds in Atambua, Kefa, Malaka, and Soe which were developed using multi-gamma irradiation and determined planting locations. At this stage, observations and measurements of several important characteristics of the parent plant are carried out such as plant height, resistance to pests and diseases, flowering age, number of pods per tree, harvest age, mass per 200 seeds, production level, water content, and so on, 2) Planting land by plowing using a tractor so that it is free from weeds and loose so that red peanuts grow well. Furthermore, beds are made with a width of 100 cm, a height of 30 cm, a distance between beds of 40 cm, a trench width for a drainage system of 25 cm, 3) A sample of selected local red peanut seeds, then irradiated with a multi gamma source at a dose of 3000 rads for 30 minutes, 4) Soak the cultivated land and allow it to dry again for two days (moist), 5) Local red peanut seeds that have been irradiated, are immediately planted in the prepared land with a spacing of 40 cm x 15 cm, a planting hole depth of 4 cm , and each hole was filled with 4 seeds, 6) Irrigation after planting was adjusted to the rainfall at the research site, 7) Observing the ability to grow and age of growth in both parent and treatment varieties. The number of seeds that did not grow in the selected sample was 100 seeds, both treatments were counted. Crop embroidery is done at the age of 7 days after planting, 8) Weeding the plants and fertilizing with NPK and Urea 2:1 so that the plants are free from weed disturbances and thrive, 9) Conduct a series of observations during plant growth, including: pest-disease resistance, adaptation to dry land, and plant selection. At the time of approaching harvest, the selection was carried out again, measuring plant height, number of pods per tree, and number of seeds per pod. Perform the 1st selection at the age of 30 days after planting, the 2nd selection when the pods begin to fill densely, before harvesting the M_n selection, and post-harvest the last selection, 10) Harvesting and drying. Harvesting is done after the pods are well ripe, which is indicated by the dense contents and the skin color of the pods is dull, brownish white and rough by pulling the pods assisted with a crowbar so that no pods are left in the soil. Count the number of pods for each tree in a randomly selected sample. After that, the pods were separated from the tree, dried the pods to dry well, 11) Post-harvest weighing the mass per 200 seeds, as well as water content analysis, 12) Comparing the physical and chemical properties of the superior mutant cultivar with the parent variety, 13 Do interpretation and draw conclusions.

The research design for the first stage (acquisition of mutant varieties) and the second stage (purification) are shown in Figure 1 and Figure 2. The focus of the research in the first phase was genetic engineering of local Timor red penaut cultivars using multi-gamma irradiation to obtain superior mutant cultivars. Other variables that were observed and measured in this study were adaptive to dry land, pest and disease resistance, growth percentage, water content, production range, average production, and percentage increase in production. The focus of the research in the second stage is purification and multi-location testing, as well as other tests which are a continuation of the first stage of research testing.

Journal of Physics: Conference Series **2165** (2022) 012001 doi:10.1088/1742-6596/2165/1/012001

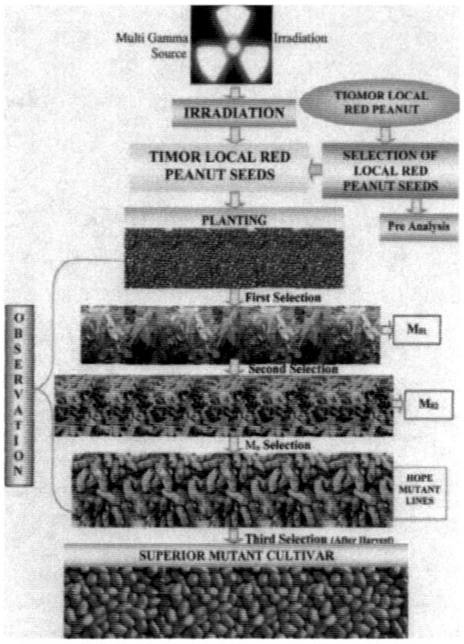

Figure 1. Research flow chart for obtaining superior mutants in the multigamma irradiation stage

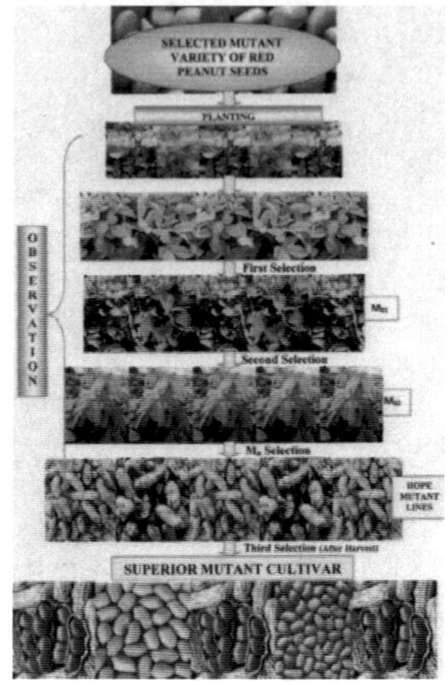

Figure 2. Research flow chart for obtaining superior cultivars in the refining stage

2.5. Statistics for Data Analysis

Some of the mathematical formulations needed are [9].

2.5.1. Percentage of growth is calculated by the equation [10, 11]:

$$PG = \frac{(T_{AS} - T_{SG})}{T_{AS}} \times 100\% \tag{1}$$

where: PG: percentage of growth (%), T_{AS}: total number of seeds planted, T_{SG}: number of seeds that did not grow.

2.5.2. The percentage increase in the average production is calculated by the formula:

$$I_{PAP} = \left(\frac{A_{PTS} - A_{PCS}}{A_{PTS}} \right) \times 100\% \tag{2}$$

with: I_{PAP}: percentage increase in average production, A_{PTS} : average production of mutant cultivars selected superior, and A_{PCS} : the average production of the control sample (parent).

2.5.3. The average production of each superior mutant cultivar in the three planting locations was calculated by the equation:

$$A_{PTS(n)} = \left(\frac{P_{L1} + P_{L2} + P_{L3}}{3} \right) \tag{3}$$

with: P_{L1}, P_{L2}, P_{L3}: production of superior and control mutant cultivars in three locations and n = 1…..3, the mean production of each selected mutant cultivar at three planting sites.

2.5.4. The average production in the control sample at the three planting sites is calculated by the equation [12]:

$$A_{PCS} = \left(\frac{P_{L1} + P_{L2} + P_{L3}}{3} \right) \tag{4}$$

2.5.5. average total production in the treatment sample is calculated by the equation:

$$T_{AP} = \left(\frac{A_{PTS(1)} + A_{PTS(2)} + A_{PTS(3)}}{3} \right) \quad (5)$$

Where $A_{PTS(n)}$ is the average production of each superior mutant cultivar at the three planting sites, and is T_{AP} the average total production in the treatment samples, n = 1, 2, 3.

2.5.6. The average total percentage increase in the production of selected superior mutant cultivars was calculated by the formulation:

$$I_{PAP} = \left(\frac{T_{AP} - A_{PCS}}{T_{AP}} \right) \times 100\% \quad (6)$$

Where I_{PAP} is the average total percentage increase in production of selected superior mutant cultivars.

3. Result and Discussion

3.1. Observation, Measurement, and Calculation Results

Multi-gamma irradiation on samples of local Timor red peanut seeds resulted in three variations of selected high yielding mutant varieties with different characteristics such as leaf color, plant height, number of pods per tree, seed size, and seed color, but the production rate was relatively the same. The results of observations of the initial growth of seeds in the treatment and control samples to calculate the percentage of growth, various physical and chemical characteristics observed, measured, and calculated, as well as estimates of production levels, are presented in Table 1, Table 2, and Table 3, respectively, as well as various photos of research results as follows.

Figure 3. Growth of Mutant Variety-1 (MV_1)

Figure 4. Growth of Mutant Variety-2 (MV_2)

Figure 5. Growth of Mutant Variety-3 (MV_3)

Figure 3, Figure 4, and Figure 5 show examples of the growth of three selected high-yielding mutant varieties (MV_1, MV_2, and MV_3) of Timor local red peanuts irradiated by multigamma at 28 days after planting (d a p) which have started flowering, while Figure 6, shows examples of parent varieties at the age of 28 days after planting.

Figure 6. Growth of parent variety

Figure 7. Pods of Mutant Variety-1

Figure 8. Pods of Mutant Variety- 2

Figure 9. Pods of Mutant Variety-3

Figure 10. Pods of Parent variety

Figure 11. Pods with 3 and 4 seeds of Mutant Variety-1

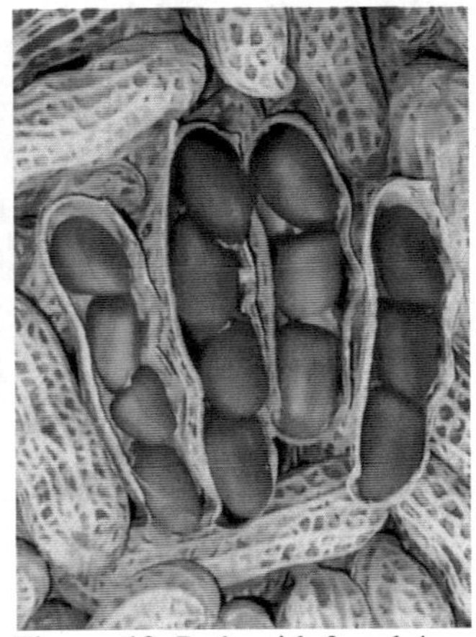

Figure 12. Pods with 3 and 4 seeds of Mutant Variety-2

Figure 7, Figure 8, and Figure 9 show examples of the number of pods per tree and the color of the pods after harvest at the age of 92 days after planting (d a p) on three varieties of selected high yielding mutant varieties resulting from multi-gamma irradiation, while Figure 10. Shows one example of parent variety pods after harvesting at 135 days after planting (d a p).

Figure 11, Figure 12, and Figure 13 show an example of pods with 3 and 4 seeds from three variations (MV_1, MV_2, and MV_3) selected high-yielding mutant varieties of Timor red peanuts produced by multigamma irradiation with relatively similar production, while Figure 14 variety of parent.

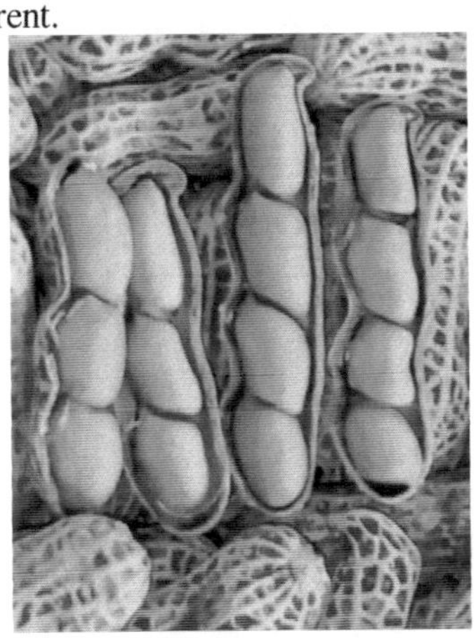

Figure 13. Pods with 3 and 4 seeds of Mutant Variety-3

Figure 14. Pods with 3and 4 seeds of Parent variety

Figure 15. Seeds of Mutant Veriety-1

Journal of Physics: Conference Series **2165** (2022) 012001 doi:10.1088/1742-6596/2165/1/012001

Figure 16. Seeds of Mutant Veriety-2

Figure 17. Seeds of Mutant Variety-3

Figure 18. Seeds of parent variety

Figure 15, Figure 16, and Figure 17 show an example of seeds from three variations (MV_1, MV_2, and MV_3) selected high yielding mutant varieties of Timor local red peanuts irradiated by multigamma with relatively similar production, while Figure 18 is the parent variety.

To calculate the percentage of growth, 5 groups were taken at random with the number of seeds observed being 100 seeds in each group. The treatment and control samples were taken in 5 groups with the treatment sample consisting of 10 variations of observations.

Table 1. Number of seeds that did not grow in control and treatment samples

No SG	Total number of seeds in each group	The number of seeds were not growth										
		Control Sample	Selected mutant varieties									
			1	2	3	4	5	6	7	8	9	10
I	100	17	3	4	2	3	5	2	1	3	2	3
II	100	14	1	2	4	1	3	1	3	2	4	2
III	100	13	4	2	3	5	2	3	5	4	3	5
IV	100	18	5	3	5	2	4	3	4	5	5	3
V	100	21	6	5	3	4	2	5	3	1	3	2
	Average = 100	16.60	3.80	2.80	3.40	3.00	3.20	2.80	3.20	3.00	3.40	3.00
.	**Total average of mutant**						3.16					
	Growth percentage (%)	83.40	96.20	97.20	96.60	97.00	96.80	97.20	96.80	97.00	96.60	97.00
	Total average of growth percentage	83.40					96.84					

SG: Sample Group

Table 2. Important physical and chemical characteristics that were observed, measured, and calculated in 3 variations of selected high yielding mutant varieties of Timor red peanuts produced by multi gamma irradiation.

No	Description	Control Sample	Treatment Samples or Mutants		
			MV_1	MV_2	MV_3
1	Average of time growth	11 d a p	6 d a p	6 d a p	6 d a p
2	Average of growth percentage	83.40 %	96.84 %	96.84%	96.84%
3	Harvest age range	39 to 48 d a p	26 to 38 d a p	28 to 40 d a p	27 to 39 d a p
4	Average age of flowering	43 d a p	35 d a p	37 d a p	36 d a p
5	Average total flowering time	43 d a p		36 d a p	
6	Leaf color	Dull green	Dark green	Yellowish green	Light green
7	Flower color	yellowish	Yellow	Yellow	Yellow
8	Plant height range	27 cm to 61 cm	46 cm to 83 cm	44 cm to 85 cm	43 cm to 84 cm
9	Average plant height	46 cm	65 cm	68 cm	67 cm
10	Harvest age range	(93 to 135) d a p	(62 to 92) d a p	(64to 95) d a p	(63 to 93) d a p
11	Average harvest age	123 d a p	83 d a p	84 d a p	83 d a p
12	Number range of pods per tree	(21 to 67) pods per tree	(49 to 123) pods per tree	(48 to 121) pods per tree	(46 to 120) pods per tree
13	Average number of pods per tree	45 pods per tree	88 pods per tree	86 pods per tree	85 pods per tree
14	Range of number of seeds per pod	1 to 4 seeds per pod	1to 5 seeds per pod	1to 5seeds per pod	1 to 5 seeds per pod
15	Average number of seeds per pod	2 (50%), 3 (35%) seeds per pod	3 (50%), 4(40%) seeds per pod	3 (50%), 4(40%) seeds per pod	3 (50%), 4(40%) seeds per pod

16	Skin color of dry pods	Dull white	Bright white	Yellowish white	Light yellow white
17	Seed color	Red	Red	Dark red	Yellowish
18	Range of mass per 200 seeds (grams)	21.6 to 38.5	38.6 to 56.5	38.5to 56.6	38.7 to 56.4
19	Average of mass per 200 seeds (grams)	32.40	48.75	48.73	48.74
20	Adaptation to less water (drought) land	Less	Adaptive	Adaptive	Adaptive
21	Tolerant to germ (disease and pest)	Less	Tolerant	Tolerant	Tolerant
22	Range of water content (%)	11 to 19	10 .2 to 12.5	9.8 to 12.2	10.1 to 12.6
23	Average of water content (%)	16.5	11.1	10.8	11.3
24	Production range (tons/ha)	(1.98 to 2.24) tons/ha	(3.97 to 4.38) tons/ha	(3.97-4.36) tons/ha	(3.98-4.34) tons/ha
25	Production average (tons/ha)	2.12 tons/ha	4.18 tons/ha	4.16 tons/ha	4.17 tons/ha
26	Percentage of production increase	-	49.28%	49.04%	49.16%
27	Average percentage increase in Production (%)	-		49.16%	
28	Potential of Maximum production	2.24 tons/ha	4.38 tons/ha	4.36 tons/ha	4.34 tons/ha

d a p: days after planting.

The average number of pods per tree for each variety was 88 (MV_1), 86 (MV_2), 85 (MV_3), and 45 for the parent variety.

Table 3. Production levels in 3 planting locations (3 provinces) in the control sample and the treatment sample (selected mutant).

No	Planting Location	Control Sample (tons/ha)	Selected Mutant Varieties (tons/ha)		
			VM_1	VM_2	VM_3
1	East Perfui Kupang NTT	2.24	4.38	4.36	4.34
2	Tanah Toraja South Sulawesi	2.15	4.18	4.15	4.19
3	Mamuju West Sulawesi	1.98	3.97	3.97	3.98
	Average production	**2.12**	**4.18**	**4.16**	**4.17**
	Average production of mutant varieties	-		**4.17**	
	Percentage increase in production of mutant varieties (%)	-	**49.28**	**49.04**	**49.16**
	Average percentage increase in production of mutant varieties (%)			**49.16**	

3.2. Statistical Calculations

The percentage of seed growth in the control and treatment (mutant) samples can be calculated based on equation (1) using the data in Table 5.1, as follows.

The number of observed seeds that were randomly selected was 100 seeds, the average seed that did not grow in the control sample was 16.60 seeds. The average seed that did not grow in the selected superior mutant variety of local Timor red peanut was 3.16 seeds.

Percentage of growth in the control sample.

$$PG = \left(\frac{T_{AS} - A_{SG}}{T_{AS}}\right) \times 100\% = \left(\tfrac{100-16.60}{100}\right) \times 100\% = 83.40\% \quad (7)$$

The growth percentage in the treated samples (treatment or seeds irradiated with multigamma sources) were:

$$PG = \left(\frac{T_{AS} - A_{SG}}{T_{AS}}\right) \times 100\% = \left(\tfrac{100-3.16}{100}\right) \times 100\% = 96.4\% \quad (8)$$

The average production of the control sample (parent variety) in the three provinces of planting locations.

$$A_{RPSC(t)} = \frac{P_{L1} + P_{L2} + P_{L3}}{3} = \frac{2.24 + 2.15 + 1.8}{3} = 2.2\, ton\,/\,ha \quad (9)$$

The average production of treated samples for each selected high yielding mutant variety of Timorese local red peanuts as a result of multi gamma irradiation was:

Journal of Physics: Conference Series **2165** (2022) 012001 doi:10.1088/1742-6596/2165/1/012001

Mutant Variety-1 (MV_1):

$$A_{RPSP(1)} = \frac{P_{L1} + P_{L2} + P_{L3}}{3} = \frac{4.38 + 4.18 + 3.7}{3} = 4.tons / ha \qquad (10)$$

Mutant variety-2 (MV_2):

$$A_{RPSP(2)} = \frac{P_{L1} + P_{L2} + P_{L3}}{3} = \frac{4.36 + 4.15 + 3.7}{3} = 4.16 tons / ha \qquad (11)$$

Mutant variety-3 (VM_3):

$$A_{RPSP(3)} = \frac{P_{L1} + P_{L2} + P_{L3}}{3} = \frac{4.4 + 4.9 + 3.98}{3} = 4.17 tons / ha \qquad (12)$$

The average total production of the selected superior mutant variety was:

$$A_{RPSP(t)} = \frac{A_{PTS(1)} + A_{PTS(2)} + A_{PTS(3)}}{3} = \frac{4.18 + 4.16 + 4.17}{3} = 4.7 tons / ha \qquad (13)$$

The percentage increase in production for each selected superior mutant variety of local Timor red peanuts as a result of multi-gamma irradiation:

Mutant variety-1 (MV_1):

$$I_{PPPVM(1)} = \left(\frac{A_{RPSP(1)} - A_{RPSC(t)}}{A_{RPSP(1)}} \right) \times 100\% = \frac{4.8 - 2.12}{4.8} \times 100\% = 49.28\% \qquad (14)$$

Mutant variety-2 (MV_2):

$$I_{PPPVM(2)} = \left(\frac{A_{RPSP(2)} - A_{RPSC(2)}}{A_{RPSP(2)}} \right) \times 100\% = \frac{4.6 - 2.12}{4.16} \times 100\% = 49.04\% \qquad (15)$$

Mutant variety-3 (MV_3):

$$I_{PPPVM(3)} = \left(\frac{A_{RPSP(3)} - A_{RPSC(3)}}{A_{RPSP(3)}} \right) \times 100\% = \frac{4.7 - 2.12}{4.7} \times 100\% = 49.16\% \qquad (16)$$

The average total percentage increase in production of selected high yielding mutant varieties of local Timor red peanuts as a result of multi gamma irradiation:

$$I_{PPPRTVM} = \left(\frac{I_{PPPVM(1)} + I_{PPPVM(2)} + I_{PPPVM(3)}}{3} \right) = \frac{49.8\% + 49.04\% + 49.16\%}{3} = 49.6\% \qquad (17)$$

3.3. Discussion

3.3.1. Growing Time, Growth Percentage, Flowering Age, and adaptation to less water (drought conditions), and Pest-Disease Resistant. The results of observations, measurements, and calculations on parent varieties and selected high yielding mutant varieties of Timor local red peanuts irradiated by multigamma showed that the growing time of the parent varieties was 11 days after planting (d a p), while the average growing time of the selected superior mutant varieties was 6 days after planting (d a p). This proves that local Timor red peanuts developed using standard multi-gamma irradiation techniques grow faster than their parent varieties.

The average percentage growth of selected superior mutant varieties was 96.84%, while the percentage of growth of the parent variety was 83.40%. The average flowering age of the selected superior mutant variety was 36 days after planting and the average flowering age of the parent variety was 43 days after planting. This indicates that the selected high yielding mutant variety as a result of multi gamma irradiation has a higher growth percentage and faster flowering than the parent variety.

Figure 1, Figure 2, and Figure 3 show three variations of selected high yielding mutant varieties of local Timor red peanuts produced by multi-gamma irradiation at the age of 28 days after planting, growing fertile and smooth leaves. This indicates that the local Timor red peanuts produced by multigamma irradiation can adapt to less water and are tolerant of pests and diseases. This statement in accordance with the results of previous research by experts [13, 14].

3.3.2. Range and Average Plant Height, Range and Average Harvesting Age of Local Timor Red Peanuts Selected and Parent Mutant Varieties. Selected high yielding mutant varieties resulting from standard multi-gamma irradiation had heights ranging from (46 to 83) cm for mutant-1 (MV_1), (44 to 85) cm for mutant-2 (MV_2) varieties, and (43 to 84) cm for the mutant-3 (MV_3) variety, while the height range of the parent variety (27 to 61) cm, with the average height respectively 65 cm (MV_1), 68 cm (MV_2), 67 cm (MV_3) and the parent 46 cm. Harvest age ranges (62 to 92) d a p (MV_1), (64 to 95) d a p (MV_2), (63 to 93) d a p (MV_3), and (93 to 135) d a p (parent). These data indicate that the selected high yielding mutant varieties of local Timor red peanuts produced by multi-gamma irradiation varied, most of which were different from the height of the parent and had a shorter harvest life than the parent variety.

3.3.3. Resistance to Pests. Figure 1, Figure 2, and Figure 3 clearly show the growth of three varieties of selected high-yielding mutant varieties of Timor red peanuts resulting from multi-gamma irradiation that grew with fertile, smooth leaves (no pest and disease attacks). In selected superior mutant varieties, leaf shape and color, plant height, plant fertility, flower color and shape, stalk and pod shape, pod growth position, were relatively uniform for each variety. These data indicate that the selected high yielding mutant varieties of local Timor red peanuts as a result of multi-gamma irradiation are resistant to pests and diseases, have relatively the same diversity in each selected mutant variety, and have high production potential (indicated by relatively large number of pods) compared to other selected mutant varieties. with the parent variety.

3.3.4. Range and Average Number of Pods, Range and Average Number of Seeds, Range and Mass per 200 Seeds. The range of the number of pods per tree in local Timor red peanuts selected high yielding mutant varieties from multigamma irradiation was (49 to 123) pods per tree (MV_1), (48 to 121) pods per tree (MV_2), and (46 up to 120) pods per tree (MV_3), while the parent variety (21 to 67) pods per tree. The average number of pods per tree for each variety was 88 (MV_1), 86 (MV_2), 85 (MV_3), and 45 for the parent variety.

The range and mean number of seeds per pod of selected high yielding mutant varieties from multi-gamma irradiation and parent varieties were: 1 to 5 (MV_1), 1 to 5 (MV_2), 1 to 5 (MV_3), and 1 to 4 for parent varieties with the average number of seeds per pod in the treatment and parent samples respectively 3 (50%), 4 (50%) for MV_1, 3 (50%), 4 (50%) for MV_2, 3 (50%), 4(50%) for MV_3, and 2 (50%), 3 (50%) for parent varieties. The range and mean mass per 200 seeds of selected superior mutant varieties and parent varieties were (38.6 to 56.5) grams (MV_1), (38.5 to 56.6) grams (MV_2), (38.7 to 56.4) grams (MV_3), and (21.6 to 38.5) grams (parent variety) with average mass: 48.75gram (MV_1), 48.73 grams (MV_2), 48.74 grams (MV_3) and 32.40 grams (parent variety).

So, it can be stated that the average mass per 200 seeds in the three selected varieties of high yielding mutant varieties which were relatively similar was higher than that of the parent variety (mutant varieties experienced a significant increase in production).

3.3.5. Production Range and Average, First Percentage of Production Material, and Water Content. The production of local Timor red peanuts for the three varieties of selected high yielding mutant varieties from multigamma irradiation ranged from (3.97 to 4.38) tons/ha (MV_1), (3.97 to 4.36) tons/ha. (MV_2), and (3.98 to 4.34) tons/ha (MV_3), as well as parent varieties (1.98 to 2.24) tons/ha. The maximum production potential is 4.38 tons/ha (MV_1), 4.36 tons/ha (MV_2), and 4.34 tons/ha (MV_3), and 2.24 tons/ha for parent varieties. The average production of each selected high yielding mutant variety for three variations was 4.18 tons/ha (MV_1), 4.16 tons/ha (MV_2), 4.17 tons/ha (MV_3), and 2.12 tons/ha (parent variety). The average total production of selected high yielding mutant varieties in the three provinces of planting locations was 4.17 tons/ha. The first percentage of production of each selected superior mutant variety for the three variations in three provinces was 49.28% (MV1), 49.04% (MV_2), and 49.16% (MV_3) with the average total percentage increase in production of the selected superior mutant variety in three provinces is 49.16%.

The range and mean water content of each selected superior mutant variety for three successive variations were 10.2% to 12.5% (MV_1), 9.8% to 12.2% (MV_2), 10 .1 % to 12.6% (MV_3) and 11.0% to 19.0% (parent variety), with mean water content: 11.1% (MV_1), 10.8% (MV_2), 11.3% (MV_3), and 16.5% (parent variety), the mean total water content of the selected mutant varieties for the three variations in three provinces was 11.07 %.

The above results indicate that the production of selected high yielding mutant varieties resulting from multigamma irradiation in the three provinces of planting locations increased significantly with the percentage of water content according to the standard.

4. Conclusion

Based on the description in the discussion section, several conclusions can be put forward as follows. The development of local Timor red peanut varieties through breeding with multigamma irradiation techniques and careful selection resulted in superior mutant cultivars that could adapt to several abiotic and biotic conditions and significantly increased production. The use of multigamma irradiation in the development of local Timor red peanuts resulted in three varieties of selected high yielding mutant varieties with relatively similar production and significantly increased compared to the parent variety. The average total percentage increase in production for the three varieties of selected high yielding mutant varieties resulting from multigamma irradiation was 49.16% with an average total dry seed moisture content of 11.07%.

Acknowledgement

This research or study was supported and funded by Decentralization research, Directorate of Research and Public Service, Department of Education and Culture of Indonesia Government and Technology Research through PNBP funds of Nusa Cendana University, on the scheme: University Leading Applied Research or Professor's Research. Especial thanks to authorized official had been funded of this research and also especial thanks to all colleagues in English Training Centre, whom have helped to me in technical writing of this manuscript. The publication fee of manuscript is included in this financial.

References

[1] Sutrisno, I and Fachrur Rozi. 2015. Local Wisdom as a Basis for Optimizing Food Crops Cultivation in East Sumba, Development of Environmentally Friendly Local Resources Potential for Strengthening National Agricultural Products with Global Competitiveness. Research Institute for Various Nuts and Tubers. Proceeding. p. p 247-255.

[2] Rahmianna, A. A, Herdina Pratiwi, dan Didik Harnowo. 2017. Peanut Cultivation, Research Institute for Various Nuts and Tubers, Monograph Balitkabi No. 13 pp. 133-169.

[3] BPS NTT. 2010. NTT in Numbers 2010. WWW://ntt.bps.go.id.

[4] Murdolelono, B. 2011. Peanut Plant Technology for Farmers in East Nusa Tenggara, Proceeding of Research Results on Various Beans and Bulbs.

[5] BPS. 2015. Peanut Production in Indonesia.
[6] Balitkabi. 2016. Description of Superior Peanut Varieties, KT.45.
[7] BPS. 2018. Peanut Production by Province, Results of Coordination Meeting in Solo, 25-27 July 2018.
[8] Marwiyah, S, Heny Purnamawati, and Pinni Sembiring. 2017. *Physical Mutation Induced by Gamma Ray Irradiation on Kidney Bean, Journal Comm. Hort. J*, February 2017, 1(4): 49-55.
[9] Pasangka, B. 2021. Genetic Mutation of Vigna Radiata L with Multigamma Irradiation, Book. International Edition, LAP Lambert Academic Publishing Germany, ISBN-13: 978-620-3-86211-9., ISBN-10: 6203862118., EAN: 9786203862119
[10] Pasangka and Refli. 2013. *Developing of Local Peanut Based on Multiculture that Tolerant to Abiotic and Biotic Conditions with Multigamma Radiation Method (Nuclear), Journal of Agricultural Science,* **5** *(5), doi.10.5539/jas.v5n5p15, URL: http://dx.doi.org/10.5539/jas.v5n5p15, May, 2013*
[11] Pasangka, B and Refli. 2013. *The Application of Multigamma Radiation as a Physical Mutagen for Breeding of Local Soybean, Journal of Agricultural Science*, **5** (6) pp. 201-210.
[12] Pasangka, B and Refli. 2016. *Development and ultivation of Local Kidney Bean (Phaseolus Vulgaris L) Through Breeding to Use Multigamma Irradiation Technique (Nuclear), American Journal of Agricultural and Biologycal Sciences*, **11** Issue. 3. Pp: 117-127.
[13] Rozi, F, Imam Sutrisno, dan A. A. Rahmianna. 2016. Peanut Development Opportunities in Dry Land of East Nusa Tenggara, Palawija Bulletin, **14** No. 2 pp. 71-77.
[14] Seran, A. H and Krisantus Tri Pambudi Raharjo. 2018. Study of Drought Stress After Flowering of Red Beans (Phaseolus Vulgaris L) in *Entisol* Soil, Journal of Dry Land Conservation Agriculture, *International Standard of Serial Number* 2477-7927, Timor University.

The effect of brain-based learning on students' mathematical communication ability viewed from creativities in the thematic subjects of science physics - mathematics

Syahrir[1] and S Prayogi[2]

[1]Universitas Negeri Jakarta, Jakarta 13220, Indonesia
[2]Universitas Pendidikan Mandalika, Mataram 83125, Indonesia

E-mail: syahrir_9913920001@mhs.unj.ac.id

Abstract. This study aims to determine the effect of Brain Based Learning on students' mathematical communication ability viewed from creativities, in the thematic subjects of science physics - mathematics. The research was carried out at SDN Pantai Harapan Jaya 01, Muara Gembong District, using experimental research on treatment with a level 2 x 2 design. The results showed that:(1) the mathematical communication ability of students who are given brain-based learning treatment is higher than students who use expository, (2) there is an interaction of learning models with creativity on students' mathematical communication ability, (3) students' mathematical communication ability with brain-based learning are morehigher than expository, this is for students who have high creativity, and (4) there is no difference in mathematical communication ability with brain based learning and expository treatment between students who have low creativity.

1. Introduction

The 2015 report from PISA (Program for International Student Assessment) with 70 countries participating in the country, Indonesia ranked 62nd[1].

Table 1.PISA results of performance in science, reading and mathematics in 2015.

No	Country	Science		Reading		Mathematics	
		Mean Score	Average in3-year	Mean Score	Average in3-year	Mean Score	Average in3-year
		Mean	*Score dif*	*Mean*	*Score dif*	*Mean*	*Score dif*
1	Singapore	556	7	535	5	564	1
2	Japan	538	3	516	-2	532	1
3	Estonia	534	2	519	9	520	2
4	Chinese Taipei	532	0	497	1	542	0
5	Finland	531	-11	526	-5	511	-10
..							
62	**Indonesia**	**403**	**3**	**397**	**-2**	**386**	**4**
..							
70	Dominican Republic	332	M	358	M	328	M

The real that occurs in the results of mathematics learning is still very low[2–6]. A report from the 2015 TIMSS (Trends in International Mathematics and Science Study) from the International Association for Evaluation of Educational Achievement (IEA), Indonesia ranked 44th out of 56 participating countries with an average score of students' mathematics scores in grade IV namely amounting to 397 below the average score of 500 [7–9]. This shows that the ability in learning mathematics in Indonesia has not shown satisfactory results.

Overcoming this statement that needs to be improved mathematical communication skills in students that is to articulate, explain, organize and consolidate mathematical thinking [10–15]. This becomes important with the "articulation" of mathematical ideas that are explained through words can be heard by students, "explain" here means that students can explain their mathematical ideas to other students, "organize" is students can coordinate their ideas to be coherent and clear so that young people are accepted by their friends. After articulating, explaining, arranging further strengthens or connects each other about mathematical ideas. So that students are able to develop students' abilities in communicating ideas with verbal speaking, notes, symbols, tables, graphics, or other media to clarify the situation or problem.

Mathematical communication skills are very important in improving the quality of education [16–19], then in learning must be improved. The right learning to use is. Brain-Based Learning is a learning approach that is more parallel with how the brain learns best naturally based on the disciplines of neurology, biology, psychology, an understanding of the relationship between learning and the brain now leads to the role of emotions, patterns, meaning, environment, body rhythm and attitude, stress, trauma, assessment, music, movement, gender, and enrichment[3,20–31]. So as to be able to create learning that is oriented towards efforts to empower brain potential.

Mathematical communication skills are strongly influenced by students 'mathematical creativity, students' mathematical creativity is an ability that must be possessed by students in the learning process with the aim of being able to produce new ideas and ideas [32–55]. So that in the learning process, especially in mathematics learning, can increase.

After identification of the problem above, the formulation of the problem in this study is 1) Are there differences in mathematical communication skills of students who use Brain-Based Learning with class students who use Expository?; 2) Are their interactions between learning models and creativity on students' mathematical communication skills?; 3) Are there differences in mathematical communication skills between students who study with Brain-Based Learning and students who study with Expository in students who have high creativity?; 4) Are there differences in mathematical communication skills between students who study with Brain-Based Learning and students who study with Expository in students who have low creativity?

The objectives to be achieved in the study are 1) Explain the differences in mathematical communication skills of students who use Brain-Based Learning with class students who use Expository; 2) Explain the interaction between learning models and creativity on students' mathematical communication skills; 3) Explain differences in mathematical communication skills between students who learn with Brain-Based Learning and students who study with Expository in students who have high creativity; 4) Explain differences in mathematical communication skills between students who learn with Brain-Based Learning and students who study with Expository in students who have low creativity.

2. Methods

In this study, the method used in measuring the results of students' mathematical communication skills is the experimental method with 2x2 factorial by level or the two-way two-levels crossed factorial design, this means that the experiment consists of 2 factors, each of which consists of 2 levels. This design is used to assess the impact of each factor and its interaction on the response variable [56]. Research design as presented in Table 2.

Table 2. Design of experimental research

Attribute Variables Creativity (B)	Treatment Variables (A)	
	Brain Based Learning (A_1)	Ekspositori (A_2)
High Creativity (B_1)	A_1B_1	A_2B_1
Low Creativity (B_2)	A_1B_2	A_2B_2

The subjects of this study were fourth-grade students at SDN 01 Pantai Harapan Jaya Bekasi with the number for the experimental class as many as 30 students and for the control class as many as 28 students. The average age of students in the experimental class and control class is ± 9 years. Where in the experimental and control class the average has a social background having a well-off life with the parents mostly fishermen and agriculture and about 6 people per class of their parents are civil servants. The average ability after the trial test has almost the same ability of the two classes.

Implementation procedures in the study as in Table 2, namely the design used by level 2 x 2 treatments (two-way two-levels crossed factorial design) [56]. Where in the experimental class will use Brain-Based Learning while in the control class using expository learning In the expository Brain-Based Learning and Learning variables, each will be viewed from the creativity of the students so that students will find students who have high creativity and low creativity.

Furthermore, after students are given a creativity test it will be taken as a result of data analysis on students who have high creativity on the score range of 33% and students who have low creativity on the score range of 33% as well, this proportion is adapted from previous studies [57]. So that it will be obtained in the experimental class that is students who use Brain-Based Learning with high creativity (A_1B_1) as many as 10 students, and students who use Brain-Based Learning with low creativity (A_1B_2) as many as 10 students. Then in the control class is students who use expository learning with high creativity (A_2B_1) as many as 10 students, and students who use expository learning with low creativity (A_2B_2) as many as 10 students.

The sampling technique was carried out by the following processes: 1) Randomly selected SDN in Muara Gembong sub-district, Bekasi Regency and selected SDN 01 Pantai Harapan Jaya, 2) Then determined the experimental class and control class, so that IVA class students were selected as the experimental class and IVB class as a control class. Data collection techniques on students' creativity and mathematical communication skills are using essay tests with 5 questions.

3. Result and Discussion

After following the learning process using Brain-Based Learning and Expository, the results of students' mathematical communication skills in the form of scores were obtained.

Table 3. Description of Student Mathematical Communication Ability

Creativity	Brain Based Learning					Ekspository				
	N	Max	Min	Average	SD	N	Max	Min	Average	SD
High	10	80	60	71.5	6.69	10	70	45	54.5	8.32
Low	10	70	50	60.5	5.99	10	70	40	60	9.43
$\sum$	20	80	50	66	8.37	20	70	40	57.25	9.1

Based on the results of the analysis of the normality test in the group of students who use Brain-Based Learning and Expository by using SPSS Version 24, it shows the significance value of students' mathematical communication skills in the group of students who use Brain-Based Learning shows Kolmogorov-Smimov[4] of 0.200 > 0.05, then Shapiro-Wilk shows that the significant value is 0.530 > 0.05. Then the Expository shows Kolmogorov-Smimov of 0.200 > 0.05, then Shapiro-Wilk shows that the significant value is 0.236 > 0.05. This means that the data of the mathematical communication values of 20 samples of students given treatment with Brain-Based Learning and Expository are normally distributed.

The results of the analysis of the homogeneity test in the group of students who use Brain-Based Learning and Expository by using SPSS Version 24, it shows that significant Levene'sstatistics 0.475 > 0.05, which means that the data value of mathematical communication ability has a varianceshomogeneous. The normality test in the group of students who use Brain-Based Learning and Expository on students who have high creativity, it shows the significance value of students' mathematical communication skills in groups of students who use Brain-Based Learning with high creativity shows Kolmogorov-Smimov[4] is 0.200 > 0.05, then Shapiro-Wilk shows that the significant value is 0.466 > 0.05. Then the Expository with high creativity shows Kolmogorov-Smimovof 0.200 > 0.05, then Shapiro-Wilk shows that the significant value is 0.325 > 0.05. This means that the data of the mathematical communication value of the 10 samples of students who were treated with Brain-Based Learning and Expository in students who have the high creativity that is a normal distribution.

Then, based on the results of the analysis of the homogeneity test in the group of students who use Brain-Based Learning and Expository on students who have high creativity, it shows that the Levene'sstatistic is significant 0.591 > 0.05 which means the data value of mathematical communication skills has a varianceshomogeneous. Furthermore, the results shows the significance value of students' mathematical communication skills in groups of students who use Brain-Based Learning with low creativity shows Kolmogorov- Smimovis 0.200 > 0.05, then Shapiro-Wilk shows that the significant value is 0.691 > 0.05. Then the Expository with low creativity showed Kolmogorov-Smimovof 0.200 > 0.05, then Shapiro-Wilk showed that the significant value was 0.190 > 0.05. This means that the data of mathematical communication values from 10 samples of students given treatment with Brain-Based Learning and Expository on students who have low creativity are normally distributed.

The results of the analysis of the homogeneity test in the group of students who use Brain-Based Learning and Expository on students who have low creativity by using SPSS Version 24, it shows that significant Levene's statistics 0.282 > 0.05, which means that the data value of mathematical communication skills has a homogeneous variances.

3.1 Metamorphic Communication Ability Students who get Higher Brain-Based Learning Treatment for Students Who Get Expository Treatment.

Based on the results of hypothesis testing using two-way ANOVA analysis which shows that students who are given the Brain-Based Learning and Expository treatment get a F_{count} value of 12.84. With a significance level of 0.05 with $df_1 = 2$ and $df_2 = 18$, when viewed from F_{table}, the value of $F_{table} = 3.55$ is obtained.

Thus, that the value of $F_{count} = 12.84 >$ value of $F_{table} = 3.55$, this means there is a rejection of H_0, which means that there is a difference in the value of mathematical communication skills between the two groups of students given different learning. In addition, the acceptance of H_1 means that the mathematical communication ability of the group of students given Brain-Based Learning treatment is higher than the group of students given Expository.

3.2 The interaction between Learning Model and creativity on students' mathematical communication skills.

Based on the results of the analysis using SPSS Version 24, there is an interaction between learning, namely Brain-Based Learning and expository with creativity towards mathematical communication skills. This shows that through the interaction column learning with creativity obtained significant values of 0.002 < 0.05 with a significance level of 0.05. This means that there is a significant interaction between learning used with creativity in mathematical communication skills.

3.3 Differences in Mathematical Communication Ability of Students Who Get Brain-Based Learning and Expository Treatment for Students Who Have High Creativity.

Based on the results of the analysis by t-test using SPSS Version 24 that students who were given treatment with Brain-Based Learning with high creativity and Expository with high creativity obtained a count of 5.04. With a significance level of 0.05 and a degree of magnitude 18, when viewed in t_{table}, it is obtained $t_{table}$of 2.10. Thus, that the value of $t_{count} = 5.04 >$ the value of $t_{table} = 2.10$, this means that there is a rejection of H_0 which means that there is a difference in the value of mathematical communication skills between students who use different learning treatments even though both have high creativity. In addition, the acceptance of H_1 means that the mathematical communication ability of students who are given treatment using Brain-Based Learning is higher than students who use the Expository treatment which both have high creativity.

3.4 Differences in Mathematical Communication Ability of Students Who Get Brain-Based Learning and Expository Treatment for Students Who Have Low Creativity.

Based on the results of the analysis by t-test using SPSS Version 24, it shows that the results of the analysis with the t-test are students who are given treatment with Brain Based Learning with low creativity and expository with low creativity get a total of 1.42. With a significance level of 0.05 and a degree of 18, when viewed from the t_{table}, the t_{table} is 2.10. So the value of $t_{count} = 1.42 < t_{table}$ value = 2.10 this means that there is acceptance of H_0 which means that there is no difference in the value of mathematical communication skills between students who use different learning treatments even though both have low creativity.

Overall, the results of this study indicate that there is an interesting relationship between students' mathematical communication ability and brain-based learning, which is also found to be related to students' creativity. This result is in line with the previous study[58], that mathematics learning that applies a brain-based learning approach contributes to the development of students' mathematical communication ability. When compared with conventional teaching, students' mathematical communication ability using Brain Based Learning are better than those using conventional learning approaches[59]. Brain based learning was also found to be effective on students' creative thinking skills and self-efficacy[60]. In the context of neuroscience, brain capacity directly impacts how students can understand and communicate in learning[61]. Mathematical communication is very important in the learning process. Through effective mathematical communication, students can organize their mathematical thinking in writing and communicate their understanding to others[59].

4. Conclusion

Based on the findings made by researchers, it was concluded that:(1) the mathematical communication ability of students who are given brain-based learning treatment is higher than students who use expository, (2) there is an interaction of learning models with creativity on students' mathematical communication ability, (3) students' mathematical communication ability with brain-based learning are morehigher than expository, this is for students who have high creativity, and (4) there is no difference in mathematical communication ability with brain based learning and expository treatment between students who have low creativity.

Acknowledgement

The authors would like to thank those who have helped in this study, the research team, teachers, students, and universities who have supported this study process.

References

[1] OECD 2017 PISA 2015 Assessment and Analytical Framework: Science, Reading, Mathematic, Financial Literacy and Collaborative Problem Solving (OECD)

[2] Saleh M, Charitas R, Prahmana I and Isa M 2018 Improving the Reasoning Ability of Elementary School Student Through the Indonesian Realistic Journal on Mathematics Education **9** 41–54
[3] Adiansha A A and Sumantri M S 2017 The Effect of Brain Based Learning Model and Creative Thinking on the Ability of Mathematics Concept of Elementary Students **5** 1195–9
[4] Heymann H W 2003 Mathematics Instruction from the Perspective of General Education Why Teach Mathematics? A Focus on General Education (Dordrecht: Springer Netherlands) pp 83–223
[5] Rachmadtullah R, Ms Z and Sumantri M S 2018 Development of computer - based interactive multimedia : study on learning in elementary education International Journal of Engineering & Technolog **7** 2035–8
[6] Sumantri M S 2016 The Effect of Formative Testing and Self- Directed Learning on Mathematics Learning Outcomes International Electronic Journal of Elementary Education **8** 507–24
[7] NCES N C for E S 2017 Highlights From TIMSS and TIMSS Advanced 2015 1–58
[8] Hole A, Grønmo L S and Onstad T 2018 The dependence on mathematical theory in TIMSS, PISA and TIMSS Advanced test items and its relation to student achievement Large-scale Assessments in Education **6** 3
[9] McComas W F 2014 Trends in International Mathematics and Science Study (TIMSS) The Language of Science Education: An Expanded Glossary of Key Terms and Concepts in Science Teaching and Learning ed W F McComas (Rotterdam: SensePublishers) p 108
[10] Ellerton N F 2018 Book Review: NCTM's Compendium: finding a balance between historical details, contemporary practices, and future resources. Jinfa Cai (Ed.) (2017) Compendium for research in mathematics education Educational Studies in Mathematics **99** 109–23
[11] Merenda R C 2000 Numeracy encounters in a book bag: Meeting the NCTM standards Early Childhood Education Journal **27** 151–7
[12] Charlesworth R 2005 Prekindergarten Mathematics: Connecting with National Standards Early Childhood Education Journal **32** 229–36
[13] Sriraman B and Pizzulli M 2005 Balancing mathematics education research and the NCTM standards ZDM **37** 431–6
[14] Oluwole Pratt E 2002 Aligning Mathematics Teacher Work Sample Content with Selected NCTM Standards: Implications for Preservice Teacher Education Journal of Personnel Evaluation in Education **16** 175–90
[15] McLeod D B, Stake R E, Schappelle B P, Mellissinos M and Gierl M J 1996 Setting the Standards Bold Ventures: Case Studies of U.S. Innovations in Mathematics Education ed S A Raizen and E D Britton (Dordrecht: Springer Netherlands) pp 13–132
[16] Hayal Yavuz Mumcu., Aktürk T 2017 An Analysis Of The Reasoning Skills Of Pre-Service Teachers In The Context Of Mathematical Thinking European Journal of Education Studies **3** 225–54
[17] Muhammad 'Azmi Nuha., S. B. Waluya. I J 2018 Mathematical Creative Process Wallas Model in Students Problem Posing with Lesson Study Approach International Journal of Instruction **11** 527–38
[18] Syahputra E 2018 Differences in Metacognition and Mathematical Communication Ability Between Students Taught Using Problem Based Learning Model and Numbered Head Together Cooperative Learning Model at SMP Kartika I-2 Medan Journal of Education and Practice **9** 30–7
[19] Englar R 2018 Tracking Veterinary Students' Acquisition of Communication Skills and Clinical Communication Confidence by Comparing Student Performance in the First and Twenty-Seventh Standardized Client Encounters Journal of Veterinary Medical Education 1–23
[20] Samat C, Saengjan P, Chaijaroen S, Kanjug I and Vongtathum P 2018 Designing of the Learning Innovation Enhance Learning Potential of the Learners Using Brain-Based Learning Innovative

Technologies and Learning ed T-T Wu, Y-M Huang, R Shadiev, L Lin and A I Starčič (Cham: Springer International Publishing) pp 196–204
[21] Chaijaroen S and Samat C 2018 Design and Development of Learning Innovation Enhancing Learning Potential Using Brain-Based Learning Innovative Technologies and Learning ed T-T Wu, Y-M Huang, R Shadiev, L Lin and A I Starčič (Cham: Springer International Publishing) pp 189–95
[22] Waree C 2017 An Increasing of Primary School Teachers? Competency in Brain-Based Learning International Education Studies **10** 176
[23] Duman B 2010 The Effects of Brain-Based Learning on the Academic Achievement of Students with Different Learning Styles Educational Sciences: Theory & Practice **10** 2077–103
[24] Inocian R B 2015 Integrated Arts-based Teaching (IAT) Model for Brain-based Learning Journal of Curriculum and Teaching **4** 130–43
[25] GÖZÜYEŞİL E and DİKİCİ A 2014 The Effect of Brain Based Learning on Academic Achievement: A Meta-analytical Study Educational Sciences: Theory & Practice **14** 642–8
[26] Suparta I N 2018 The Effect Of Brain Based Learning On Second Grade Junior Students' Mathematics Conceptual Understanding On Polyhedron Journal on Mathematics Education **9** 145–56
[27] Shabatat K and Al-Tarawneh M 2016 The Impact of a Teaching-Learning Program Based on a Brain-Based Learning on the Achievement of the Female Students of 9th Grade in Chemistry Higher Education Studies **6** 162
[28] A. Al-Balushi K and Al-Balushi S 2018 Effectiveness of Brain-Based Learning for Grade Eight Students' Direct and Postponed Retention in Science International Journal of Instruction **11** 525–38
[29] Atta A and Safein M 2017 Engaging ESP Students with Brain-Based Learning for Improved Listening Skills, Vocabulary Retention and Motivation English Language Teaching **10**
[30] Yasar D M 2017 Brain Based Learning in Science Education in Turkey: Descriptive Content and Meta Analysis of Dissertations Journal of Education and Practice **8** 161–8
[31] Koşar G 2018 European Journal of Education Studies Improving Knowledge Retention Via Establishing Brain-Based Learning Environment European Journal of Education Studies **4** 208–18
[32] Sriraman B, Haavold P and Lee K 2014 Creativity in Mathematics Education ed S Lerman Encyclopedia of Mathematics Education 109–15
[33] Mazzola G, Guitart R, Ho J, Lubet A, Mannone M, Rahaim M and Thalmann F 2017 Mathematical Models of Creativity The Topos of Music III: Gestures: Musical Multiverse Ontologies (Cham: Springer International Publishing) pp 1001–82
[34] Sriraman B, Yaftian N and Lee K H 2011 Mathematical Creativity and Mathematics Education The Elements of Creativity and Giftedness in Mathematics ed B Sriraman and K H Lee (Rotterdam: SensePublishers) pp 119–30
[35] Schindler M, Joklitschke J and Rott B 2018 Mathematical Creativity and Its Subdomain-Specificity. Investigating the Appropriateness of Solutions in Multiple Solution Tasks Mathematical Creativity and Mathematical Giftedness: Enhancing Creative Capacities in Mathematically Promising Students ed F M Singer (Cham: Springer International Publishing) pp 115–42
[36] Sriraman B 2008 The characteristics of mathematical creativity ZDM **41** 13
[37] Krummheuer G, Leuzinger-Bohleber M, Müller-Kirchof M, Münz M and Vogel R 2013 Explaining the mathematical creativity of a young boy: an interdisciplinary venture between mathematics education and psychoanalysis Educational Studies in Mathematics **84** 183–99
[38] Shriki A 2010 Working like real mathematicians: developing prospective teachers' awareness of mathematical creativity through generating new concepts Educational Studies in Mathematics **73** 159–79
[39] Glas E 2002 Klein's Model of Mathematical Creativity Science & Education **11** 95–104

[40] Hoshino R 2018 Supporting Mathematical Creativity Through Problem Solving Teaching and Learning Secondary School Mathematics: Canadian Perspectives in an International Context ed A Kajander, J Holm and E J Chernoff (Cham: Springer International Publishing) pp 367–75
[41] Sorge V 2014 Combining Systems for Mathematical Creativity (Invited Talk) Artificial Intelligence and Symbolic Computation ed G A Aranda-Corral, J Calmet and F J Martín-Mateos (Cham: Springer International Publishing) pp 7–8
[42] Muir A 1988 The psychology of mathematical creativity The Mathematical Intelligencer **10** 33–7
[43] Ervynck G 1991 Mathematical Creativity Advanced Mathematical Thinking ed D Tall (Dordrecht: Springer Netherlands) pp 42–53
[44] Carayannis E G 2013 Mathematical Creativity Encyclopedia of Creativity, Invention, Innovation and Entrepreneurship (New York, NY: Springer New York) p 1228
[45] Pitta-Pantazi D, Kattou M and Christou C 2018 Mathematical Creativity: Product, Person, Process and Press Mathematical Creativity and Mathematical Giftedness: Enhancing Creative Capacities in Mathematically Promising Students ed F M Singer (Cham: Springer International Publishing) pp 27–53
[46] Sriraman B, Haavold P and Lee K 2013 Mathematical creativity and giftedness: a commentary on and review of theory, new operational views, and ways forward ZDM **45** 215–25
[47] Haylock D W 1987 A framework for assessing mathematical creativity in school chilren Educational Studies in Mathematics **18** 59–74
[48] Goldin G A 2017 Mathematical creativity and giftedness: perspectives in response ZDM **49** 147–57
[49] Luria S R, Sriraman B and Kaufman J C 2017 Enhancing equity in the classroom by teaching for mathematical creativity ZDM **49** 1033–9
[50] Lev M and Leikin R 2017 The Interplay Between Excellence in School Mathematics and General Giftedness: Focusing on Mathematical Creativity Creativity and Giftedness: Interdisciplinary perspectives from mathematics and beyond ed R Leikin and B Sriraman (Cham: Springer International Publishing) pp 225–38
[51] Wessels H M 2017 Exploring Aspects of Creativity in Mathematical Modelling Mathematical Modelling and Applications: Crossing and Researching Boundaries in Mathematics Education ed G A Stillman, W Blum and G Kaiser (Cham: Springer International Publishing) pp 491–501
[52] Kattou M, Kontoyianni K, Pitta-Pantazi D and Christou C 2013 Connecting mathematical creativity to mathematical ability ZDM **45** 167–81
[53] Van Harpen X Y and Sriraman B 2013 Creativity and mathematical problem posing: an analysis of high school students' mathematical problem posing in China and the USA Educational Studies in Mathematics **82** 201–21
[54] Leikin R and Lev M 2013 Mathematical creativity in generally gifted and mathematically excelling adolescents: what makes the difference? ZDM **45** 183–97
[55] Samsudin N S, Samsuddin I and Yusof A F 2018 Creativity in Mathematical Thinking Through Constructivist Learning Approach for Architecture Students Proceedings of the Second International Conference on the Future of ASEAN (ICoFA) 2017 -- Volume 2 ed R Saian and M A Abbas (Singapore: Springer Singapore) pp 133–42
[56] Harrar S W, Ronchi F and Salmaso L 2019 A comparison of recent nonparametric methods for testing effects in two-by-two factorial designs Journal of Applied Statistics **46** 1649–70
[57] Suhirman S, Prayogi S and Asy'ari M 2021 Problem-Based Learning with Character-Emphasis and Naturalist Intelligence: Examining Students Critical Thinking and Curiosity INT J INSTRUCTION **14** 217–32
[58] Triana M, Zubainur C M and Bahrun B 2019 Students' Mathematical Communication Ability through the Brain-Based Learning Approach using Autograph JRAMathEdu (Journal of Research and Advances in Mathematics Education) **4** 1–10

[59] Iski N 2019 Penerapan Model Brain Based Learning untuk Meningkatkan Kemampuan Komunikasi Matematis Siswa Kelas VIII pada Siswa MTs/SMP Al Khawarizmi: Jurnal Pendidikan dan Pembelajaran Matematika **3** 33–43
[60] Imanuel I, Waluya S B and Mariani S 2019 The Effectiveness of Brain Based Learning Assisted by Schoology towards Student's Creative Thinking and Self-Efficacy Journal of Primary Education **8** 274–81
[61] Darling-Hammond L, Flook L, Cook-Harvey C, Barron B and Osher D 2020 Implications for educational practice of the science of learning and development Applied Developmental Science **24** 97–140

What are the dynamics of hydrometeorological parameters on peatlands during the 2019 extreme dry season?

M Irfan[1], S Safrina[2], E Koriyanti[1], K Saleh[1], N Kurniawaty[1], I Iskandar[1]

[1]Department of Physics, Faculty of Mathematics and Natural Sciences, Sriwijaya University, Indonesia;
[2]Department of English Education, Faculty of Teacher Training and Education Sriwijaya University, Indonesia.

Email: irfplg@yahoo.com

Abstract. Peatlands in Indonesia are generally flammable, including peatlands in Kalimantan. The Indonesian government has made efforts to prevent fires on peatlands in Indonesia. One of them is by establishing several stations for measuring hydrometeorological parameters which are managed by the Peat Restoration Agency (BRG). This study utilizes BRG data, especially rainfall and groundwater levels, combined with hotspots data from MODIS satellite measurements during the 2019 extreme dry season at two research locations, namely at BRG stations in Central Kalimantan and West Kalimantan. The objectives of this research is to study the dynamics of these three parameters and find their relationship during the 2019 extreme dry season. To achieve the objectives of this study, an analysis of the time series graphs of these three parameters was carried out during the 2019 extreme dry season. It is hoped that the results of this research will help to prevent severe fires on peatlands, especially in Kalimantan. The results of this study indicate that during this period there was a decrease in rainfall, a decrease in groundwater level, and the emergence of a large number of hotspots. It can be concluded that the hotspots that appear are closely related to the low groundwater level due to low rainfall.

1. Introduction

Indonesia has very large peatlands. The area of peatland in Indonesia is about 20 million hectares or about half of the total peatland area in the world. These peatlands are located in Papua, Kalimantan, Sumatera and Sulawesi. Peatlands in Indonesia are flammable, so efforts should be made to prevent this recurrence[1–6].

The Indonesian government has done many things to prevent severe fires on peatlands. One of them is by establishing an integrated measurement station for hydrometeorological parameters, where these parameters are thought to be closely related to the occurrence of fires in peatlands. The station is managed by the Peat Restoration Agency (BRG). This research makes use of the data from the measurement results of the BRG station, especially the rainfall and groundwater level data combined with the hotspots data from NASA's MODIS satellite measurements.

In 2019 a natural phenomenon called positive Indian Ocean Dipole (IOD +) occurred which caused minimal rainfall in several regions in Indonesia, including Kalimantan. This minimal rainfall has resulted in severe drought and fires on the peatlands. This period is called the extreme dry season which occurred from July 2019 to December 2019 [7–11]. It is interesting to examine how the dynamics of the hydrometeorological parameters are estimated to be associated with fires on peatlands, including rainfall, groundwater levels and the number of hotspots during this period as an effort to mitigate fire disasters on peatlands.

The purpose of this study is to study changes in rainfall, groundwater levels, hotspots, and the relationship between these three parameters on peatlands in Kalimantan during the extreme dry season in 2019. This research is expected to be able to answer the following questions, namely: (a) What is the pattern of changes in the hydrometeorological parameter values?; (b) How is the relationship between rainfall and groundwater level?; (c) How is the relationship between the groundwater level and the number of hotspots?.

2. Methods

2.1. Data

The data used in this research are hourly rainfall and hourly groundwater levels which are downloaded from the SESAME website at the following address: https://web.sesame-system.com. SESAME (Sensory Data Transmission Service Assisted by Midori Engineering Laboratory) is an in-situ integrated measurement system of several hydrometeorological parameters on peatlands in Indonesia, including in Kalimantan. SESAME was built on the cooperation between the Indonesian government and the Japanese government. SESAME is managed by an institution called the Peat Restoration Agency (BRG) [12]. The data of this research are data from measurement results at 2 BRG stations in West Kalimantan (WK) and in Central Kalimantan (CK). This study also uses hotspots data from NASA's MODIS satellite measurements [5]. Hotspots data is available on the NASA website at the following address: https://firms.modaps.eosdis.nasa.gov. Details about data and stations coordinates are shown in Table 1 and Table 2.

Table 1. Data used in this study

Parameters	Source of Data	Type of data
Rainfall	BRG	In-situ, hourly
Average groundwater level	BRG	In-situ, hourly
Number of hotspots	MODIS	Remote sensing, daily, 0.5°x 0.5°

Table 2. Data of study locations

Station Name	Coordinate
Central Kalimantan (CK)	-3.492, 104.978
West Kalimantan (WK)	-2.911, 105.082

2.2. Data Analysis

To study the dynamics of the parameters of rainfall, groundwater level, and hotspots, time series graphs for these three parameters were made during the 2019 extreme dry season, from July to October. After that, it is analyzed to find out when the extreme values of the three parameters occur. Analysis of the relationship between the three parameters is also carried out by making an overlay graph between these parameters.

3. Result and Discussion

3.1. The Time Series of Changes in Hydrometeorological Parameters

Based on the time series graph of rainfall shown in Figures 1 (a) and (b), it can be seen that very low rainfall occurs from July to September 2019. In these months the peak of extreme dryness occurs in Kalimantan. If we compare the rainfall that occurs at stations in Central Kalimantan with those in West Kalimantan, the rainfall in Central Kalimantan is lower. This is also evident in Table 1, where the total monthly rainfall in Central Kalimantan is lower than the monthly total rainfall in West Kalimantan.

Rainfall in Central Kalimantan is lower than West Kalimantan, resulting in a lower groundwater level in Central Kalimantan compared to the groundwater level in West Kalimantan. This can be seen in the graphs in Figures 1 (c) and (d). The lowest groundwater level in Central Kalimantan reached -

Journal of Physics: Conference Series **2165** (2022) 012003 doi:10.1088/1742-6596/2165/1/012003

1.46 m occurred in October, while in West Kalimantan it only reached -0.98 m occurred in July. For the average monthly groundwater level, in Central Kalimantan the lowest value was -1.35 m occurred in October and in West Kalimantan was -0.90 m occurred in August as shown in Table 3.

Lower groundwater levels in Central Kalimantan mean that the number of hotspots that appear in Central Kalimantan is more than the number of hotspots in West Kalimantan, as shown in Figures 1 (e) and (f). The highest total number of monthly hotspots was in Central Kalimantan, reaching 1062 which occurred in September, and in West Kalimantan it was 269 which also occurred in September. It can be concluded that the highest number of hotspots in Kalimantan occurred in September.

Table 3. Monthly values of hydrometeorological parameters

	Total of Rainfall (mm)		Average of GWL (m)		Total of Hotspots	
Month	CK	WK	CK	WK	CK	WK
July	6.5	45.0	-0.82	-0.71	143	16
August	52.0	113.5	-1.13	-0.90	535	242
September	7.0	101	-1.32	-0.73	1062	269
October	93.5	258	-1.35	-0.55	126	2

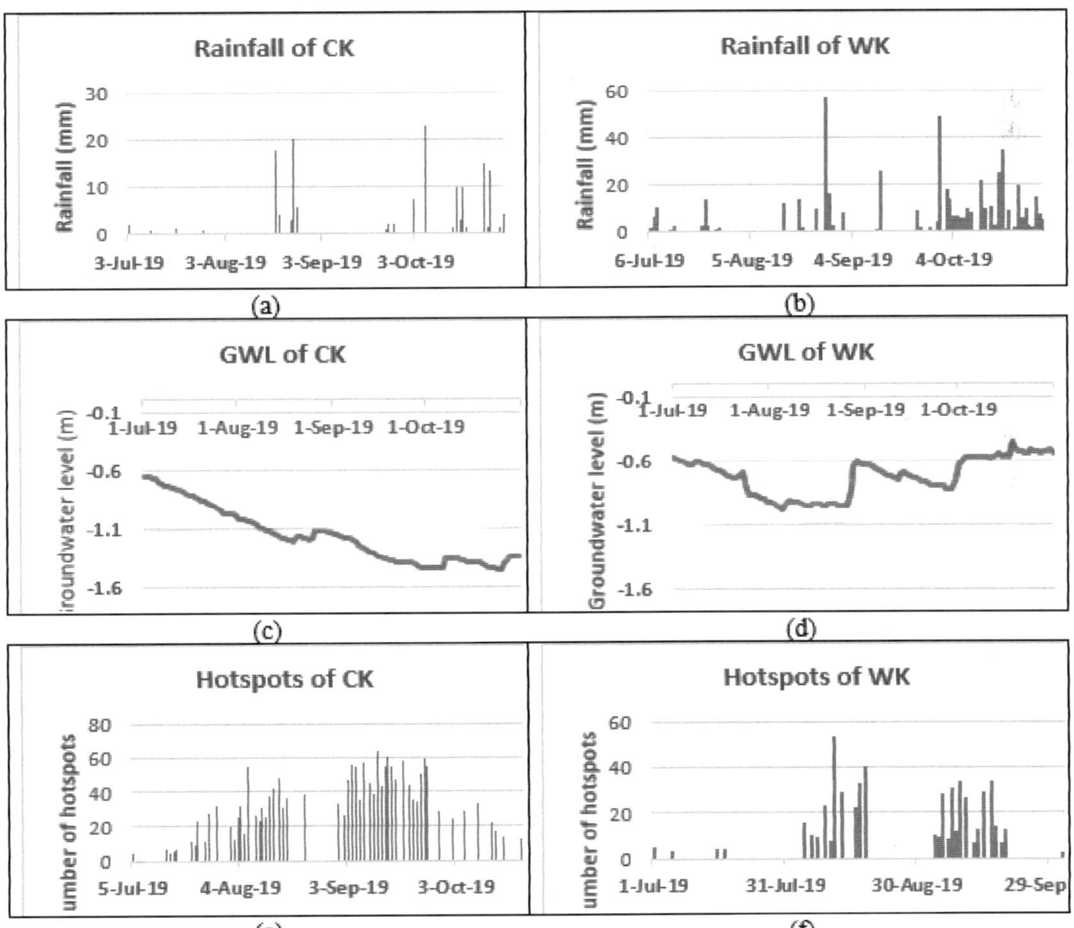

Figure 1. Time series graph of change in hydrometeorological parameters

Journal of Physics: Conference Series **2165** (2022) 012003 doi:10.1088/1742-6596/2165/1/012003

The extreme dry season in 2019 that occurred in several areas in Indonesia, including in Kalimantan, was the result of a natural phenomenon called positive Indian Ocean Dipole (IOD +). IOD occurs due to anomalous sea surface temperature in the Indian Ocean which can have an impact on excess rain in one place and lack of rain in another. In 2019, parts of Indonesia were affected by a lack of rain, which resulted in an extreme dry season [13,10].

3.2. Relationship between Rainfall and Groundwater Level

The graphs in Figures 2 (a) and (b) show the relationship between rainfall and groundwater levels. It appears that if there is no rain, the groundwater level tends to fall and vice versa. In fact, the rise and fall of the groundwater level in a peatland, apart from being influenced by rainfall, is also influenced by the condition of peatland cover, the types of plants that grow on the peat surface, the maturity of the peatlands, the porosity of the peat soil layer, the susceptibility of the peat soil layer, and the presence of canals near the peatland[14-17].

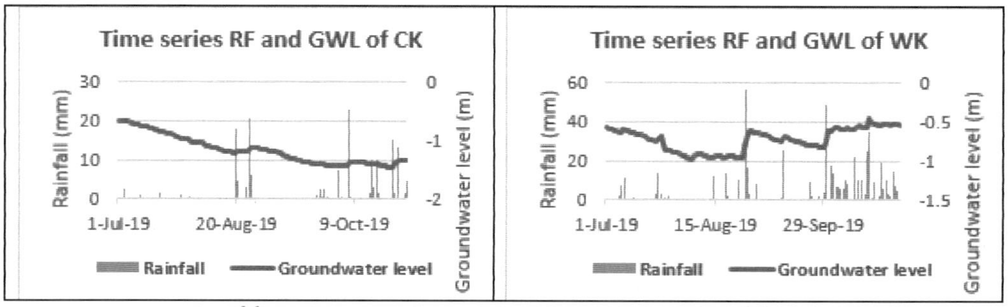

(a) (b)

Figure 2. Overlay graph of rainfall and groundwater level

3.3. Relationship between Groundwater Level and Hotspots

Figures 3 (a) and (b) show a graph of the relationship between the groundwater level and the number of hotspots that appear on the peatlands around the Central Kalimantan and West Kalimantan stations. In these graphs, it appears that when the groundwater level tends to rise, no hotspots will appear and vice versa. Then it appears that the number of hotspots at the Central Kalimantan station is more than the number of hotspots at the West Kalimantan station, while the groundwater level in Central Kalimantan is lower than that in West Kalimantan. Based on this, it can be said that there is a relationship between the groundwater level and the number of hotspots that appear on a peatland. The relationship is that the lower the groundwater level, the more hotspots that appear and vice versa. This is understandable because the lower the groundwater level, the drier the surface of a peatland is, making it easier for hotspots to appear on the peatland [18,19,20].

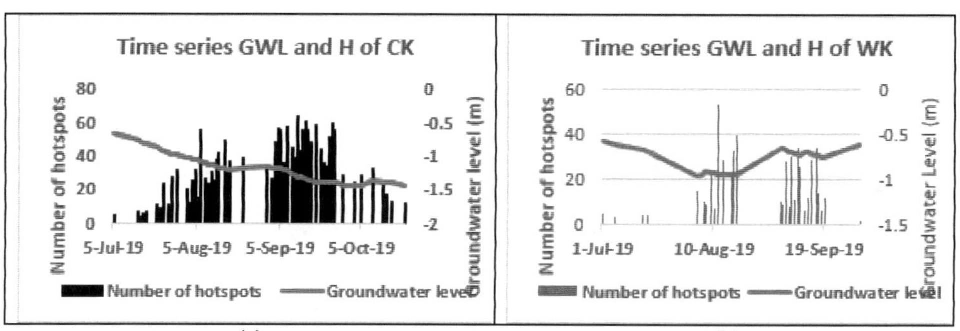

(a) (b)

Figure 3. Overlay graph of groundwater level and hotspots

4. Conclusion

In the extreme dry season of 2019 on the Kalimantan peatlands, there was low rainfall, low groundwater levels, and the emergence of a large number of hotspots. It was also found that there is a relationship between the rainfall and the groundwater level where the lower the rainfall, the lower the groundwater level and vice versa. There is also a relationship between the groundwater level and hotspots, where the lower the groundwater level, the more hotspots that appear and vice versa.

Acknowledgements

We thank to Sriwijaya University for supporting this research activity through the *Hibah Unggulan Kompetitif* 2021 for the first author and a *Hibah Unggulan Profesi* 2021 for the last author. We also thank to the Agency for Peatland Restoration (BRG) for providing rainfall and groundwater level data.

References

[1] Hugron S, Guêné-Nanchen M, Roux N, LeBlanc M C, and Rochefort L 2020 Plant reintroduction in restored peatlands: 80% successfully transferred – Does the remaining 20% matter? *Glob. Ecol. Conserv.* **22**

[2] Cahyarini S Y and Henrizan M 2018 Coral based-ENSO/IOD related climate variability in Indonesia: A review *IOP Conf. Ser. Earth Environ. Sci.* **118** (1)

[3] Leng L Y, Ahmed O H, and Jalloh M B 2019 Brief review on climate change and tropical peatlands *Geosci. Front.* **10** (2) pp. 373–380.

[4] Murdiyarso D, Lilleskov E, and Kolka R 2019 Tropical peatlands under siege: the need for evidence-based policies and strategies*Mitig. Adapt. Strateg. Glob. Chang.* **24** (4) pp. 493–505.

[5] Park H, Takeuchi W, Ichii K 2020 Satellite-based estimation of carbon dioxide budget in tropical peatland ecosystems*Remote Sens.* **12** (2), pp. 1–21.

[6] Uda S K, Hein L, and Atmoko D 2019 Assessing the health impacts of peatland fires: a casestudy for Central Kalimantan, Indonesia*Environ. Sci. Pollut. Res.* **26** (30) pp. 31315–31327.

[7] Hendrawan I G, Asai K, Triwahyuni A, and Lestari D V 2019 The interanual rainfall variability in Indonesia corresponding to El Niño Southern Oscillation and Indian Ocean Dipole*Acta Oceanol. Sin.* **38** (7) pp. 57–66.

[8] Irfan M, Satya O C, Arsali, Ariani M, Sulaiman A, and Iskandar I 2021 What is the rate of groundwater level decline on peatlands in South Sumatera during the 2019 extreme dry season? *J. Phys. Conf. Ser.* **1816** (1) p. 012008.

[9]Irfan M, Virgo F, Khakim M Y N, Ariani M, Sulaiman A, and Iskandar I 2021 The dynamics of rainfall and temperature on peatland in South Sumatra during the 2019 extreme dry season*J.Phys. Conf. Ser.* **1940** (1), p. 012030.

[10] Huang P, Zheng X T, and Ying J 2019 Disentangling the changes in the Indian Ocean dipole-related SST and rainfall variability under global warming in CMIP5 models*J. Clim.* **32** (13)pp. 3803–3818.

[11] Sankar S*et al.* 2019 The influence of tropical Indian Ocean warming and Indian Ocean Dipole on the surface chlorophyll concentration in the eastern Arabian Sea*Biogeosciences Discuss.*no. June, pp. 1–23.

[12] Irfan M, Mardiansyah W, Khakim M Y N, Ariani M, Sulaiman A, andIskandar I 2019 Some insight into direct observation of hydrological parameters in peatland area of the south sumatera *Int. J. Geomate* **17** (60) pp. 124–129.

[13] Doi T, Behera S K, and Yamagata T 2020 Predictability of the Super IOD Event in 2019 and Its Link With El Niño Modoki*Geophys. Res. Lett.* **47** (7) pp. 1–9.

[14] Dohong A, Aziz A A, and Dargusch P 2017 A review of the drivers of tropical peatland degradation in South-East Asia*Land use policy* **69**no. September, pp. 349–360.

[15] Hooijer A*et al.* 2013 Deep instability of deforested tropical peatlands revealed by fluvial organic carbon fluxes*Nature***493** (7434) pp. 660–663.
[16] Kajita R 2019 Historical precipitation data in Sumatra and Kalimantan from 1879 to 1900, by using Dutch colonial materials*IOP Conf. Ser. Earth Environ. Sci.***361** (1).
[17] Sloan S, Locatelli B, Wooster M J, and Gaveau D L A 2017 Fire activity in Borneo drivenby industrial land conversion and drought during El Niño periods, 1982–2010 *Glob. Environ. Chang.***47** no. November 2016 pp. 95–109.
[18] Ijumulana J, Ligate F, Bhattacharya P, Mtalo F, and Zhang C 2020 Spatial analysis and GIS mapping of regional hotspots and potential health risk of fluoride concentrations in groundwater of northern Tanzania*Sci. Total Environ.***735**p. 139584.
[19] Multsch S, Pahlow M, Ellensohn J, Michalik T, H. Frede H G, andBreuer L 2016 A hotspot analysis of water footprints and groundwater decline in the High Plains aquifer region, USA *Reg. Environ. Chang.***16** (8) pp. 2419–2428.
[20] Wallis I, Prommer H, Berg M, Siade A J, Sun J, and Kipfer R 2020 The river–groundwater interface as a hotspot for arsenic release *Nat. Geosci.***13** (4) pp. 288–295.

Analysis of college students misconceptions in astronomy using four-tier test

S N Azizah[1], H Akhsan[1], M Muslim[1] and MAriska[1]

[1]Physics education study program, FKIP Sriwijaya University

E-mail: mellyariska@fkip.unsri.ac.id

Abstract. This study aims to find out the misconceptions experienced by college students in Astronomy material using four-tier Diagnostic test through the google form. The population of this study were all 5th semester students of the Sriwijaya University Physics Education Study Program. This research is a quantitative descriptive research. Descriptive research here is intended to identify the physics misconceptions experienced by college students in material Astronomy using four-tier diagnostic test. Quantitative research is intended to obtain data from research samples analyzed according to the statistical methods used. This research produces a description of the phenomenon of Astronomical misconceptions that occur. The instrument used is a multiple choice four-tier diagnostic test consisting of 10 questions. Based on the results of research and data analysis showed that overall percentage, college students experienced misconceptions of 57.34%, so that college students' misconceptions were categorized as moderate level misconceptions.

1. Introduction

According to Law No. 23 of 2003 concerning National Education System, Education is a conscious and planned effort to create a learning atmosphere and learning process so that students actively develop their potential to have extraordinary strength, self-control, intelligence, noble character, and the skills they need, society, nation and state. One of the fields of knowledge in the educational process is Natural Sciences (Science), one of the branches of science is physics [1].

To create science education including quality physics learning is influenced by five things, namely understanding concepts, process skills, creativity, developing attitudes, and using concepts in everyday life. In the implementation of the five things above, of course, it cannot be separated from the role of an educator [2,3]. An educator of physics subjects based on Permendiknas No. 16 of 2007 is required to have four competencies, namely pedagogical competence, personality competence, social competence and professional competence. One of the above competencies, namely Professional competence, requires educators to have mastery skills in terms of material, structure, concepts, and scientific mindsets that support the learning process.In the process of learning Physics, understanding concepts are needed for students, in this case the role of educators is needed. Therefore, prospective Physics Educators should understand the concept according to the actual scientific concept. However, sometimes college students as propective of physics subject educators still experience errors in understanding the concept of physics. The source of errors in understanding concepts can come from the wrong initial interpretation, if the wrong interpretation is allowed, there will be misconceptions [4].

Misconception is a concept that is not in accordance with the concept recognized by experts. Misconception refers to a concept that is not in accordance with the scientific understanding or understanding accepted by experts in that field [5,6]. when a person experiences a misconception it means he cannot master the concept as a whole, even mixing between the actual concept with the preconception can cause a lot of physics difficulties. A prospective educator should correct the misconceptions he experiences.To correct these misconceptions, of course college students must know how far the misconceptions they have experienced [7].

Based on the material studied, physics is divided into three states, namely micro physics, visual physics and universe physics. One of the three states of physics is the physics of the universe, which is a physical science that studies the earth and space whose size is so large that it cannot be seen clearly with the naked eye [8]. One of the topics in the physics of the universe is Astronomy. Astronomy is a material that studies celestial bodies.As is well known, this astronomical material is included as one of the fields tested in the OSN (National Student Olympiad)at the school, there are no special subjects taught about astronomy, this makes it difficult for students to explore this astronomical material and even causes students to lack understanding of concepts or experience misconceptions. If improvements are not made by educators and students themselves, misconceptions will spread.Misconceptions can be built through a learning process with the surrounding environment or called constructivism, besides that misconceptions can come from educators, textbooks, teaching methods, contexts, and others [9]. As mentioned earlier, educators are one source of misconceptions, so it is important to identify the misconceptions of prospective educators, namely college students so that the misconceptions experienced by prospective educator students can be corrected [10,11].

In identifying misconceptions, diagnostic tests are needed for college students. Diagnostic tests are tests carried out to find out the weaknesses student, so that from those weaknesses the appropriate treatment [12]. There are many instruments that have been carried out in science education to identify misconceptions such as interviews, open-ended questions, concept maps and multiple choice tests all of which have advantages and disadvantages. The limitations of the previously written instruments encourage researchers to create more effective instruments in identifying misconceptions such as two-tier test, three-tier test and four-tier test [13,14].

Previous research related to the identification of misconceptions about astronomical material uses the CRI (Certainty Of Response Index) instrument as done by Sekar Rachmawati,she identified a misconception in the matter of the solar system.Then there is Taufik Ramlan Ramalis who identified the misconception of IPBA using the CRI (Certainty of Response Index) instrument.Another study related to Astronomy Misconceptions was conducted by Melly Ariska, the instrument used was a two-tier diagnostic test and it was concluded that students' understanding of concepts for each item was still diverse. Overall, the percentage of college students experiencing misconceptions was 15% [15–17].

In its development, the two-tier test instrument has drawbacks because it cannot determine the cause of someone who really experiences misconceptions, in this instrument all incorrect answers are considered misconceptions. Then this weakness is corrected by the presence of a three-tier test instrument which adds a tier of confidence that is placed after the first two levels to ensure one's confidence in answering the questions of the previous two tier [18]. However, this instrument also has limitations, namely when someone fills in at one level feeling unsure of the answer he chooses but inevitably has to choose sure because there is only one level of confidence in the question. For this reason, the present Four-tier test instrument adds two levels of confidence which are located after the question and after the reason. This four-tier diagnostic test instrument is considered to know more about a person's concept understanding[19,20]. Based on the description above, this study will analyze college students' misconceptions about astronomy using a four-tier diagnostic test based on google form.

Journal of Physics: Conference Series **2165** (2022) 012004 doi:10.1088/1742-6596/2165/1/012004

2. Methods

This type of research is a quantitative descriptive study with a Google Form based Four-Tier Diagnostic Test instrument.The test was carried out on 64 students of the physics education study program. the first tier of this diagnostic test,College Students must choose one of the answers provided, at the second tier college students must choose the tier of confidence in the previous answer.At the third tier, college students must choose one of the reasons that have been provided as a reason for the previous answer. And at the fourth tier, college students must choose the tier of belief for the reasons that have been chosen [21,22]. The four-tier Diagnostic Test can be easier and more detailed in distinguishing between understanding the concepts, not understanding the concepts and misconceptions. The confidence tierl in this diagnostic test is high if the selected scale is 4-6, then the confidence level is low if the selected scale is 1-3.

The results of this four-tier diagnostic test were analyzed with the first step, namely grouping college student answers based on the categories of understanding concepts (UC), not understanding concepts (NUC) and misconceptions (M) according to the criteria in Table 1.

Table1. Interpretation of four-tier diagnostic test results

Category	Tier 1	Tier 2	Tier 3	Tier 4
UC	True	High	True	High
NUC	True	Low	True	Low
	True	High	True	Low
	True	Low	True	High
	True	Low	False	Low
	False	Low	True	Low
	False	Low	False	Low
	True	High	False	Low
	False	Low	True	High
M	True	Low	False	High
	True	High	False	High
	False	High	True	Low
	False	High	True	High
	False	High	False	Low
	False	Low	False	High
	False	High	False	High

The second step is calculates the percentage of answers for each category. To calculate the percentage of college students who understand the concept, do not understand the concept and misconceptions can using the equation as follows[9]:

$$P = \frac{f}{n} x\ 100\% \quad (1)$$

P= percentage value
f= answer frequency
n=total number of college students

The third step is to make a graph of the percentage of misconceptions, and the fourth step is to describe the students' misconceptions for each question and are grouped based on the level of misconceptions according to Table 2 [1,23].

Table 2. Categories for the percentage of misconceptions

Percentage	Category
0 - 30%	Low
31% - 60%	Medium
61% - 100%	High

3. Result and Discussion

This study was attended by 64 students of Semester 5. The diagnostic test given through this google form consists of 10 multiple-choice four-tier questions equipped with options and the level of confidence in the answers and reasons..For the range of confidence levels, it includes only guess (1), very unsure (2), not sure (3), sure (4), very sure (5) and very very sure (6) [9,11].

The results of this diagnostic test are grouped and the number of college students calculated according to the categories Understand the concept (UC), Do not understand the concept (NUC) and Misconception (M), so that the results are as shown in Table 3.

Table 3. The number of college students based on the category of understanding concepts,not understanding concepts and misconceptions

Question Number	Number of college students		
	UC	NUC	M
1	8	12	44
2	0	8	56
3	5	29	30
4	3	16	45
5	8	27	29
6	0	19	45
7	6	30	28
8	10	14	40
9	9	28	27
10	2	39	23
Total	51	222	367
Percentage	7.96%	34.68%	57.34% (Medium)

From the data in table 3, it can be calculated the percentage of each category of college student concept understanding and classifying the level of student misconceptions based on the percentage as shown in Table 4.

Table 4. The results of the analysis of college student concept understanding

Sub Concepts	Percentage (%)			Category (misconceptions)
	UC	NUC	M	
Moon Phase	12.5	18.75	68.75	High
Outer Space Gravity	-	12.50	87.50	High
Relationship beetween planetary mass and body weight	7.81	45.31	46.87	Medium
Center Universe	4.68	25.00	70.31	High
Stars	12.50	42.18	45.31	Medium
Changing Seasons	-	29.68	70.31	High
Sun	9.37	46.87	43.75	Medium
Planets	15.62	21.87	62.50	High
Black Hole	14.06	43.75	42.18	Medium
Influence the mass of stars	3.12	60.93	35.93	Medium

Based on Table 4, it can be seen that in the concept of the moon phase contained in question number 1, the results of college students experiencing misconceptions are 68.75%.In the concept of outer space gravity contained in question number 2, it was found that students experienced misconceptions of 87.5%.In the concept of the relationship between planetary mass and body weight in question number 3, the results showed that college students had misconceptions of 46.87%. on the concept of the center of the universe contained in question number 4, the results of students' misconceptions are 70.31%. In the concept of stars contained in question number 5, it was found that students experienced a misconception of 45.31%.In the concept of changing seasons contained in question number 6, it was found that college students experienced misconceptions of 70.31%.In the concept of the sun contained in question number 7, it was found that college students experienced a misconception of 43.75%. In the concept of planets contained in question number 8, the results showed that college students had misconceptions of 62.5%. In the concept of black holes contained in problem number 9, the results showed that college students had misconceptions of 42.18%. and lastly, on the concept of the influence of the mass of stars contained in question number 10, it was found that college students experienced a misconception of 35.93%.

Based on the results of manual data analysis, it is known that college students experience varying levels of misconceptions in each question, the largest percentage of misconceptions (70.31%) is in items number 4 and 6 and is included in the medium level misconception. The lowest percentage of misconceptions (35.93%) is in question number 10 and is included in the moderate level of misconceptions.The percentage of misconceptions for each question is briefly explained in Figure 1.

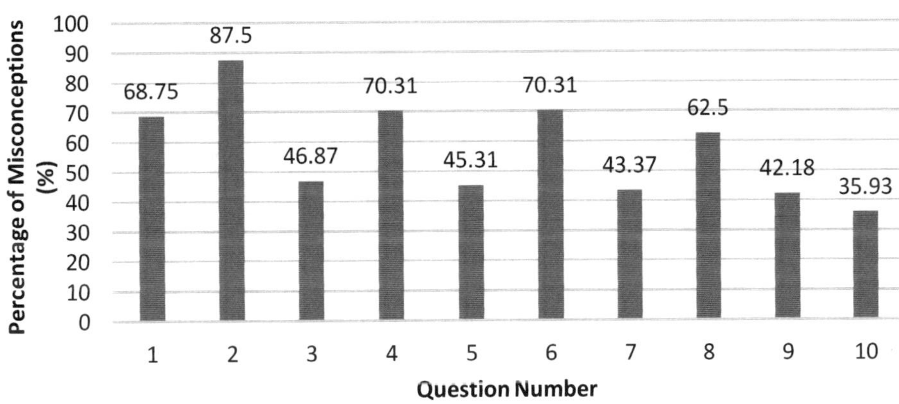

Figure1. Graph of the percentage of college students' misconceptions for each question

Based on the results of the analysis of test data, on the concept of the moon phase, college students experienced a high level of misconception,college Students assume that a solar eclipse occurs when the moon is in its full phase.and college students have a conceptual view that a solar eclipse occurs when the sun, earth and moon form a straight line, it is true that the three form a straight line but the position of the moon in the middle is not the earth.So the correct answer is that a solar eclipse occurs when the moon is in a new phase and a solar eclipse occurs when the moon's position is in the middle between the sun and the earth or the moon can be said to be a bridge between the sun and the earth.

The concept of outer space gravity, college students experience a high level of misconception, college students assume that what causes astronauts to float in spacecraft is because there is no gravitational force in outer space,while in reality the spacecraft has a gravitational force but it is smaller than the earth's gravitational force, this small gravitational force is called the microgravity force. This microgravity force causes astronauts to float in space.

The concept of the relationship between planetary mass and body weight, college students experience a moderate level of misconception, students assume that human weight is related to the weight of the planet they inhabit, but in reality the weight of the human does not depend on the weight of the planet but depends on the mass of the planet inhabited. Concept of the center of the universe, college students experience a high-level misconception, students assume that the center of the universe is the sun, when in fact the universe has no center, because the big bang theory which is considered the beginning of the formation of the universe does not occur in one place, but occurs in all over the place. Therefore the universe has no center.

The concept of stars, college students experience a moderate level of misconception, there are still students who think that the hottest star is red, in fact the hottest star is blue, and the concept of seasons, college students experience a high level of misconception, students assume that the change of seasons only depends on the revolution of the earth, but in fact the existence of the earth's revolution is to form various types of seasons that differ in each region, while the change of seasons itself depends on the tilt of the axis of rotation. which is the angle of the sun's position with respect to the equator formed during the earth's orbit about the sun.

The concept of the sun, college students experience a moderate level of misconception, in this concept there are still college students who think that solar energy is formed due to a reaction in which two atomic nuclei combine to form one or more small atomic nuclei, but in fact in the sun there is a proton-proton chain reaction which emits enormous energy.The concept of planets, college students experience a high level of misconception, students assume that the planet with the highest temperature is Mercury because it receives a large amount of solar energy, but in fact the planet with the highest temperature is Venus because the planet with the highest temperature is covered by thick clouds coated with carbon dioxide,This carbon dioxide prevents heat from the sun from escaping back into space.

In the concept of a black hole, college students experience a moderate level of misconception, in this concept some students think that the black hole's escape speed has reached the speed of light, but more precisely, a black hole is a giant hole that moves through space that sucks everything in its path. The concept of stellar mass, college students experience a moderate level of misconception, in this concept some students assume that the mass of the star is directly proportional to the age of the star, but the correct concept is that the mass of the star is inversely proportional to the age of the star.

This study shows results that are in line with previous research by Melly Ariska (2021) entitled Overview of the Initial Understanding of Physics Education Students' Concepts on Celestial Objects in Astrophysics Lectures using a two-tier diagnostic test instrument,where in this study it was found that college students experienced misconceptions about astronomical material by 15%.Another studyentitled Understanding the concepts and misconceptions of prospective teacher college students in Kepler law using a two-tier test instrument, where the results of students experiencing misconceptions of 33.82% [6,24]. Another studyentitled Identification of misconceptions about IPBA material in high school using CRI (Certainty Response Index) in an effort to improve the order of giving IPBA material on KTSP, where the results obtained are that there are still many students who have misconceptions and do not know the concepts in the material the IPBA [6,25].

4. Conclusion

Based on the results of the analysis with the four-tier multiple choice diagnostic test, it can be concluded that college students experience the highest misconception of 87.5% regarding the concept of the outer space gravity.Overall, the percentage of college students who understand concepts is 7.96%, college students who do not understand concepts are 34.68% and college students who experience misconceptions are 57.34%, college students' misconceptions on this astronomical material can be categorized as moderate level misconceptions.

Acknowledgement
The author would like to thank the 64 college students of the physics education study program who have become the samples of this research.

References
[1] M. Ariska, H. Akhsan, M. Muslim, and N. Azizah, "Pemahaman Konsep Awal Mahasiswa Pendidikan Fisika Terhadap Materi Benda-Benda Langit dalam Perkuliahan Astrofisika pembelajaran fisika merupakan ilmu yang digunakan untuk mengetahui dan memahami konsep-konsep dasar fisika yang ada di alam semesta (Aviyan," **5**, (3), 2021.
[2] J. Nirahua, J. Taihuttu, and V. Sopacua, "Pengembangan Bahan Ajar Berbasis Blended Learning Dan Critical Thinking Skill Pada Mata Kuliah Astrofisika Dalam Menyongsong Era Revolusi Industri 4.0," *Jambura Phys. J.*, **2**(1), pp. 24–36, 2020, doi: 10.34312/jpj.v2i1.6869.
[3] U. Kanli, "A study on identifying the misconceptions of pre-service and in-service teachers about basic astronomy concepts," *Eurasia J. Math. Sci. Technol. Educ.*, **10**, (5), pp. 471–479, 2014, doi: 10.12973/eurasia.2014.1120a.
[4] M. Ariska, "Studi Pemahaman Konsep Siswa Pada Sub Konsep Rangkaian Listrik Arus Searah Di Kelas Xi Sma Negeri 1 Palembang," *J. Inov. dan Pembelajaran Fis.*, **2**, (2), pp. 147–154, 2015, doi: 10.36706/jipf.v2i2.2616.
[5] J. Febrianti, H. Akhsan, and M. Muslim, "Analisis Miskonsepsi Suhu Dan Kalor Pada Siswa Sma Negeri 3 Tanjung Raja," *J. Inov. dan Pembelajaran Fis.*, **6**, (1), pp. 90–102, 2019, doi: 10.36706/jipf.v6i1.7819.
[6] Syuhendri, "Konsepsi Alternatif Mahasiswa Pada Ranah Mekanika: Analisis Untuk Konsep Impetus Dan Kecepatan Benda Jatuh," *J. Inov. dan Pembelajaran Fis.*, **1**(1), pp. 56–67, 2014, [Online]. Available: http://ejournal.unsri.ac.id/index.php/jipf/ article/view/1265.
[7] Supardi, L. Leonard, H. Suhendri, and Rismurdiyati, "Pengembangan Media Pembelajaran dan Minat Belajar Terhadap Hasil Belajar Fisika," *J. Form.*, **2**, (1), pp. 71–81, 2012, [Online]. Available: http://journal.lppmunindra.ac.id/index.php/Formatif/article/view/86/84.
[8] M. Ariska, H. Akhsan, and M. Muslim, "Dynamic Analysis of Tippe Top on Cylinder's Inner Surface with and Without Friction based on Routh Reduction," in *Journal of Physics: Conference Series*, 2020, **1467**(1), doi: 10.1088/1742-6596/1467/1/012040.
[9] B. Junedi, I. Mahuda, and J. W. Kusuma, "Optimalisasi keterampilan pembelajaran abad 21 dalam proses pembelajaran pada Guru MTs Massaratul Mut'allimin Banten," *Transform. J. Pengabdi. Masy.*, **16**(1), pp. 63–72, 2020, doi: 10.20414/transformasi.v16i1.1963.
[10] W. Rusli, A. Haris, and A. Yani, "Pokok Bahasa Gerak dan Gaya Three & Tier Test 5."
[11] A. Pujianto, "Analisis Konsepsi Siswa Pada Konsep Kinematika Gerak Lurus," *JPFT (Jurnal Pendidik. Fis. Tadulako Online)*, **1**(1), pp. 16–21, 2013, doi: 10.22487/j25805924.2013.v1.i1.2370.
[12] S. Nurfadila, I. Kaniawati, and W. Liliawati, "Identifikasi Miskonsepsi dan Penyebabnya Menggunakan Tes Diagnostik Pada Siswa SMA Kelas 11 Materi Gelombang Mekanik," **0**, pp. 99–107, 2020.
[13] D. Jumadi, stepanus, "MenggaliMiskonsepsi Siswa SD Tentang Tata SuryaSecara Lisan dalam Bahasa Dayak.," *J. Pendidik. Dan Pembelajaran Khatulistiwa*, **7** (5), 2018.
[14] M. Ariska, H. Akhsan, and M. Muslim, "Utilization of physics computation based on maple in determining the dynamics of tippe top," in *Journal of Physics: Conference Series*, 2019, **1166**(1), doi: 10.1088/1742-6596/1166/1/012009.
[15] D. R. Ningsih, T. R. Ramalis, and U. Purwana, "Pengembangan Tes Keterampilan Berpikir Kritis Berdasarkan Analisis Teori Respon Butir," *WaPFi (Wahana Pendidik. Fis.*, **3**(2), p. 45, 2018, doi: 10.17509/wapfi.v3i2.13730.
[16] H. Akhsan, K. Wiyono, M. Ariska, and N. E. Melvany, "Development of Higher order Thinking Test Instrument on Fluid Material for Senior High School Students," in *Journal of Physics:*

Conference Series, 2020, **1467**(1), doi: 10.1088/1742-6596/1467/1/012046.

[17] M. Ariska, H. Akhsan, and M. Muslim, "Vector Fields of the Dynamics of Non-Holonomic Constraint System With Elliptical Configuration Space," vol. 513, pp. 738–744, 2020.

[18] N. Novrianti, "Pengembangan Computer Based Testing (Cbt) Sebagai Alternatif Teknik Penilaian Hasil Belajar," *Lentera Pendidik. J. Ilmu Tarb. dan Kegur.*, **17**(1), pp. 34–42, 2014, doi: 10.24252/lp.2014v17n1a3.

[19] R. Mahmudah, Y. Pramudya, and D. Sulisworo, "Analisis Validitas Butir Soal Certainty of Respons Index (Cri) Untuk Identifikasi Miskonsepsi Materi Tata Surya Dan Fenomena Astronomi," *Seminar Nasional Pendidikan Sains 2016*. pp. 1–15, 2016.

[20] A. Fauziah and Y. Darvina, "Analisis Miskonsepsi Peserta Didik Dalam Memahami Materi Gerak Lurus Dan Gerak Parabola Pada Kelas X Sman 1 Padang," *Pillar Phys. Educ.*, **12** (1), pp. 73–80, 2019, doi: 10.1017/CBO9781107415324.004.

[21] W. Abdullah, "Seberapa Luas Sih Alam Semesta Itu?(Sebuah Tinjauan Singkat Beberapa Miskonsepsi Tentang Alam Semesta Yang Mengembang …," *Academia.Edu*, pp. 1–23, 2014, [Online]. Available: https://www.academia.edu/download/40466305/Seberapa_Luas_Sih_Jagat_Raya_Itu.pdf.

[22] PADU, "Malaysia Education Blueprint 2013 - 2025," *Education*, **27**, (1), pp. 1–268, 2013, doi: 10.1016/j.tate.2010.08.007.

[23] H. H. Kamaluddin dan Fihrin, P. Studi Pendidikan Fisika FKIP Universitas Tadulako Jl Soekarno Hatta Km, and K. Bumi Tadulako Tondo Palu – Sulawesi Tengah, "Analisis Pemahaman Konsep Gerak Lurus pada Siswa SMA Negeri di Kota Palu," *J. Pendidik. Fis. Tadulako*, **4**(3), pp. 3–5, 2338.

[24] A. L. Ball and B. L. Garton, "Modeling Higher Order Thinking: The Alignment Between Objectives, Classroom Discourse, And Assessments," *J. Agric. Educ.*, **46**(2), pp. 58–69, 2005, doi: 10.5032/jae.2005.02058.

[25] B. Limbach and W. Waugh, "Developing higher level thinking."

The influence of the hybrid learning model using the schoology application in basic physics lectures to improve process skills in the industrial 4.0 period

E S Nasution[1], Y Rhamyanti[1], R U Lubis[1], D G S Harahap[2], A N Harahap[2], Y Sara[3], R S Sikumbang[3]

[1] Lecture Department of Physics Education, faculty of teaching and education, Graha Nusantara University Padangsidimpuan, Indonesia
[2]Lecture Department of Mathematics Education, faculty of teaching and education, Graha Nusantara University Padangsidimpuan, Indonesia
[3]Student Department of Physics Education, faculty of teaching and education, Graha Nusantara University Padangsidimpuan, Indonesia

E-mail: enisumanti.nst@gmail.com, yunirhamayantiugnp@gmail.com

Abstract. The development of science and technology in the industrial period 4.0 needs to be improved through the improvement of human resources. The learning process in the industrial period 4.0 is required for the world of education to develop, especially in the field of technology. One of the learning models used is the Hybrid learning model by using schoology applications to assist students in improving process skills in basic physics course materials. Therefore, one of the objectives of this study is to find out the influence of hybrid learning models using schoology applications in basic physics courses in improving process skills in the face ofindustry 4.0. The population of this study is all students of the Faculty of Teacher Training and Education. Sampling techniques are used cluster random sampling where students as experimental classes that use hybrid learning models using schoology and math students as control classes using e learning models. The instrument used is a matterof process skills test. The resultof the study for the average value of the control class on pretes was 37.17 and the average postes score was 66.94. The average pretest score in the training experiment classwas 36.83 and the average postes score was 78.61, this suggests there is an influenceof hybrid learning models using schoology applications on process skills in basic physics courses.

1. Introduction

In early 2020, the world was shocked by the spread of a new type of coronavirus (SARS-Covid) and a disease called Coronavirus disease 2019 (COVID) originated in Wuhan, China. Found now it has been confirmed that there are 65 countries that have contracted this virus [1].

Indonesia has contracted covid-19 in March 2020. With this spread, the government made a policy that everything is done at home. This makes no difference in the world of education. Based on the policy carried out by the government, the learning process also applies at the Faculty of Teacher Training and Education, Graha Nusantara University so that the learning process in the faculty of teaching and education through home learning.

However, the process of implementing distance learning did not run smoothly. Many obstacles were experienced as well as complaints submitted by students, parents, and also teachers. For example, about too many tasks, limited devices and internet access for learning, to saturation in carrying out the learning process.

Based on the results of a survey conducted with pandemic conditions by conducting online learning in students having difficulty in terms of capturing lecture materials, especially in basic physics course materials, this is because there are still many students who are lacking in technology, resulting in students' process skills in the learning process in low basic physics courses. besides that also the low skills of the student process due to distance learning used through online learning still using whatssapp application media is commonly used and utilized by students so that when students do the process of skills where in the process skills students have to go through the stages of low-skill student process because in the learning used during the pandemic using the online learning model in utilizing whatsapp so that when displayed in the learning students still have difficulty in understanding the basic physics lecture materials.

Basically, with the advancement of technology today, it has an impact on aspects of the whole world, one of which is the learning process [2,3], so to overcome the problems experienced by students in online learning applied today, hybrid learning is used. In this hybrid learning, students participate in face-to-face learning and online learning. In this hybrid learning, learning is generally a process of social activities that take place in online learning where there are interactions.

In contrast to distance learning which is carried out fully online, hybrid learning is a learning system that combines face-to-face learning and online learning. In the classroom, the teacher teaches two groups of students, those who attend directly to school and those who join online via video conferencing applications. Well, like other learning models, the application of hybrid learning during a pandemic cannot be separated from the pros and cons that arise in the community.

Hybrid learning is basically a social process that will be compromised if all the processes that take place learning away from human interaction[4,5,6] state that with hybrid learning has a different meaning in each person. This hybrid learning model is a blend that combines face-to-face learning in the classroom and online learning. In this hybrid learning students are faced with structural problems, then students work in groups to formulate problems. Teachers facilitate students in learning, organize learning tasks, mediate problem formulations, and formulate hypotheses [8].

The many uses of applications used in online learning is one that is used in improving the skills of this process is to use schoology applications. Schoology is one of those social web sites where it offers learning just like in the classroom for free and easy to use like facebook. Through schoology, learning is very easy. Schoology is also almost as functional as other websites such as Web CT and Blackboard and in it, it offers Lecturers to upload up all the important coursework as well as learning materials required by students in their subjects [9].

Schoology is a Learning Management System (LMS) for schools where visual and functional is easy to use such as social media facebook, services that can be used in the form of attendance records, online gradebook (facilities to manage grades), tests and quizzes, and homework. Schoology is one of several types of Social Learning Networks (SLNs) circulating in the world wide web. E-learning media by using schoology as a learning medium that may be able to help the effective learning process. With this media is expected to attract the attention of students sabegai new learning media and support the interest of students to learn [10].

2. Methods

2.1. Types of Research and Research Design

This qualitative research is a measurement that determines the number or data in the form of numbers [11]. This type of research is quasi experiments that have treatment, and do not use random placement. The design used in this study is *Nonequivalent Posttest Only Control Group Design.*

X

Treatment Group ———————— O_1

Control Group O_2

Figure 1. *Nonequivalent Posttest Only Control Group Design* [12,13]

The population used is all students of the faculty of teaching and education. The sample is for experimental classes using physics students in semester 1 and control classes of mathematics students in semester 1.

2.2. Methods of Data Collection and Data Analysis

This research uses instruments about process skills. The skill of this process consists of several indicators namely 1) observing, 2) interpreting, 3) asking questions, 4) formulating hypotheses, 5) using tools and materials, 6) applying concepts, 7) communicating [14]. The form of test used is to use a description test.

This study was given pretest first before being treated to both classes after it was conducted similarity test, then after that given post- test by giving treatment [15] where the control class using online learning model and experimental class using hybrid learning model using schoology application.

The data analysis used at the time of the pretest in determining similarity tests in two classes is the independent samples T-test. This T-test is done after the data obtained is normally distributed and has the same variations or homogeny data. After that, an analysis of variance was conducted to see the posttest in that class. The posttest value is used to see how the hybrid learning model influences using schoology applications.

3. Result and Discussion

3.1. Result

The results of the research obtained after pretest and posttest in experimental class and control class. The average value in experimental classes where here using hybrid learning models using schoology applications and experimental classes using online learning models or e learning. The results of process skills in both classes are as in figure 2:

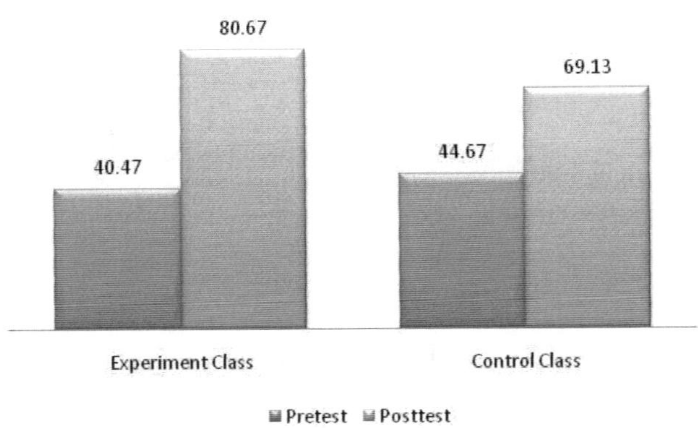

Figure 2. Average Results of Process Skills in Experimental Classes and Control Classes

Based on figure 2 above obtained that the average value of process skills increased whereas in the control class obtained an average score from 40.47 to 69.13 and in the experimental class from the pretest score of 40.47 in the pretest and 80.67 in the posttest.

Once obtained the pretest and posttest values in the experiment class are then looked for whether both classes have normally distributed data. Here using SPSS software in finding data whether it is normal distribution or not. As for the condition if significant > 0.05 then the data is distributed normally. The data to see if it is normally distributed in the experiment class and control class on pretest and posttest can be found in Table 1.

Tabel 1. Test The Normality of The Experimental Class and control class.

	Kolmogorov-Smirnov[a]		
	Statistic	df	Sig.
Pre-test Control Class	.146	15	.200*
Pre-test Experimental Class	.186	15	.171
Post-test Control Class	.101	15	.200*
Post-test Experimental Class	.163	15	.200*

From the data in Table 1 above shows the control class at the time of pretest 0.200 > 0.05 and posttest 0.200 > 0.05 and in the experiment class pretest 0.171> 0.05 and posttest 0.200>0.05. Based on these results both control classes and experiments at the time of pretest and posttest are normally distributed. Furthermore, after knowing that the data obtained is normal distribution then continued with whether the data is in the same variance. The test is a homogeneity test at the time of pretest and posttest. The data is in Table 2.

Tabel 2. Homogeneity Test of Pre-test and Post-test

	Levene Statistic	df1	df2	Sig.
Pre-test	.001	1	28	.972
Post-test	.631	1	28	.434

The requirement for homogeneity test is significantly greater than 0.05. From table 2 above shows in pretest 0.972>0.05 indicates that both classes have the same variance and in the posttest class 0.434 > 0.05 indicates that both classes have the same variance. The next step that needs to be done is to know the t test at the time of pretest and posttest. The hypothesis is that if it is significantly larger than 0.05 then there is no difference in both classes and if it is significantly smaller than 0.05 then there is a difference in both classes. The data on the t test is contained in Table 3.

Table 3. Test Independent Paired T test at Pretest and Post-test

	t-test means	Df	Sig. (2-tailed)	Description
Pre-test	4.20000	28	.324	No differences in control classes and experimental classes
Post-test	-11.53333	28	.008	There are differences in control classes and experiments

In Table 3 here it is obtained that by conducting an independent paired T test that by using significantly stated greater than 0.05 i.e., 0.324>0.05 with it can be concluded when pretest initial ability in experimental class and control class is the same. When the posttest significant is smaller than 0.05 i.e., 0.008<0.05 can be inferred to have differences in capabilities in the experimental class and control class.

Furthermore, to see the influence is to use analysts of variance of its use with SPSS application that is if significantly smaller than 0.05 i.e., 0.008 then it can be concluded that there is an influence of hybrid learning model by using Schoology application in improving student process skills as contained in Table 4.

Table 4. Variance Analysis Test

	Sum of Squares	df	Mean Square	F	Sig.
Between Groups	997.633	1	997.633	8.122	.008
Within Groups	3439.067	28	122.824		
Total	4436.700	29			

3.2. Discussion

Based on the results of the above research where in the experimental class there was an increase from 40.67 to 80.67 it can be shown that there is an influence in the hybrid learning model using schoology this is due to students becoming active and independent in learning activities so as to improve student achievement as well as students facilitate in solving problems and evaluating them [16,17,18]. In addition, students in the process of learning psychomotor abilities or student process skills become improved because students more easily understand the concept by using this hybrid learning model [19].

Hybrid learning using schoology also improves students' process skills in basic physics materials because by using schoology, lecturers and students build a collaborative learning. The schoology app can also be used to take advantage of time outside of learning hours [22].

Hybrid learning is also by using schoolgy at the time of Online learning conducted before face-to-face learning to observe phenomena and determine the hypotheticals of the problem. This is done by looking at the videos that have been given by teachers on the schoology platform and to hypothesize it is done in a discussion in the comments field. Through activities conducted by observation and giving conjecture can improve the ability to think critically [23]. In addition, with schoology application can also improve students' learning outcomes when face-to-face experiments are conducted, the activity students can find the facts of the truth of the problem and interact socially with their peers. Through experimental methods students learn to find answers to problems and through student's social interaction students are able to improve students' critical thinking skills [24,25].

Based of the research in average value of process skills increased whereas in the control class obtained an average score from 40.47 to 69.13 and in the experimental class from the pretest score of 40.47 in the pretest and 80.67 in the posttest, from here show the influence hybrid learning to improved process skills, if compare when compared with previous research conducted by [26] the result show that the effect by hybrid learning has had an impact interactive environment and can learn to be independent (for upgrade them skill) including interaction with teacher or peers. Limitation and future research are discussed is the practicality of Hybrid Learning and an average score of 91.

4. Conclusion

Based on the research data, it can be concluded that there is an influence of hybrid learning model by using schoolgy application in improving process skills in basic physics lecture materials this is seen by the increasing learning outcomes where there is a significant difference of $0.008<0.05$ or process skills for posttest higher experimentation (80.67) than the control class (69.13).

Acknowledgement

The author thanks the Faculty of Teacher Training and Education of Graha Nusantara University Padangsidimpuan for providing facilities in conducting this research. Thanks also to the physics education courses and mathematics courses with permission to conduct this research.

References

[1] S. Kannan, P. Shaik Syed Ali, A. Sheeza, and K. Hemalatha, "COVID-19 (Novel Coronavirus 2019) - recent trends," *Eur. Rev. Med. Pharmacol. Sci.*, **24**, (4), pp. 2006–2011, 2020, doi: 10.26355/eurrev_202002_20378.

[2] A. Benešová and J. Tupa, "Requirements for Education and Qualification of People in Industry 4.0," *Procedia Manuf.*, **11** June, pp. 2195–2202, 2017, doi: 10.1016/j.promfg.2017.07.366.

[3] Afrianto, "Being a professional teacher in the era of industrial revolution 4.0: opportunities, challenges and strategies for innovative classroom practices," *English Lang. Teach. Res.*, **2** 1, p. 3, 2018, [Online]. Available: http://ejournal.unp.ac.id/index.php/eltar/article/view/102675.

[4] R. Donnelly, "Blended problem-based learning for teacher education: Lessons learnt," *Learn. Media Technol.*, **31**, (2), pp. 93–116, 2006, doi: 10.1080/17439880600756621.

[5] G. Pool and A. S. Byatt, "Chronicle of Higher Education," *Women's Rev. Books*, **13**, (10/11), p. 35, 1996, doi: 10.2307/4022491.

[6] A. A. Okaz, "Integrating Blended Learning in Higher Education," *Procedia - Soc. Behav. Sci.*, **186**, 600–603, 2015, doi: 10.1016/j.sbspro.2015.04.086.

[7] B. F. Melton, H. Bland, and J. Chopak-Foss, "Achievement and Satisfaction in Blended Learning versus Traditional General Health Course Designs," *Int. J. Scholarsh. Teach. Learn.*, **3**, no. 1, 2009, doi: 10.20429/ijsotl.2009.030126.

[8] Slavin R E, *Educational psychology theory and practice Eight edition*. 2009.

[9] E. Yuliyanto and M. Imaduddin, "Panduan Pembelajaran E-Learning," 2014.

[10] Beny Hari Firmansyah, "[JOURNAL BM-Beny Hari Firmansyah] Pengembangan Blended Learning berbasis Schoology," *Teknol. Pembelajaran*, **109**, (1), pp. 5–10, 2015.

[11] P. V Young, *Scientific Social Surveys and Research, 3rd ed*, 3rd ed. New York: Prentice-Hall, 1960.

[12] C. Prof. Dr. H. M. Wahyudin Zarkasyi, *Penelitian Pendidikan Matematika*, 1st ed. Kerawang: PT Refika Aditama, 2015.

[13] C. R. Kothari, *Research Metodology Methods And Techniques*, Second Rev. New Delhi: New Age International Limeted, 2004.

[14] K. L. Ting and N. M. Siew, "Effects of Outdoor School Ground Lessons on Students' Science Process Skills and Scientific Curiosity," *J. Educ. Learn.*, **3**, (4), pp. 96–107, 2014, doi: 10.5539/jel.v3n4p96.

[15] J. W. Creswell, *Research Design Pendekatan Kualitatif, Kuantitatif, dan Mixed*, 4rd ed. SAGE Publication Inc, 2012.

[16] P. Indarto, M. Fatoni, and U. M. Surakarta, "Model Pembelajaran Hybrid Learning Pada Mata," pp. 4–5, 2018.

[17] T. Ramdhani, I. G. P. Suharta, and I. G. P. Sudiarta, "Pengaruh Model Pembelajaran Hybrid Learning Berbantuan Schoology Untuk Meningkatkan Prestasi Belajar Matematika Siswa Kelas Xi Sman 2 Singaraja," *J. Pendidik. Mat. Undiksha*, **11**, (2), pp. 2613–9677, 2020, [Online]. Available: https://ejournal.undiksha.ac.id/index.php/JJPM/article/view/24967.

[18] F. Alonso, G. López, D. Manrique, and J. M. Viñes, "An instructional model for web-based e-learning education with a blended learning process approach," *Br. J. Educ. Technol.*, **36**, (2), pp. 217–235, 2005, doi: 10.1111/j.1467-8535.2005.00454.x.

[19] Y. Sunaryo, I. Nuraida, and N. E. Zakiah, "Pengaruh Model Pembelajaran Hybrid Tipe Traditional Clasess-Real Workshop Terhadap Kemampuan Pemahaman Matematik Ditinjau Dari Self-Confidence Siswa," *Teorema*, **2**, (2), p. 93, 2018, doi: 10.25157/.v2i2.1071.

[20] S. Biswas, "Schoology-Supported Classroom Management: A Curriculum Review," *Northwest J. Teach. Educ.*, **11**, (2), pp. 0–10, 2013, doi: 10.15760/nwjte.2013.11.2.12.

[21] W. S. Ambar Ningsih, W. Suana, and N. Maharta, "Pengaruh Penerapan Blended Learning Berbasis Schoology Terhadap Kemampuan Berpikir Kritis Siswa," *Konstan - J. Fis. Dan Pendidik. Fis.*, **3**, (2), pp. 85–93, 2018, doi: 10.20414/konstan.v3i2.16.

[22] I. Dobre, "Learning Management Systems for Higher Education - An Overview of Available Options for Higher Education Organizations," *Procedia - Soc. Behav. Sci.*, **180**. November 2014, pp. 313–320, 2015, doi: 10.1016/j.sbspro.2015.02.122.

[23] P. D. A. Putra and Sudarti, "Pengembangan sistem e-learning untuk meningkatkan keterampilan berpikir kritis mahasiswa pendidikan fisika," *J. Fis. Indones.*, **19** (55), pp. 45–48, 2015.

[24] L. S. Vygotsky, *Mind in Society Development of Higher Psychological Processes*. New York:, 1994.

[25] A. Ardila, "L.S. Vygotsky in the 21st century," *Psychol. Russ. State Art*, **9**, (4), pp. 4–15, 2016, doi: 10.11621/pir.2016.0401.

[26] Disa Hediansah , Herman Dwi Surjono. Hybrid Learning Development to Improve Teacher Learning Management. JKTP **3(**1) Februari (2020): 1-9

Synthesis of $BaFe_{12-x}Cr_xMn_xO_{19}$ material based on Telindung beach iron sand using coprecipitation method

H A Fatari[1], A Doyan[1,2*], M Makhrus[1,2], Susilawati[1,2*] and Y Taryana[3]

[1]Master of Science Education Program, University of Mataram, Lombok, West Nusa Tenggara, Indonesia
[2]Physics Education, FKIP, University of Mataram, Lombok, West Nusa Tenggara Indonesia
[3]Research Center for Electronics and Telecommunication, National Research and Innovation Agency, Indonesia

E-mail: halfatari@gmail.com, aris_doyan@unram.ac.id, susilawatihambali@unram.ac.id

Abstract. The production of $BaFe_{12-x}Cr_xMn_xO_{19}$ material has been carried out based on iron sand at Telindung Beach, East Lombok Regency, using the coprecipitation method. The manufacture of $BaFe_{12-x}Cr_xMn_xO_{19}$ material has three stages: the separation stage, the magnetic material separation stage, and the synthesis stage. The separation stage aims to separate the material from non-magnetic materials to obtain the maximum total Fe content. The results obtained are 43.50 grams of magnetic material from 50 grams of natural sand using a permanent magnet. The next stage is separating the magnetic materialused as the basic ingredients in the synthesis of $BaFe_{12-x}Cr_xMn_xO_{19}$. The fundamentalcomponents used in the synthesis process were $BaCO_3$, Fe_3O_4 obtained from the separation, $MnCl_2$ and CrO_3 as doping, NH_4OH as a precipitate, and 37% HCl as a solvent. The dissolution process was carried out using a hot plate magnetic stirrer for 30 minutes at 80°C. Synthesis Barium M-Hexaferrite based on natural iron sand by using doping ion variations x= 0.0; 0.4; 0.8; 1.0 and calcination temperature of 200 °C, 600 °C and 1000 °C. The synthesis results made the colour change of $BaFe_{12-x}Cr_xMn_xO_{19}$ powder darker. This happens because the higher the mole fraction of doping used, the darker the powder obtained. Moreover, the darker colour of the powder indicates that the solvent has evaporated due to the calcination process.

1. Introduction

Iron sand is a natural material that is superabundant and has the main components of iron oxide, namely magnetite (Fe_3O_4), hematite (α-Fe_2O_3), and maghemite (γ-Fe_2O_3), several other oxide compounds such as Al_2O_3, MgO, and SiO_2 [1]. The difference in mineral content is caused by the structural geology and mineralization process in each region. These minerals are 88% magnetic and 12% without magnetic properties. Geologically, iron sand on the Lombok in West Nusa Tenggara province, Indonesia can be found along 2,333 km and certainly has natural iron sand that has been potentially developed [2]. One of the beaches with prospective natural iron sand is Telindung beach, located in Mudung hamlet, Anggaraksa village, Pringgabaya sub-district.

Telindung Beach is one of the tourist areas with a stretch of black sand beach in the 90s. However, now the tourist area is no longer used because there has been a beach abrasion caused by erosion. When viewed from these characteristics, Telindung Beach has very black, dense and smooth sand. Black iron sand reveals a lot of iron oxide content, and the magnetic properties increase proven by the substantial amount of sample is pulled with a permanent magnet [3]. The size of the sampling distance also influences the increasing value of Fe. This affects the resistivity value of a sample. The higher the resistivity value of a material, the more difficult for the material to be electrified. Factors that increase the sample resistivity are the colour and the grain size of iron sand. The grain size of the iron sand influences the resistivity value because the smaller the grain size, the more grain boundaries. The number of grain boundaries reduces the average free path of charge carriers, or the electrons are difficult to move; hence, it will have a greater resistivity [4]. The mineral content of iron sand can be applied in various fields so that experts use iron sand as a study material in microwave absorption research.

One of the microwave absorption material applications is the manufacture of RAM (Radar Absorbing Material), which can be used in the military. This material is able to dampen reflections or absorb microwaves so that objects coated with RAM are not detected by (RADAR). The microwave absorber consists of two components, a dielectric absorber and a magnetic absorber. Therefore, it is necessary to develop materials that can absorb both dielectric and magnetism. However, the material-based iron rock form is still a perfect choice for microwaves' magnetic absorbing material [5].

The material used in microwave absorption must have low coercivity, high resistivity and high magnetic saturation to produce a strong wave absorption material [6]. One material that is quite ideal for use as a microwave absorbing material is Barium hexaferrite ($BaFe_{12}O_{19}$). Barium ferrite magnets have been developed since 1960, which are used as an alternative to metal magnets. This magnet belongs to hard magnetic ceramics with a hexagonal structure ($BaFe_{12}O_{19}$). The advantages of Barium hexaferrite are good chemical stability, relatively inexpensive, and easy production. Therefore, it is very suitable for technological equipment in a reasonably wide range [6]. In addition, Barium hexaferrite, which has lossy material properties, has a high dielectric loss and magnetic loss factor, so the material has good properties for electromagnetic wave absorption [7]. However, Barium hexaferrite magnetic material only has a relatively small and district width of microwave absorption frequency. Therefore, special treatment is needed to increase the width of microwave absorption frequency.One solution that can be done is adding additives in the metal form to widen the frequency of microwave absorption by $BaFe_{12}O_{19}$ [8]. The addition of additives in this research used a mixture of Cr and Mn metal ions doping.

The substitution carried out in this study used a combination of Mn-Cr transition metal elements. Double ion substitution aims to combine the good properties of each doping component to reduce the strength of the coercivity field significantly. Double ion substitution of Co-Zn, Co-Mn, and Mn-Ni successfully modified the magnetic properties [9]. Mn ion substitution can reduce the coercivity field by 0.197 T at a calcination temperature of 800°C [10]. This research uses a combination of transition metal elements Mn-Cr to form the $BaFe_{12-2x}Cr_xMn_xO_{19}$ phase because these two elements have a radius and ionic configuration similar to Fe^{3+}. The electrical and magnetic properties of the substitution of Barium M-Hexaferrite depend on the synthesis process [7].

Several methods are able to synthesize Barium M-Hexaferrite, such as thermal decomposition method, solvothermal method, hydrothermal method, sonochemical method, microemulsion method, coprecipitation method, salts method, and ionic liquid at room temperature. Of the several types of methods, the coprecipitation method is used in the synthesis of iron sand into magnetite nanoparticles because it has the advantages of short reaction time, can be dissolved in water solvents, has a low economical price, is simple in the synthesis process and takes place at low temperatures (< 100°C) [11].

Based on the description above, this research concentrates on the production of $BaFe_{12-x}Cr_xMn_xO_{19}$ powder based on $BaFe_{12}O_{19}$of natural iron sand doped with Mn and Cr metals. The doping variation of x=

Journal of Physics: Conference Series **2165** (2022) 012006 doi:10.1088/1742-6596/2165/1/012006

0.0; 0.4; 0.8 and 1.0 and with calcination temperatures of 200°C, 600°C and 1000°C. This treatment is expected to produce magnetic materials with ideal electrical and magnetic properties as microwave absorption materials.

2. Methods

This research is experimental research on a laboratory scale and carried out at the Advanced Chemistry Laboratory, Faculty of Mathematics and Natural Sciences, University of Mataram. The method used is the coprecipitation method to synthesize Barium M-Hexaferrite based on natural iron sand at Telindung Beach, East Lombok Regency, with mixed doping of Mn and Cr ions. This synthesis aims to produce $BaFe_{12-x}Cr_xMn_xO_{19}$ material with a doping variation of x = 0.0; 0.4; 0.8 and 1.0 and with calcination temperatures of 200°C, 600°C and 1000°C.Calcination process is carried out for 2 hours. This synthesis process is carried out in 3 stages: the separation stage, the magnetic material separation stage, and the synthesis stage, as shown in Figure 1.

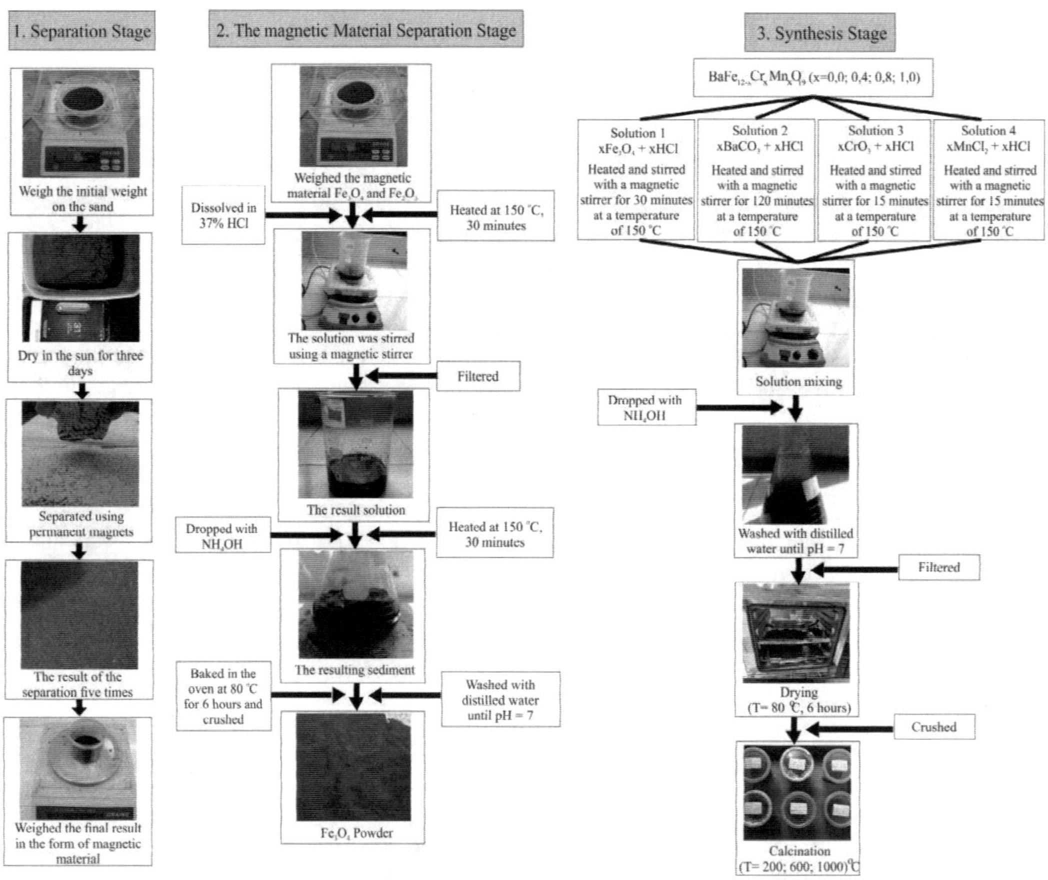

Figure 1. Stages of synthesis $BaFe_{12-x}Cr_xMn_xO_{19}$ material using the coprecipitationmethod (1) Separation stage; (2) Magnetic material separation stage; (3) Synthesis Stage.

3. Result and Discussion

The natural iron sand of Telindung Beach, East Lombok Regency, is the essential ingredients for the synthesis of $BaFe_{12-x}Cr_xMn_xO_{19}$ material. Magnetic material from natural sand is separated five times using a permanent magnet. Hence, the obtained results were43.50 grams or 87% magnetic material from 50 grams of iron sand at Telindung Beach, East Lombok Regency. After obtaining the magnetic material, it is separated from the impurities for processing to the synthesis stage. Before carrying out the synthesis stage, chemical reactions are equalized according to the required composition. The reaction equation is balanced based on the concentration of doping ions used: x= 0.0; 0.4; 0.8 and 1.0. Thereaction equalization for each compound composition is as follows:

Reaction equation on the composition of $BaFe_{12}O_{19}$ (x = 0.0)

$$4Fe_3O_4 + 32HCl \rightarrow 4FeCl_2 + 8FeCl_3 + 16H_2O$$
$$BaCO_3 + 2HCl \rightarrow BaCl_2 + H_2CO_3$$

$$4FeCl_2 + 8FeCl_3 + 16H_2O + BaCl_2 + H_2CO_3 + 34NH_4OH \rightarrow BaFe_{12}O_{19} + 34NH_4Cl + 32H_2O + 2H_2 + CO_2$$

Reaction equation on the composition of $BaFe_{11.2}Cr_{0.4}Mn_{0.4}O_{19}$ (x = 0.4)

$$3.733Fe_3O_4 + 29.866HCl \rightarrow 3.733FeCl_2 + 7.467FeCl_3 + 14.933H_2O$$
$$BaCO_3 + 2HCl \rightarrow BaCl_2 + H_2CO_3$$
$$0.4CrO_3 + 0.8HCl \rightarrow 0.4CrCl_2 + 0.4H_2O + 0.4O_2$$
$$0.4MnCl_2 + H_2O \rightarrow 0.4MnCl_2 + H_2O$$

$$3.733FeCl_2 + 7.467FeCl_3 + 14.933H_2O + BaCl_2 + H_2CO_3 + 0.4CrCl_2 + 0.4H_2O + 0.4O_2 + 0.4MnCl_2 + H_2O + 33.4NH_4OH \rightarrow BaFe_{11.2}Cr_{0.4}Mn_{0.4}O_{19} + 33.4NH_4Cl + 40.03H_2O + CO_2$$

Reaction equation on composition $BaFe_{10.4}Cr_{0.8}Mn_{0.8}O_{19}$ (x = 0.8)

$$3.466Fe_3O_4 + 27.78HCl \rightarrow 3.466FeCl_2 + 6.934FeCl_3 + 13.864H_2O$$
$$BaCO_3 + 2HCl \rightarrow BaCl_2 + H_2CO_3$$
$$0.8CrO_3 + 1.6HCl \rightarrow 0.8CrCl_2 + 0.8H_2O + 0.8O_2$$
$$0.8MnCl_2 + H_2O \rightarrow 0.8MnCl_2 + H_2O$$

$$3.466FeCl_2 + 6.934FeCl_3 + 13.864H_2O + BaCl_2 + H_2CO_3 + 0.8CrCl_2 + 0.8H_2O + 0.8O_2 + 0.8MnCl_2 + H_2O + 32.8NH_4OH \rightarrow BaFe_{10.4}Cr_{0.8}Mn_{0.8}O_{19} + 32.8NH_4Cl + 39.06H_2O + CO_2$$

Reaction equation on composition $BaFe_{10}Cr_{1.0}Mn_{1.0}O_{19}$ (x = 1.0)

$$3.333Fe_3O_4 + 26.667HCl \rightarrow 3.333FeCl_2 + 6.667FeCl_3 + 13.333H_2O$$
$$BaCO_3 + 2HCl \rightarrow BaCl_2 + H_2CO_3$$
$$CrO_3 + 2HCl \rightarrow CrCl_2 + H_2O + O_2$$
$$MnCl_2 + H_2O \rightarrow MnCl_2 + H_2O$$

$$3.333FeCl_2 + 6.667FeCl_3 + 13.333H_2O + BaCl_2 + H_2CO_3 + CrCl_2 + H_2O + O_2 + MnCl_2 + H_2O + 32.4NH_4OH \rightarrow BaFe_{10}Cr_{1.0}Mn_{1.0}O_{19} + 32.4NH_4Cl + 38.53H_2O + CO_2$$

Based on the results of the reaction, the composition of the compounds for each doping variation is as follows:

Table 1. Reaction equalization Results

No	Compound Composition	X	Fe_3O_4 (gr)	$BaCO_3$ (gr)	CrO_3 (gr)	$MnCl_2$ (gr)
1	$BaFe_{12}O_{19}$	0.0	24	5.10799	-	-
2	$BaFe_{11.2}Cr_{0.4}Mn_{0.4}O_{19}$	0.4	24	5.47334	1.10937	1.39617
3	$BaFe_{10.4}Cr_{0.8}Mn_{0.8}O_{19}$	0.8	24	5.89498	2.38967	3.00747
4	$BaFe_{10}Cr_{1.0}Mn_{1.0}O_{19}$	1.0	24	6.13021	3.10628	3.90934

After obtaining the composition of the compounds in Table 1, a synthesis process was carried out based on the process shown in Figure 1(3). The final result obtained from this process is $BaFe_{12-x}Cr_xMn_xO_{19}$ powder which is shown in Figure 2. The powder has gone through a calcination process using temperature variations that are 200 °C, 600 °C, 1,000 °C. The calcination process aims to eliminate residual stresses so that the powder is not easily cracked and reduces the pores. This happens because, during the calcination process, it is possible to change the particle morphology [12].

Figure 2. Synthesis Results (a) Temperature 200 °C (x= 0.0; 0.4; 0.8 and 1.0); (b) Temperature 600 °C(x= 0.0; 0.4; 0.8 and 10); Temperature 1000 °C (x= 0.0; 0.4; 0.8 and 10)

Based on Figure 2, it can be seen that the colour difference reveals in each sample. The colour change occurs becausethe mole fraction of doping increases, making the powder darker or closer to blackish brown [12]. The colour change that occurs in the doping variation is light brown to blackish brown. This happens because the colour formed is a combination of dark red Cr doping and white Mn doping, so that mixing the solution will produce a dark colour. This colour change is influenced by the addition of NH_4OH solution as a precipitator; the volume used is higher as the mole fraction of the doping ion increases. In addition, the colour change becomes darker because of the calcination process [9]. As shown in Figure 2, the colour of the sample gets darkeras the temperature increases. When viewed at 200 °C each doping variation has an almost similar sample colour. This is because the endothermic reaction identified the release of crystal water. Crystal water is water bound to crystals. At 600 °C, the mass shrinkage to a temperature of 962.5 °C and no significant mass loss. At 1000 °C, the colour of the sample becomes dark black; at this temperature, an exothermic reaction happens, which identifies crystallization [13,14].

4. Conclusion

Based on the results of the synthesis in the production of $BaFe_{12-x}Cr_xMn_xO_{19}$ material using the coprecipitation method with ion doping variations x= 0.0; 0.4; 0.8 and 1.0, and calcination temperature: 200, 600, and 1,000 °C. Calcination process is carried out for 2 hours.It can be inferred that there was an effect of doping and calcination temperature in the formation of $BaFe_{12-x}Cr_xMn_xO_{19}$ material. It can be seen that there is a colour change in each sample powder from light brown to blackish brown due to doping variations and increasing temperature.

Acknowledgement
The author would like to thank the Advanced Chemistry Laboratory University of Mataram, Basic Chemistry Laboratory Mataram of University and all those who helped the realization of this article.

References

[1] Gustomo, F&Suwadi, S. 2013. Analisa Penggunaan Sinyal Radar Bentuk Pulsa dan Gelombang Kontinyu untuk Target Bergerak dengan Model Clutter Terdistribusi Rayleigh. *J. Tek. ITS 2*: A235–A240.
[2] Taryana, Yana., Manal Azwar., Sudrajat, Nanang., & Wahyu, Yuyu. 2019. Material Penyerap Gelombang Elektromagnetik Jangkauan Frekuensi Radar. *Jurnal Keramik dan Gelas Indonesia* **28** (1): 1-29.
[3] Saville, Paul. 2005. *Review of Radar Absorbing Materials*. Canada: Defence R&D Canada - Atlantic.
[4] Pratitajati, E & Manaf, A. 2013. Microstructural Characterisation and Microwave Absorption Characteristics of $La_{(1-x)}Ba_xFe_{0,25}Mn_{0,5}Ti_{0,25}O_3$. *In Advanced Materials Research* **789**: 161-166.
[5] Ristiani, D & Zainuri M. 2016. Desain Double Layer Radar Absorbing Materials dengan Metode Dallenbach Layer Berbasis Bahan Magnetik Alam Tanah Laut dan Polianilin. *Jurnal Sains dan Seni 5* (2).
[6] Susilawati, Khairunnisa, dan Doyan A. 2015. Sintesis Bahan M-Hexaferrites dengan Doping Logam Co Menggunakan FTIR. *Jurnal Pendidikan Fisika dan Teknologi* **1**(3): 180 – 184.
[7] Doyan, A., Khalilurrahman, Susilawati. 2015. Sintesis Dan Uji Ftir Barium M-Hexaferrite Dengan Doping Logam Mn. *Jurnal Pendidikan Fisika dan Teknologi* **1**(4) : 235-238.
[8] Sulistyo, S., Marhaendrajaya, I., & Priyono, P. Sintesis dan Karakterisasi Material Magnetik Barium Hexaferrite Tersubstitusi Menggunakan Teori Sol-Gel untuk Aplikasi Serapan Gelombang Mikro pada Frekuensi X-Band. *Berkala Fisika* **15** (2): 63-68.
[9] Simbolon, S., Tetuko, A.P., Sebayang, P., Sebayang k., &Ginting H. Sintesis Dan Karakterisasi Barium M-Heksaferit Dengan Doping Ion Mn Dan Temperatur Sintering. *Jurnal Sains Materi Indonesia.*
[10] Setiadi, E.A., Sari, F.P., Sari, A.Y., Ramlan & Sebayang, P. 2016. Mikrostruktur dan Sifat Kemagnetan $BaFe_{12}O_{19}$ dengan Aditif MgO-Al_2O_3. *Widyariset* **2** (1): 1-8.
[11] Rahmawati, D.T. 2015. Variasi Penambahan Fe_3O_4 pada Paduan PANi/Fe_3O_4 Sebagai Bahan Penyerap Gelombang Mikro. *Jurnal Fisika* **04** (1): 1-5.
[12] Ahmeda, Okashab, Kershi. 2008. Influence of Rare-earth Ions on The Structure and Magnetic Properties of Barium W-type Hexaferrite. *Journal ofMagnetism and Magnetic Materials320,* hal.

1146–1150.
[13] Sudrajat, N dan Idayanti,N. Karakterisasi Pembentukan Magnet Barium Ferit Nano Partikel dengan X-Ray Diffraction. Jurnal Sains Materi Indonesia,Edisi Khusus, 2009, hal. 71-73
[14] Pangga, D. 2011. Pengaruh Substitusi Ion Dopan Co/Zn Terhadap Struktur Kristal Barium M-Hexaferrit BaFe12O19. *Tesis*. Surabaya: ITS.

Journal of Physics: Conference Series **2165** (2022) 012007 doi:10.1088/1742-6596/2165/1/012007

The process of making BaM ($BaFe_{12}O_{19}$) samples based on natural iron sand doped with metal (Co-Cu-Zn) using the coprecipitation method

Z Fatimah[1], Susilawati[1,2*] and A Doyan[1,2*]

[1]Master of Science Education Program, University of Mataram, Lombok, West Nusa Tenggara, Indonesia
[2]Physics Education, Faculty of Teacher Training and Education, University of Mataram, Lombok, West Nusa Tenggara, Indonesia

E-mail: ziadatulfatimah96@gmail.com, aris_doyan@unram.ac.id, susilawatihambali@unram.ac.id

Abstract. $BaCO_3$ with natural iron sand is the basic material for making samples of Barium M-Hexaferrite (BaM) which is doped to make the utilization of BaM more effective by adding metal ions that have almost the same size as cobalt (Co), copper (Cu), and Zinc (Zn) and using the coprecipitation method. The process of making Barium M-hexaferrite is carried out in the advanced and basic chemistry laboratory, FMIPA University of Mataram. In this study, variations in concentration and temperature variations during the calcination process were used, namely (x = 0.0; 0.4; 0.8; 1.0) (T = 200; 600; and 1000 ^{0}C). These variations greatly affect the sample, where the greater the concentration and temperature of calcination, the color produced in the sample will be darker and this is by the purpose of the calcination process carried out, namely the greater the temperature at the time of calcination, the more water content in the sample will be. reduced because the solvent in the sample evaporation occurs. In addition to these materials, the solvent used also plays an important role in the sample-making process, namely using HCl37% to dissolve, NH_4OH 65% so that the deposition process is faster and aquades as a neutralizer.

1. Introduction

Electromagnetic waves are one of the uses that have been widely applied both in the field of information and others. The influence of electromagnetism itself greatly affects human health and defense. So that scientists develop a wave-absorbing material as one of the materials used to break down excessive impacts that can harm humans. One of the materials used in making wave absorbing materials is using Barium M-Hexaferrite (BaM) ($BaFe_{12}O_{19}$) material which is known as ferric-magnetic which has stable chemical properties, anti-corrosive, and has high anisotropy [1].

BaM it self is a magnet whose composition consists of iron (Fe) as a synthetic material. Elemental iron is obtained from one of the natural resources found along the Indonesian coastline, especially the island of Lombok [2]. Natural black sand has a lot of content consisting of magnetite (Fe_3O_4) nanoparticles which is an attractive material and has a very wide application. In the spinel group, magnetite (Fe_3O_4) is one type of iron oxide mineral phase, in addition to maghemite (γ-Fe2O3) and hematite (αFe_2O_3) [3].

BaM has high magnetic strength so that in its application as a wave absorbing the material it is very less effective [4]. Therefore, Fe^{3+} ions are doped with other metal ions to reduce the hard magnetic properties of BaM[5]. The metal ions that can be used are those that have almost the same size as Cobalt (Co)[6,7], Copper (Cu) [6], and Zinc (Zn)[6,8].The doping was carried out to reduce the anisotropic properties of the Barium M-Hexaferrite (BaM) type without changing the structure. The synthesis process has various methods, such as sol-gel [9][10], powder metallurgy method [11], and coprecipitate method [12].

2. Methods

In this research, the method used is a pure experimental method, which includes three stages, namely: the stage of separating natural iron sand, the stage of separating Fe_3O_4 compounds from magnetic materials, and the synthesis stage. The three stages used the coprecipitation method based on $BaCO_3$ and Fe_3O_4 doped with Co, $CuSO_4$ and $ZnCl_2$ using 37% HCl solvent, distilled water to form a precipitate using NH_4OH solution. Fe_3O_4 is obtained from the process of separating natural iron sand, where the natural iron sand used is obtained from one of the beaches on the island of Lombok, namely Cemara Beach located in Lembar District, West Lombok Regency which is then washed and dried in the hot sun, where the purpose of drying is drying. The sand is to make it easier to separate magnetic material from natural sand which is then separated again to get the desired Fe_3O_4 as the sample base material for synthesis. The synthesis process uses several concentrations and temperature variations, namely (x = 0.0; 0.4; 0.8 and 1.0) and (T = 200; 600 and 1000 °C). The steps in the coprecipitation method are as shown in Figure 1.

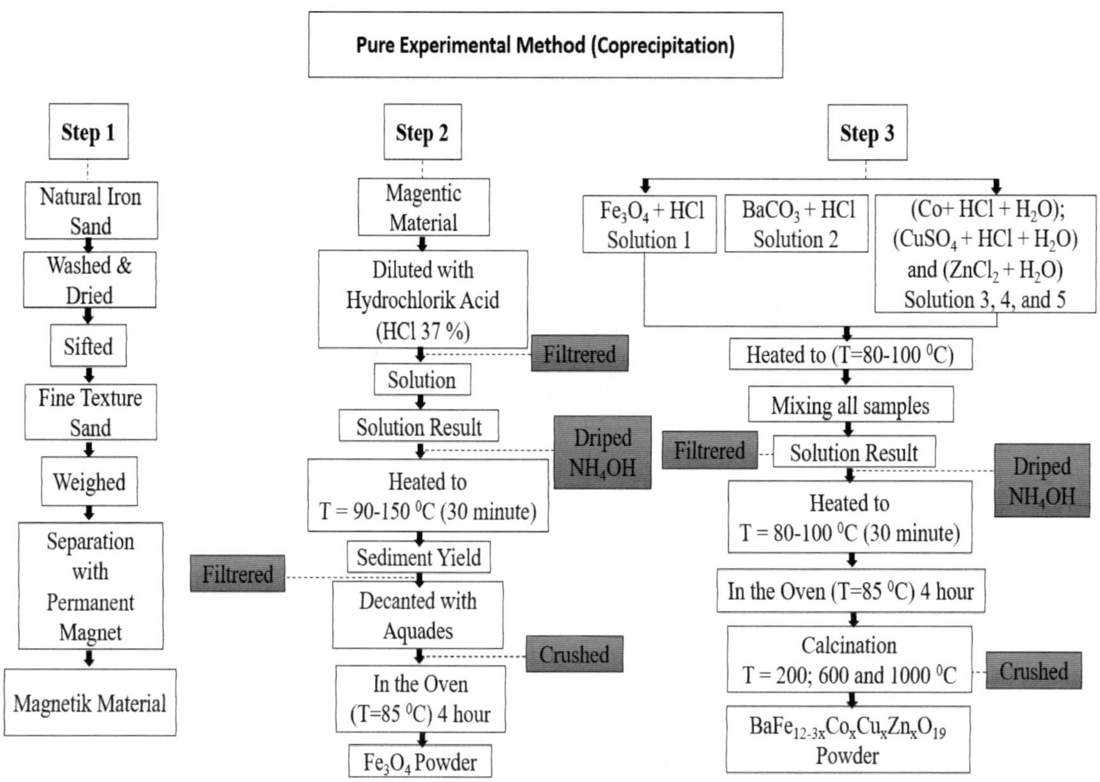

Figure 1. Step 1 (sample separation process); step 2 (separation of Fe_3O_4 from magnetic materials); and step 3 (syenthesis process) with coprecipitation method

3. Result and Discussion

The synthesis of $BaFe_{12-3x}Co_xCu_xZn_xO_{19}$ was carried out using the coprecipitation method in which the basic ingredients used were natural iron sand (Fe) obtained from one of the beaches in West Lombok, namely Cemara beach and $BaCO_3$ and doped with metal Co-Cu-Zn. The materials is then weighed according to the composition obtained in the chemical reaction equation below [13].

Dopant chemical reaction equation x = 0

Equation A

$$4Fe_3O_4 + 32HCl \rightarrow 4FeCl_2 + 8FeCl_3 + 16H_2O$$

Equation B

$$BaCO_3 + 2HCl \rightarrow BaCl_2 + H2CO_3$$

Equation A + B

$$4FeCl_2 + 8FeCl_3 + BaCl_2 + H_2CO_3 + 34NH_4OH \rightarrow BaFe_{12}O_{19} + 34NHCl + 32H_2O + 2H_2 + CO_2$$

The chemical equation for the dopant isx (12-3x = 0,2; 0,4; 0,6; 0,8 dan 1,0)

Equation A

$$(12\text{-}3x)Fe_3O_4 + (32\text{-}1,6)HCl \rightarrow FeCl_2 + FeCl_3 + H_2O$$

Equation B

$$BaCO_3 + 2HCl \rightarrow BaCl_2 + H_2CO_3$$

Equation C + D + E

$$(x)\text{doping }(Co/Cu/Zn) + 2HCl \rightarrow (x)\text{doping }(Co/Cu/Zn)l_2 + H_2$$

Equation A + B + C + D + E

$$(12\text{-}3x)FeCl_2 + FeCl_3 + BaCl_2 + H_2CO_3 + 0,2CoCl_2 + H_2 + 0,2CuCl_2 + H_2SO_4 + 0,2ZnCl_{2(aq)}$$

$$33,2NH_4OH \rightarrow BaFe_{12\text{-}3x}Co_xCu_xZn_xO_{19} + 33,2NHCl + H_2O + 2H_2 + CO_2$$

Table 1. The required compound composition

x	Natural iron sand (gr)	$BaCO_3$ (gr)	Co (gr)	$CuSO_4$ (gr)	$ZnCl_2$ (gr)	HCl (ml)	The resulting compound
0.0	20.79	4.43	-	-	-	63.38	$BaFe_{12}O_{19}$
0.2	19.51	4.37	0.26	0.70	3.67	66.9	$BaFe_{11,4}Co_{0,2}Cu_{0,2}Zn_{0,2}O_{19}$
0.4	18.76	4.32	0.51	1.39	1.19	64.57	$BaFe_{10,8}Co_{0,4}Cu_{0,4}Zn_{0,4}O_{19}$
0.6	18.00	4.26	0.763	2.07	1.77	62.3	$BaFe_{10,2}Co_{0,6}Cu_{0,6}Zn_{0,6}O_{19}$
0.8	17.31	4.21	1.00	2.73	2.33	60.19	$BaFe_{9,6}Co_{0,8}Cu_{0,8}Zn_{0,8}O_{19}$
1.0	17.02	4.19	1.19	2.87	2.62	60.39	$BaFe_9CoCuZnO_{19}$

The resulting solution that has been heated (Fe_3O_4, $BaCO_3$, Co, $CuSO_4$, and $ZnCl_2$+ HCl 37 %) is then dripped with NH_4OH to form the desired precipitate[14](as shown in Figure a). The precipitate is then filtered and washed using distilled water until it gets a neutral pH (as shown in Figure b). The precipitate was then placed in the oven for four hours at a temperature of 85 ^0C (as shown in Figure c).

(a)

(b)

(c)

Figure 2. a) dripped NH_4OH; b) washing process using distilled water to get a neutral pH; c) oven process

The results of the oven are then ground and calcined at temperatures of 200, 600, and 1000°C, as shown below:

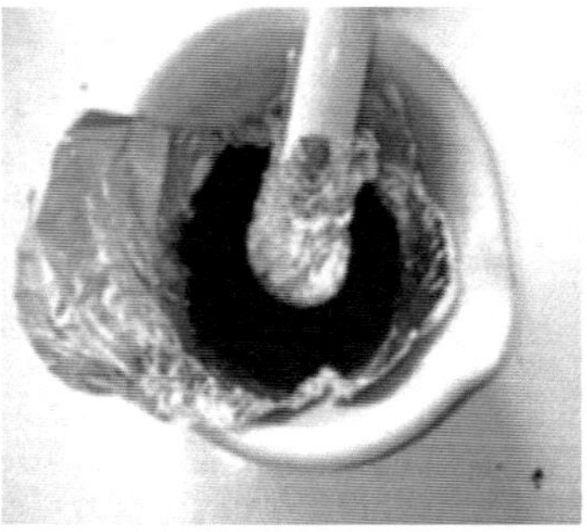

(a)

(b)

Figure 3. a) grinding process; b) calcination process

The calcination process aims to purify the sample by removing the water content, solvent, and impurities from doping such as (Co-Cu-Zn), natural iron sand, and barium carbonate ($BaCO_3$) after the dissolving process [15]. The powder used for calcination must have been in the oven so that the texture of the powder can easily be ground using a mortar. The calcination process also affects the color of the sample as shown in Table 2.

Journal of Physics: Conference Series **2165** (2022) 012007 doi:10.1088/1742-6596/2165/1/012007

Table 2. Effect of calcination temperature on powder color change $BaFe_{12-3x}Co_xCu_xZn_xO_{19}$

Calcination Temperature	Concentration Variation (x)			
	0	0.4	0.8	1.0
200 ^{0}C				
600 ^{0}C				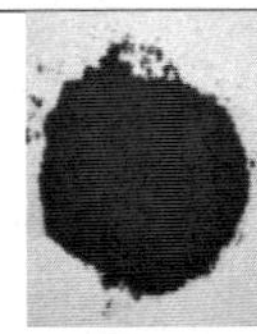
1000 ^{0}C				

In Table 2 the color difference in the sample is very clear. The changes that occur are due to doping which plays a role in the dissolution process which is dripped with NH_4OH as a solution that makes the sample easy to settle [16]. The higher the mole fraction of the metal as doping and the volume of NH_4OH, the darker the color of the resulting sample [17]. This indicates that the greater the temperature during calcination, the water content in the sample will decrease because the solvent in the sample evaporates [18].

HCl and NH_4OH solvents are the best media because they are easily removed from the synthesis results (an integration of two or more existing elements that produce a new product) [19]when washed with distilled water or by heating at low temperatures. So it can be concluded that the doping used is very influential in the sample making process or the synthesis process using the coprecipitation method in the formation of BaM powder with higher temperatures, the darker the resulting color will be.

The color change that occurs at a temperature of 200 ^{0}C does not produce a very dark color and is fairly consistent with the changes that occur. This is because the water content and impurities in the sample are still high and different at temperatures of 600 ^{0}C and 1000 ^{0}C. The changes that occur are quite visible with the difference between without doping which is still light in color and using doping, where the higher the doping, the darker the color of the sample. The difference that occurs between temperature variations is also in accordance with the calcination theory, namely heating carried out at high temperatures.

The results of this study are supported by research that has been carried out which used a calcination temperature of 80 to 800 ^{0}C and there was an increase in this study using a higher temperature of 1000 ^{0}C which adjusted to the capacity of existingequipment in the laboratory [20,21].

4. Conclusion

Based on the results and discussion above, it can be concluded that Barium M-Hexaferrite ($BaFe_{12}O_{19}$) is based on natural iron sand at the Cemara beach, West Lombok with metallic cobalt (Co), Copper (Cu), and zinc (Zn) using a soluble coprecipitation method using hydrochloric acid (HCl) which produces a darker color with higher calcination temperatures of 200, 600, and 1000°C. The darker the color of the sample is also influenced by the doping used and NH_4OH which plays an important role in the process of deposition and neutralization of the sample using distilled water. So that in the process of making $BaFe_{12-3x}Co_xCu_xZn_xO_{19}$ samples, it can be said to be successful by seeing the color of the powder produced according to the calcination temperature from low to high and the amount of doping given.

Acknowledgment
Thank you to all parties who have taken part in the process of research and writing this article. Advanced chemistry laboratory and basic chemistry laboratory, FMIPA, University of Mataram, and other parties who cannot be mentioned one by one, gladly provide input for the improvement of this paper.

References

[1] Purnama S, Purwanto P, and Ari W A 2015 Pembuatan Bahan Magnetik Barium Heksaferit ($BaFe_{12}O_{19}$) *Jurnal Kimia dan Kemasan* **37**(2)pp95-102DOI:http://dx.doi.org/10.24817/jkk.v37i2.1818
[2] Susilawati, Doyan A, Taufik M, Wahyudi, Gunawan E R, Kosim, Fithriyani A, and Khair H 2018 Identifikasi Kandungan Fe Pada Pasir Besi Alam di Kota Mataram*Jurnal Pendidikan Fisika dan Teknologi* **4**(1) pp 105-110 DOI: http://dx.doi.org/10.29303/jpft.v4i1.571
[3] Saragi T, Permana B, Saputri M, Safriani L, and Iman Rahayu R 2018 Karakteristik Optik dan Kristal Nanopartikel Magnetit *JIIF (Jurnal Ilmu dan Inovasi Fisika)* **2**(1) pp 53-56DOI: https://doi.org/10.24198/jiif.v2i1.15438
[4] Sahlam S, Doyan A, and Susilawati S 2018 Sintesis Bahan M-Heksaferit Substitusi Logam Kobalt-Mangan dengan Metode Kopresipitasi *Jurnal IPA &Pembelajaran IPA* **2**(2) pp 64-68 DOI: https://doi.org/10.24815/jipi.v2i2.12073
[5] Seta P D C 2017 Pengaruh Dopan Fe Pada Sifat Kemagnetan Nanopartikel ZnO Hasil Kopresipitasi *(Doctoral dissertation, Institut Teknologi Sepuluh Nopember).*
[6] Miessler G L, Fischer P J, and Tarr D A 2014*Inorganic Chemistry Fifth Edition* US: Pearson Education, Inc.
[7] Susilawati, Doyan A 2021 Synthesis and Characterization Barium M-Hexaferrites (BaFe12-2xCoxMnxNixO19) as a Microwave Absorbent Material*In Solid State Phenomena* **317**, *pp. 46-52) Trans Tech Publications Ltd.*DOI: https://doi.org/10.4028/www.scientific.net/SSP.317.46
[8] Doyan A, Susilawati, and Halik I 2015 Characterization of Electrict and Magnetict Priperties of Barium M- Hexaferrites Doped with Zinc Procedia ICMSE University of Mataram Lombok Island Indonesia Nov. 4-5DOI: https://doi.org/10.29303/jppipa.v7i1.461
[9] Murti F, Ramli R, and Darvina Y 2017 Analisis Sifat Listrik Lapisan Tipis Fe_3O_4 yang Dipreparasi dari Pasir Besi Pantai Tiram Kabupaten Padang Pariaman Sumatera Barat dengan Metode Sol-gel Spin Coating *Pillar of Physics* **10**(1)pp31-38.
[10] Siswoyo S W, Winatapura D S, and Sukirman E 2017 Pengaruh Penambahan Aditif Bi2O3 Terhadap Karakteristik Barium Heksaferit Hasil Sintesis dengan Metode Sol-Gel *Jurnal Kimia dan Kemasan* **39**(2) pp 61-68.
[11] Hayati R, Budiman A, and Puryanti D 2016 Karakterisasi Suseptibilitas Magnet Barium Ferit yang Disintesis dari Pasir Besi dan Barium Karbonat Menggunakan Metode Metalurgi Serbuk *Jurnal Fisika Unand* **5**(2) pp 187-192 DOI: https://doi.org/10.25077/jfu.5.2.187-192.2016
[12] Winatapura D S, Sukirman E, and Ridwan R 2019 Sintesis Nanopartikel Heksaferit $MFe_{12}O_{19}$ (M=Ba,Sr) dengan Metode Kopresipitasi *Jurnal Sains Materi Indonesia* **13**(1) pp 39-43 DOI: 10.17146/jusami.2011.13.1.5397
[13] Munib, Susilawati, and Savalas T 2016 Sintesis Barium M-Hexaferrite ($BaFe_{12-x}Ni_xO_{19}$) Doping

Logam Nikel dengan Metode Kopresipitasi*Jurnal Penelitian Pendidikan IPA* **4**(1) pp 9-17DOI: https://doi.org/10.29303/jppipa.v2i1.28

[14] Nasution N and Aida F 2018 Sintesis Nanopartikel TiO2 Fasa Rutile dengan Metode Kopresipitasi*FISITEK: Jurnal Ilmu Fisika dan Teknologi* **2**(2) pp 18-25.

[15] Susilawati S, Doyan A, and Muliyadi L2021 Synthesis of M-Hexaferrites Material Based on Natural Iron Sand with Metal Co Doping Using the Coprecipitation Method *JurnalPenelitian Pendidikan IPA* **7**(1)pp 1-4 DOI: https://doi.org/10.29303/jppipa.v7i1.461

[16] Susilawati, Doyan A, and Sahlam 2017 Synthesis and Characterization Materials M-Barium Hexaferrite Doping Ions Co-Mn Nano Particle*IOP Conf. Series: Materials Science and Engineering* 196 012016.

[17] Susilawati, Khairunnisa, andDoyan A 2018 SintesisBahan M-Hexaferrites dengan Doping Logam Co Menggunakan FTIR*JurnalPendidikan Fisika dan Teknologi* **1**(3) pp 180-184DOI: https://doi.org/10.29303/jppipa.v4i1.112

[18] Susilawati, Doyan A, Taufik M, Wahyudi, Gunawan E R, Fithriyani A, and Nazarudin 2019 Characterization of Barium M-Hexaferrite with Doping Zn and Mn for Microwaves Absorbent *Materials Science Forum* **966** pp 282-289DOI: https://doi.org/10.4028/www.scientific. net/MSF.1028.32

[19] Wulandari I O 2021 Sintesis dan Karakterisasi Nanopartikel Fe_3O_4 termodifikasi Biokompatibel Polimer serta Potensinya sebagai Penghantar Obat *The Indonesian Green Technology Journal* **10**(1) 1-8.

[20] Halik, I., Doyan, A., & Hakim, A. 2016. Sintesis Barium M-heksaferit dengan Doping Logam Zn Menggunakan Metode Kopresipitasi. *Jurnal Penelitian Pendidikan IPA*, *2*(1).

[21] Sahlam, S., Doyan, A., & Susilawati, S. 2018. Sintesis Bahan M-Heksaferit Substitusi Logam Kobalt-Mangan dengan Metode Kopresipitasi. *Jurnal IPA & Pembelajaran IPA*, **2**(2), 64-68.

Synthesis $BaFe_{12}O_{19}$ based on natural iron sand with Cu and Ni doping using coprecipitation method

B U P Elinda[1], Susilawati[1,2*], Kosim[1,2], A Doyan[1,2*]

[1]Master of Science Education Program, University of Mataram, Mataram, Indonesia
[2]Physics Education, FKIP, University of Mataram, Mataram, Indonesia

E-mail: ulfaputrielinda@gmail.com, aris_doyan@unram.ac.id, susilawatihambali@unram.ac.id

Abstract. Synthesis of $BaFe_{12}O_{19}$ based on natural iron sand with Cu and Ni doping using the coprecipitation method has been done. This research was conductedat the Chemistry Laboratory, Faculty of Teacher Training and Education, Mataram University. The synthesis was carried out through the variation of doping ions and calcination temperature. The concentration of doping ion used is x = 0.0; 0.4; 0.8; and 1.0 with variations in calcination temperature of 200, 600, and 1000 ^{0}C. The materials used are $BaCO_3$, Fe_3O_4, $CuSO_4$, and $NiCl_2.6H_2O$ powder. This synthesis process aims to obtain $BaFe_{12-2x}Cu_xNi_xO_{19}$ powder that will later be used as a microwave absorbent material. To obtain sample powder of $BaFe_{12-2x}Cu_xNi_xO_{19}$, thisresearch is carried out through three stages: natural sand separation, magnetic minerals separation, and synthesisand calcination. The separation is to attach magnetic materials from the sand using a permanent magnet. After that, the magnetic mineral is detached from the separated sand to obtain Fe_3O_4 powder, which will later be used for synthesis. The results of synthesis and calcination showed that the more doping ions applied, the more concentrated the resulting powder. Similarly, according to the temperature variation, the higher the temperature, the more black the sample powder is.

1. Introduction

Electromagnetic wave (EM)-based technology devices are currently widely used in various fields, including military defense. The application of EM waves in this field is Radio Detection and Ranging (Radar) [1]. In military defense, there is a technology used to make ships or aircraft undetectable by radar, namely stealth technology. Stealth technology has two methods that are the design method and the coating method. In the device's design, the aircraft is modified in an angular shape. This modification aims to engineer the reflected wave (scattering) not to lead to the radar detector so that the plane cannot be detected [2]. However, this method requires an enormous cost. Therefore, another efficient alternative technique to develop is the coating method [3].

The coating method with Radar Absorbing Material (RAM) can reduce reflections and absorb microwaves (absorber) so that the material coated with RAM is not detected by radar [4]. The material used as an absorber must have high permeability (μ), high permittivity (ε), low coercivity (Hc) [5], and high magnetic saturation (Ms) [6]. Generally, the material used as an absorber is Barium M-Hexaferrite (BaM) [7]. BaM is known as a permanent magnet that has hardmagnetic properties [8]. Because BaM is hardmagnetic, this will cause the absorption of BaM to be weak, making it less practical to implement as an absorber. However, the material used as an absorber must be soft-magnetic [9]. Therefore, for this reason, doping with other metal cations with ionic radii quite similar to Fe, such as Co, Cu, Zn, Ni, and Mn, is required. [10]. The doping used in this research is Copper (Cu) and Nickel (Ni) because they have almost the same ionic radius as iron (Fe). In addition, based on their electron configurations, Cu and Ni have 1 and 2 unpaired electrons, which means that the doped metal has fewer electrons than Fe metal. Fe metal has four unpaired electrons, so it will be doped using a metal with fewer unpaired electrons. As it is known that the more unpaired electrons a metal has, the stronger its magnetism.

The synthesis of $BaFe_{12}O_{19}$ using Cu and Ni doping unveiled by the formula $BaFe_{12-2x}Cu_xNi_xO_{19}$ can be carried out using several methods: the coprecipitation method [11], the sol-gel method [12], the mechanical alloying method [13], and the mechanical milling method [14]. $BaFe_{12-2x}Cu_xNi_xO_{19}$ was synthesized using natural iron sand from the Ijobalit Beach, Labuhan Haji District, East Lombok. Barium Hexaferrite is classified according to its structure into six types, namely $BaFe_{12}O_{19}$ (type M), $BaMe_2Fe_{16}O_{27}$ (type W), $Ba_2Me_2Fe_{28}O_{46}$ (type X), $Ba_2Me_2Fe_{12}O_{22}$ (type Y), $Ba_3Me_2Fe_{24}O_{41}$ (type Z) and $Ba_4Me_2Fe_{36}O_{60}$ (type U) where Me is a divalent ion from the first transition series [15]. Type W-, X-, Y-, Z- is a type that is less economical because the process is relatively intricated. Thus, type M is preferable to use because it has uniaxial magnetic properties at room temperature and has an anisotropic field of 17.5 KOe. The synthesis in this study was carried out to produce $BaFe_{12-2x}Cu_xNi_xO_{19}$ powder, which would then be calcined to reduce the moisture content of the sample powder.

2. Methods

The method used in this research is a laboratory experiment. This experiment is conducted at the Chemistry Laboratory, Faculty of Teacher Training and Education, the University of Mataram. The fundamental ingredients are $BaCO_3$, Fe_3O_4 taken from natural sand from Ijobalit beach, East Lombok, $CuSO_4$, $NiCl_2.6H_2O$, 37% HCl, NH_4OH, and H_2O. In the forming of barium ferrite, $BaFe_{12-2x}Cu_xNi_xO_{19}$ was carried out using four mole fractionvariations i.e. $x = 0.0$; 0.4; 0.8; and 1.0.

The process to obtain $BaFe_{12-2x}Cu_xNi_xO_{19}$ sample powder is through three stages, the first stage is to separate natural sand, then the second stage is to separate magnetic minerals, and the third stage is synthesis. The separation is done by attaching magnetic materials from natural iron sand using apermanent magnet, as shown in Figure 1. After that, separation and purification are carried out to obtain Fe_3O_4 powder. Afterward, Fe_3O_4 powder will be synthesized using the coprecipitation method by settling more than one substance together through the steps shown in Figure 2 [16]. The following process is calcination, as shown in Figure 3.

(a) (b) (c)

Figure 1. Natural sand separation stage. (a) Natural sand; (b) Separation using permanent magnet; (c) Magnetic minerals

Journal of Physics: Conference Series **2165** (2022) 012008 doi:10.1088/1742-6596/2165/1/012008

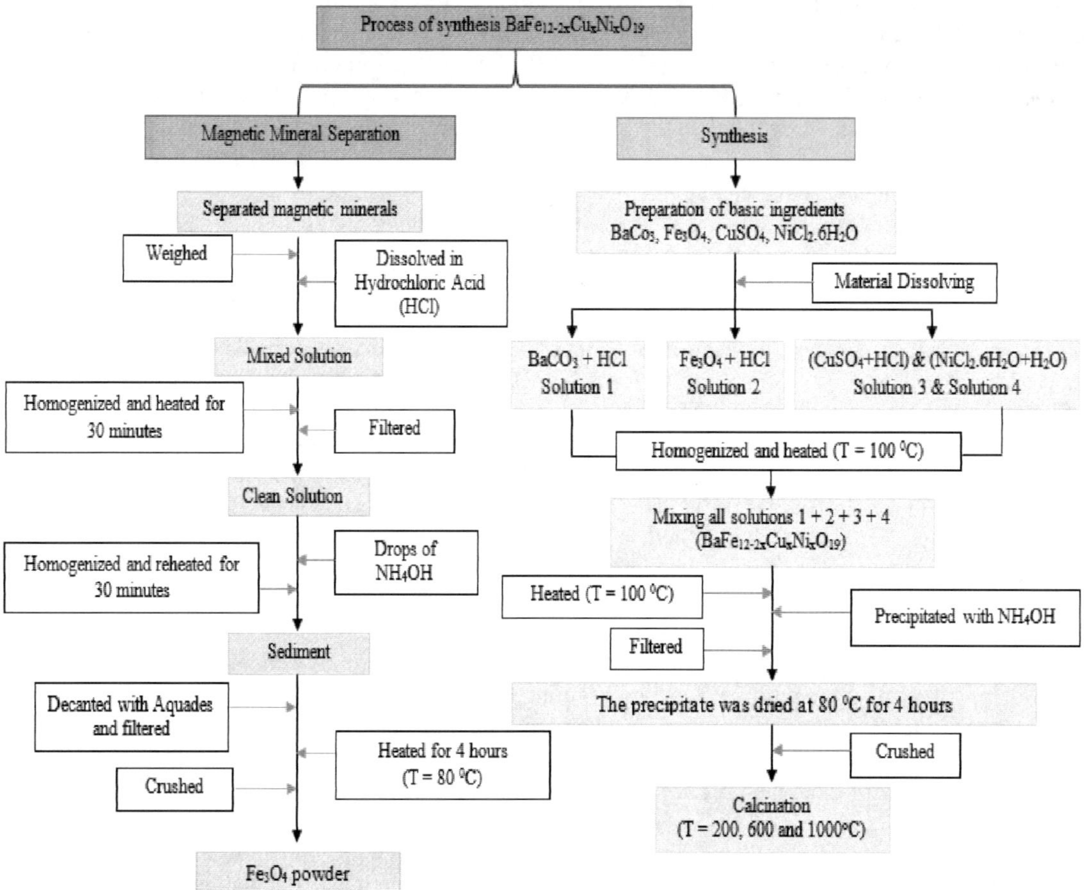

Figure 2. The process of synthesis $BaFe_{12-2x}Cu_xNi_xO_{19}$ powder

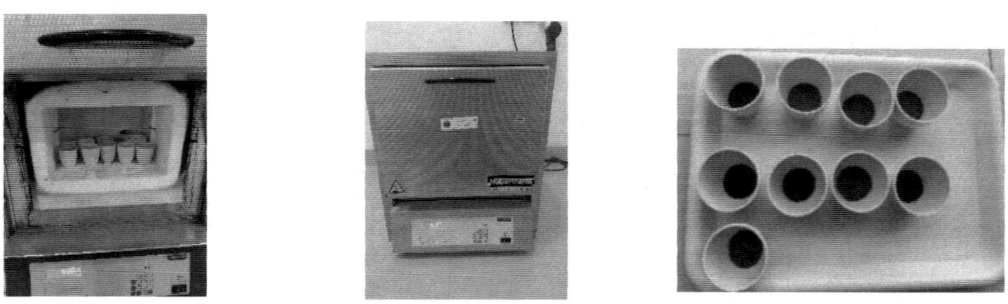

(a) (b) (c)

Figure 3. Calcination stage. (a)put the sample into the kiln; (b) heated at various temperatures; (c) calcination results

3. Result and Discussion

The synthesis of Barium M-Hexaferrite begins with the chemical reaction equalization of the materials used to adjust the composition required in this process. The reaction equation is calculated based on the amount of doping ion concentration used, namely x = 0.0; 0.4; 0.8 and 1.0.

The reaction equation for the compound without doping

Equation A

$$BaCO_3 + 2HCl \rightarrow BaCl_2 + H_2CO_3$$

Equation B

$$4Fe_3O_4 + 32HCl \rightarrow 4FeCl_2 + 8FeCl_3 + 16H_2O$$

Equation A + B

$$BaCl_2 + H_2CO_3 + 4FeCl_2 + 8FeCl_3 + 16H_2O + 34NH_4OH \rightarrow BaFe_{12}O_{19} + 34NH_4Cl + 32H_2O + 2H_2 + CO_2$$

The reaction equation for the compound with doping x = 0.4 ($BaFe_{11.2}Cu_{0.4}Ni_{0.4}O_{19}$)

Equation A

$$BaCO_3 + 2HCl \rightarrow BaCl_2 + H_2CO_3$$

Equation B

$$3.733Fe_3O_4 + 29.866HCl \rightarrow 3.733FeCl_2 + 7.467FeCl_3 + 14.933H_2O$$

Equation C

$$0.4CuSO_4 + 0.8HCl \rightarrow 0.4CuCl_2 + 0.4H_2SO_4$$

Equation D

$$0.4NiCl_2.6H_2O + H_2O \rightarrow 0.4NiCl_2 + 7H_2O$$

Equation A+B+C+D

$$BaCl_2 + H_2CO_3 + 3.733FeCl_2 + 7.467FeCl_3 + 14.933H_2O + 0.4CuCl_2 + 0.4H_2SO_4 + 0.4NiCl_2 + 7H_2O + 33.4NH_4OH \rightarrow BaFe_{11.2}Cu_{0.4}Ni_{0.4}O_{19} + 33.4NH_4Cl + 40.03H_2O + CO_2$$

Based on the equation above, it can be seen that the required compound is accordingto the value of Mr (relative atomic mass), M (Molarity), and n (molality) of each compound used. The following table shows the results of the compound calculation needed in the synthesis process.

Table 1. Calculation results of compounds for synthesis

No	Compound	X	Fe_3O_4	$BaCO_3$	$CuSO_4$	$NiCl_2.6H_2O$
1	$BaFe_{12}O_{19}$	0	20,8354	4,43448		
2	$BaFe_{11,2}Cu_{0,4}Ni_{0,4}O_{19}$	0,4	19,3724	4,41800	1,42934	2,12858
3	$BaFe_{10,4}Cu_{0,8}Ni_{0,8}O_{19}$	0,8	17,9202	4,40165	2,84810	4,24141
4	$BaFe_{10}Cu_{1,0}Ni_{1,0}O_{19}$	1,0	17,2007	4,39351	3,55355	5,29196

After gaining the required compound composition, the next step is to carry out the synthesis process according to the steps shown in Figure 2.

The powder produced from the synthesis process will then be calcined at different temperatures: 200, 600, and 1000 ^0C. Calcination aims to reduce the water content in the sample and shrink the pores during the calcination process. The results of the calcination of the $BaFe_{12-2x}Cu_xNi_xO_{19}$ powder obtained underwent a color change, as shown in Table 2.

Table 2. Effect of doping ion variation and temperature variation of $BaFe_{12-2x}Cu_xNi_xO_{19}$ powder calcination

$BaFe_{12-2x}Cu_xNi_xO_{19}$ X = 0.0 (T = 200⁰C)	$BaFe_{12-2x}Cu_xNi_xO_{19}$ X = 0.4 (T = 200 ⁰C)	$BaFe_{12-2x}Cu_xNi_xO_{19}$ X = 0.8 (T = 200 ⁰C)	$BaFe_{12-2x}Cu_xNi_xO_{19}$ X = 1.0 (T = 200⁰C)
	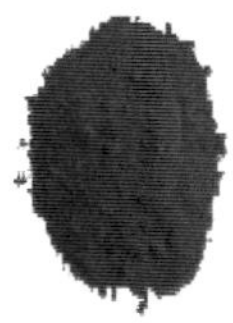		
$BaFe_{12-2x}Cu_xNi_xO_{19}$ X = 0.0 (T = 600⁰C)	$BaFe_{12-2x}Cu_xNi_xO_{19}$ X = 0.4 (T = 600⁰C)	$BaFe_{12-2x}Cu_xNi_xO_{19}$ X = 0.8 (T = 600 ⁰C)	$BaFe_{12-2x}Cu_xNi_xO_{19}$ X = 1.0 (T = 600⁰C)

$BaFe_{12-2x}Cu_xNi_xO_{19}$ X = 0.0 (T = 1000 ⁰C)	$BaFe_{12-2x}Cu_xNi_xO_{19}$ X = 0.4 (T = 1000⁰C)	$BaFe_{12-2x}Cu_xNi_xO_{19}$ X = 0.8 (T = 1000⁰C)	$BaFe_{12-2x}Cu_xNi_xO_{19}$ X = 1.0 (T = 1000 ⁰C)

As elaboratedin Table 2, it can be seen that the color of the sample powder produced varies with the increase in the calcination temperature and the addition of doping use. The more doping ions mixed, the more concentrated the color of the sample powder. At 200 ⁰C, the sample powder appears to produce a similar color because at 100-300 ⁰C only H_2O is released, and the sample will be bound to O-H [17]. Next, at 600 ⁰C, the sample has been seen a color difference from brick red to dark brown. This occurrence happens because, at 600 ⁰C, there is a release of CO_2 gas and a significant reduction in mass weight. Finally,at the last temperature of 1000 ⁰C, a color change from brown to black indicated that a crystal structure had formed. Dark brown to black powder demonstrates that all the H_2O and HCl content of 100% has been used up [18]. This reveals that in the heating process, the basic ingredients are completely dissolved [19].

The calcination process on the color change of $BaFe_{12-2x}Cu_xNi_xO_{19}$ powder indicates crystallization, influenced by temperature in the calcination process. The $BaFe_{12-2x}Cu_xNi_xO_{19}$ powder will form hexagonal crystals at a temperature of at least 600 ^{0}C. Based on this research, it was found that the color change of the $BaFe_{12-2x}Cu_xNi_xO_{19}$ powder will be darker as the calcination temperature increases and the doping ion is used.

4. Conclusion

Based on the synthesis of $BaFe_{12-2x}Cu_xNi_xO_{19}$ using the coprecipitation method with variations in doping ions and calcination temperature. The more doping ions used, the more concentrated the resulting powder. Likewise, with the temperature variations used, the higher the temperature,the blacker the sample powder is. The color of the sample indicates that the sample has formed crystals or not.

Acknowledgment

The author would like to thank the Basic Chemistry Laboratory, Faculty of Teacher Training and Education, the University of Mataram for the research facilities provided in completing this research. Thank you to the Research and Technology Directorate of Higher Education who has assisted the funding of this research through the research of Dra. Susilawati, M.Sc., Ph.D on PTUPT research at the University of Mataram 2021.

References

[1] Gustomo, F&Suwadi, S. 2013. Analisa Penggunaan Sinyal Radar Bentuk Pulsa dan Gelombang Kontinyu untuk Target Bergerak dengan Model Clutter Terdistribusi Rayleigh. *J. Tek. ITS* **2**: A235–A240.

[2] Taryana, Yana., Manal Azwar., Sudrajat, Nanang., & Wahyu, Yuyu. 2019. Material Penyerap Gelombang Elektromagnetik Jangkauan Frekuensi Radar. *Jurnal Keramik dan Gelas Indonesia* **28** (1): 1-29.

[3] Saville, Paul. 2005. *Review of Radar Absorbing Materials*. Canada: Defence R&D Canada - Atlantic.

[4] Susilawati, Doyan. A, & Khalilurrahman. 2017. Synthesis and Characterization of Barium Hexaferrite with Manganese (Mn) Doping Material as Anti-Radar. *AIP Conference Proceedings*. 040007-1.

[5] Susilawati & Doyan, A. 2021. Characteristics of Barium M-Hexaferrite with Dopping Mn and Ni in X-Band Frequency for Microwave Absorption. *Materials Science Forum* (1028); 32-37.

[6] Susilawati, Doyan, A., &Sahlam. 2017. Synthesis and Characterization Materials M-Barium Hexaferrite Doping Ions Co-Mn NanoParticle. *IOP Conf. Series: Materials Science and Engineering,* (012016): 196

[7] Susilawati & Doyan, A. 2013. Sintesis dan Studi Pendahuluan Struktur Bahan Hexaferrite Untuk Aplikasi Anti Radar. *Proseding Seminar Nasional Penelitian, Pembelajaran Sains dan Implementasi Kurikulum 2013.* Program Studi Magister Pendidikan IPA, Program Pascasarjana Universitas Mataram: Mataram.

[8] Ristiani, D & Zainuri M. 2016. Desain Double Layer Radar Absorbing Materials dengan Metode Dallenbach Layer Berbasis Bahan Magnetik Alam Tanah Laut dan Polianilin. *Jurnal Sains dan Seni* **5** (2).

[9] Susilawati & Doyan, A. 2021. Synthesis and Characterization Barium M-Hexaferrites ($BaFe_{12-2x}Co_xMn_xNi_xO_{19}$) as a Microwave Absorbent Material. *Solid State Phenomena.* (317): 46-52.

[10] Susilawati, Khairunnisa, dan Doyan A. 2015. Sintesis Bahan M-Hexaferrites dengan Doping Logam Co Menggunakan FTIR. *Jurnal Pendidikan Fisika dan Teknologi* **1**(3): 180 – 184.

Journal of Physics: Conference Series **2165** (2022) 012008 doi:10.1088/1742-6596/2165/1/012008

[11] Doyan, A., Khalilurrahman, Susilawati. 2015. Sintesis dan Uji Ftir Barium M-Hexaferrite Dengan Doping Logam Mn. *Jurnal Pendidikan Fisika dan Teknologi* **1**(4) : 235-238.

[12] Sulistyo, S., Marhaendrajaya, I., & Priyono, P. Sintesis dan Karakterisasi Material Magnetik Barium Hexaferrite Tersubstitusi Menggunakan Teori Sol-Gel untuk Aplikasi Serapan Gelombang Mikro pada Frekuensi X-Band. *Berkala Fisika* **15** (2): 63-68.

[13] Simbolon, S., Tetuko, A.P., Sebayang, P., Sebayang k., &Ginting H. Sintesis Dan Karakterisasi Barium M-Heksaferit Dengan Doping Ion Mn Dan Temperatur Sintering. *Jurnal Sains Materi Indonesia.*

[14] Setiadi, E.A., Sari, F.P., Sari, A.Y., Ramlan & Sebayang, P. 2016. Mikrostruktur dan Sifat Kemagnetan $BaFe_{12}O_{19}$ dengan Aditif MgO-Al_2O_3. *Widyariset* **2**(1): 1-8.

[15] Ahmeda, Okashab, Kershi. 2008. Influence of Rare-earth Ions on The Structure and Magnetic Properties of Barium W-type Hexaferrite. *Journal ofMagnetism and Magnetic Materials* **320**, 1146–1150.

[16] Susilawati, Doyan, A., Taufik, M., Wahyudi, Gunawan, E.R., Fithriyani, A, & Nazarudin. 2019. Characterization of Barium M-Hexaferrite with Doping Zn and Mn for Microwaves Absorbent. *Materials Science Forum* (966): 282-289.

[17] Sudrajat, N dan Idayanti,N. Karakterisasi Pembentukan Magnet Barium Ferit Nano Partikel dengan X-Ray Diffraction. Jurnal Sains Materi Indonesia,Edisi Khusus, 2009, hal. 71-73

[18] Munib, Susilawati, & Savalas, T. 2016. Sintesis Barium M-Hexaferrite ($BaFe_{12-x}Ni_xO_{19}$) Doping Logam Nikel dengan Metode Kopresipitasi. Jurnal Penelitian Pendidikan IPA, **4**(1): 9-17.

[19] Pangga, D. 2011. Pengaruh Substitusi Ion Dopan Co/Zn Terhadap Struktur Kristal Barium M-Hexaferrit BaFe12O19. *Tesis*. Surabaya: ITS.

Journal of Physics: Conference Series **2165** (2022) 012009 doi:10.1088/1742-6596/2165/1/012009

Structure and optical properties of Titanium Dioxide thin film with mixed Fluorine and Indium doping for solar cell components

A Doyan[1,2], Susilawati[1,2*], I K Mahardika[3], D R Rizaldi[1], Z Fatimah[1]

[1]Magister of Science Education Program, University of Mataram, Mataram, West Nusa Tenggara, Indonesia
[2]Physics Education Program, Faculty of Teacher Training and Education, University of Mataram, Mataram, West Nusa Tenggara,Indonesia
[3]Magister of Science Education Program, Faculty of Teacher Training and Education, University of Jember, Jember, East Java, Indonesia

Email: aris_doyan@unram.ac.id

Abstract. Research has been carried out in the form of the synthesis of a thin film of Titanium Dioxide with a mixture of Fluorine and Indium doping. The purpose of this research is to obtain the characteristics of a thin filmphotocatalyst with a crystal structure and optical properties that can be applied to solar cell components. This research was carried out in two stages, namely the synthesis stage and the thin filmcharacterization stage. The TiO_2 thin film synthesis stage is through the substrate preparation process, sol-gel solution preparation, thin film deposition, and heating process, while the thin-film characterization stage uses X-Rays Diffraction (XRD) to identify crystal properties and UV-VisSpectrophotometer to determine optical properties of a thin film of TiO_2 and TiO_2:(F+In). Based on data analysis and discussion, XRD test results showed that the thin film of TiO_2:(F+In) has an Anatase phase with a tetragonal shape, and has an average crystal size that is getting smaller from 23.14 to 12.17 nm. Based on the results of research, data analysis, and discussion, it was found that the sample of TiO_2 : (F+In) has characteristics that can be used as materials in the development of solar cell components. These characteristics are like 1). havenano-sized particles, 2). have a low transmittance value and high absorbance in absorbing sunlight, and 3). have a low activation energy and energy gap (direct and indirect).

1. Introduction

The development of thin films is a very interesting research study to do. This is because the thin film can be applied to various technologies that can facilitate human activities. One area of study that uses a lot of thin films is the energy field, which is related to the development of solar cell components [1, 2]. Solar cell components such as absorbent layers are parts developed by humans to absorb sunlight by utilizing the properties of semiconductor materials [3]. The absorbed sunlight will be converted into alternative electrical energy by utilizing the principle of the photovoltaic effect [4,5,6].

Titanium Dioxide (TiO_2) is a type of semiconductor material that is often used in developing various thin films to be used as components of solar cells [7]. This is because the energy bandgap of the thin film of TiO_2 material is relatively small, ranging from 3.4 to 3.7 eV [8]. The smaller the energy band gap possessed by a thin film, the better the resulting characteristics. However, the problem with Titanium Dioxide is its ability to absorb sunlight with a very small wavelength (ultraviolet region) [9]. These limitations make the intensity of light that can be captured will be less.

To increase the ability of Titanium Dioxide in the absorption process, a doping process is carried out using a mixture of Fluorine and Indium. The addition of a mixture of Fluorine and Indium doping is expected to increase the photocatalytic activity of visible light radiation [10], reduce the energy band gap and activation energy [11,12,13], increase chemical stability, absorption, and reduce the resistance of the thin film so that it has high sensitivity [14].

2. Methods

This research was carried out in two stages, namely the synthesis process and the thin film characterization which can be seen in Figure 1.

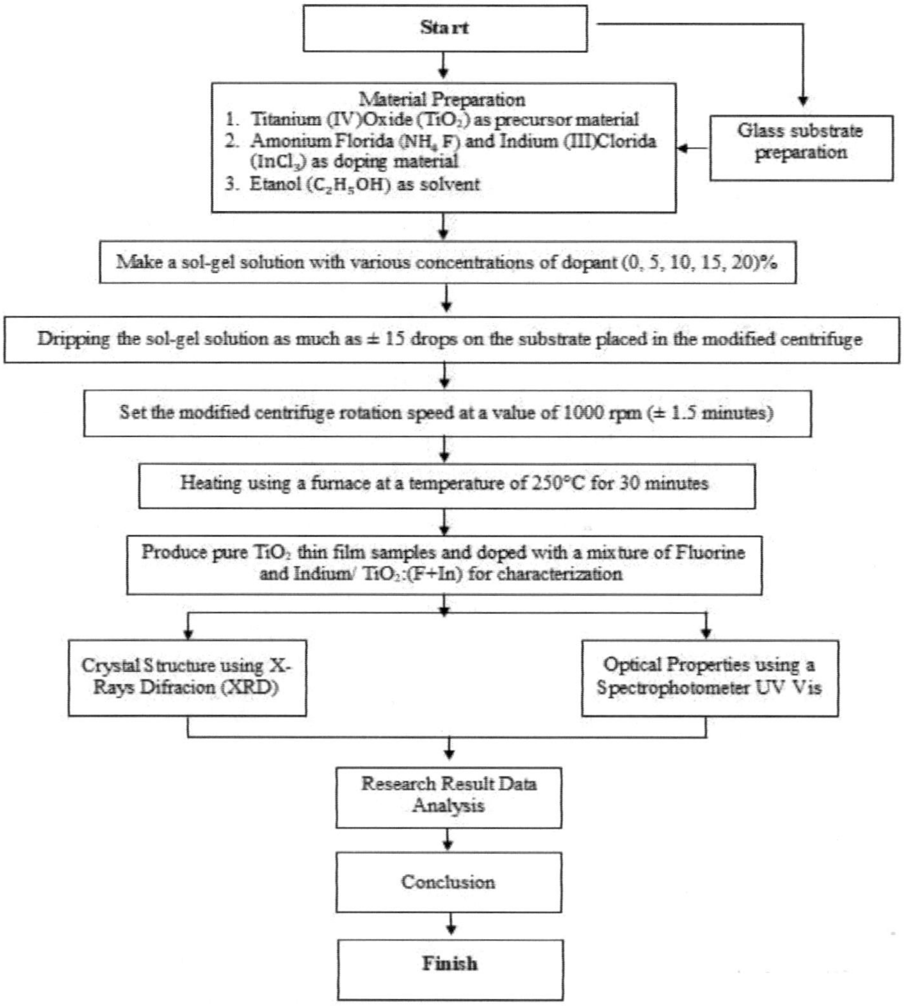

Figure 1. Procedure for Synthesis and Characterization of Titanium Dioxide Thin Films with Fluorine and Indium Doping Mixture [15,16]

Journal of Physics: Conference Series **2165** (2022) 012009 doi:10.1088/1742-6596/2165/1/012009

3. Result and Discussion

Based on the thin film synthesis process carried out, it can be seen that the TiO_2 thin film sample with a mixture of Fluorine and Indium doping has a darker appearance than the pure TiO_2 sample as shown in Figure 2.

(a) (b) (c) (d) (e)

Figure 2. TiO_2:(F+In) Thin Film Samples at Doping Concentrations: (a) 0%, (b) 5%, (c) 10%, (d) 15%, and (e) 20%

XRD testing in this study using X-Rays Diffraction (XRD) Shimadzu type 7000 Maxima to study the crystal structure formed in thin films of TiO_2 and TiO_2:(F+In). X-ray diffraction measurements were carried out at an angle range of 20°-80° because peaks in TiO_2 compounds tended to appear in that angle range, especially at an angle of 25° [17].

Based on the results of the XRD test by comparing the data from the computer analysis (using the Origin 8) application in Figure 3 with the JCPDS (Joint Committee on Powder Diffraction Standard) data, it shows that the two thin-film samples, namely TiO_2 and TiO_2:(F+In) with a concentration of 20% have anatase phase. So when viewed from the reference data, it is known that the crystal structure formed in the two samples is tetragonal.

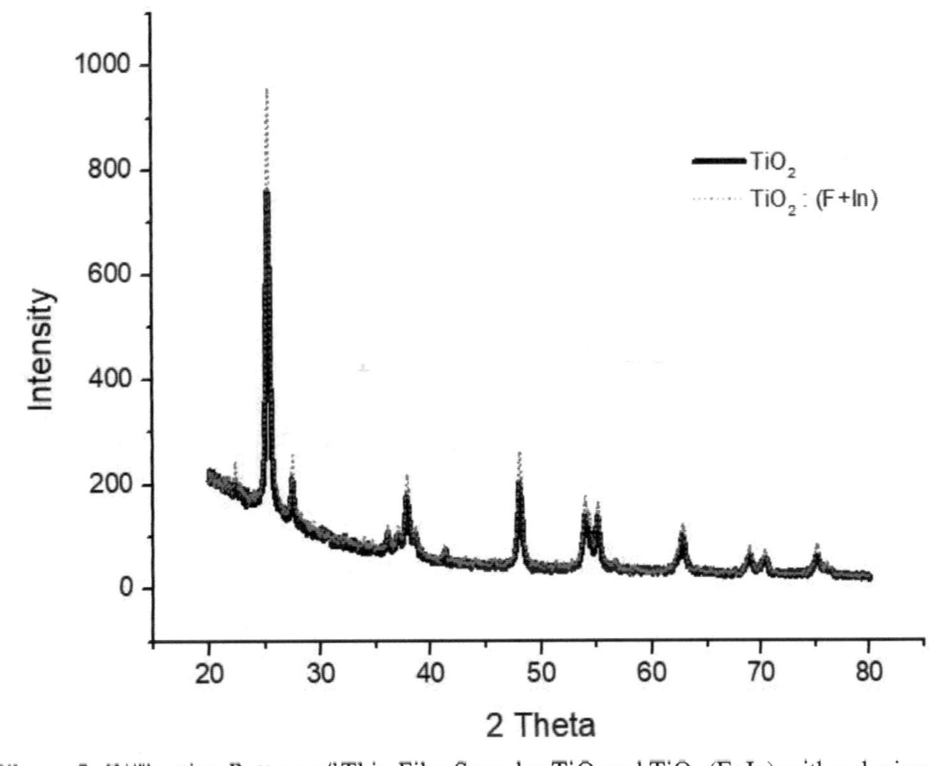

Figure 3. Diffraction Pattern of Thin Film Samples $TiO_2$and TiO_2:(F+In) with a doping concentration of 20%

The intensity of the two samples is quite high, especially in the field (101) with the maximum intensity value in the TiO_2 sample is 782 nm while the TiO_2:(F+In) sample is 960 nm. This corresponds to the characteristic principal peak of anatase phase TiO_2 which appears in the plane (101) at an angle of $2\theta = 25°$. The intensity value on the XRD measurement results is proportional to the level of crystallinity of a material, where the higher the intensity of the diffractogram peak, the more perfect the crystal form formed on the material. The process of calculating the value of the distance between the crystal lattice (d_{hkl}) using the Bragg equation [18], while obtaining the value of the crystal grain size (D) using the Debye-Scherrer equation [19] and obtained data as shown in Tables 1 and 2.

Table 1. XRD Test Result Data for TiO_2 Thin Film(100:0)%

Scattered Angle (2θ)	Miller Index (hkl)	Crystal Lattice Parameters				Lattice Distance (d_{hkl})(Å)	Crystal Grain Size (D) (nm)
		a = b (Å)		c (Å)			
		Standard Value	Calculation Value	Standard Value	Calculation Value		
25.30	101	3.7848	3.7832	9.5124	9.4984	3.51	30.39
37.86	004					2.37	32.44
48.06	200					1.89	31.90
54.53	105					1.68	10.05
62.67	204					1.48	16.90
62.70	204					1.48	17.18
Average							23.14

Table 2. XRD Test Result Data for TiO_2:(F+In) Thin Film(80:20)%

Scattered Angle (2θ)	Miller Index (hkl)	Crystal Lattice Parameters				Lattice Distance(d_{hkl})(Å)	Crystal Grain Size (D) (nm)
		a = b (Å)		c (Å)			
		Standard Value	Calculation Value	Standard Value	Calculation Value		
25.31	101	3.7848	3.7821	9.5124	9.4909	3.51	12.85
27.44	101					3.25	15.16
37.89	004					2.37	11.29
48.07	200					1.89	12.99
54.05	105					1.69	9.82
55.14	211					1.66	10.92
Average							12.17

Based on the calculation process contained in Tables 1 and 2, it is found that the average grain size of the TiO_2:(F+In) sample is smaller than that of the TiO_2 sample. The decrease in crystal grain size is thought to be due to the use of a mixture of Fluorine and Indium dopant which can inhibit crystal growth by being at the crystal boundary. TiO_2 compounds will experience a reduction in crystal grain size after doping with metal [20].

In the application of photocatalyst solar cells, the grain size of the nanometer-scale TiO_2 has the advantage of being able to accommodate more sunlight energy because the smaller the crystal size, the more cavities formed [18]. This can increase the amount of current flowing in a solar cell circuit. The addition of dopants in the synthesis process can affect the physical properties of TiO_2 compounds such as the degree of crystallinity process, surface area, and smaller crystal grain size [21]. In addition, nano-sized photocatalysts in the 1-100 nm range will provide high catalytic activity [22].

The UV-Vis Spectrophotometry test was carried out to know the optical properties of the TiO_2:(F+In) thin-film samples at all doping concentrations (0, 5, 10, 15, and 20)%. The optical properties of the investigated thin films include the properties of transmittance, absorbance, energy bandgap, and activation energy. The graph of the transmittance and absorbance values of TiO2:(F+In) thin films in all doping concentrations can be seen in Figure 4.

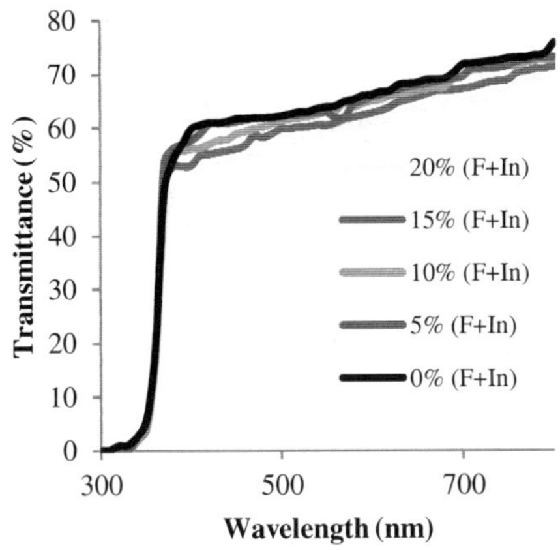

(a)

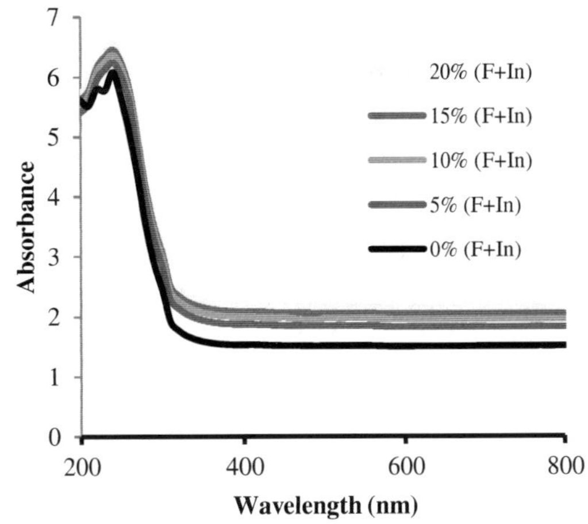

(b)

Figure 4. (a) TiO_2:(F+In) Thin Film Transmittance Value with Doping Concentration 0, 5, 10, 15, and 20%, (b) TiO_2:(F+In) Thin Film Absorbance Value with Doping Concentration 0, 5, 10, 15, and 20%

Based on Figure 4a.it can be seen that the relatively significant difference between TiO_2:(F+In) samples at all concentrations tends to be observed in the visible light absorption region with a wavelength range of 350–700 nm. The maximum transmittance value of the TiO_2:(F+In) thin film at concentrations of 0, 5, 10, 15, and 20% in the visible light region, respectively was 73.76; 73.01; 72.78; 71.01; and 69.55%. The occurrence of changes in the transmittance value of TiO_2 which is quite low in the wavelength range of ultraviolet light indicates a high enough optical absorption so that it certainly has the lowest transmittance value [23].

The maximum absorbance value of the TiO_2:(F+In) thin-film sample seen in Figure 4b occurs in the region with a wavelength range of 220-250 nm or is in the ultraviolet region. The maximum absorbance values for TiO_2:(F+In) films at concentrations of 0, 5, 10, 15, and 20% were 6.08; 6.25; 6.37; 6.46; and 6.57. This is supported by the characteristics of TiO_2 compounds in the anatase phase which absorb wavelengths with small values (close to the ultraviolet region) [23, 24]. The addition of a mixture of Fluorine and Indium doping causes a shift in absorption towards visible light with a wavelength range from 320-700 nm (Figure 4b). With the increase in energy absorption in the visible light wavelength range, it is expected to increase the photocatalytic effect of the thin film [25, 26].

The value of the energy band gap can be determined using the Tauc Plot graph method which describes the relationship between $h\nu$ and $(ah\nu)^n$ [27]. Determination of the energy bandgap from the graph of the relationship between $h\nu$ and $(ah\nu)^n$ to intersect the $h\nu$ axis. The energy band gap values obtained in this study consist of a direct energy bandgap (n = 1/2) and an indirect energy bandgap (n = 2). Based on Figure 5. it is obtained that the direct energy band gap of TiO_2:(F+In) thin films at concentrations of 0, 5, 10, 15, and 20%, respectively are 3.23; 3.19; 3.07; 3.02; and 2.97 eV, while in Figure 6. the indirect energy band gap values of TiO_2:(F+In) thin films at concentrations of 0, 5, 10, 15, and 20% are 3.83; 3.80; 3.79; 3.79; and 3.75 eV.

Journal of Physics: Conference Series **2165** (2022) 012009 doi:10.1088/1742-6596/2165/1/012009

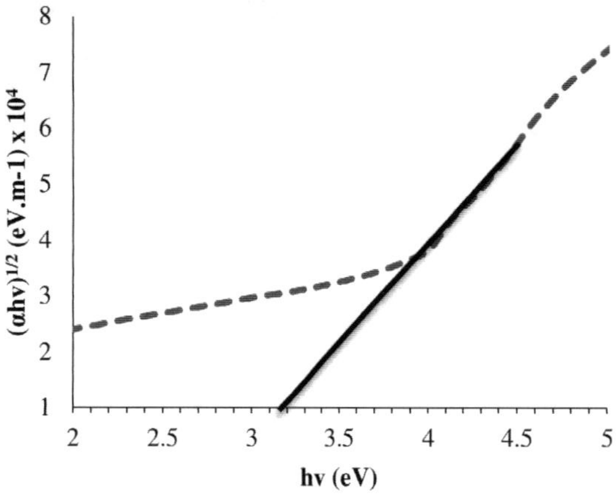

(a)

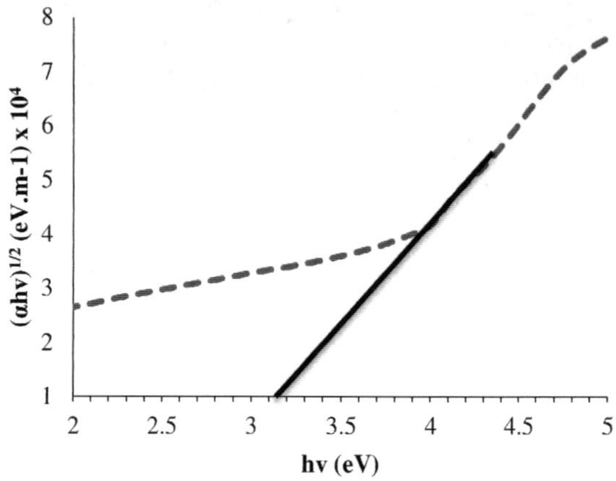

(b)

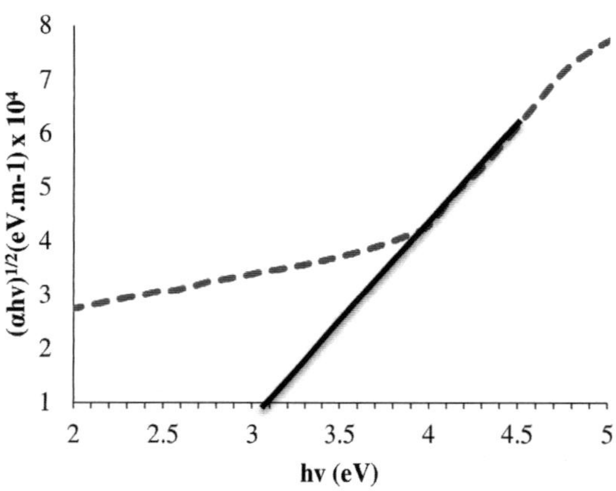

(c)

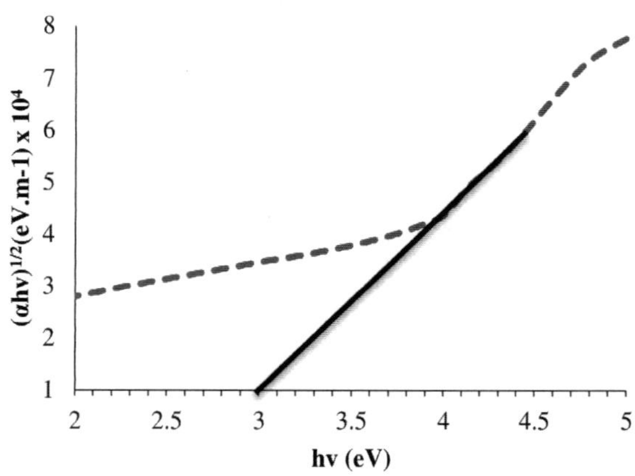

(d)

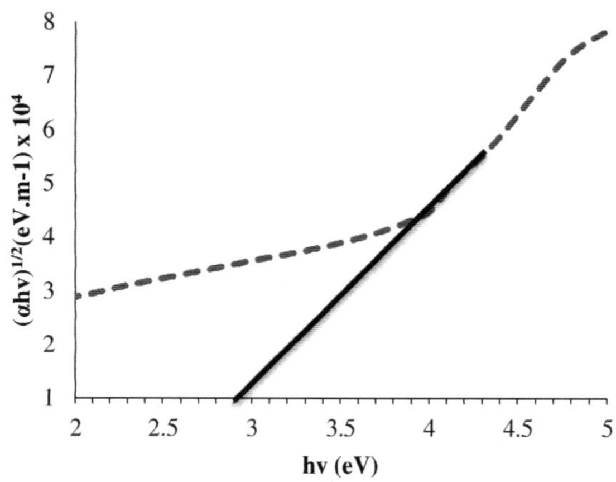

(e)

Figure 5. Graph of Direct Energy Bandgap TiO_2:(F+In) Thin Film at Doping Concentration (a) 0%, (b) 5%, (c) 10%, (d) 15%, and (e) 20%

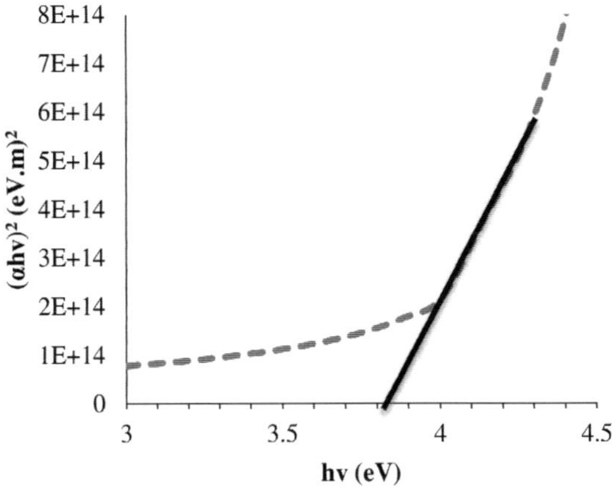

(a)

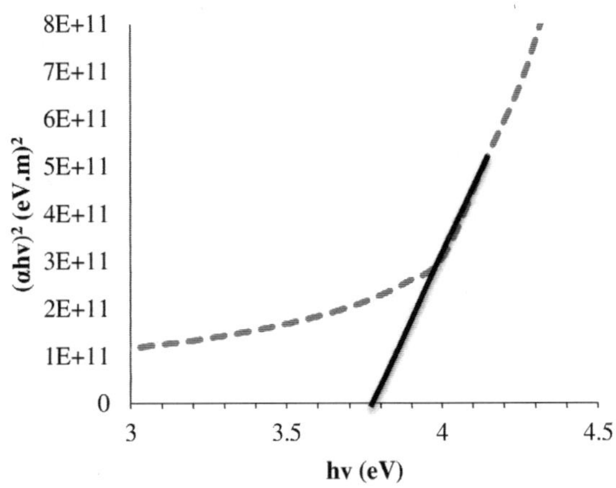

(b)

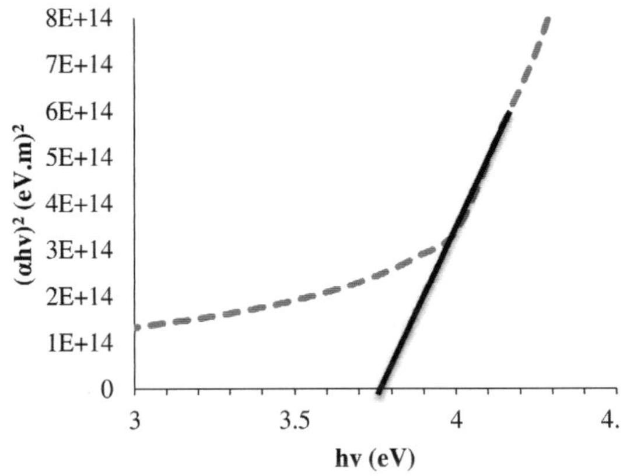

(c)

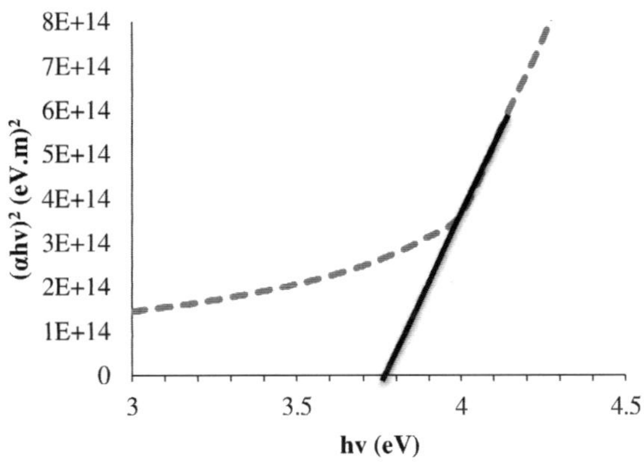

(d)

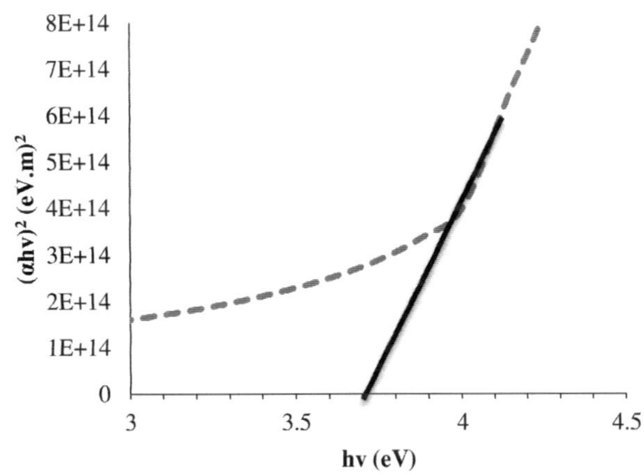

(e)

Figure 6. Graph of Indirect Energy Bandgap TiO_2:(F+In) Thin Film at Doping Concentration (a) 0%, (b) 5%, (c) 10%, (d) 15%, and (e) 20%

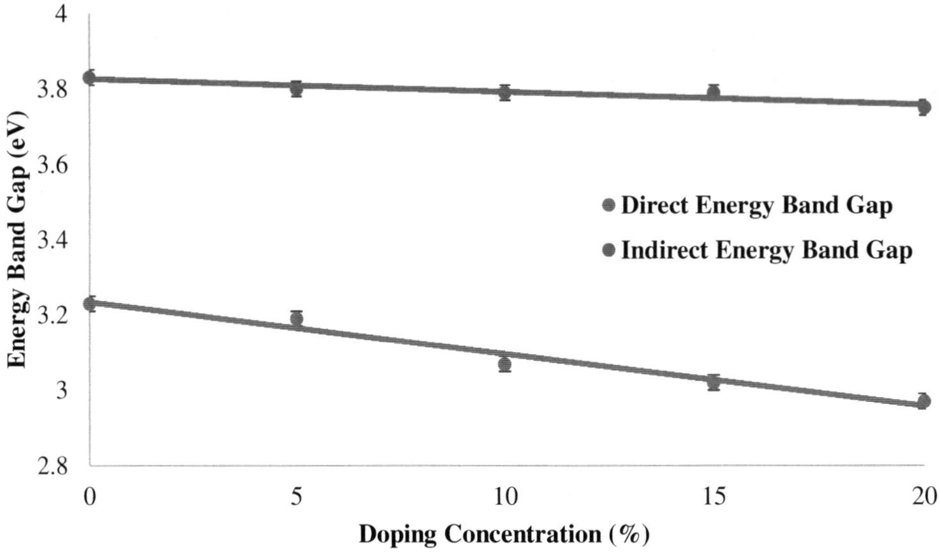

Figure 7. Graph of Relationship between Direct and Indirect Energy Band Gap TiO_2:(F+In) Thin Films at Doping Concentration (a) 0%, (b) 5%, (c) 10%, (d) 15%, and (e) 20%

Based on Figure 7, it can be seen that the value of the energy bandgap is getting smaller, the more doping concentrations of a mixture of Fluorine and Indium are added. The decrease in the energy band gap value will certainly have an impact on the ability of the thin film in the photocatalysis process [17, 28]. The smaller the energy bandgap will increase the photocatalytic activity [29]. Therefore, by decreasing the energy bandgap, there is a shift in the optical properties of TiO_2 towards the visible spectrum region (ranging from 320-700 nm). This condition allows the photocatalytic process to occur even though it uses visible light for the activation process. The decrease in the energy band gap is to facilitate the process of excitation of electrons from the valence band to the conduction band [30].

Figure 7 also shows that the value of the indirect energy bandgap tends to be greater than the direct energy bandgap because electrons in the conduction band cannot recombine with holes or the opposite condition in the excitation process [31,32,33]. This is because electrons and holes are not at the same crystal momentum in the Brillouin zone. So that for the recombination or excitation process to occur in the indirect energy bandgap, it must involve a phonon absorption or emission process, where the phonon momentum must be equal to the difference between the electron and hole momentum [34, 35,36].

If viewed from the aspect of activation energy, based on Figure 8, the activation energy values for TiO_2:(F+In) thin films at concentrations of 0, 5, 10, 15, and 20%, respectively are 3.96; 3.83; 3.73; 3.57; and 3.15 eV. The greater the concentration of doping used, the smalleractivation energy produced [37]. The smaller the value of the activation energy possessed by a layer, the less energy is needed for a chemical reaction to occur.The use of Fluorine and Indium doping in this study seems to be able to affect the activation energy of the Titanium Dioxide thin film. The addition of Fluorine (F) doping can reduce the activation energy of the thin film from 4.04 to 1.99 eV [12], while the addition of Indium doping can reduce the activation energy of the thin film from 2.91 to 2.35 eV [38,39].

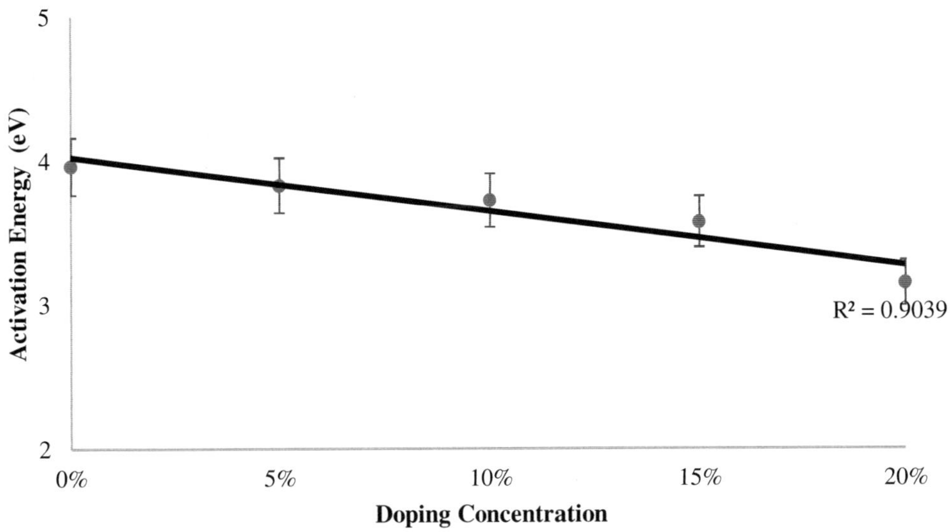

Figure 8. TiO_2:(F+In) Thin Film Activation Energy Graph at Doping Concentration (a) 0%, (b) 5%, (c) 10%, (d) 15%, and (e) 20%

4. Conclusion

The process of synthesizing a thin film of TiO_2 with a mixture of Fluorine and Indium doping has been carried out and produces a thin film sample with a surface appearance that tends to get darker with the addition of doping. The results of the X-Rays Diffraction (XRD) test showed that the samples of TiO_2 and TiO_2:(F+In) had an anatase phase and a tetragonal shape with the crystal grain size getting smaller from 23.14 to 12.17 mm. The results of the UV-VisSpectrophotometertest showed that the transmittance value was very low in the ultraviolet region but increased in the wavelength range of 350–700 nm (visible light region). The highest absorbance value is in the ultraviolet region in the range of 220-250 nm and the absorption shifts towards the visible light range with a wavelength range of 320-700 nm. The values of the direct energy bandgap, indirect energy bandgap, and the obtained activation energy decreased with the values being respectively 3.23 to 2.97 eV; 3.83 to 3.75 eV; and 3.96 to 3.15 eV.

Acknowledgments

The authors would like to thank Postgraduates of the University of Mataram, Laboratory of Basic Chemistry and Organic Chemistry at the University of Mataram, Hydrogeology &Hydrogeochemistry Laboratory of the Bandung Institute of Technology, Mataram City Environmental Laboratory and the Calibration Testing Health Laboratory of Mataram City, and all those who helped in completing this article.

References

[1] Astuti A and Ningsi S 2017Sintesis Dan KarakterisasiNanopartikel Titanium Dioksida (TiO2) MenggunakanMetodeSonokimia *Jurnal Ilmu Fisika Universitas Andalas* **9**(1) pp 26-32 DOI: 10.25077/jif.9.1.26-32.2017

[2] Chen Z,Guo X,Guo H, Ma C,Qiu J, Yuan N, and Ding J 2019 Fabrication of a semi-transparent thin-film Sb_2Se_3 solar cell *Materials Letters* **236** pp 503-505 DOI: 10.1016/j. matlet.2018.10.173.

[3] Buwarda S 2019 AZTS denganMetode Sol-Gel sebagai Lapisan Buffer Bebas Cadmium padaSel Surya CZTS*Jurnal Keteknikan dan Sains (JUTEKS)* **2**(1) pp 1-7.

[4] Bredas J L, Sargent E H, and Scholes G D 2017 Photovoltaic Concepts Inspired by Coherence Effects in Photosynthetic System*Nature Materials* **16**(1) pp 35-44 DOI: 10.1038/nmat4767.

[5] Husain AA, Hasan W Z W, Shafie S, Hamidon M N, and Pandey SS 2018 A Review of Transparent Solar Photovoltaic Technologies *Renewable and Sustainable Energy Reviews* **94** pp 779-791 DOI: 10.1016/j.rser.2018.06.031.

[6] Nayak P K, Mahesh S, Snaith H J, and Cahen D 2019 Photovoltaic Solar Cell Technology: Analysing the State of The Art*Nature Reviews Materials* **4**(4) pp 269 DOI: 10.1038/s41578-019-0097-0.

[7] Voloshin R A, Bedbenov V S, Gabrielyan D A, Brady N G, Kreslavski V D, Zharmukhamedov SK, ... and Allakhverdiev S I 2017 Optimization and characterization of TiO2-based solar cell design using diverse plant pigments *International Journal of Hydrogen Energy* **42**(12) pp 8576-8585 DOI: 10.1016/j.ijhydene.2016.11.148

[8] Bedghiou D, Reguig F H, and Boumaza A 2019 Novel High/Ultrahigh Pressure Structures of TiO_2 with Low Band Gaps *Computational Materials Science* 166 pp 303-310 DOI: 10.1016/j. commatsci.2019.05.016

[9] Manurung D N and Deswardani F 2021PengaruhPenggunaan Dye Tunggal dan Dye CampuranAntosianin-KlorofilTerhadapEfisiensiKerja Dye Sensitized Solar Cell (DSSC) *JPFT (JurnalPendidikanFisikaTadulako Online)* **9**(1) pp 83-88.

[10] Dewi Y S 2019. *Sintesis dan Karakteristik Foto katalis TiO_2 Mesopori Terdoping Galium (III) dengan Variasi Konsentrasi Dopan Ga^{3+} Menggunakan Metode Sonikasi* Tesis S2 Universitas Islam Negeri Maulana Malik Ibrahim.

[11] Hakim S, Doyan A, Susilawati S, and Muliyadi L 2019 Synthesis Thin Film SnO_2 with Doping Indium by Sol-Gel Spin Coating*JurnalPenelitianPendidikan IPA* **5**(2) pp 171-174 DOI: 10.29303/jppipa.v5i2.254.

[12] Muliyadi L, Doyan A, Susilawati S, and Hakim S 2019 Synthesis of SnO2 Thin Layer with a Doping Fluorine by Sol-Gel Spin Coating Method *Jurnal Penelitian Pendidikan IPA* **5**(2) pp 175-178 DOI: 10.29303/jppipa.v5i2.257.

[13] Munandar H, Doyan A, and Susilawati S 2020 Synthesis of SnO2 Thin Coatings by Indium and Aluminum Mixed Doping using the Sol-Gel Spin-Coating Technique *Jurnal Penelitian Pendidikan IPA* **6**(2) pp 152-156 DOI: 10.29303/jppipa.v6i2.391.

[14] Ivanova T, Harizanova A, Koutzarova T,Vertruyen B, and Stefanov B 2018 Structural and morphological characterization of sol-gel ZnO: Ga films: Effect of annealing temperatures *Thin Solid Films* **646** pp 132-142 DOI: 10.1016/j.tsf.2017.11.042.

[15] Rizaldi D R, Doyan A, and Susilawati S 2021 Sintesis Lapisan Tipis TiO2:(F+ In) pada Substrat Kaca Dengan Metode Spin-Coating Sebagai Bahan Sel Surya *Orbita: JurnalKajian, Inovasi dan Aplikasi Pendidikan Fisika* **7**(1) pp 219-224 DOI: 10.31764/orbita.v7i1.4655.

[16] Susilawati S, Doyan A, Muliyadi L, and Hakim S 2020 Growth of Tin Oxide Thin Film by Aluminum and Fluorine Doping Using Spin Coating Sol-Gel Techniques *Jurnal Penelitian Pendidikan IPA* **6(1)** pp 1-4 DOI: 10.29303/jppipa.v6i1.264.

[17] Subalakshmi K and Senthilselvan J 2018 Effect of Fluorine-Doped TiO_2 Photoanode On Electron Transport, Recombination Dynamics, and Improved DSSC Efficiency*Solar Energy* **171** pp 914-928 DOI: 10.1016/j.solener.2018.06.077.

[18] Hastuti E 2011AnalisaDifraksi Sinar X TiO_2 dalam Penyiapan Bahan Sel Surya Tersensitisasi Pewarna *Jurnal Neutrino* **4**(1) pp 93-100 DOI: 10.18860/neu.v0i0.2416.

[19] Dahlan D and Fahyuan H D 2014 Pengaruh Beberapa Jenis Dye Organik Terhadap Efisiensi Sel Surya Dye Sensitized Solar Cell *Jurnal Sains Materi Indonesia* **15**(2) pp 74-79 DOI: 10.17146/jsmi.2014.15.2.4359.

[20] Ganesh I, Gupta A K, Kumar P P, Sekhar P S C, Radha K, Padmanabham G, and Sundararajan G 2012 Preparation and Characterization of Ni-Doped T i O 2 Materials for Photocurrent

and Photocatalytic Applications *The Scientific World Journal* 2012 DOI: 10.1100/2012/127326.

[21] Tan Y N, Wong C L, and Mohamde A R 2011 An Overview On the Photocatalytiv Activity of Nano-Doped-TiO_2 in the Degradation of Organic Pollutans*ISRN Materials Science* 34 DOI: 10.5402/2011/261219.

[22] Naimah S and Ernawati R 2011EfekFotokatalisis Nano TiO_2terhadapMekanisme Anti Mikroba*E Coli* dan *Salmonela Jurnal Riset Industri* **5**(2) pp 113-120.

[23] Perdana R and Dahlan D 2013Sifatoptiklapisan tipis Tio2 yang disintesismenggunakanmetode sol gel spin coating *Jurnal Ilmu Fisika| Universitas Andalas* **6**(1) pp 18-24 DOI: 10.25077/jif.6.1.18-24.2014.

[24] Listanti A, Taufiq A, Hidayat A, and Sunaryono S 2018InvestigasiStrukturdanEnergi Band Gap Partikel Nano TiO2 HasilSintesisMenggunakanMetode Sol-Gel *JPSE (Journal of Physical Science and Engineering)* **3**(1) pp 8-15 DOI: 10.17977/um024v3i12018p008.

[25] Al-Kdasi A, Idris A, Saed K, and Guan C T 2004 Treatment of textile wastewater by advanced oxidation processes—a review *Global nest: the Int. J.* **6**(3) pp 222-230.

[26] Agus A, Fahyuan H D, and Damris M 2019PenentuanNilai Energy Gap Lapisan Tipis TiO2/C Dengan Menggunakan Metode Touc Plot *JIFP (Jurnal Ilmu Fisika dan Pembelajarannya)* **3**(2) pp 63-67.

[27] Efelina V 2017PreparasidanPenentuanEnergi Gap Fim Tipis TiO_2:Cu yang Ditumbuhkan Menggunakan Spin Coating *JurnalPendidikanFisikadanKeilmuan* **3**(1) pp19-27.

[28] Tahir M and Amin N S 2015 Indium-doped TiO2 nanoparticles for photocatalytic CO2 reduction with H2O vapors to CH4 *Applied Catalysis B: Environmental* **162** pp 98-109 DOI: 10.1016/j.apcatb.2014.06.037.

[29] Salim and Sudaryanto 2016 Penambahan N Pada TiO_2danPengaruhnyaPadaEnergi Band Gap TiO_2Sebagai Bahan Pengelola Limbah *Prosiding Seminar Nasional Teknologi dan Aplikasi Reaktor Nuklir PRSG Tahun 2016* pp59-64.

[30] Berli A U, Dahlan D, and Umar A A 2016 Pengaruh Lama Penumbuhan Titanium DioksidaDidoping Copper TerhadapEnergi Gap *Jurnal Ilmu Fisika| Universitas ANDALAS* **8**(2) pp 60-63 DOI: 10.25077/jif.8.2.60-63.2016.

[31] Yuan L D, Deng H X, Li SS, Wei S H, and Luo J W 2018 Unified theory of direct or indirect band-gap nature of conventional semiconductors *Physical Review B* **98**(24) pp 245203 DOI: 10.1103/PhysRevB.98.245203.

[32] Doyan A, Imawanti Y D, Gunawan E R, and Taufik M 2018 September Characterization Thin Film Nano Particle of Aluminum Tin Oxide (AlTO) as Touch Screen *In Journal of Physics: Conference Series* **1097**, No. 1, p. 012009 IOP Publishing.

[33] Doyan A, Muliyadi L, Hakim S, and Taufik M 2019 December Characteristics and Optical Properties of Fluorine Doped SnO2 Thin Film Prepared by a Sol–Gel Spin Coating *In Journal of Physics: Conference Series* **1397** (1), p. 012003 IOP Publishing.

[34] Gu L,Srot V, Sigle W, Koch C, Van Aken P, Scholz F, ... and Rühle M 2007 Band-gap measurements of direct and indirect semiconductors using monochromated electrons *Physical Review B* 75(19) 195214 DOI: 10.1103/PhysRevB.75.195214.

[35] Kocevski V, Eriksson O, and Rusz J 2013 Transition between direct and indirect band gap in silicon nanocrystals *Physical Review B*. **87** (24) pp 245401 DOI:10.1103/PhysRevB.87.245401.

[36] Doyan A, Harjono A, Azzahra S, and Taufik M 2019 Characterization of Tin Oxide Doping Antimony Thin Layer With Sol-Gel Spin Coating Method for Electronic Device*In Materials Science Forum* **966**, pp. 30-34) Trans Tech Publications Ltd.

[37] Doyan A, Taufik M, Hakim S, and Muliyadi L 2021 The Optical Properties of Thin Films Tin Oxide with Triple Doping (Aluminum, Indium, and Fluorine) for Electronic Device*In Solid State Phenomena* **317**, pp. 177-182) Trans Tech Publications Ltd.

[38] Doyan A and Munandar H 2021 March Optical Characteristics of Tin Oxide Thin Films Doped with Indium and Aluminum Using the Sol-Gel Spin Coating Technique*In 7th International Conference on Research, Implementation, and Education of Mathematics and Sciences (ICRIEMS 2020)* (pp. 396-403) Atlantis Press **(39)**.

[39] Doyan A, Hakim S, Muliyadi L, and Taufik M 2019 The Effect of Indium Doped SnO_2 Thin Films on Optical Properties Prepared by Sol-gel Spin-Coating Technique*Journal of Physics: Conference Series.* **1397**(2019) pp 012005.

Journal of Physics: Conference Series 2165 (2022) 012010 doi:10.1088/1742-6596/2165/1/012010

Development of digital handout of quantum mechanics no spin and quantum mechanics with spin using professional pdf flip app

D Nurvita[1], H Akhsan[1], M Muslim[1] and M Ariska[1]

[1]Physics Education, FKIP, Sriwijaya University, South Sumatra, Indonesia

E-mail: hamdiakhsan@gmail.com

Abstract. This study aims to produce a digital handout on the subject of mechanics quantum mechanics without spins and quantum mechanics with spins are valid and practical. Development This digital handout uses a combination of Rowntree's development model and evaluation model Tessmer formative. The Rowntree development model consists of three stages, namely: planning, development and evaluation. The evaluation stage consists of four stages, namely: self evaluation , expert review , one-to-one evaluation , small group evaluation. Technique data collection in this study using walkthroughs and questionnaires. Research subject This is a student who has taken the introductory quantum physics class of class 2017 Indralaya class for the Physics Education Study Program, Sriwijaya University. Obtained the average scorethe average that meets the very valid category with a score on the content aspect of 4.92, the design aspect 4.46, and the language aspect 4.5. And meet the very practical category with an average score of 4.43at the one-to-one evaluation stage and 4.52 at the small group evaluation stage.

1. Introduction

The faster the progress of science knowledge and technology in the field of information and new innovations in technology make it easy life human. Advances in science and technology make an impact on the world education, one of which is to increase quality of education so as to create quality human and answer market demand for labor quality [1]. Quality of education and learning outcomes influenced by many things, including educator skills and learning resources which will be used during the learning process teaching [2]. Source learning has a very important role in solving problems in learning, therefore learning resources are very important to improve the quality of education. One type of learning resource that is very often used are teaching materials.

Teaching materials are learning resources which serves to convey content or learning materials/lectures for encourage teacher efficiency and improve student performance [3]. With teaching materials to make learning more interesting, practical, and realistic. In addition, the use of teaching materials in learning allows good teachers or students can participate actively and make learning more effective. Teaching materials can provide space to acquire knowledge and skills, develop confidence self-actualization and self-actualization of students [3].

According to material steach has several types based on one of the technologies used namely handouts [4]. Handout is one of the types of printed teaching materials containing a summary of the material prepared by educators for make it easier for students to follow learning [5]. Handouts load the required parts for learning which consists of; destination, learning or competence, prerequisite namely learning materials that support or need to be studied first beforehand, learning procedures, theory learning that arranged systematic, exercises or tasks and questions evaluation questions [6].

Great technological developments quickly create a paperless culture or paperless in the form of teaching materials/handouts print using paper and ink, as if immersed in the discovery of teaching material electronic (Digital Handout), so that more varied and innovative teaching material scan be equipped with hyperlinks, images as well as videos assisted with use a professional PDF flip application to get a digital display that sophisticated.

Development technology especially high technology has very close relationship with development of science, especially physics quantum. Main concept Introduction to Quantum Physics can used to develop or create new technology if studied furthermore. However, some references learning resources used in teaching and learning process for subjects it's dominant in English so that if some terms are direct translated can give meaning doubles as well as references used too only focused on understanding the concept only the material so it's very minimal to help improve literacy student technology. Therefore, many students experience difficulty understanding the concept Introduction to Quantum Physics especially on its application in the field of technology.

Based on the results of filling out the questionnaire online with google from by 41 student class 2017 physics education Indralaya, then the results obtained are 43.9% of students who have difficulty caused by book references that used in language learning english. 61% stated strongly agree and39% agree if done similar development research.

Research and development of teaching materials has been carried out by several researchers in UNSRI, including 1) [7] with products in the form of digital handouts based on technological literacy in courses modern physics the subject of relativity and particle wave dualism [8] with a product in the form of a textbook introduction to STEM-based quantum physics in the spin sub-topic. 3) [9] with products in the form of -based modules STEM subject matter variation potential in the introductory course quantum physics. While in research in this case, researchers will develop digital handout on introductory physics course quantum subject of quantum mechanics without spin and quantum mechanics with technology literacy-based spin because the lack of development of teaching materials accessible electronics and materials quantum mechanics without spin and mechanics quantum with integrated spin with technological literacy.

2. Methods

Research this done in Education Study Program Laboratory Sriwijaya University Physics with use method research Rown tree model product development. According to [10] model Rown tree development consists of 3 stages, that is step planning, step development and evaluation stage. On evaluation stage using the evaluation model Tessmer formative consisting of (1) self evaluation ; (2) expert review ; (3) one to one evaluation ; (4) small group evaluation and (5) Field Test [11]. But stage the fifth is not done because the purpose is to be achieved in this research only to produce digital handout products valid and practical.

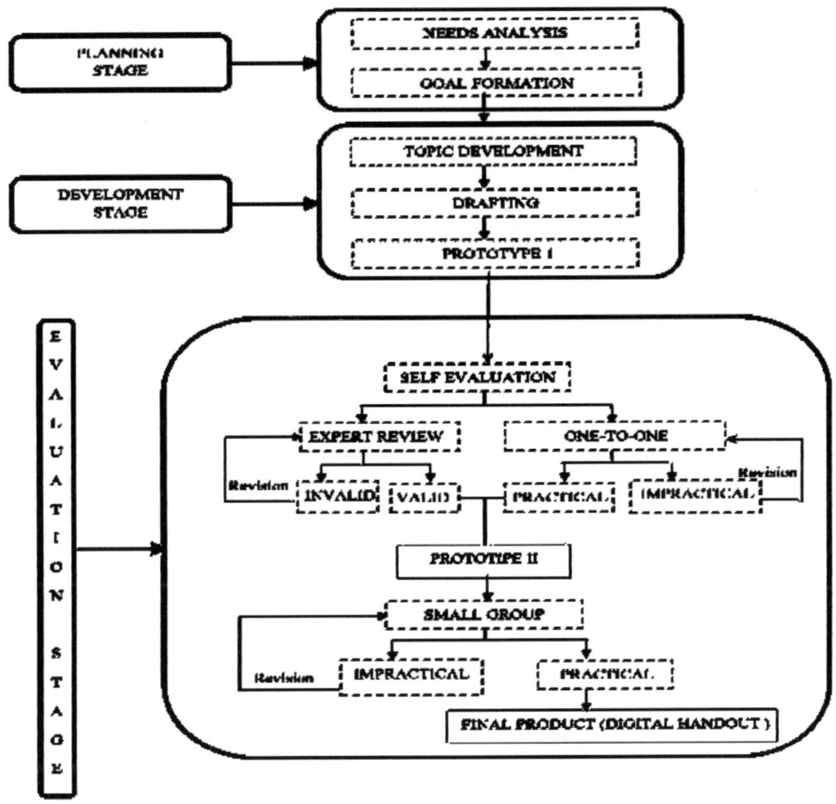

Figure 1. Research Development Procedure (Rowntree and Tessmer)

Then for the gathering technique the data uses a walkthrough to see the level of product validity and questionnaires to see the level of practicality of the product. In walkthrough data collection techniques the data collection tool is in the form of sheets validation given to experts who includes content, design, and language validation. With guided validation indicator son the Guide to Development of Teaching Materials by the Ministry of National Education (2008). Indicators that validated for the content aspect, namely suitability material with achievements and indicators learning, the truth of the substance of the material, suitability to student needs, conformity with digital needs handouts, benefits to add insight knowledge and detailed study of material concepts about its application in technology. Validated indicators for design aspects consists of completeness of information, sequence presentation, use of fonts (type and size), Illustration, image, audio, video, linkweb, layout and display design. Validated indicators for language aspect consists of legibility, clarity of information, conformity with the rules of the Indonesian language good and correct as well as the use of language effectively and efficiently.

For data collection techniques a questionnaire in the form of a student response questionnaire given at the one-to-one stage evaluation and small group evaluation that involving 41 physics education students Indralaya class 2017. Sheet validation and questionnaire given later made in the form of a Likert scale. Scale The likert used is made in the form of checklist with five categories of answers such as the table below which is useful for analyze the results of the walkthrough and technique questionnaire.

Table 1. Criteria for Scoring Validation and Questionnaire [12].

Category Answer	Score Statement
Very good	5
Good	4
Doubtful	3
Not good	2
Not very good	1

Validation results and questionnaire data will be described in tabular form. Then in calculate the average score with the formula [13].

$$\bar{X} = \frac{\Sigma x}{n} \quad (1)$$

Description: $\overline{X}$= average score, Σ_x= total score of each indicator, n = total statement.

The average value of validator and customized questionnaire data with the criteria set out in table 2 below:

Table 2. Criteria for Expert Validation Scores and Student Practical Response [14].

Average Score	Category
$4 \leq \overline{X} \leq 5$	Very valid
$3 \leq \overline{X} < 4$	Valid
$2 \leq \overline{X} < 3$	Not valid
$1 \leq \overline{X} < 2$	Invalid

Suggestions/comments received on the validation sheet and the questionnaire is used as input to fixdigital handouts.

3. Result and Discussion

3.1. Planning Stage

At this planning stage, it begins by conducting a needs analysis using a questionnaire via google form, the result is 100% students have difficulty in learn Introduction to Quantum Physics especially regarding the main concepts introduction to quantum physics on its .application in technology. These difficulties are due to: (1) as many as 43.9% of students experienced trouble caused by book reference used in learning speak English, (2) 31.7% of student shaving trouble due to lack of additional book references regarding the discussion of the main concepts introduction to quantum physics on applications in the field of technology, (3) 24.4% of students have difficulty caused by the material presented is not associated with the application of it s application in technology field. Another reason why this research was conducted based on previous research there are no researchers who develops course handouts introduction to quantum physics in particular the subject of quantum mechanics without spin and quantum mechanics with a spin that integrated with technological literacy as well as no researchers have packaged handouts in digital form.

Next after analysis needs, researchers formulate learning objectives that produce learning objectives and achievements from the results material analysis based on RPS analysis introductory quantum physics course. Materials presented on digital handouts this will focus on understanding the concept introduction to quantum physics on application in technology.

3.2. Development Stage

At this stage, the researcher conducts topic development, drafting and producing prototypes. First on the development of the topic the researcher conducts Creating an Outline of Contents for a Digital Handout (GBIDH) based on the topic of discussion developed based on destination learning that has been formulated. Then the drafting is carried out by specifying the components which will be written in the digital handout. The next step is prototype production1 or the initial product. Digital handout that produced in the form of supporting teaching materials with material focused on application its application in the field of technology on the concept of quantum mechanics without spin and mechanics quantum with spin.

Here's a digital handout display in image form. In figure 2 is a digital handout cover that contains the title, name of the researcher and the institution.

Figure 2. Cover

Prior to discussing the material, there is figure 3 which shows when hyperlink is clicked in the list section contents and other posts that go directly to additional reading resources and so on. There is also a page view that has figures and videos are shown on the figure4, when the image is clicked then as in display figure 5, while typing video rotated as shown in figure 6. Next on the practice questions at the end discussion there is a barcode that can scanned to make it easier for users to fill in the answer to the question, which is displayed on figure 7.

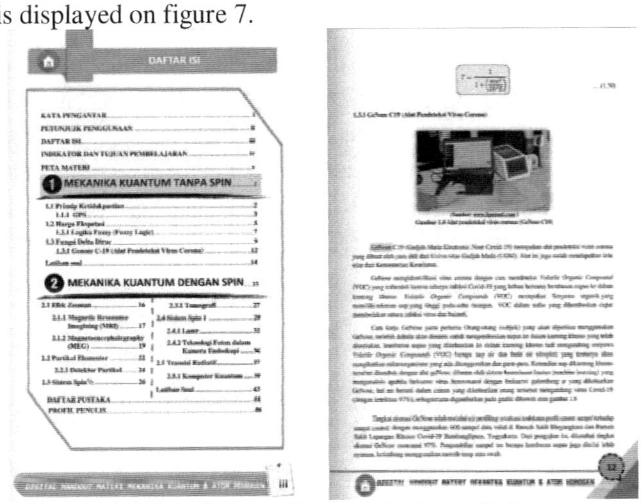

Figure 3. Page display with hyperlinks

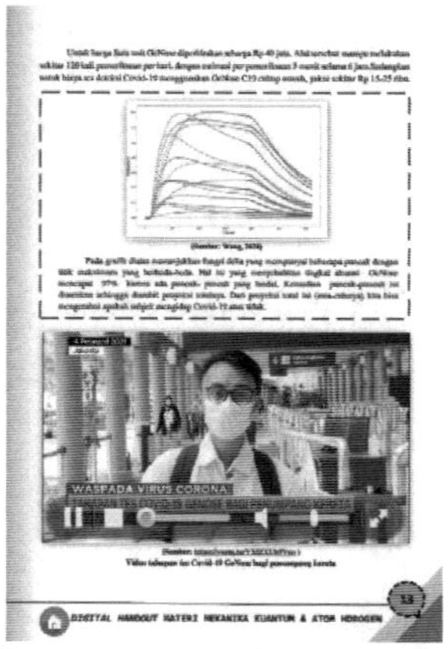

Figure 4. Page view with images and videos

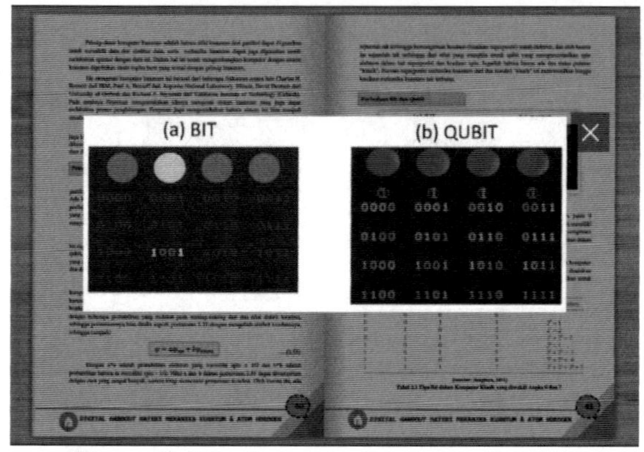

Figure 5. Display when the image is clicked

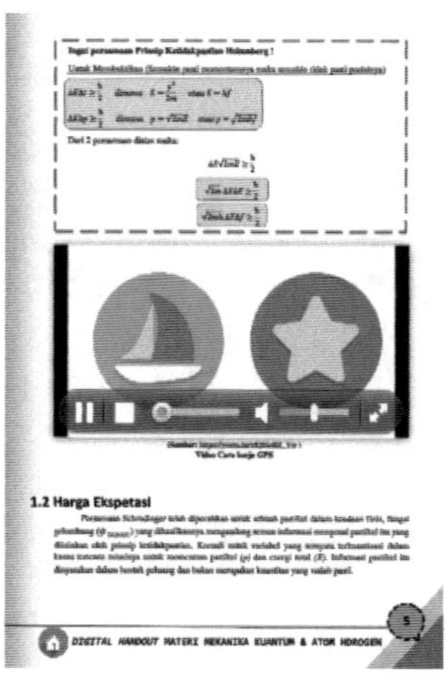

Figure 6. Display when the video is playing

Figure 7. Display the barcode of the question

3.3. Evaluation Stage

After prototype 1 is generated then the next stage is evaluated based on Tessmer's formative evaluation stages consists of four stages. First stage self evaluation that is self-examination to the initial prototype draft that has been developed and checked by lecturers supervisor, as for enter the provided in the form of cover repairs, addition of sub chapters and additions using hyperlinks in digital handouts to make it easier to use. The second is expert review stage conducted by experts (Lecturer of Physics Education Study Program UNSRI) to see the level of validity in terms of content , design and language prototype 1.

Table 3. Recapitulation of Research Results Phase Expert Review

Aspects that validated	Score validity	Category
Contents	4.92	Very valid
Design	4.46	Very valid
Language	4.50	Very valid

Based on the table above, it can be it is known that the average yield obtained from the expert review stage on each aspect has fulfilled very valid category, so digital handouts that have been developed already relevant to learning outcomes, aspects of the design is already attractive and adequate for used, and the language aspect has been consistent.

The third stage is one-to-one evaluation involving three people UNSRI physics education student class of 2017 Indralaya class who have take an introductory physics course quantum.

Table 4. Research Results Phase One-to-one evaluation

Indicator	Average Score	Category
1	4.67	Very Practical
2	3.60	Practical
3	4.30	Very Practical
4	5.00	Very Practical
5	5.00	Very Practical
6	4.00	Practical
7	4.67	Very Practical
8	4.67	Very Practical
9	4.00	Practical
10	4.17	Very Practical
11	4.67	Very Practical
Score Average	4.43	Very Practical

Based on table 4, it is known that the resulting prototype 1 has practical, with suggestions / comments obtained from the stage of expert reviews and one-to-one this be input for revising the prototype 1 so as to produce a prototype 2 which will then be tested at the stage small group evaluation.

The fourth stage is small group the ninth trialed evaluation UNSRI physics education students class of 2017 Indralaya class who have take an introductory physics course quantum. This evaluation stage is the final stage of research for produce prototype 3 (Digital handout based on technological literacy of the subject quantum mechanics without spin and mechanics quantum with a valid and practical spin). The result data from this evaluation can be seen in table 5.

Table 5. Small Group Stage Research Results Evaluation

Indicator	Average Score	Category
1	4.60	Very Practical
2	4.60	Very Practical
3	4.20	Very Practical
4	4.40	Very Practical
5	4.40	Very Practical
6	4.40	Very Practical
7	4.60	Very Practical
8	4.50	Very Practical
9	4.60	Very Practical
10	4.60	Very Practical
11	4.80	Very Practical
Score Average	4.52	Very Practical

Based on the results obtained in the table above, it is known that the prototype 2 is already practical. Practical means that the resulting product is easy used, have clear instructions, and is in accordance with linguistic standards [15]. At this stage there is also suggestions/comments given are only in the form of fix one of the pictures the resolution is a bit low. From enter it produces prototype 3 as the end product is a digital handout on the subject of quantum mechanics without spin and quantum mechanics with a spin that valid and practical.

Digital Handout on the subject quantum mechanics without spin and mechanics quantum with developed spin with this professional PDF flip application has advantages, namely, 1) Teaching materials that don't just stick to writing, but with this Professional PDF flip app hyperlinks can be inserted as well as various multimedia content such as audio, animation, and videos [16]. 2) Format the output or output is flexible, such as html, exe, zip, Mac App, mobile version (IOS/Android) and inserted into the CD, making it more efficient to carry and used [17]. While the weakness of this digital handout is that every three months must be updated or uploaded back, with a maximum of one flip account Professional PDF upload three times product [18] and can't yet observed the effectiveness of this digital hand out broad because it has not been tested field test.

4. Conclusion

Digital has been successfully developed quantum mechanics handout without spin and quantum mechanics with spin in Physics Education Study Program a valid and practical UNSRI. Average score the validity of the product meets the category very valid with each score on the aspect content validation is 4.92, design aspect is 4.46, and language aspects 4.50. Product practicality meet the very practical category with the overall results of the questionnaire assessment of the three students with an average score of 4.43 for the one-to-one evaluation stage and at the stage small group evaluation with results overall assessment of the questionnaire from the nine students with an average score 4.52.

References

[1] Mustari, M., & Rahman, M. T. (2014). Manajemen pendidikan. Jakarta: Raja Grafika Persada.

[2] A, B., Sjarif, E. (2010). Pendekatan konstruktivistik dan pengembangan bahan ajar pada sistem pendidikan jarak jauh. Jurnal Pendidikan Terbuka dan Jarak Jauh. **11**(2):117-128.

[3] Akhsan, H., Muslim, M., Ariska, M., & Rianti, S., (2020). Pengembangan handout digital pada materi relativitas umum dan relativitas khusus menggunakan aplikasi 3D page flip. Jurnal Ilmu Fisika dan Pembelajarannya. **4**(2) 43-51.

[4] Asrizal, Festiyad, Sumarmin, R. (2017). Analisis kebutuhan pengembangan bahan ajar ipa terpadu bermuatan literasi era digital untuk pembelajaran siswa SMP kelas VIII. Jurnal Eksakta Pendidikan . 1 (1): 1-8.

[5] Nisa, H. A., Mujib., & Putra, R. W. Y. (2020). Efektivitas *e-modul* dengan flip PDF professional berbasis gamifikasi terhadap siswa SMP. Jurnal Pendidikan Matematika Raflesia. 5(2): 13-25.

[6] Depdiknas.(2008). *Panduan Pengembangan Bahan Ajar*. Jakarta: Depdiknas.

[7] Prawiradilaga, D.S. (2009). *Prinsip desain pembelajaran (instructional design principles).* Jakarta: Kencana Prenada Media Group.

[8] Sugiyono. (2019). *Metode penelitian pendidikan kuantitatif, kualitatif, kombinasi, R&D dan penelitian pengembangan.* Bandung: Alfabeta.

[9] Rianti, S. (2020). *Pengembangan digital handout fisika modern berbasis literasi teknologi di program studi pendidikan fisika universitas sriwijaya. Skripsi.* Inderalaya: FKIP UNSRI.

[10] Destiani, D., Ismet., & Wiyono, K. (2017). Pengembangan bahan ajar IPA berorientasi framework science Pisa untuk Sekolah Menengah Pertama. *Prosiding Seminar Nasional Pendidikan IPA 2017 STEM untuk Pembelajaran SAINS Abad 21.* 654-663.

[11] Hermawan, A.H., Permasih, H., & Dewi, L. (2012). *Pengembangan Bahan Ajar.* Bandung: UPI.

[12] Himmah, E., F. (2019). Pengembangan e-modul menggunakan flip pdf professional pada materi

suhu dan kalor. *Skripsi*. Lempung: FKIP UIN Raden Intan.
[13] Lestari, L., Alberida, H., & Rahmi, Y. L. (2018). Validitas dan Praktikalitas Lembar Kerja Peserta Didik (LKPD) Materi Kingdom Plantae Berbasis Pendekatan Saintifik untuk Peserta Didik Kelas X SMA/MA. Jurnal Eksata Pendidikan. 2(2): 170-177.
[14] Murtafi'ah, S., S. (2019). Pengembangan e-modul pembelajaran pada kompetensi dasar menerapkan pembuatan website kelas xi bisnis daring dan pemasaran di smk negeri 1 jombang. *Jurnal* Pendidikan Tata Niaga (JPTN). **7**(2): 493-498.
[15] Sudjana, N. (2017). Penilaian Hasil Proses Belajar Mengajar. Bandung: PT Remaja Rosdikarya.
[16] Tessmer, M. (1993). *Planning and conducting formative evaluation*. Routledge: London.
[17] Putra, G., S. (2018). *Pengembangan modul pendahuluan fisika kuantum berbasis STEM pada sub pokok bahasan variasi potensial di program studi pendidikan fisika universitas sriwijaya*. *Skripsi*. Inderalaya: FKIP UNSRI.
[18] Zakiyah, S. (2018).Pengembangan buku ajar pendahuluan fisiaka kuantum berbasis STEM pada sub pokok bahasan spin. *Skripsi*. Inderalaya: FKIP UNSRI.

The effect of reciprocal teaching-learning model on the mastery of physics concepts and creativity of senior high school

A Sandopa[1], A Doyan[1,2] and J 'Ardhuha[1]

[1]Physics Education, FKIP, University of Mataram, Lombok, West Nusa Tenggara Indonesia
[2]Master of Science Education Program, University of Mataram, Lombok, West Nusa Tenggara, Indonesia

E-mail: albaenasandopa18@gmail.com, aris_doyan@unram.ac.id

Abstract. This research form to preposition examines the effect of the reciprocal teaching-learning model on the mastery of physics concepts and the creativity of students. This type of research is a quasi-experiment with a non-equivalent control group design. The independent variable of this study is the reciprocal teaching-learning model, while the dependent variable is mastery of physics concepts and creativity. The research population was all students of XI MIA MAN 2 Mataram. The sampling technique of this research uses purposive sampling so that there are two classes is XI MIA 1 as the experimental class and XI MIA 2 as the control class. The experimental class was given treatment using the reciprocal teaching-learning model and the control class was given treatment in the form of a conventional learning model. The instrument used to measure students 'mastery of physics concepts was 10 multiple choice questions, while the instrument used to measure students' creativity was in the form of 5 items in the description. Hypothesis testing of this research is MANOVA. The result of the MANOVA test shows a significant 0,00 so H_0 is rejected and Ha is accepted. The results of the study using reciprocal teaching-learning models showed that the value of mastery of students' physical concepts increased from 37.67 to 83.67 and the creativity of learners increased from 38.83 to 63.83.

1. Introduction

Natural science is not only about the mastery of knowledge in the form of facts, concepts, or principles, but also about the process of discovery. Science or science learning emphasized the direct experience experienced by students. Based on this learning, learners must use scientific methods that are to explore knowledge through investigation or research. One of the branches of science is physics also requires direct experience experienced by students. Physics is not only about facts, concepts, or principles, as well as the process of discovery, but also further development in applying them in everyday life [1]. In addition, in the 21st-century learners are required to be able to have high-level thinking skills that not only receive information from teachers but also be able to convey information they know or process information that has been provided into new information. So that through scientific discovery learners can project their knowledge instead of just imagining it. It is hoped that through scientific discoveries, learners will not only understand the concepts of physics but can also develop the concepts they acquire into new concepts [2].

Based on observations that have been made in the school the physics value of students is low. The low value indicates that learners still have difficulty in learning. The cause of the low mastery of concepts and creativity of learners in physics learning according to previous studies, namely in the learning of physics concepts obtained directly by learners without going through the process of discovery made by the learners themselves, physics learning is only presented as a collection of formulas that must be memorized by learners so that important concepts in physics are often overlooked, the learning process in the classroom is more inclined to the textbooks containing a summary of materials and practice questions, questions asked by teachers to learners are more dominant in the aspect of memory and not based on problems, teachers rarely associate learning materials with the phenomena of students' daily lives as well as tasks given more dominantly on the task of working on textbooks that are generally only mathematical. So the mastery of the concept of physics of learners is low and the creativity of learners can not develop [3,4,5].

To solve these problems, a learning model is needed that can not only improve the mastery of the concept of learners but also can develop the creativity of learners. One of the appropriate learning models applied is the reciprocal teaching-learning model. The reciprocal teaching-learning model is a learning model that applies four strategies of self-understanding, namely concluding teaching materials, drawing up questions and solving them, re-explaining the knowledge they have gained, then predicting the next question of the problem given to the learner. Reciprocal teaching is one of the learning models implemented so that learning objectives are achieved quickly through the process of self-learning and students can present them in front of the class so that it is expected that the learning objectives are achieved and the ability of learners in self-learning can be improved [6].

The reciprocal teaching-learning model is a learning model that applies four strategies of self-understanding, namely concluding the materials, compiling questions and trying to answer them, re-explaining the knowledge they have gained to their friends, then predicting the next question of the problem given by the teacher [7]. The reciprocal teaching-learning model has four phases: summarizing, compiling questions and solving them, clarifying, and resectioning. In the summarizing phase, the teacher will provide the text of the material and ask the learners to summarize it. In the process of compiling questions and solving them, the teacher will help students to make inquiries from summaries that students have made about what they do not understand from the material. In the clarifying phrase, the learners will conduct experiments and it is expected that by experimenting the learners can answer the previous questions. As for the predicting phase, students will predict the level of more difficult questions and try to solve them with the help of the teacher [8,9].

2. Methods

The type of research used is quasi-experimental (pseudo experiment), with the research design is a non-equivalent control group design. The variables in this study are free in the form of reciprocal teaching-learning models, bound variables in the form of mastery of concept and creativity, and control variables in the form of the same teacher, temperature and calorific materials, and 3 meetings. The test instrument used to measure the mastery of the concept of physics is a multiple-choice test with a total of 10 questions, for creativity using a description test with a total of 5 questions [10,11].

This research was conducted at SMAN 8 Mataram. This study used two classes, namely class XI MIA 1 and 2 which were selected by random sampling cluster technique. Class XI MIA 1 as experimental class and XI MIA 2 as control class. Experimental classes use reciprocal teaching-learning models, and control classes use conventional learning models. Data mastery of concepts and creativity tested prerequisites, namely homogeneity test, and normality test. Then continued with the hypothesis test using the MANOVA test. Manova test has 2 test conditions namely the Box'M test and Leneve's test. THE MANOVA test can be performed if both tests are met i.e. if the significance obtained for both tests is more than 0.050 [12].

Journal of Physics: Conference Series **2165** (2022) 012011 doi:10.1088/1742-6596/2165/1/012011

3. Result and Discussion

The average value of the pre-test control class is 52.12, then increased to 64.29 in the post-test. While in the experimental class the average pre-test score of learners was 37.67 and in the post-test increased to 83.67. Mastery of the concept of physics learners for experimental classes and control classes before being given treatment are still many students with low grades. Low scores for experimental classes and control classes in pre-tests are due to both classes not being treated and have not been taught about temperature and calorific material.

After being given experimental class treatment has a higher average compared to the control class this is in line with the benefits of the reciprocal teaching-learning model that students learn by understanding and not easily forget what has been learned. The highest score of learners was obtained on temperature conversion materials for the experimental class. This happens because the difficulty of temperature conversion is still in the easy level of C2 or understand. It is also supported by the syntax on the reciprocal teaching-learning model that is the first phase when students are shared reading text and asked to read it, learners are required to find and understand the concepts contained in the reading text. Then in the second phase, students draw up questions about concepts they do not yet understand. The third phase is the experiment, after the learner conducts the experiments, the learner is asked to teach the knowledge they have got to his friends, by teaching his friends the learner becomes not easy to forget the concept that has been learned. The lowest score was obtained by learners on substance change material. This is due to the difficulty of changing the form of substances has a high level of difficulty that is at the level of C3, C4, and C5. This is also because when learning students do not practice much to work on problems that have a high level of difficulty.

The average value of the pre-test control class is 36.07, then increased to 62.68 in the post-test. While in the experimental class the average pre-test value of learners was 38.83 and in the post-test increased to 63.83. In the data of the results of the second pre-test class, The average creativity of learners in experimental classes and control classes can be categorized as less creative so it can be said that the learning applied previously emphasized less on the creativity of learners. The cause of low creativity in both classes is also caused because both classes have not studied temperature and heat material and have not been treated.

Learning using reciprocal teaching-learning models can increase students' creativity. The first and second phases require learners to develop creativity i.e. learners learn to find concepts in reading texts shared by teachers, then ask problems related to temperature and color then formulate answers to the problem to train the creativity of learners, especially in verbal indicators. The third phase is that students conduct experiments and collect information from other sources such as books to solve the problems formulated so that figural, numerical, and procedural indicators can be seen and improved. The highest score that students get is on the verbal indicator and the lowest score is obtained on the figural indicator. This is because in the verbal indicators the learners mention the concepts of temperature and heat that they know. This is supported by the first sentencing, which summarizes when summarizing the learners will read about the concepts in the temperature and heat material and the third syntax that is when the learner explains back to his friends the learners do not easily forget about the concepts they have learned. While the lowest score on the figural indicator, this is because in the figural indicators students are asked to describe concepts about temperature and heat. Students still struggle to relate the concepts they receive to the application of these concepts to their daily lives.

Based on the Canova test that has been conducted, it is obtained significantly smaller than 0.050 so it can be concluded that there is an influence of reciprocal teaching-learning model on the mastery of physics concepts and creativity of learners. The results obtained from previous research ware concluded that the application of the reciprocal teaching-learning model can improve the understanding of the concept of students of grade X Senior High School [13]. That students who get learning with a reciprocal teaching-learning model can improve the mastery of student concepts. Descriptively, the mastery of the concept of science for experimental classes is higher than the mastery of the IPA concept for the control class based on the average value obtained [14]. The results obtained from previous research, concluded that and implementing reciprocal teaching-learning strategies can increase student creativity [15].

4. Conclusion

The reciprocal teaching-learning model is one of the learning models that can improve the ability of independent learning learners. The reciprocal teaching-learning model is very suitable for the subjects who experimented, because the third phase in this learning model, clarifying, can be done by conducting experiments. Through experiments that learners do learners can find concepts regarding physical materials and can also develop the creativity of learners.Thanks to the principal of Senior High School 8 Mataram and the Physics teachers who have allowed the author to research and assist in this research process.

Acknowledgments

Thank you to the principal of SMAN 8 Mataram and the physics teacher who has allowed researchers to conduct research and guide in the research process.

References

[1] Gunawan. 2015. Model Pembelajaran Sains Berbasis ICT. Mataram: FKIP UNRAM. Indonesian
[2] Gunawan. 2017. Keterampilan Berpikir dalam Pembelajaran Sains. Mataram: Arga Puji Press.
[3] Niami, K. Kosim, K. dan Gunawan, G. 2018. Model Problem Based Learning Berbatuan Simulasi Komputer untuk Meningkatkan Penguasaan Konsep pada Materi Alat-alat Optik. Jurnal Pendidikan Fisika dan Teknologi. **5**(2): 220-225. doi: http://dx.doi.org/10.29303/jpft.v4i2.850 [Indonesian]
[4] Salsabillah, S. 2018. Analisis Penguasaan Konsep-Konsep Fisika Pokok Bahasan Gelombang Elektromagnetik pada Siswa Kelas XII SMA. Seminar Nasional Pendidikan Fisika 2018. **3**(3): 259-267. doi: https://repository.unej.ac.id/handle/123456789/87349 [Indonesian]
[5] Slameto. 2003. Belajar dan Faktor-Faktor yang Menpengaruhinya. Jakarta: Rineka Cipta.
[6] Trianto. 2009. Model-model Pembelajaran Inovatif Berorientasi Konstruktivistik. Jakarta : Perpustakaan Nasional.
[7] Adhani, A. 2014. Pengaruh Pembelajaran Reciprocal Teaching dan Kemampuan Akademik terhadap Aktivitas Lisan dan Hasil Belajar Kognitif Biologi. Jurnal Pendidikan Sains, **2**(3) : 148-158. doi: https://doi.org/10.24042/ajpm.v9i1.2567 [Indonesian]
[8] Efendi, N. 2013. Pendekatan Pengajaran Reciprocal Teaching Berpotensi Meningkatkan Ketuntasan Hasil Belajar Biologi Siswa SMA. Pedagogis, **2**(1): 84-97. doi: https://doi.org/10.21070/pedagogia.v2i1.49 [Indonesian]
[9] Argikas, T. B. & Khuzaini, N. 2016. Penerapan Model Pembelajaran Reciprocal Teaching untuk Meningkatkan Pemahaman Konsep Matematika Siswa Kelas VII SMP Negeri 2 Depok. Jurnal Metematika, 1(1) : 67-79. doi: https://doi.org/10.22460/jpmi.v2i5.p229-236 [Indonesian]
[10] Sugiyono. 2017. Metode Penelitian Pendidikan. Bandung: Alfabeta. [Indonesian]
[11] Arikunto, S. 2013. Manajemen Penelitian. Jakarta: PT. Rineka Cipta. [Indonesian]
[12] Rahayu, S. 2014. Diktat Ststistik Dasar. Mataram: UNRAM Press. [Indonesian]
[13] Prisiani, D. E. 2016. Pengaruh Pendekatan Reciprocal Teaching Terhadap Kemampuan Pemahaman Konsep pada Pembelajaran Matematika di Kelas X SMA Muhammadiyah 2 Palembang. Jurnal Pendidikan Matematika, **2**(2): 80-85. doi: https://doi.org/10.26877/aks.v11i1.3749 [Indonesian]
[14] Putra, I Gst. Ngr. D., Suarjana, Md. dan Arcana, I Nym. 2012. Pengaruh Model Pembelajaran Reciprocal Teaching terhadap Penguasaan Konsep Siswa. Jurnal Pembelajaran IPA. **1**(1): 1-10.

[15] Saleh, Hairus. 2016. Penerapan Strategi Pembelajaran Terbalik (Reciprocal Teaching) untuk Meningkatkan Kreativitas Mahasiswa pada Matakuliah Analisis Real. Sigma. **1**(2): 13-18. doi: http://dx.doi.org/10.0324/sigma.v2i1.71 [Indonesian]

Exploring pre-service teachers' views of science process skills

L O Nursalam[1], Syarifuddin[1], Z Sailan[1], Saifullah[2], A R Hakim[3], A Rosadi[4], M Suhardi[5], M Asy'ari[5], S Prayogi[5], and M R Bilad[5]

[1]University of Halu Oleo, Indonesia
[2]College of Teacher Training and Education (STKIP) Bima, Indonesia
[3]College of Teacher Training and Education (STKIP) Taman Siswa Bima, Indonesia
[4]College of Social and Political Sciences (STISIP) Mbojo Bima, Indonesia
[5]Mandalika University of Education, Indonesia

E-mail: 79syarifuddin@gmail.com

Abstract. The school curriculum in Indonesia requires students to master science comprehensively through problem-based learning with the integration of science process skills (SPS) or scientific methods. This study aims to identify the responses of pre-service teacher students about the importance of SPS components. The data collection was done via a survey with samples of 120 pre-service teachers. The response of pre-service teacher to SPS was collected using 14 items in the questionnaire. The responses were analyzed qualitatively and statistically to find out the frequency of the strongest responses towards the SPS components. The results showed that pre-service teachers stated that problem-solving/critical thinking skills (63 voters), basic understanding answers (59 voters), hypotheses (41 voters), and making conclusions (39 voters) were key components in SPS. Based on these findings, the application of learning that can improve the science process skills of pre-service teachers is important to do.

1. Introduction

Thinking skills are essential in order to track and edit knowledge and control cognitive efforts in problem-solving [1], where process skills are the key to mastering these thinking skills [2]. The process skills are skills acquired through training mental, physical and social abilities that underlie higher mastery of abilities [3] according to the demands of constructivism learning [4] so that it is essential to learn [5]. These skills include observing, interpreting observations, predicting, using tools and materials, applying concepts, research planning, and communicating [6,7]. On the other hand, state that science learning emphasizes the use of science process skills (SPS) such as observation, calculation, experimentation, and reasoning to understand the phenomena of science [8]. Further explained that SPS is essential to be mastered in developing capacity and problem solving, where problem-solving skills and research are life skills and are the highest learning achievements that students must have [9].

SPS can be classified into two, namely basic standalone SPS and integrated SPS[10–12]. SPS is classified because of differences in local, regional and national science education standards tailored to the needs[13]. SPS is classified into three by Long field [14], namely basic, intermediate and advance. Whereas, do not classify SPS [15,16]. Basic SPS indicators include observation, comparison, classification, measuring, and communication; intermediate SPS includes inference and prediction;

while integrated/advance SPS includes hypothesis and define and control variables [17,18]. On the other hand, who classified SPS into basic and integrated SPS stated that basic SPS is a skill in observing, inferring, classifying, communicating, measuring, and predicting [19]. While integrated SPS is student skills in controlling variables, defining operationally, formulating hypotheses, interpreting data, experimenting, and formulating models.

Science learning in Indonesia are more oriented towards memorizing sciences' concepts and facts [19] and does not accommodate the demands of students' SPS development [20,21]. Such condition is thought to be the reason for students' low SPS [20,22]. The pre-service teachers' SPS is generally in moderate category, with hypothesis and defining variables operationally indicators as the lowest score [23]. Those findings indicate the low level of students' high-level thinking skills: critical thinking, problem solving, creative thinking, decision making and meta cognition [24]. Recommend that critical thinking dispositions of pre-service teachers need to be improved.On the other hand [25], found that critical thinking skills were still low [26,27]. Furthermore stated that 81% of pre-service teachers have critical thinking skills and critical analysis with criteria that have not yet appeared or are still underdeveloped, while only 19% are categorized as developing or developing well [28], whereas Gelder found that students' metacognition abilities were generally in low category [29].Furthermore, recommends that some components of metacognition awareness such as information management system (IMS), monitoring (M), evaluation (E), and debugging (D) must be maximized in the learning process [30]. The high-level thinking skills described were very important to be mastered by students in understanding science comprehensively and SPS was claimed to be a gateway to mastering these skills [6,8,23].

Exploring the pre-service teachers' views on the importance of SPS is one of the preliminary step in solving the problems. The results of this study are also expected to be a solution so that learning is more directed and in accordance with students' interest in learning. This becomes important because learning skills outside the context are not meaningful and cannot develop students' conceptual understanding [31]. On the other hand, Molefe and Stears [32] which emphasizes the integration of SPS in learning is still questionable, considering the low students' SPS [20]. Asy'ari et al [23] found that the pre-service teachers' SPS based on gender were significantly different only on observation, inference, and analysis indicators with the moderate category. This study aims to explore pre-service teachers' views on the importance of SPS in Universitas Pendidikan Mandalika. The pre-service teachers' views on the importance of selected SPS indicators and these indicators relevance to selected study programs were also evaluated in this study.

2. Methods

This research is survey research using quantitative-qualitative research design to get the pre-service teachers' views towards SPS. Instruments with 14 SPS items in the form of questionnaires with five Likert scales adapted from Coil et al [33] were used to collect the pre-service teachers' views of the importance of SPS.Initial knowledge of pre-service teachers about science process skills was also identified in this study. The summary of questionnaire is shown in Table 1.

Table 1. Science process skills questionnaire

Scientific Skills	Very Important	Important	Moderately Important	Of Little Importance	Unimportant
Interpreting data: graphs and tables					
Understanding basic statistics					
Questioning: raising questions that are testable, measurable and repeatable					
Observing (and comparing): proficiency in describing patterns and ordering and sequencing events					
Interpreting data: ability to construct an argument from					
Ability to create a testable hypothesis					
Measuring: understanding concepts of accuracy and precision					
Ability to design an experiment: identifying and controlling variables					
Problem solving/critical thinking					
Recording and communicating information: understanding forms of information or data representation (i.e., verbal, written, pictorial and mathematical forms)					
Classification: grouping and organising objects or attributes					
Predicting: forecast future observations on the basis of present trends or previous knowledge					
Being able to infer plausible reasons for failed experiments					
Creating appropriate graph from data					

Data from the results of subsequent studies were analyzed descriptive-statistic to determine the mean and frequency of the pre-service teachers' views. The correlation between teachers' perceptions of the most important and least important SPS indicators was also analyzed using the Spearman correlation test with the help of IBM SPSS Statistics 23 to determine the consistency of the pre-service teachers' choice regarding the most important and less important indicators in SPS.

This research was conducted at Universitas Pendidkan Mandalika Mataram with 120 pre-service teachers as samples. The sample was determined using a purposive sampling technique, where the sample is a pre-service teacher who has conducted Basic Science courses, as such SPS had been introduced to them.

3. Result and Discussion

SPS has long been the basis for developing higher-order thinking skills (HOTs). The Indonesian national curriculum explicitly emphasizes the integration of SPS components as both a learning approach and a learning product. The results of this study are not only a portrait of how the pre-service teachers' perceptions and knowledge about SPS, more than that, the results of this study are very useful as a reference for the integration of SPS components to be learned based on the perception of the most important SPS components to be taught to pre-service teachers based on the program selected study especially in physics, mathematics, biology, and chemistry education. The pre-service teachers' views on the 14 SPS items contained in the research questionnaire are presented in Table 2 which is sorted by the highest to lowest views.

Table 2. The pre-service teachers' views on the importance of SPS.

Indicators	Very important	Important	Moderately Important	Less important	Unimportant	M	SD
Problem-solving/critical thinking	78	35	5	2	0	3.25	0.28
Recording and communicating information	34	67	19	0	0	2.79	0.23
Data interpretation: graphs and tables	28	65	26	1	0	2.71	0.22
Observing and comparing	35	62	20	2	1	2.58	0.21
Understand the basic calculation of research data	61	55	4	0	0	2.54	0.26
Questioning	46	61	13	0	0	2.54	0.23
Classifying	27	60	28	5	0	2.50	0.20
Creating the appropriate graph from data	60	46	11	0	3	2.50	0.23
Ability to design experiments	39	57	23	1	0	2.38	0.20
The ability to make hypotheses that can be tested	56	54	8	2	0	2.33	0.24
Measuring: understanding the concepts of accuracy and precision	48	56	16	0	0	2.33	0.22
Being able to infer plausible reasons for failed experiments	46	56	18	0	0	2.33	0.22
Predicting	37	54	23	6	0	2.25	0.18
Data interpretation: the ability to construct arguments from data	47	50	21	1	1	2.08	0.20
Percentage	0.38	0.46	0.14	0.01	0.003		

Table 2 shows the pre-service teachers' views about the importance of the SPS indicator based on the relevance of these indicators to the study program taken. It shows that indicators of *problem-solving/critical thinking* (78) is considered to be the most important, followed by *recording and*

Journal of Physics: Conference Series **2165** (2022) 012012 doi:10.1088/1742-6596/2165/1/012012

communicating information (34), *interpretation of data: graphs and tables* (28), and *observing and comparing* (35). In general, the pre-service teacher considers all SPS indicators to be very important (0.38) and important (0.46).

The views of the importance of SPS indicators as presented in Table 1 were verified by asking respondents to select each of the four indicators that are considered the most important and less important as presented in Table 3.

Table 3. The SPS indicators that considered the most important and less important

Indicators	Most important	Less important
Problem-solving/critical thinking	63	17
Understand the basic calculation of research data	59	20
Being able to infer plausible reasons for failed experiments	41	30
The ability to make hypotheses that can be tested	39	16
Observing and comparing	35	29
Data interpretation: the ability to construct arguments from data	34	27
Questioning	33	29
Data interpretation: graphs and tables	29	34
Recording and communicating information	29	28
Predicting	22	32
Creating the appropriate graph from data	22	47
Ability to design experiments	21	21
Measuring: understanding the concepts of accuracy and precision	19	31
Classifying	11	53

Table 3 shows the ranking of SPS indicators that are considered the most important and less important according to the pre-service teacher in Universitas Pendidikan Mandalika. The SPS indicator that most often chosen based on the pre-service teachers' views of the most important indicator was problem solving/critical thinking (63) followed by understand the basic calculation of research data (59), being able to infer plausible reasons for failed experiments (41) , and the ability to make hypotheses that can be tested (39). The correlation test result also showed a very strong negative correlation between the pre-service teachers' views towards the SPS indicator which is considered the most important and less important (Spearman's correlation coefficient = -0.664), which indicates that respondents are consistent with the choice of selected indicators.

Initial knowledge of teacher candidates related to SPS was identified based on the written response in the open questionnaire given at the end of the SPS questionnaire. Some examples of responses from pre-service teachers who do not understand SPS comprehensively such as *(*1) skills are an interesting activity where those skills are traded in various communities can perform optimally when doing activities or events, (2) process skills are the ability to process something broader, (3) process skills are thinking or conceptualizing something carefully and carefully so that mistakes do not occur, (4) skills in understanding the process, (5) in the process we must be skilled, each person has different skills to achieve the intended goal."

A better response was conveyed to pre-service teachers related to their initial knowledge about SPS such as (1) process skills are skills acquired from the exercise of basic mental, physical and social abilities as movers of higher abilities, (2) process skills are skills that can train students to solve problems, such as formulating problems, make hypotheses, collect data and be able to draw conclusions, (3) approaches that aim for students to be skilled in managing information obtained through teaching and learning processes that provide broader opportunities for students to observe, classify, interpret, research and communicate learning outcomes, and (4) approach that aims to train students' way of thinking, mental skills, measuring skills, observing skills, interpreting skills, communicating, conducting experiments through the learning process.

The written responses of pre-service teachers presented indicate the possibility that (1) pre-service teachers do not yet understand SPS comprehensively, and (2) lecturers have not comprehensively taught and integrated the components of SPS in the classroom learning activity. This statement is reinforced by the finding that the frequency of response of pre-service teachers who do not understand SPS comprehensively (F = 97) > the frequency of pre-service teachers who gave better responses related to initial knowledge about SPS (F = 23).

The problem-solving/critical thinking is the most important SPS indicators that have to achieve according to the pre-service teachers' views and relevant to the study program being selected. These results are relevant to many study results which state that problem solving/critical thinking is an essential skill that must be possessed and learned in 21st-century learning [33–35]. On the other hand, predicting become the SPS indicator that considered less important by pre-service teachers in Universitas Pendidikan Mandalika. Based on the perspective of cognitive learning theory and information processing, pre-service teachers psychologically require learning that meaningful oriented through the association of new information to relevant concepts contained in one's cognitive structure [36] and this can be done through integrated SPS learning. Predicting which is an SPS intermediate indicator [17,18] and has been taught in high school [32] less relevant to higher education learning orientation that emphasizes integrated SPS learning.

The authentic learning [37], inquiry learning [38,39] can be implemented as an effort to teach SPS. Result study of Bilgin [40] shows that hand-on learning activities with a cooperative approach significantly impact students' SPS and scientific attitudes. Inquiry learning that emphasizes the scientific experience of students through problem formulation activities, hypothesizing, testing hypotheses, developing critical thinking, and self-regulation to comprehensively understand the concepts and/ theories [29,39] very relevant to be used to improve students' SPS so that it has an impact on abilities in critical thinking, logical reasoning, and problem solving [34,41].

The response of pre-service teachers shows that initial knowledge about SPS is not yet comprehensive. These findings reinforce the statement that pre-service teachers SPS are in the low and moderate categories[20,23]which has implications for the low level of pre-service teachers' critical thinking, critical analysis skills [28], scientific attitudes [22], designing experiment in physics learning that need physics factual and conceptual knowledge [42].

4. Conclusion

The pre-service teachers' views in Universitas Pendidikan Mandalika shows the importance of SPS being taught. Problem-solving/critical thinking is the most important SPS indicator to be mastered and relevant to the study program taken while predicting is an SPS indicator that is considered less important based on the response given. Apart from these findings, in general, the pre-service teacher in Universitas Pendidikan Mandalika assumes that all SPS indicators are important to be taught as the provision to become competent teachers in the future. SPS indicators such as the *ability to design experiments* only get 21 points by the pre-service teacher and are ranked low based on the SPS indicator which was considered the most important. On the other hand, The ability to design experiments is one of the most important but most challenging indicators in SPS so that all SPS indicators are basically important to learn. These findings also contribute to the lecturers' reference in teaching SPS, the preparation of authentic teaching materials by integrating SPS indicators. The results of this study can also be the initial reference for lecturers to determine the SPS indicators that are a priority for learning using innovative students oriented learning models such as inquiry learning, problem-based learning, and project-based learning that empirically can teach students' SPS.

Acknowledgement

We would like to thank to the Dean of Faculty of Science, Engineering, and Applied Science of Mandalika University of Education and all respondents that would to spend their time to filled out the questionnaire used in this study.

References

[1] Gürses A, Cuya Ş, Güneş K and Doğar Ç 2014 Determination of the Relation between Undergraduate Students' Awareness Levels Regarding Their Scientific Process Skills and Application Potentials *Am. J. Educ. Res.* **2** 250–6
[2] Kaya V H, Bahceci D and Altuk Y G 2012 The Relationship Between Primary School Students' Scientific Literacy Levels and Scientific Process Skills *Procedia - Soc. Behav. Sci.* **47** 495–500
[3] Akinbobola A O and Afolabi F 2010 Analysis of Science Process Skills in West African Senior Secondary School Certificate Physics Practical Examinations in Nigeria **5** 234–40
[4] Asy'ari M and Fitriani H 2017 Literatur Reviu Keterampilan Proses Sains sebagai Dasar Pengembangan Keterampilan Berpikir Tingkat Tinggi *Prisma Sains J. Pengkaj. Ilmu Dan Pembelajaran Mat. Dan IPA IKIP Mataram* **5** 1
[5] Dahlia Yuliskurniawati I, Ika Noviyanti N, Rosyadah Mukti W, Mahanal S and Zubaidah S 2019 Science Process Skills Based on Genders of High School Students *J. Phys. Conf. Ser.* **1241** 012055
[6] Nur M 2011 *Modul Keterampilan-keterampilan Proses dan Hakikat Sains* (Surabaya: PSMS Unesa)
[7] Dahar R W 2003 *Teori-teori belajar* (Jakarta: Gelora Aksara Prima)
[8] Saido G A M, Siraj S, DeWitt D and Al-Amedy O S 2018 Development of an instructional model for higher order thinking in science among secondary school students: a fuzzy Delphi approach *Int. J. Sci. Educ.* **40** 847–66
[9] Ibrahim M 2010 *Dasar-Dasar Proses Belajar Mengajar* (Surabaya: Unesa University Press)
[10] Bulent A 2015 The investigation of science process skills of science teachers in terms of some variables *Educ. Res. Rev.* **10** 582–94
[11] Turiman P, Omar J, Daud A M and Osman K 2012 Fostering the 21st Century Skills through Scientific Literacy and Science Process Skills *Procedia - Soc. Behav. Sci.* **59** 110–6
[12] Padilla M J 1986 *The Science Process Skills. Research Matters...To the Science Teacher*
[13] Ash D 2000 The Process Skills of Inquiry **2** 12
[14] Longfield J 2002 Science process skills
[15] Rustaman N 2005 *Strategi belajar mengajar biologi* (Malang: UM Press)
[16] Mechling K, Smith B, Oliver D, Kepler L and Ires N 1985 *A Recommended Science Competency Continuum for Grades K-6 for Pennsylvania Schools* (Harrisburg: Pennsylvania Department of Education)
[17] Charlesworth R 2016 *Math and science for young children* (Boston, MA: Cengage Learning)
[18] Meador K S 2003 Thinking Creatively about Science Suggestions for Primary Teachers *Gift. Child Today* **26** 25–9
[19] Prayitno B A, Corebima D, Susilo H, Zubaidah S and Ramli M 2017 Closing the Science Process Skills Gap between Students with High and Low Level Academic Achievement *J. Balt. Sci. Educ.* **16** 266–77
[20] Andini T E, Hidayat S and Fadillah E N 2018 Scientific process skills: Preliminary study towards senior high school student in Palembang *J. Pendidik. Biol. Indones.* **4**
[21] Irwanto, Rohaeti E, Widjajanti E, and Suyanta 2017 Students' science process skill and analytical thinking ability in chemistry learning THE 4TH INTERNATIONAL CONFERENCE ON RESEARCH, IMPLEMENTATION, AND EDUCATION OF MATHEMATICS AND SCIENCE (4TH ICRIEMS): Research and Education for

Developing Scientific Attitude in Sciences And Mathematics (Yogyakarta, Indonesia) p 030001
[22] Mirawati B and Royani I 2019 Pengembangan LKS Biologi SMA Berbasis Praktikum dengan Model Pembelajaran Langsung untuk Meningkatkan Keterampilan Proses Sains Siswa *J. Penelit. Dan Pengkaj. Ilmu Pendidik. E-Saintika* **3** 88
[23] Asy'ari M, Fitriani H, Zubaidah S and Mahanal S 2019 The Science Process Skills of Prospective Biology Teachers in Plant Cell Material Based on Gender *Int. J. Emerg. Technol. Learn. IJET* **14** 168
[24] OCED 2016 *Country Note – Results from PISA 2015*
[25] Fitriani H, Asy'ari M, Zubaidah S and Mahanal S 2018 Critical Thinking Disposition of Prospective Science Teachers at IKIP Mataram, Indonesia *J. Phys. Conf. Ser.* **1108** 012091
[26] Qing Z, Jing G and Yan W 2010 Promoting preservice teachers' critical thinking skills by inquiry-based chemical experiment *Procedia - Soc. Behav. Sci.* **2** 4597–603
[27] Guest K 2000 Introducing Critical Thinking to 'Non-standard' Entry Students. The Use of a Catalyst to Spark Debate *Teach. High. Educ.* **5** 289–99
[28] Fitriani H, Asy'ari M, Zubaidah S and Susriyati M 2019 Exploring the Prospective Teachers' Critical Thinking and Critical Analysis Skills *J. Pendidik. IPA Indones.* **8**
[29] Gelder T van 2005 Teaching Critical Thinking: Some Lessons From Cognitive Science *Coll. Teach.* **53** 41–8
[30] Muhali M, Yuanita L and Muslimin L 2019 The Validity and Effectiveness of the Reflective-Metacognitive Learning Model to Improve Students' Metacognition Ability in Indonesia *Malays. J. Learn. Instr.* **16**
[31] Jegede∗ O J and Okebukola P A 1991 The relationship between African traditional cosmology and students' acquisition of a science process skill *Int. J. Sci. Educ.* **13** 37–47
[32] Molefe L and Stears M 2014 Rhetoric and Reality: Science Teacher Educators' Views and Practice Regarding Science Process Skills *Afr. J. Res. Math. Sci. Technol. Educ.* **18** 219–30
[33] Coil D, Wenderoth M P, Cunningham M and Dirks C 2010 Teaching the Process of Science: Faculty Perceptions and an Effective Methodology ed M Grossel *CBE—Life Sci. Educ.* **9** 524–35
[34] Fuad N M, Zubaidah S, Mahanal S and Suarsini E 2017 Improving Junior High Schools' Critical Thinking Skills Based on Test Three Different Models of Learning *Int. J. Instr.* **10** 101–16
[35] Mahanal S and Zubaidah S 2017 Model Pembelajaran Ricosre yang Berpotensi Memberdayakan Keterampilan Berpikir Kreatif *J. Pendidik. Teori Penelit. Dan Pengemb.* **2** 676–85
[36] Arends R 2012 *Learning to teach* (Dubuque, Iowa: McGraw-Hill)
[37] Keys C W and Bryan L A 2001 Co-constructing inquiry-based science with teachers: Essential research for lasting reform *J. Res. Sci. Teach.* **38** 631–45
[38] Ergül R 2011 The Effects of Inquiry-Based Science Teaching on Elementary School Students' Science Process Skills and Science Attitudes *Bulg. J. Sci. Educ. Policy* **5** 48–68
[39] Eggen P D and Kauchak D P 2012 *Strategies and models for teachers: teaching content and thinking skills* (Boston: Pearson)
[40] Bilgin I 2006 The Effects of Hands-On Activities Incorporating a Cooperative Learning Approach on Eight Grade Students' Science Process Skills and Attitudes toward Science *J. Balt. Sci. Educ.* **5** 27–37
[41] Prayogi S, Yuanita L and Wasis 2018 Critical Inquiry Based Learning: A Model of Learning to Promote Critical Thinking Among Prospective Teachers of Physic *J. Turk. Sci. Educ.* **15** 43–56
[42] Hodosyová M, Útla J, MonikaVanyová, Vnuková P and Lapitková V 2015 The Development of Science Process Skills in Physics Education *Procedia - Soc. Behav. Sci.* **186** 982–9

Journal of Physics: Conference Series 2165 (2022) 012013 doi:10.1088/1742-6596/2165/1/012013

PIMCA: A New Alternatives to Physics Learning Model

C Poluakan[1] and D Katuuk[2]

[1]Physics Education, Faculty of Mathematics and Natural Sciences, Universitas Negeri Manado, North Sulawesi, Indonesia
[2]Primary Teacher Education, Faculty of Educational Science, Universitas Negeri Manado, North Sulawesi, Indonesia

e-mail: cosmaspoluakan@unima.ac.id

Abstract. The general development of knowledge in physics particulary, requires the ability to understand the right and obey the principles. This study purposed to carried out the use of learning models to ensure a quality learning process with the right concepts and avoid misconceptions. The learning model is also adaptive to the 4.0 era demands and the special situation of the pandemic era. The learning model is PIMCA (presentation-idea mapping-conceptualization-assessment formative). The PIMCA learning model is based on the main theories of educational psychology, such as behavioristic, cognitive, constructivist, and humanistic. It was conducted to students in the physics education department at Manado State University, used experimental study in various fields of physics studies such as waves, optics, electricity, and magnetism, it was found that the average achievement of learning progress from the pre-test and post-test was 73.14%. Likewise, in terms of conceptual understanding of physics, namely from idea maps and concept maps, an average increase of up to 84% is obtained. The major findings of this study indicate that the PIMCA learning model can be effectively used as an alternative in learning physics. Study using PIMCA learning models needs to be continued to be applied to the learning of physics in the field of mechanics and modern physics, as well as in the learning of mathematics, science, including biology-chemistry-geology and astronomy, and engineering.

1. Introduction

The education world in the 21st century is marked by the disruption phenomenon. This phenomenon causes changes in the learning world. Learning becomes a continuum of learning according to interests and needs. Learning activities tend to be more informal, in addition to requiring learners to be self-motivated learners. Learning resources through the MOOC (massive open online course) application are becoming increasingly popular. The global community is forced to adapt to the use of IoT (Internet of Things) in almost all aspects of life. Educational institutions are forced to carry out the online learning process. As a result, many problems arise in teaching and learning activities, including in terms of how to communicate, strategies, methods, and learning models that suit the needs. Therefore, a comprehensive approach is needed in the learning process through a learning model that is adaptive to the demands of learning in the 21st century and is also relevant to learning when the pandemic era continues. This paper reports research on the use of the PIMCA learning model as an alternative in physics learning and STEM (Science-Technology Engineering-Mathematics) learning in general. The PIMCA learning model is based on the principles of educational psychology and integrated learning theory, namely behavioristic, cognitive, constructivist and humanistic theories. The PIMCA model was created as an alternative learning model to answer the challenges of learning in the 4.0 era, which is to stimulate learners to be able to think critically, be creative, have high-level thinking skills and be adaptive to the use of ICT, in addition to being able to collaborate and communicate well.

1.1. Literature Review

The ability to think critically, logically, and creatively from a person is a necessity that must be developed within him [1]. Education and learning through STEM, among others, can be a way for someone to sharpen their minds more intelligently, creatively, and logically. The Industrial Revolution 4.0 is mostly based on mathematics and natural sciences and these two things are the basis of human logical thinking [2]. Therefore, in the learning process, a learning model that is suitable to the demands of the times is needed. The PIMCA model is an alternative new learning model that is adaptive to the needs of the times. Why? The PIMCA model provides opportunities for the use of multimedia and information technology. The multi-representation (MR) based PIMCA model, in the first stage, the presentation will condition teachers and students to optimize the use of multimedia through youtube shows, animated images, simulations, virtual labs, demonstrations, film projects, graphic info, etc. With the help of this multimedia, students can construct knowledge by building conceptual ideas related to ongoing learning materials. This is where the initial demands of students' critical thinking begin to be formed. In this phase, learners enter the second stage of the PIMCA model which is called idea mapping. Furthermore, the concept that has been built is deepened through the third stage of conceptualization. In this third stage, the concepts of ideas that were built previously are straightened out so as not to form a wrong concept understanding (misconceptions) and then these concepts are directed into a network of related and meaningful concepts as a complete concept map. The concept of knowledge that has been fully formed is verified through the formative assessment stage.

From the perspective of educational psychology, the PIMCA model is part of the instructional learning model as well as other learning models MOMBI (the model of model base instruction) which uses the theoretical basis of the mental model. Learning this model starts from the presentation stage where the main emphasis is based on the use and application of multiple-representation (MR) which is based on behavioristic learning theory that prioritizes the process of providing stimulus. The impact of the stimulus through the use of MR is the emergence of a response in the learners, in the form of a mental process of constructing concepts. This process is known as the idea mapping stage. According to cognitive psychology, the construction process of this concept is explained as a form of perceptual change that does not always appear in the form of behavior. However, the constructed concept is the initial part of the constructivist learning theory called by Driver and Oldham (1994) as orientation, elicitation, and restructuring of ideas. This is in accordance with Vygotsky's opinion, namely the formation of concepts through self-regulation requires assistance in the Zone Proximal Development stage [3]. Thus, the representation and idea mapping stages in the PIMCA model provide optimal contributions to the formation of critical and creative thinking skills. This is clarified by the results of research using cognitive psychology theory and the theory of multimedia learning, showing that the use of Multiple Representations (MR) can provide tremendous benefits to student learning [4-10]. According to Mayer, the use of multiple sources and multiple representations, from the perspective of The Cognitive Theory of Multimedia Learning (CTML), and by multimedia principles: "students learn better from words and pictures than from words alone", and with CTML, "using multiple representations can foster learning" [11].

The interaction that occurs in the teaching and learning process is the absorption and deposition of information, according to The Theory of Mental Models, that "learning is a form of information processing" [12]. This is confirmed according to Behavioral Psychology which states, successful learning involved the mental acquisition of a 'copy of the information being taught [13]. Concepts constructed at the idea mapping stage are clarified in the next stage called conceptualization. At this conceptualization stage, concepts are planted through experiments (laboratory practices), virtual lab experiments, demonstrations, project assignments, discussions, and others. At this stage intellectual development enactive-iconic-symbolic (Bruner) is applied through three stages, namely: the information stage where learners who are learning obtain much information about the material being studied; then in the transformation stage, the information that has been obtained is analyzed, changed or transformed into an abstract or conceptual form; Finally, at the evaluation stage, the learners themselves assess the extent to which the information that has been transformed can be used to understand the symptoms or problems they are facing [10]. With a process like the one above, learners can find and conclude the right concept which can then be shown in a concept map. The continuation from the conceptualization stage of the PIMCA model is formative assessment. At this stage, the learners are fully assessed as a whole human beings. Krathwohl (2002) developed and revised Bloom's taxonomy by stating that the categories of cognitive dimensions should include from simple to complex and knowledge dimensions from concrete to abstract. The assessment model through cognitive dimension-based measurements includes the domains of remembering - remember (C1), understand - understand (C2), apply - apply (C3), analyze - analyze (C4), evaluate - evaluate (C5), and create - create (C6); and the knowledge dimension which includes factual knowledge, conceptual knowledge, procedural knowledge and metacognitive knowledge [14]. With the psychological foundation as stated above, the PIMCA learning model is proposed as shown in Figure 1.

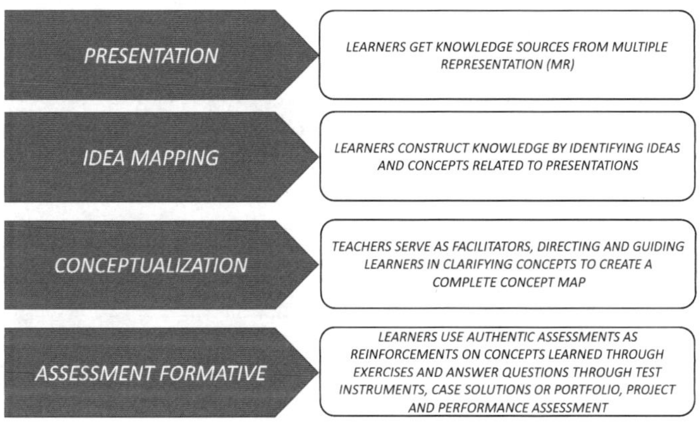

Figure 1. PIMCA Learning Model

2. Methods

Research on the use of the PIMCA model (Figure 1), begins with a reference study of the theoretical basis of educational psychology, learning theory as well as multimedia theory, and information technology. To ensure that the PIMCA model is effectively used as a learning model in the field of science, a trial was carried out in the field of physics on several selected teaching materials with students as teaching models and students as the target sample learners. The research was conducted at the Physics department of Manado state university. The syntax of the model was validated by experts in the fields of psychology and pedagogy, while the physics materials as teaching materials were validated by experts in the fields of physics. Measurement of achievement of learning outcomes is indicated by pre-test and post-test scores, as shown in Table 1.

To check the effectiveness of the formation or construction of concepts from the presentation and idea-mapping stages to the conceptualization stage, it is necessary to check the number of concepts found by the learners. The concepts are written in the idea maps and concept maps sheets. The data on the calculation of the number of concepts found and written by the learners are shown in Table 2.

Table 1. Achievement of learning outcomes scores (PIMCA learning model, N=24)

No	Teaching Material	Average Score		Increase (%)
		Pre-Test	Post-Test	
1	Refraction of Light (Dido, 2021) [15]	15.94	60.22	73.50
2	Lup–Magnifying Glass (Hartati, 2021) [16]	27.86	75.00	62.90
3	Convex Lens (Koming, 2021) [17]	23.33	78.96	70.50
4	Microscope (Lampeang, 2021) [18]	45.00	92.50	51.40
5	Optical Instruments (Mamengko, 2021) [19]	43.75	77.08	43.00
6	Phyphox (Mayampoh, 2021) [20]	11.33	18.11	37.40
7	Electric Charge (Nasra, 2021) [21]	18.88	75.60	75.00
8	Magnetic Field, (Patol, 2021* [22]	17.50	55.00	68.00
9	Magnetic Field (Patol, 2021** [22]	5.20	55.00	89.00
10	Lorentz Style (Rettob, 2021) [23]	7.44	37.44	80.10
11	Convex Mirror (Reskin, 2021) [24]	20.69	72.44	71.40
12	Doppler Effect (Tokolang, 2021) [25]	1.25	7.13	82.50
		Mean score increase (%)		73.15

** Small group, ** Large Group; Sig*.000*

To check the effectiveness of the formation or construction of concepts from the presentation and idea-mapping stages to the conceptualization stage, it is necessary to check the number of concepts found by the learners. The concepts are written in the idea maps and concept maps sheets. The data on the calculation of the number of concepts found and written by the learners are shown in Table 2.

Table 2 Achievement of concept formation through the PIMCA learning model

	Content	N	Idea Maps		Concept Maps		%
			Incorrect	Correct	Incorrect	Correct	
1	Refraction of light	24	2	1	0	7	85.00
2	Lup-Magnifying Glass	28	1	1	0	6	83.00
3	Convex lens	24	2	1	0	8	87.50
4	Microscope	24	3	3	1	11	72.00
5	Optical-Eye Tools	24	3	1	0	7	85.00
6	Electrical charge	22	3	0	0	11	100.00
7	Magnetic field	30	2	3	0	12	75.00
8	Lorentz forces	30	3	1	1	10	90.00
9	Convex mirror	29	2	1	0	8	87.50
10	Doppler effect	24	2	3	2	12	75.00
			Total concept formation achievement				84.00

3. Result and Discussion

The findings of the research show that the PIMCA learning model is quite effective as a learning model with an average score increase of 73.15% *(Sig*.000)*. Table 1 also shows that there is no significant difference between teaching materials in the field of optics and waves, electricity and magnetism. For this, it is necessary to test with other materials in the field of physics such as mechanics, thermal physics, and modern physics. Regarding the formation or construction of concepts, there is a tendency that the use of Multiple Representations in the presentation stage is not optimal. This is indicated by the tendency to write down the wrong concept or not a physical concept (eg force, momentum-physical concept, flight-not a concept) at the idea mapping stage. However, the effectiveness of using PIMCA as a learning model, in terms of concept construction at the conceptualization stage, reached an average of 84%. The fact that the research results show that the function of the conceptualization phase is to clarify and straighten out previously formed concepts or ideas, effectively minimize misconceptions and enrich the formation of appropriate concepts according to the topic of teaching materials. Table 2 shows, only 1 wrong concept out of 11 concepts found in Lorentz style teaching materials and 2 wrong concepts out of 14 concepts found in Doppler effect material. In addition, all the concepts that are found by the students are the correct concepts. These findings reinforce previous studies which suggest that learners have difficulty understanding vector-related physics concepts.

4. Conclusion

PIMCA turned out to be effectively used as a physics learning model. The PIMCA learning model which consists of presentation, idea mapping, conceptualization, and formative assessment can improve the ability to master physics material, especially it can also trigger learners to construct concepts correctly. PIMCA learning model research needs to be continued in STEM fields. In addition, the PIMCA model can be developed into a research roadmap in the field of educational psychology and learning theories as well as educational assessments and evaluations.

Acknowledgement

The authors would like to thank the Directorate of Research and Community Service, Ministry of Research Technology and Higher Education, who funded this research, the Rector of Manado State University (Unima) and Chair of the Institution of Research and Community Service (LPPM) Unima for facilitating this research, and the Dean of the Faculty of Mathematics and Natural Sciences (FMIPA) Unima who has supported the implementation of this research, and merits to all the research team colleagues and students in physics education who participated in this study.

References

[1] A reference Gleason NW 2018 *Higher Education in the Era of the Fourth Industrial Revolution* (Singapore: Palgrave Macmillan-Springer Nature Singapore Pte Ltd)
[2] Handoko L T 2018 Higher Education: Applying Mathematics and Natural Sciences as a Solution to Daily Problems, *Kompas*, p. 12.
[3] Vygotski, Lev Semenovich 1978 *Mind in Society: The Development of Higher Psychological Processes* (Harvard University Press Cambridge)
[4] Penprase B E 2018 *The Fourth Industrial Revolution and Higher Education,* (Ed. Nancy W. Gleason), Higher Education in the Era of the Fourth Industrial Revolution, (p. 217). Singapore: Palgrave Macmillan-Springer Nature Singapore Pte Ltd
[5] Cook M P 2006 Visual representations in science education: the influence of prior knowledge and cognitive load theory on instructional design principle science Education. DOI 10.1002/sec. 20164 pp 1073-1094
[6] Olympiou G, Zacharias Z, deJong T 2013 Making the invisible visible: enhancing students' conceptual understanding by introducing representations of abstract objects in a simulation *Instructional science* **41** 3 Page: 575-596 DOI 10.1007/s11251-012-9245-2
[7] Waldrip B, Prain V and Sellings P 2013 Explaining newton's laws of motion: using student reasoning through representations to develop conceptual understanding *Instructional Science* 41 p 165–189
[8] Savinainen A, Makynen A and Nieminen P 2015 The effect of using a visual representation tool in a teaching-learning sequence for teaching newton's third Law Science Education **47** pp 119-135
[9] Rau M A and Matthews P G How to make 'more' better? Principles for effective use of multiple representations to enhance students' learning about fractions ZDM *Math. Edu.* **49** p 531
[10] Sirait J et al 2018 Students' understanding of forces: Force diagrams on horizontal and inclined plane *Journal of Physics: Conferences Series* **997** 012030
[11] Opfermann M, Schmeck A, Fischer and Hans E 2017 Multiple Representations in Physics and Science Education – Why Should We Use Them?. In Multiple Representations in Physics Education, Models and Modeling in Science Education, DF Treagust et al. (eds) Springer

International Publishing AG 2017

[12] Mayer R E 2009 *Multimedia Learning (2nd ed.)* (Cambridge, MA: Cambridge University Press) in the book: Multiple Representations in Physics Education - Models and Modeling in Science Education, by Treagust, DF, Duit, R., Fischer HE, eds. (2017). Springer International Publishing AG, Switzerland.

[13] Hanke U 2008 *Realizing Model-Based Instruction - The Model of Model-Based Instruction, In D. Ifenthaler, P. Pirnay-Dummer ve JM Spector (Eds.), Understanding Models for Learning and Instruction* (pp. 175-186) Springer Science+Business Media, LLC.

[14] Krathwohl D R 2002 A Revision of Bloom's Taxonomy: An Overview *Theory into Practice* **41** 4, pp. 212-218) Autumn 2002 © 2002 College of Education, The Ohio State University

[15] Dido, et al. 2021 Analysis of student difficulties in learning refraction of light *Journal of Physics: Conference Series* **1968** (2021) 012033 doi:10.1088/1742-6596/1968/1/012033, IOP Publishing

[16] Hartati, et al. 2021 PIMCA learning models to improve student learning outcomes in optic magnifying glass tool *Journal of Physics: Conference Series* **1968** (2021) 012036 doi:10.1088/1742-6596/1968/1/012036, IOP Publishing

[17] Koming, et al. 2021 Teaching convex Lens materials with a PIMCA model *Journal of Physics: Conference Series*, **1968** (2021) 012037 doi:10.1088/1742-6596/1968/1/012037, IOP Publishing

[18] Lampeang, et al. 2021 Use of the four-tier diagnostic test with PIMCA model on learning of microscope *Journal of Physics: Conference Series* **1968** (2021) 012039 doi:10.1088/1742-6596/1968/1/012039, IOP Publishing

[19] Mamengko, et. al. 2021 The use of the MR-SR based PIMCA learning model in eye as optical tools subject *Journal of Physics: Conference Series* **1968** (2021) 012043 doi:10.1088/1742-6596/1968/1/012043, IOP Publishing

[20] Mayampoh, et.al. 2021 Phyphox application with PIMCA learning model, *Journal of Physics: Conference Series* **1968** (2021) 012042 doi:10.1088/1742-6596/1968/1/012042, IOP Publishing

[21] Nasra, et.al. 2021 Teaching and learning of electric charge with PIMCA model, *Journal of Physics: Conference Series,* **1968** (2021) 012032 doi:10.1088/1742-6596/1968/1/012032, IOP Publishing

[22] Patol, et.al. 2021 Development of PIMCA learning model on magnetic field, *Journal of Physics: Conference Series,* **1968** (2021) 012034 doi:10.1088/1742-6596/1968/1/012034, IOP Publishing

[23] Rettob, et.al. 2021 Students learning difficulties in understanding the Lorentz force, *Journal of Physics: Conference Series* **1968** (2021) 012041 doi:10.1088/1742-6596/1968/1/012041, IOP Publishing

[24] Reskin, et.al. 2021 Implementation of the PIMCA model to learning convex mirror, *Journal of Physics: Conference Series* **1968** (2021) 012040 doi:10.1088/1742-6596/1968/1/012040, IOP Publishing

[25] Tokolang, et.al. 2021 The application of multiple-choice in learning the doppler effect with PIMCA model *Journal of Physics: Conference Series* **1968** (2021) 012038 doi:10.1088/1742-6596/1968/1/012038, IOP Publishing

Synthesis TiO_2 using sonochemical method and responses the CO_2 gas of the nanoparticle TiO_2 layers on the QCM sensor surfaces

L Mardiana[1,2*], A Y P Wardoyo[1], Masruroh[1*], and H A Dharmawan[1]

[1]Physics Department, Brawijaya University, Jl. Veteran 65145 Malang, East Java, Indonesia
[2]Physics Department, University of Mataram, Jl. Majapahit 83125 Mataram, West Nusa Tenggara, Indonesia

E-mail: ruroh@ub.ac.id, lailimardiana@student.ub.ac.id

Abstract. TiO_2synthesis using sonochemical methods and their coating on the surface of the Quartz Crystal Microbalance (QCM) sensor utilizing the spin coating method has been carried out. The synthesis started by mixing 7 ml of titanium (IV) isopropoxide (TTIP) (97%) in 70 ml of ethanol (96%). The mixture of TTIP solution was stirred for 30 minutes. After ward, the sonication was carried out for 4 hours at a frequency of 40 kHz and heated for 16 hours at 80 ^oC. To produce TiO_2 powder, the precursor was calcined at 500^oC for 3 hours. The TiO_2 nanoparticle was analyzed using XRD, and the particle size was observed using SEM/Eds. Based on the XRD results, the crystal structure of Anatases was obtained at the 2θ (deg) of 25.503°(101), 29.56° (103), 36.20° (004), 37.977° (112), 48. 25.9° (200), 54.100° (105), 55.272^o (211), 62.89°. The particle size from the SEM results showed the average particle size in the range of ±7μm-±54 nm with the majority in ±96 nm. The obtained $TiO_2$was dissolved with ethanol then the solvent was deposited on the QCM surface using the spin coating method. A homogeneous and porous surface structure is obtained from the SEM observation. The coated QCM was tested for its ability to adsorb CO_2 gas. QCM was exposed to CO_2 gas for 100 seconds with a constant exposure discharge. A positive response was obtained for the adsorbability of the coated QCM to adsorb CO_2 gas with time variations. TiO_2's response against CO_2 gas adsorption indicates that TiO_2 can be further investigated as a CO_2 gas sensor. The measuring response of the TiO_2 layer on the QCM sensor was conducted to adsorb CO_2 gas.

1. Introduction

Titanium dioxide (TiO_2) is an n-type semiconductor material constituted as transition metals. It has three types of crystal structures: anatase, rutile, and brookite; these three crystal phases have different energy gaps of 3.2, 2.96, and 3.02 eV, respectively [1]. Titanium is beneficial as it has the same high refractive index, dielectric constant, and chemical stability [2]. Operationally, it is a low-cost synthesis with low oxidation potential, non-toxic, and environmentally friendly [3,4].

With its many advantages and potential, titanium is widely used in various applications such as photovoltaics [5], optoelectronics [6], photocatalysis [7] as well as across multiple sensor applications [1].

TiO_2 as a sensor material is used as a Quartz Crystal Microbalance (QCM) sensor. QCM is a mass sensor with high sensitivity and fast response. The mass sensitivity that can be responded to is in the nanogram scale in each 1 Hz change of the oscillation frequency [8,9]. QCM's sensitivity is highly dependent on its surface response when a stimulus is given in the form of solids, liquids, or gases. Therefore the QCM sensor can be modified into various sensors, such as gas sensors. Previous studies show that TiO_2 nanoparticles were coated on the surface of the QCM sensor for the mega-adsorption of CO and NO_2 gases [10].

There is huge potential for TiO_2 nanoparticles and QCM sensors for the future development of gas sensors. Thus, TiO_2 must be easily accessible as a potential sensor material. Many methods have been used to obtain titanium in the preparation of TiO_2, such as the sol-gel method [11], the hydrothermal method [12,13], the precipitation process [14], and sonochemistry [15,16]. The sonochemical method is very attractive as it is cheap, simple, capable of producing uniform morphology and high crystallinity structure. The important thing is that it has a high synthesis success rate [16]. Therefore, in this study, sonochemical methods were used in the preparation of TiO2 nanoparticles. In this study, the synthesis process of TiO_2 nanoparticles using sonochemical methods was carried out. The resulting TiO_2 nanoparticle powder will be used as the coating material for the surface of the QCM sensor- which will then be tested in the coating's adsorbability to the CO_2 gas. This research is hoped to be beneficial for developing CO_2 gas sensors based on QCM sensors in the future.

2. Methods

2.1. Preparation Particle TiO_2

The precursor used in this synthesis was Titanium (IV) isopropoxide (TTIP) (97%) from Sigma Aldrich. The sonochemical method is used in this synthesis process. A total of 7 ml of TTIP was mixed with 70 ml of ethanol (96%). This mixture was stirred using a magnetic stirrer to produce a homogeneous solution. This solution was then ultrasonicated at a frequency of 40 kHz for 4 hours. The sonicated solution was then heated at 80 °C for 16 hours producing a gel mixture. Furthermore, the calcination process was carried out at a temperature of 500°C for 3 hours and produced TiO_2 powder. TiO_2 powder was characterized using XRD and SEM to determine the crystalline phase formed and the particle size.

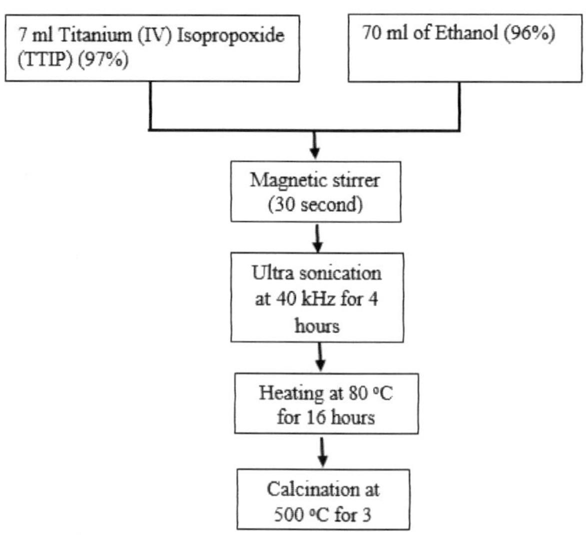

Figure 1.The sample preparation flow chart

2.2. QCM Sensor Coating

TiO_2 particles resulting from this sonochemical synthesis were used to coat the surface of the QCM sensor. A 1 molar solution of TiO_2 and aquades as a solvent were made, and 1 μl was dripped on the surface of the QCM in the spin coating machine. A rotating frequency of 500 rpm for 5 seconds was done, then rotated at a frequency of 3000 rpm for 60 seconds was used. The coated QCM is dried at room temperature, which will then be observed by SEM to determine the surface morphology produced on the coated surface of the QCM sensor.

2.3. Experimental Setup

The QCM used in this experiment was first coated with TiO_2 Particles. The QCM's initial frequency was first measured using a frequency counter. QCM has two electrodes connected to the oscillator. The oscillator's output will be connected to the microcontroller, which calculates each count of changes in the QCM oscillation frequency due to the addition of CO_2 gas exposed to the QCM sensor. The experimental set can be seen in Figure 2. This experimental set consists of a gas storage box, a gas inlet pump, an oscillator, a set of frequency counters, and a CO_2 gas sensor.

The CO_2 gas is filled in the 30 cm x 20 cm x 20 cm storage box where the CO_2 gas sensor has been placed. The storage box is connected to a pump with a constant discharge of 0.02 liters/second, this pump out is directly connected to the QCM. QCM is exposed to CO_2 gas for 100 seconds. Then each value of the change in the oscillation frequency is noted using a frequency counter.

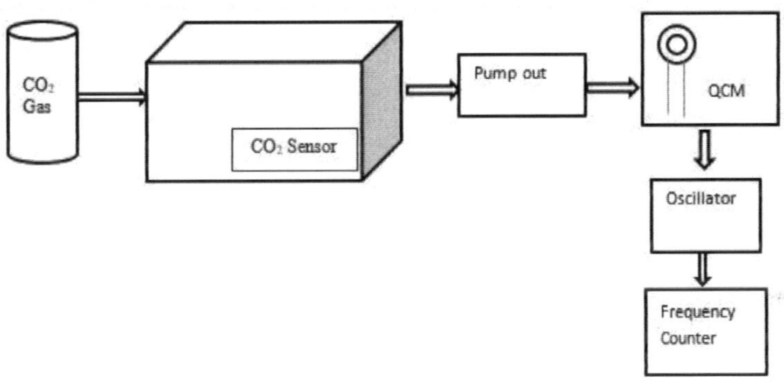

Figure 2. Diagram block of the experimental setup for responses of QCM Sensor

3. Result and Discussion

Figure 3 shows the results of XRD characterization of the TiO_2 particles synthesis with the sonochemical method. The characterization results showed that the phase formed from this synthesis was an anatase crystal structure. This is confirmed based on the Committee on Powder Diffraction Standard (JCPDS card no. 21-1272) which corresponds to a value of 2θ(deg) of 25.503° (101), 29.56° (103), 36.20° (004), 37.977° (112), 48. 259° (200), 54.100° (105), 55.272° (211), 62.89° (204). Figure 3 shows the conformity of the peaks formed from the XRD characterization results with the patterns of the anatase crystal structure (JCPDS card no. 21-1272). An anatase crystal structure is a TiO_2 crystal structure widely used in the sensor industry, one of which is gas sensors [17,18] .

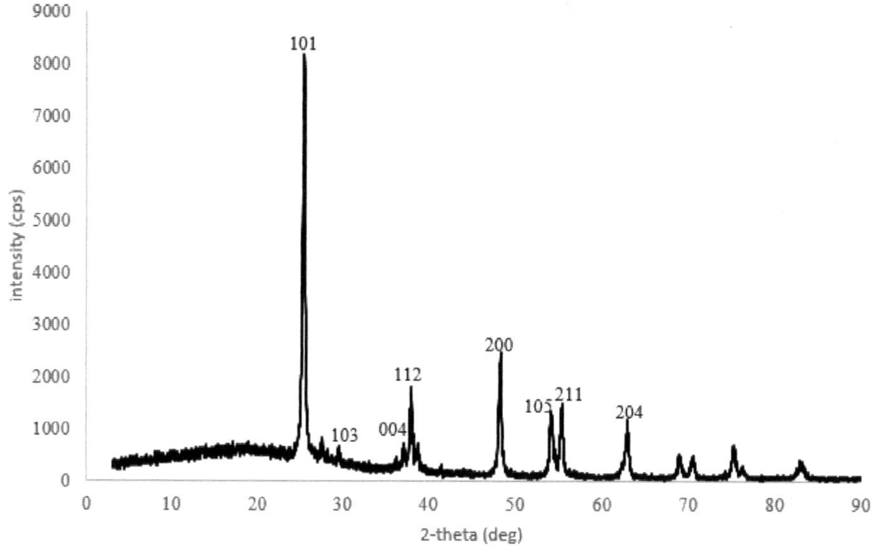

Figure 3. X-ray diffraction (XRD) patterns of TiO_2 particles using sonochemical method

Characterization was carried out on the surface morphological features using a scanning electron microscope. Figure 4 shows the obtained image. Characterization and analysis of the surface morphology showed varied sizes of ±54 nm - ±7 μm. The image indicates that the particles are dominated by an average size of ±96 nm. Because the particles are dominated in nano-size, these TiO_2 particles can be used as gas sensor materials. Due to their small size and high surface-to-volume fraction, TiO_2 nanoparticles have a high ability to adsorb and interact [19].

Figure 4. SEM image of TiO_2 particles using sonochemical method

The characterized synthesized particles from the sonochemical method are deposited on the surface of the QCM sensor. The coated QCM sensor is measured for its natural frequency before testing its adsorbability against CO_2 gas. This measurement also aims to identify the QCM sensor's ability to oscillate after the coating process. If the coating process of the QCM sensor is not right due to the layer being too thick, it will interfere with the oscillation of the sensor. The frequency counter will show a value of 0, which indicates that the QCM is unable to oscillate. The results show the frequency of the QCM sensor after being coated is 4998.464 kHz.

The QCM sensor is tested for its initial frequency value and placed in the experimental set, as shown in Figure 2. The QCM sensor is exposed to CO_2 gas for 100 seconds with a constant pump discharge and placed in the chamber. The test results obtained a positive response in the QCM sensor's adsorbability of CO_2 gas, as shown in Figure 5. Each addition of a given concentration of CO_2 gas causes a decrease in the oscillation frequency of the QCM sensor. This condition follows Sauerbrey's equation in equation 1. The addition of gas concentration indicates an increase in mass, resulting in a decrease in the oscillation frequency of the QCM sensor [20].

$$\frac{\Delta f}{f_0} = \frac{-2f_0}{A\sqrt{\rho_q \cdot \mu_q}} \Delta m \qquad (1)$$

Where A is the active area of the QCM electrode in cm^2, f_0 is the resonant frequency of the QCM in hertz (Hz), m is the change in the oscillating mass in grams (g), ρ_q is the density of quartz, and μ_q is the shear wave velocity in the quartz.

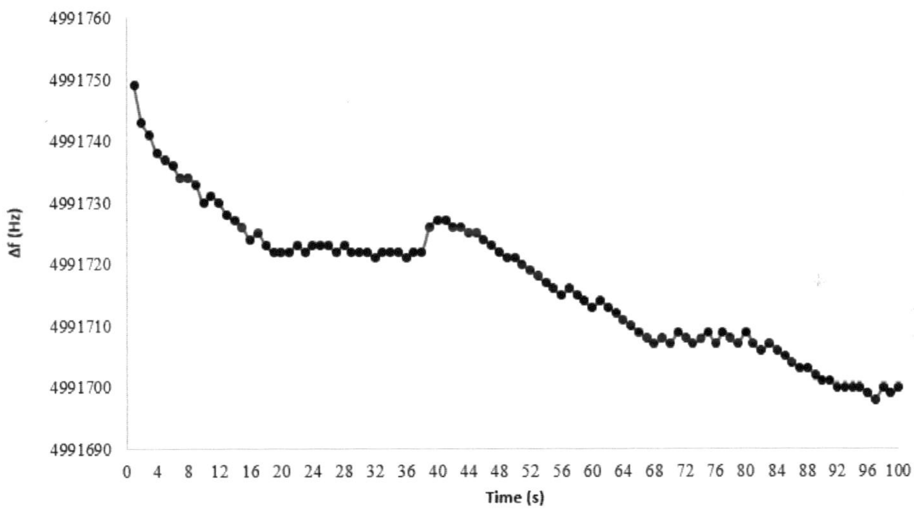

Figure 5. QCM oscillation frequency response to exposure of CO_2 gas for 100 seconds.

The response of the QCM sensor to the exposed CO_2 gas for 100 seconds is very typical. During this period, the QCM sensor experienced three phases: the active working phase, steady state, and self-recovery. At 1 to 20 seconds, the sensor has a very good response; this can be described by the value of the coefficient of determination which shows $R^2 = 0.9455$. As shown in Figure 6 (a). At 21 to 38 seconds, the QCM sensor oscillations are steady, as shown in Figure 5. At 39 to 42 seconds, the QCM sensor has an increase in the oscillation frequency value, as indicated by the self-recovery process in QCM. After 42 seconds, the sensor returns to normal conditions, as shown in Figure 6 (b), until the next 20 seconds. The condition in Figure 6 (b) is the best response from the QCM sensor because it has the highest linearity, as shown by the R^2 value of 0.9792. Figure 6 (c) has the same trend as Figures 6 (a) and 6 (b), where the sensor experiences a decrease in response as it only has 15 seconds of corresponding active work, namely at 83 seconds to 97 seconds .

This study has indicated positive response of the QCM sensor's adsorbability, which has been coated with TiO_2 nanoparticles to adsorb CO_2 gas. For further development of this sensor, the stability of the QCM sensor in responding to gas which is strongly influenced by the uniformity of the coated surface, the thickness of the layer, and the size of the nanoparticles to be deposited, need to be investigated further

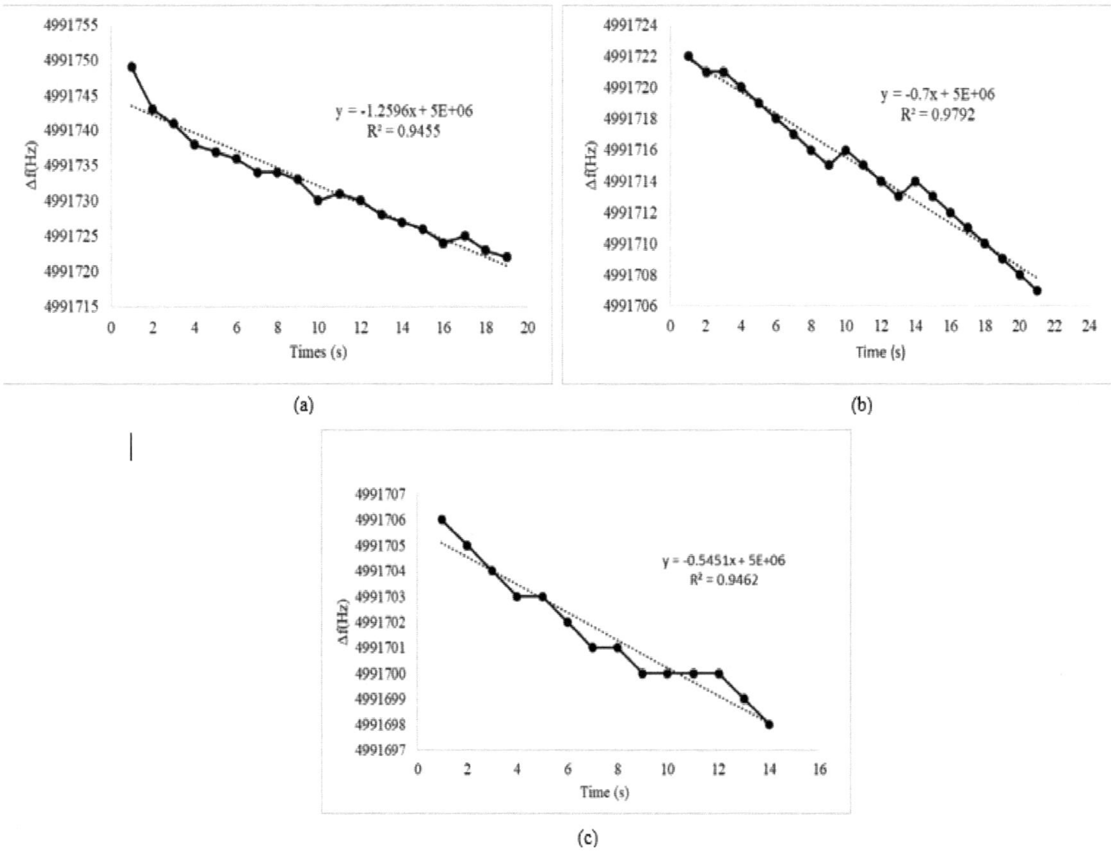

Figure 6. Linearity graph of QCM sensor oscillation frequency for time exposure of CO_2 gas.

4. Conclusion

It can be concluded that the TiO2 nanoparticles synthesized using the sonochemical method were successfully deposited on the surface of the QCM sensor and were able to adsorb CO_2 gas with a very good response, as indicated by the coefficient of restitution R^2 of 0.9792. It also suits the principle from the Saurbrey equation, which states that every increase in mass concentration on the surface of the QCM sensor results in a decrease in the oscillation frequency of the QCM sensor. This research is an initial study of the ability of the QCM sensor to be coated with TiO_2 nanoparticles as a CO_2 gas sensor.

Acknowledgments

A part of this research was funded by University of Mataram for financial supporting and laboratory fascilities and also supported by Ministry of research and Technology (BRIN) through LPPM University of Brawijaya Malang with contrat number 438.19 /UN10.C10/TU/2021.

References

[1] Bai J, Zhou B 2014Titanium dioxide nanomaterials for sensor applications *Chem. Rev.***114**10131–76.

[2] Raza MA, Habib A, Kanwal Z, Hussain SS, Iqbal MJ, Saleem M, et al Optical 2018 CO_2 Gas Sensing Based on TiO_2 Thin Films of Diverse Thickness Decorated with Silver Nanoparticles. *Adv. Mater .Sci. Eng.*

[3] Van Quy N, Minh VA, Van Luan N, Hung VN, Van Hieu N 2011 Gas sensing properties at room temperature of a quartz crystal microbalance coated with ZnO nanorods *Sensors.Actuators. B Chem***153**188–93.

[4] Swapna MV, Haridas KR 2016 An easier method of preparation of mesoporous anatase TiO_2 nanoparticles via ultrasonic irradiation. *J. Exp. Nanosci***11**540–9.

[5] Bai Y, Mora-Seró I, De Angelis F, Bisquert J, Wang P 2014 Titanium dioxide nanomaterials for photovoltaic applications *Chem. Rev***114**10095–130.

[6] Sébastien S 2012 Les soignants face à la mort. *Rev. Infirm.* **180** 39–41

[7] Liu Y, Li Z, Green M, Just M, Li YY, Chen X 2017 Titanium dioxide nanomaterials for photocatalysis *J. Phys. D. Appl .Phys.***50**(19)

[8] Mecea VM 2005 From quartz crystal microbalance to fundamental principles of mass measurements *Anal. Lett***38**753–67

[9] Liu G, Zhang G 2013 Basic Principles of QCM-D 1–8

[10] Addabbo T, Fort A, Mugnaini M, Tani M, Vignoli V, Bruzzi M 2017 Quartz crystal microbalance sensors based on TiO_2 nanoparticles for gas sensing *IEEE. Int. Instrum. Meas. Technol .Conf Proc.*

[11] Fatmawati D, Aritonang AB, Nurlina 2019 Synthesis and Characterization TiO_2 -Kaolin Using Sol Gel Method. *J.Kim .Khatulistiwa* **8**15–21.

[12] Sofyan N, Ridhova A, Yuwono AH, Udhiarto A 2018 Preparation of anatase TiO_2 nanoparticles using low hydrothermal temperature for dye-sensitized solar cell. *IOP Conf Se. Mater . Sci Eng* **316**.

[13] Nanozarah S, Dioksida T, Ester M2010 Synthesis of Titanium Dioxide Nanoparticles via Sucrose Ester Micelle-Mediated Hydrothermal Processing Route **39** 261–5.

[14] Buraso W, Lachom V, Siriya P, Laokul P 2018 Synthesis of TiO_2 nanoparticles via a simple precipitation method and photocatalytic performance. *Mater.Res.Express***5**0–10.

[15] Xu H, Zeiger BW, Suslick KS 2012 Sonochemical synthesis of nanomaterials . *Chem Soc Rev*

[16] Hinman JJ, Suslick KS 2017 Nanostructured Materials Synthesis Using Ultrasound. *Top.Curr. Chem*1–36.

[17] Boyadjiev S, Georgieva V, Vergov L, Baji Z, Gáber F, Szilágyi IM 2014 Gas sensing properties of very thin TiO_2 films prepared by atomic layer deposition (ALD) *J. Phys. Conf. Ser***559**(1).

[18] Zakrzewska K, Radecka M 2017 TiO_2-Based Nanomaterials for Gas Sensing—Influence of Anatase and Rutile Contributions *Nanoscale. Res. Lett***12**(1).

[19] Saranya SKSKS, Padil VVT, Senan C, Pilankatta R, Saranya SKK, George B, et al 2018 Green synthesis of high temperature stable anatase titanium dioxide nanoparticles using gum kondagogu: Characterization and solar driven photocatalytic degradation of organic dye. *Nanomaterials* **8**(12).

[20] Alassi A, Benammar M, Brett D2017 Quartz crystal microbalance electronic interfacing systems: A review. *Sensors (Switzerland)***17**(12) 1–41.

Journal of Physics: Conference Series **2165** (2022) 012015 doi:10.1088/1742-6596/2165/1/012015

Natural disaster mitigation using geophysical methods

A Susilo[1], Sunaryo[1], E A Suryo[2], T Rachmawati[3] and S Rosyid[4]

[1]Geophysics Engineering, Universitas Brawijaya, Malang, Indonesia
[2]Civil Engineering, Universitas Brawijaya, Malang, Indonesia
[3]Urban Planning Engineering, Universitas Brawijaya, Malang, Indonesia
[4] Physics Department, Universitas Indonesia, Jakarta, Indonesia

E-mail:adisusilo@ub.ac.id

Abstract. The research has been carried out about natural disasters (landslides, ground motion, drought) using geophysical methods.This research was conducted in several places, in East Java, i.e .land slides and ground motion occurred in Malang and Blitar areas and drought occurred in theTulungagung and Malang areas. This research uses resistivity and self-potential. The active method is resistivity, which means that an electric current is injected into the earth, after which the resulting potential difference is recorded, then processed. Self-potential and geomagnetic methods, are passive methods. In the self-potential method, the potential difference that occurs between two points, which have a certain distance, is recorded (the middle between those two points becomes a point measurement). The data of several points of measurement, that have been made within a certain distance, then is processed for determining underground river tracing. The equipment used is the OYO Mac Ohm resistivity meter. The results obtained are, an artesian well in the Tulungagung area, at a depth of 108 m, an indication of new faults in the Blitar area, and ground motion cracks in the Malang area.

1. Introduction

Indonesia is one of the countries that often experience hydrometeorological disasters. These hydrometeorological disasters include landslides, ground movements, and drought disasters.East Java is one of the provinces that frequently experiences this disaster [1]. As an illustration, throughout 2017 there were 2,372 disaster events in the territory of Indonesia. About 99 percent are hydrometeorological disasters, which are influenced by weather and runoff [2].

One of the areas in East Java that was affected by the hydrometeorological disaster was Sri Mulyo Village, Dampit Sub District, Malang Regency. On January 24, 2006, a landslide occurred in the village, which destroyed 1 house, and 14 houses had cracks in the walls. The landslide also cut off access to the village area. After doing research, it turns out that the area is composed of layers of weathered breccia, which are on top of limestone. This landslide occurred in an area where the slope was not steep. This indicates that the landslides weretriggered by the presence of normal faults [3].

Hydrometeorological disasters, in the form of drought, also occurred in Tulungagung Regency.During the dry season, rice fields in the local area do not get irrigation water supply, so that paddy fields cannot be planted.Apart from the agricultural sector, the use of clean water also comes from households.Based on data from the Tulungagung Regional development planning agency, it was recorded that three Sub-Districts in Tulungagung experienced drought: Pakel, Bandung, and Besuki.

As an effort to mitigate hydrometeorological disasters in the three areas, it is necessary to analyze the potential for disasters using geophysical methods.In this study, resistivity was used as a geophysical method to detect the presence of fractures in Pohgajih Village, Blitar;detecting the presence of slip plane in Srimulyo Village, Malang Regency, and looking for groundwater locations as a solution to drought in the dry season in Tulungagung Regency.Based on previous research, this resistivity method is commonly used in the fields of hydrology [4,5], engineering [6], archeology [7], environment [8], and disaster mitigation [9,10,11].

2. Methods

The purpose of the resistivity survey is to get a picture of the subsurface resistivity based on measurements on the surface [12].Therefdre, Direct Current is injected into the ground.Next, the measured potential difference will be obtained.Current injection is carried out with an OYO McOhm resistivity meter, which can amplify the current.An electric current of the order of several milliamperes (20 mA), can penetrate a certain depth, so that the potential difference can be measured.

There are several configurations used in resistivity surveys.In the case of analyzing the fault in Pohgajih Village and detecting the slip plane in Sri Mulyo Village, a resistivity method with a dipole-dipole configuration was used (Figure 1).Meanwhile, in finding the location of groundwater in Tulungagung, the configuration used is Wenner-Schlumberger.

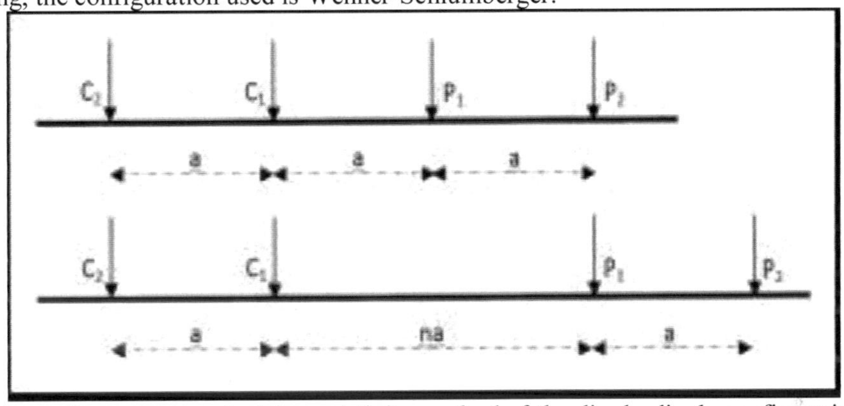

Figure 1. Electrode arrangement of the resistivity method of the dipole-dipole configuration [13]

Where C1 and C2 are the current electrodes, and P1 and P2 are the potential electrodes.The depth and accuracy that can be achieved depend on the distance between C1 and P1. The farther the distance, the deeper the depth that is achieved. While the accuracy is dependent on the distance between C2-C1 and also P1-P2.

Due to the layered and heterogeneous nature of the subsurface medium, the resistivity obtained is still an apparent resistivity [14].Based on the value of current I and potential difference ΔV, the apparent resistivity value ρ_a can be calculated using equation (1):

$$\rho_a = \frac{k\Delta V}{I} \quad (1)$$

The value of the geometric factor k, depending on the configuration used. For the dipole-dipole configuration, it is expressed by equation (2):

$$k = \pi n(n+1)(n+2)a \quad (2)$$

Meanwhile, the Wenner-Schlumberger configuration is stated by equation (3):

$$k = \pi n(n+1)a \quad (3)$$

where n is a discrete number, and a is the space between the electrodes.Furthermore, the field measurement data is processed with inversion software, to obtain the actual resistivity cross-sectionρ which is below the surface.

3. Result and Discussion

3.1 Results of Identification of Slip Plane in Sri Mulyo Village

One of the results of resistivity measurements, to mitigate landslides/subsidence, in Srimulyo village is as follows (Figure 2):

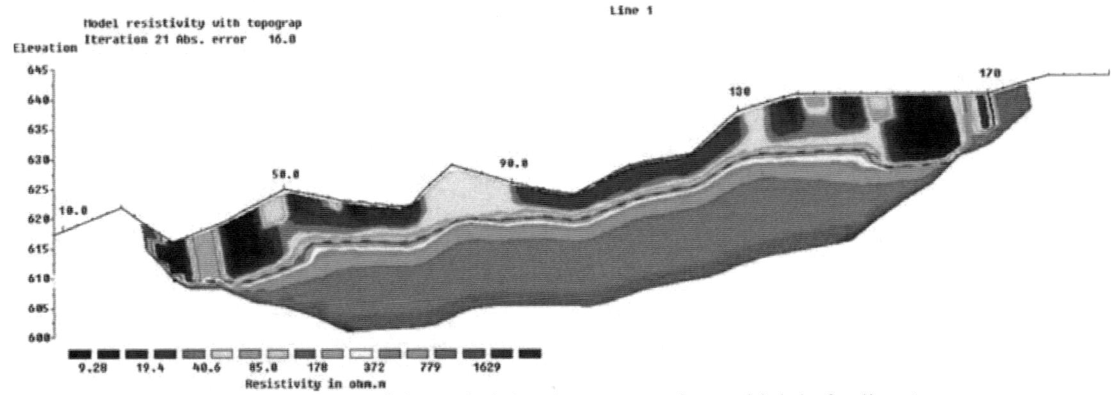

Figure 2. One of the resistivity data processing, which is for line 1

Based on the results of processing resistivity data and information on the Geological Map of the Turen Sheet [15], it can be seen that Sri Mulyo Village, Dampit sub-District is composed of three rock layers. The lithology is composed of clay (9.3-85.8) Ω.m, tuff (178-779) Ω.m, and breccias (≥ 1629 Ω.m). Figure 2 is a 2D cross-section of the result of resistivity data processing, where the dashed line shows the estimated landslide slip area at the research location. The landslide slip area in this study is the boundary between clay and tuff.

3.2 Fault Analysis Results in Pohgajih Village.

The second area is Pohgajih village, Selorejo sub-district, Blitar district. Measurement with Dipole-Dipole configuration, resistivity method in this village produces 5 paths. One of the results of the five paths is line 1, as shown in Figure 3.

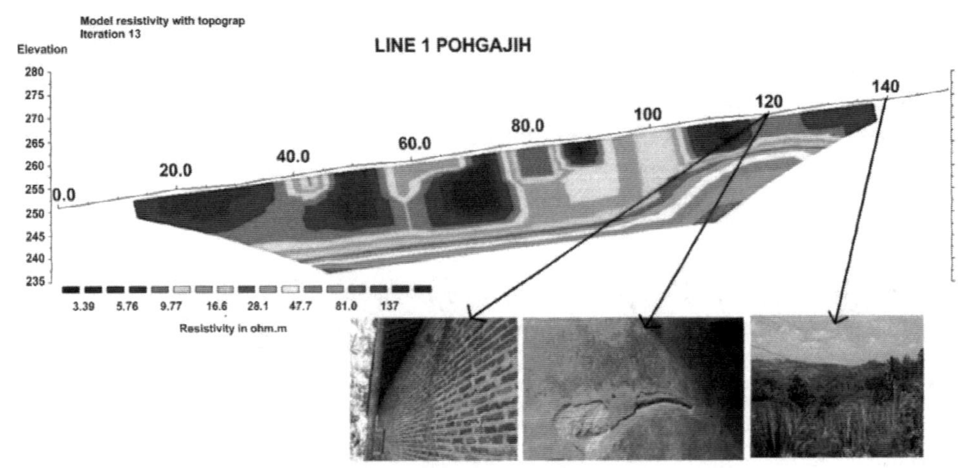

Figure 3. Interpretation results for line 1. It can be seen that there are cracked houses in the area traversed by line 1

In this figure, it can be seen that on line 1, at the 120 meter measurement point, there is a house with cracked walls and floors, with resistivity values ranging from (5.8-9.8) Ωm.

Data from the five measurement trajectories were correlated to estimate the direction of fault continuity at the study site. Based on the results of resistivity data, the discontinuity of high resistivity with low resistivity (clay) vertically is interpreted as a fault [16]. The results of the correlation of faults in the study area are shown in Figure 4. Based on the correlation, it is concluded that the faults in the study area are strike slip-faults that have northwest to southeast orientation. This is supported by information from the geological map sheet Blitar [17] which shows a fault with a northwest-southeast orientation (see red circle in the map inset Figure 4). The results of the correlation also found the presence of other faults that were far to the north of the study site.

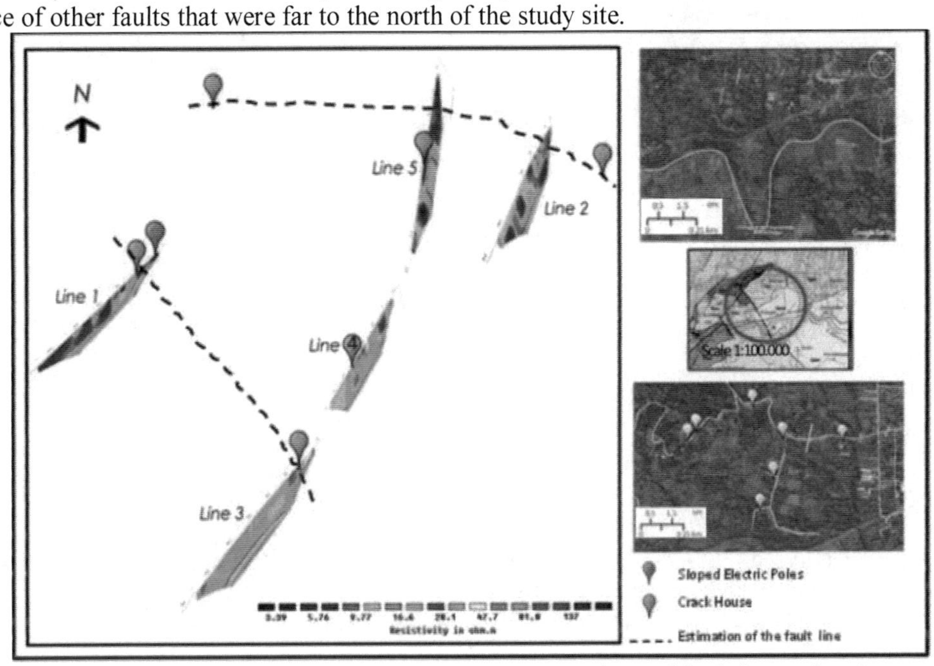

Figure 4. Correlation results of five resistivity measurement paths, and fault line estimation (dashed line).

3.3 Aquifer Exploration for Drought Disaster Mitigation in Tulungagung

There are 13 sounding measurement points, which were carried out in a number of sub-districts in Tulungagung Regency.Data from the sounding points were interpolated into resistivity map. Figure 5 shows the resistivity map for a depth of 115 meters.Based on Figure 5, it can be seen that there is a low resistivity zone which were interpreted as an aquifer.By reviewing the geology of the area that was formerly a swamp, the rock constituents in the area are terrigenous sedimentary rocks [18].Furthermore, rocks with resistivity between (0-200)Ωmwere gravel and sands), 200-800 Ωm were alluvium and > 800 Ωmwere sandstone).

The interpretation of the map in Figure 5 shows severalopen aquifers at the sounding points of Sanan, Bangun Jaya, Bono, Ngunggahan, Tanggul kundung, and Besuki villages.The depth of this aquifer varies from 6 to 21 meters below the surface.In addition, confined aquifers were also found at the Sanan, Bono, Singgit, Nguploadan, and Tanggulkundung villages at a depth of (90-115) meters below the surface.

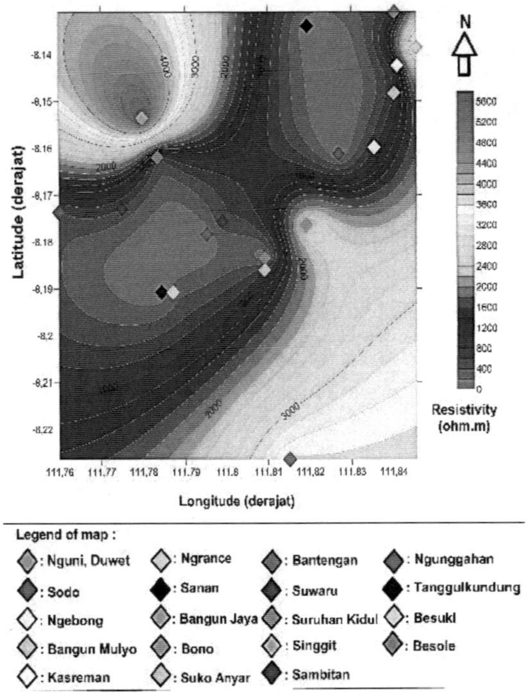

Figure 5. The resistivity map at a depth of 115 meters. Confined aquifers are still found in the Tanggulkundung, Sanan and Ngunggahan areas

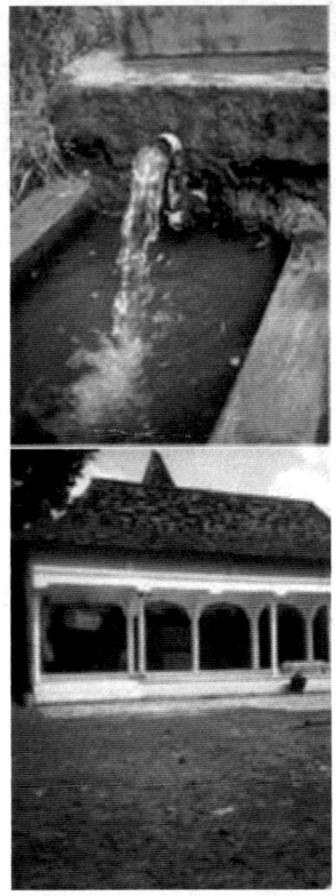

Figure 6. Artesian well and Darussabilillah Mosque.The mosque is located north of the artesian well.

Drilling is carried out at points that are accessible by drilling tools and roads.Taking into account the settlements in the surrounding area, drilling was carried out in Gempolan Village, Pakel District in front of the Darussabilillah Mosque (Figure 6).The artesian well flowedafter drilling at a depth of 108 meters from the surface.This result showed that resistivity method was useful for exploring aquifer as an effort to mitigate drought disaster, especially in Tulungagung.

Three case studies which we have been discussed showed that resistivity method was approriate for mitigating hydrometeorological disaster. For further analysis, we comparing withresistivity data from landslide disaster area in Pulung Subdistrict, Ponorogo in 2017 [19]. From 12 resistivity measuring points in the area, it was identified rock with resistivity ranging(1.4-67.5)Ωm which were interpreted as clay, tuff lapili, volcanic breccia, and andesite lava. From the resistivity cross section we concluded, thelandslide area begins at depth of 8–35 m below the surface. The thickness of thelandslidematerial andthe slope is 400 meters which supportsthe occurrence of prominent landslide. Again,the resistivity method can provide a good overview for conducting natural disaster analysis.

4. Conclusion

The geophysical method, in this case the resistivity method, can be used to mitigate hydrometeorological disasters, i.e.land movement associated with fault, landslides, and drought disaster.Fault analysis in Pohgajih Village based on resistivity map found the presence of a strike slip fault with a northwest-southeast orientation.Cracked house buildings in the measurement path are characterized by a resistivity range of (3.4-9.8)Ω.m which is clay.From the survey in Sri Mulyo Village, it is known that the landslide slip area is characterized by the difference in resistivity between clay (9.3-85.8)Ω.m and tuff (178-779) Ω.m layers. Meanwhile, aquifers with low resistivity values were found in Tulungagung Regency.Based on the resistivity map at a depth of (15-115) meters, it was found that confined aquifers were found. This is supported by the presence of artesian wells drilled at a depth of 108 meters.

References

[1] Harto A and Pandia A S 2017 *Ancaman Hidrometeorologi di Jatim Kian Menakutkan* Kompas.id https://www.kompas.id/baca/utama/2017/03/28/ancaman-hidrometeorologi-di-jatim-kian-menakutkan/ retrieved 14 September 2021

[2] BNPB 2017 *Info Bencana BNPB edisiDesember 2017* https://www.bnpb.go.id/uploads/24/infobencana/2017/info-bencana-desember-2017.pdf retrivied 14 September 2021

[3] Wasis, Adi S and Sunaryo 2011 Perunutan Jalur SesarLokal di Desa Sri Mulyo Kecamatan Dampit, Kabupaten Malang berdasarkan data geofisika*Natural B Jur. Lingkungan dan Kesehatan***1(1)** 41-50.

[4] Choudhury J, Kumar KL, Nagalah E, Sonkamble S, Ahmed S and Kumar V 2017 Vertical electrical sounding to delineate the potential aquifer zones for drinking water in Niamey city, Niger, Africa. *Jour. of Earth System Science***126(6)** 1-13

[5] Mohamaden M I I, Hamouda AZ and Mansour S 2016 Application of electrical resistivity method for groundwater exploration at the Moghra area, Western Desert, Egypt. The Egyptian*Jour. of Aquatic Research***42(3)** 261-268

[6] Sunaryo, Susilo A and Fina F 2017 Seepage zone identification at Sutami Dam by means of geoelectrical resistivity data.IOP Conf. Series: *Earth and Env. Sci.***75** 012011

[7] Susilo A,Sunaryo, Kosmas I and Rusli 2017 Investigation of Jabung Temple Subsurface at Probolinggo, Indonesia using Resistivity and Geomagnetic Method *Int. J of Geomate***13** 74-80

[8] Susilo A, Sunaryo, Alexander TS, Fina F and Hasan MFR 2017 Identification of Underground River Flow in Karst Area Using Geoelectric and Self-Potential Methods in Druju Region, Southern Malang, *Indonesia. Int. J of Applied Engineering* Research**12 (12)** 10731-10738

[9] Perrone A, Sabatino P and Vincenzo L 2017 Electrical Resistivity Tomographies for Lanslided Monitoring: *A review. Berichte Geol.***93** 129-134. ISSN 1007-8880.

[10] Pilecki Z 2017. Basic Principles for The Identification of Lanslide using geophysical methods *3rd International Conf. on Applied Geophysics*. 24 1-8. DOI: 10.1051/e3sconf/20172401001.

[11] Susilo A, Sunaryo and Fina F 2018 Fault Identification For Human Settlement in Pohgajih Village, Blitar, Indonesia Using Geoelectrical Resistivity Method for Hazard Mitigation. *Int. J of Geomate***14** 111-118

[12] Loke M H 1999 *Electrical imaging surveys for environmental and engineering studies. A practical guide* 2 70

[13] Geotomo Software 2003 *Res2Dinv Manual* (www.geoeletrical.com).

[14] Telford W M, Geldart LP and Sheriff RE1990 *Applied geophysics* (Cambridge: CambridgeUniversityPress).

[15] Sujanto R, Hadisantono R, Kusnama, Chaniago R and Baharuddin R 1992 *Geological Map of The Turen Quadrangle, Jawa* (Bandung: Geological Research and Development Centre)

[16] Olorunfemi MO, Oni AG, Bamidele OE, Fadare TK and Aniko OO 2020 Combined geophysical investigations of the characteristics of a regional fault zone for groundwater development in a basement complex terrain of South-west Nigeria *SN Applied Sci.***2(6)** 1-17

[17] Sjarifuddin M Z and Hamidi S 1992 *Geological Map of the Blitar Quadrangle, Jawa* (Bandung: Geological Research and Development Centre)

[18] Samodra H, Suharsono, Gafoer S and Suwarti T 1992 *Geological Map of Tulungagung, Jawa* (Bandung: Geological Research and Development Centre)

[19] Susilo A, Fina F, Sunaryo, Rachmawati T A R and Suryo E A 2020 Analysis of landslide area of Tulung subdistrict, Ponorogo, Indonesia in 2017 using resistivity method. *Smart and Sustainaeble Built. Env.*

Journal of Physics: Conference Series **2165** (2022) 012016 doi:10.1088/1742-6596/2165/1/012016

Thermodynamics of regular Kerr-Sen black hole

B N Jayawiguna

Departemen Fisika, FMIPA, Universitas Indonesia, Depok, 16424, Indonesia

E-mail: byon.nugraha91@ui.ac.id

Abstract. By evaluating the corresponding Killing vector, we study the thermodynamics quantity of regular black hole in the low energy string theory. From our result, we can infer that the regularity of a black hole enables the Hawking temperature to have local optima. The stability of a black hole from the heat capacity is also discussed.

1. Introduction

The gravitational field around rotating charged black hole in the low energy string theory. The metric solution is parameterized by their mass, charged, and angular momentum [1]. To get the solution, Sen use the well known Hassan-Sen transformation [2] that requires us to have some seed solution whose stationary and axisymmetric (invariant under time and rotation). The idea is that if we have a vacuum solution (zero electric charge, nontrivial dilaton, and 2-form tensor field) as an input, we have a metric solution in string frame with non zero electric charged, nontrivial dilaton, and 2-form tensor field. Thus, the Einstein frame can be easily obtained by transforming the string frame with conformal technique. As a concrete example, He use Kerr black hole as a seed solution [3]. Some aspect including event horizon(s), ergosphere, conserved quantity (ADM formulation), surface area, angular velocity, and Hawking temperature are discussed in the literature. On the other hand, Switching off the electric charge parameter makes the solution goes back to the well known Kerr, whereas for vanisihing spin combined with some rescale, $r \to r - \frac{Q^2}{M}$, the solution reduces to the charged static black hole in the low energy string theory, or it can be referred as Einstein-Maxwell-dilaton (GHS) black hole [4]. Most of them can be done from the black hole with ring singularity under the main assumption from validity of the energy condition. Since collapsing matter forms a spacetime singularity at the origin [5].

Inspired by Bardeen [6], regular (nonsingular) black hole have been considered that has a spherically symmetric, static, asymptotically flat spacetime, and regular at the center. In 2015, Ghosh employ Newman-Janis trick to static regular black hole into rotating version [7]. These solution characterized by the same parameter as Kerr, but with the free parameter, k, which determinde the regularity of the black hole. The rotating version in the low energy string theory has been studied in [8]. In this paper, we use the metric solution in [8] to discussed some of the semiclassical aspect about Hawking temperature and its stability. The paper is structured as follows. As a quick review, we present an Einstein frame metric and the corresponding matter fields in section 2. In section 3 we showed novelty aspect such as Hawking temperature and the stability from specific heat. In section 4, we conclude some of our statement.

2. Regular Kerr-Sen

The Einstein frame action describing black hole in the low energy string theory is as follows [1]

$$I = \int d^4x \sqrt{-g}\left[R(g) - \frac{1}{2}(\nabla\Phi)^2 - \frac{e^{-\Phi}}{8}F^2 - \frac{e^{-2\Phi}}{12}H^2\right], \tag{1}$$

Where I is the action of a low energy string, Φ is a scalar field, g is a determinant of metric tensor, R is a Ricci scalar, the F squared is a Maxwell strength tensor, and H squared is

$$H_{\kappa\mu\nu} = \partial_\kappa B_{\mu\nu} + \partial_\nu B_{\kappa\mu} + \partial_\mu B_{\nu\kappa} - \frac{1}{4}\left(A_\kappa F_{\mu\nu} + A_\nu F_{\kappa\mu} + A_\mu F_{\nu\kappa}\right). \tag{2}$$

is a 3-form tensor field which contain the antisymmetric 2-form tensor field and Chern-Simons term. The corresponding indices run from 0 to 3. As a review, we discussed only the Einstein frame and the corresponding matter fields from [8]. The solution reads

$$ds_E^2 = -\left(1 - \frac{2Mre^{-k/r}}{\varphi^2}\right)dt^2 - \frac{4aMre^{-k/r}\sin^2\theta}{\varphi^2}dtd\phi + \varphi^2\left[\frac{dr^2}{\Xi} + d\theta^2\right]$$
$$+\left[r^2 + a^2 + 2bre^{-k/r} + \frac{2Mra^2e^{-k/r}\sin^2\theta}{\varphi^2}\right]\sin^2\theta d\phi^2 \tag{3}$$

and the matter fields are

$$\mathrm{A} = \frac{Qre^{-k/r}}{\varphi^2}(dt - a\sin^2\theta d\phi),\ B_{t\phi} = \frac{2brae^{-k/r}\sin^2\theta}{\varphi^2},\ and\ e^{-2\Phi} = \frac{\varphi^2}{\rho^2}, \tag{4}$$

where $\varphi^2 = r(r + 2be^{-k/r}) + a^2\cos^2\theta$, $\Xi = r(r + 2be^{-k/r}) + a^2 - 2Mre^{-k/r}$, $\rho^2 = r^2 + a^2\cos^2\theta$, and $b = Q^2/2M$. These metric are parameterized by mass M , charge Q, rotation parameter a, and free parameter k. Switched off the free parameter makes the solution reduces to the well known Kerr-Sen black hole [1]. Next, we discussed some of the semiclassical aspect about the black hole thermodynamics: Hawking temperature and its stability.

3. Hawking Temperature and Stability

In this section we examine the Hawking temperature and stability which can be extract from the heat capacity. These history [9,10,11] reached a climax when Stephen Hawking discovered that black holes were not black in the sense that black hole radiates and can undergo phase transitions. By using surface gravity as fix quantity on the horizon,

$$\kappa = \sqrt{-\frac{1}{2}\nabla^\mu\xi^\nu\nabla_\mu\xi_\nu}\,|_{r\to r_+}, \tag{5}$$

we get the Hawking temperature

$$T_H = \frac{\kappa}{2\pi} = \frac{r_+ - r_-}{8\pi M r_+} = \frac{e^{2k/r_+}\sqrt{(2M^2-Q^2)^2 + 4aMe^{2k/r_+}}}{4\pi M\left[2M^2 - Q^2 + \sqrt{(2M^2-Q^2)^2 + 4aMe^{2k/r_+}}\right]}, \tag{6}$$

where M is the mass of black hole evaluated on the horizon. One can see that as $k \to 0$, the temperature reduces to that of Kerr-Sen black hole (for example, see [1]). However, to study the nature and the regularity contribution in Hawking temperature extensively, we plot the T_H vs r_+ with several fix rotational parameter and charge, a and Q. In this plot, we vary the free parameter, k, with the value: k=0 and k=0.1. It is shown that, when k=0, the temperature has one local optima indicating the instability. When the free parameter switched on, the nature of the regularity enables the Hawking temperature to have two local optima. We shall discussed the optima in the heat capacity.

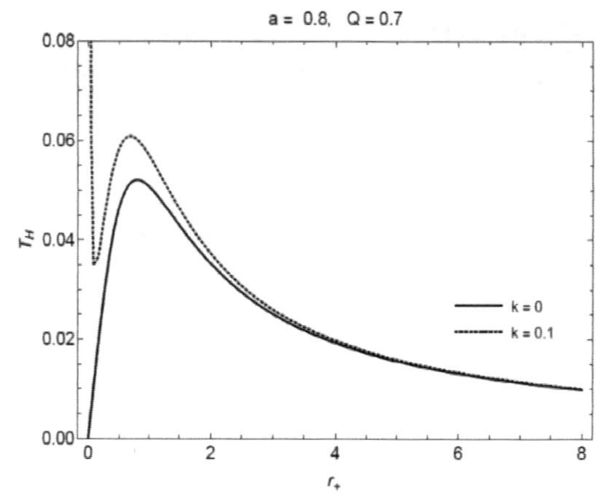

Figure 1. Profile of Hawking temperature versus radius with a = 0.8, and Q=0.7.

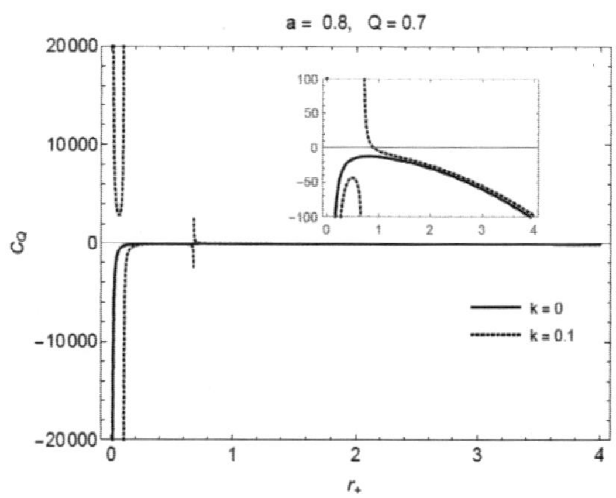

Figure 2. Profile of specific heat versus radius with a = 0.8, and Q= 0.7.

It is tempting to investigate the characteristic further with the heat capacity which can be written as

$$C_Q = \frac{\partial M}{\partial T_H} = -\frac{\chi_1}{\chi_2}, \tag{7}$$

where

$$\chi_1 = 4\pi\Delta r^2 e^{-\frac{2k}{r}} M^2 M'(\Delta + 2M^2 - Q^2)^2, \tag{8}$$

$$\chi_2 = 4r^2M^4M'\left(4a^2e^{\frac{2k}{r}} + \Delta - 3Q^2\right) - 2r^2M^2M'\left(-2a^2\Delta e^{\frac{2k}{r}} - 3Q^4 + 2\Delta Q^2\right) + 8kM^5\left(3a^2e^{\frac{2k}{r}} + \Delta - 3Q^2\right) - 4kM^3(2\Delta - 3Q^2)\left(Q^2 - a^2e^{\frac{2k}{r}}\right) + 2kQ^4M(\Delta - Q^2) + 16kM^7 + Q^4r^2(\Delta - Q^2)M' + 8r^2M^6M', \tag{9}$$

and

$$\Delta = \sqrt{4a^2 e^{\frac{2k}{r}} M^2 + (Q^2 - 2M^2)^2}, \tag{10}$$

where $M' = \frac{dM}{dr}|_{r \to r_+}$. In detail, we have plotted the heat capacity versus radius in Figure 2 with the same fix parameter as Hawking temperature. It is shown, from the plot, when k=0.1 we have two phase that have positiveC_Q and one phase that have negative sign. The positivity of the heat capacity (C_Q>0) indicate the stability and the other one indicate instability. From the figure, we can also infer that the regularity parameter enables the existence of (outer) unstable black hole to have larger radius than its Kerr-Sen (k=0) counterpart. Lastly, we complete the study by obtaining the area A and the entropy S of a regular Kerr-Sen black hole which can be written as

$$A = \int_0^{2\pi} d\phi \int_0^{\pi} \sqrt{g_{\theta\theta} g_{\phi\phi}}\, d\theta = 4\pi\, (r_+^2 + a^2)\sqrt{1 + \frac{2br_+ e^{-k/r_+}}{r_+^2 + a^2}}, \tag{11}$$

$$S = \frac{A}{4} = \pi(r_+^2 + a^2)\sqrt{1 + \frac{2br_+ e^{-k/r_+}}{r_+^2 + a^2}}. \tag{12}$$

As we can see, when b=0 and k=0, the equations reduce to the Kerr black hole's area and entropy.

4. Conclusion

We obtained the semiclassical aspect from the Einstein frame metric of regular black hole in the low energy string theory. The surface gravity of the Hawking temperature can be obtained by evaluating the corresponding Killing vector which obeys the invariant time and rotation quantity. In our thermodynamical analysis, the Hawking temperature reduces to the well known Kerr-Sen temperature. The regularity of a black hole enables the Hawking temperature to have local optima. From the heat capacity, we can extract the stability condition. We can also infer that the regularity parameter enables the existence of (outer) unstable black hole to have larger radius than its Kerr-Sen (k = 0) counterpart.

References

[1] Sen A 1992 *Phys. Rev. Lett.* **69** 1006-1009
[2] Hassan S and Sen A 1992 *Nucl. Phys. B***375** 103-118
[3] Kerr R 1963 *Phys. Rev. Lett.* **11** 237
[4] Garfinkle D, Horowitz G and Strominger A 1991 *Phys. Rev. D***43** 3140
[5] Hawking S and Ellis G *1973 The Large Scale Structure of Space-Time (Cambridge Monographs:MathematicalPhysics)*
[6] Bardeen J 1968 *Conference Proceedings of GR5*, p. 174.
[7] Ghosh S, *2015 Eur. Phys. J. C***75** 532
[8] Jayawiguna B 2020 *Preprint gr-qc* 2010.14612
[9] Bekenstein J 1972 *Lett. Nuovo Cim.* **4** 737-740
[10] Bekenstein J 1974 *Phys. Rev. D***9** 3292-3300
[11] Hawking S 1976 *Phys. Rev. D***13** 191-19

Comparison between the magnetic method data of pseudogravity transformation with gravity anomaly data from satellite imagery in the surrounding of the Sutami Dam to identify subsurface formations

M Sutasoma[1], A Susilo[1], Sunaryo[1], E A Suryo[2], S Minardi[3] and R H D Cahyo[4]

[1]Department of Physics, Faculty of science, Brawijaya University, Jl. Veteran, Malang, Jawa Timur, Indonesia
[2]Department of Civil Engineering, Faculty of Engineering, Brawijaya University, Jl. Veteran, Malang, Jawa Timur, Indonesia
[3]Department of Physics, Faculty of science, Mataram University, Jl. Majapahit, Mataram, Nusa Tenggara Barat, Indonesia
[4]Department of Physics, Faculty of Science, Billfath University, Jl. Pucuk, Lamongan, Jawa Timur, Indonesia

E-mail: msutasoma@student.ub.ac.id

Abstract: Research has been carried out to determine the subsurface formations in the Sutami dam area and its surroundings using the magnetic method of pseudogravity transformation and satellite imagery gravity anomaly data. This study aims to compare the subsurface formations in the area around the Sutami Dam between the data of the pseodu-gravity transformation magnetic method and the gravity anomaly data of satellite imagery. Data acquisition using the pseudogravity magnetic transformation method was carried out using the Proton Precession Magnetometer (PPM) Scientrex Model G-8 with a spacing of 300 meters. Satellite imagery gravity anomaly data was taken from the Gravity Model Plus (GGM plus) with a spacing of 220 meters. The radius of geomagnetic data acquisition was 15 km. The number of data for the magnetic method of pseudogravity transformation was 1,372 measurement points and satellite imagery gravity anomaly data was 3,000 measurement points. The results showed that the rock formations from the magnetic method of pseudogravity transformation and satellite gravity anomaly data were compatible. There are 4 types of subsurface formations in the study area, namely soil ($\Delta\rho$ = 1.6 g/cm^3), Butak Volcanic Product ($\Delta\rho$ = 2 g/cm^3), Tuff Deposit ($\Delta\rho$ = 2.1 g/cm^3) and Campurdarat Formation ($\Delta\rho$ = 2.6 g/cm^3).

1. Introduction

Sutami Dam is a dam located in Karangkates Village, Sumberpucung Districs, Malang Regency, East Java Province. The dam was built in 1964-1973 and started operating in 1977 [1]. Sutami Dam has a peak length of about 800 meters and a peak width of 13.7 meters. Based on previous research, the Sutami dam is located in an area prone to soil movement [2] and upstream and downstream of the dam there are fracture zones [3]. In the western part of the dam, there is a weak zone indicated by the presence of local pohgajih shear fractures, local selorejo faults, and limestone-andesite contact areas of selorejo [4]. In addition, there is a fracture zone that spreads in the upstream and downstream part of the Sutami Dam peak highway which is indicated by a low resistivity value. This shows that the Sutami Dam is very vulnerable to ground movement [5]. So it is necessary to research with further geophysical methods to determine subsurface formations.

Geophysical methods are used to determine the type of rock in the earth by measuring its physical parameters from above the surface. The purpose of this study is to determine the correlation between the magnetic anomaly data of pseudogravity transformation with satellite imagery gravity data. The data of the pseudogravity transformation magnetic method used is the data that the author published in International Journal of GEOMATE, Aug., 2021, Vol.21, Issue 84, pp.188-193 [6] but without slicing and separation anomalies while the gravity anomaly data from satellite imagery is obtained from the Gravity Model Plus (GGM plus) [7]. This research hypothesizes that there is a relationship between Pseudogravity Transformation Magnetic Data and gravity anomaly data from satellite imagery.

2. Regional Geology

Based on the geological map in Figure 1 shows that The Sutami Dam area is located in the tuff deposit area. The types of rocks in this area are lapilli tuff, pumice tuff and lava [8]. The location of the dam is also closed with the Pohgajih Local Shear Fault, the Selorejo Local Fault, and the Selorejo limestone-andesite contact area [2]. In addition, The rock formations in the research area area can be classified as follows:

a. The Wuni Formation (Tmw) is dominated by ancient volcanic eruption rocks and deposited above the Campurdarat formation. This formation consists of andesite-basalt breccia, breccia lava, andesite lava, and tuffaceous sandstone inserts.
b. The Campurdarat Formation (Tmcl) is above the tuff member of the Mandalika Formation. The Campurdarat formation consists of crystalline limestone and claystone inserts.
c. The Nampol Formation (Tmn) consists of tuffaceous-limestone sandstone, claystone, and marl.
d. Butak Volcanic Products (Qpkb) consisting of lava, volcanic breccia, tuff breccia, and sandy tuff.
e. Tuff deposits (Qptm) consist of lapilli tuff, pumice tuff and lava.

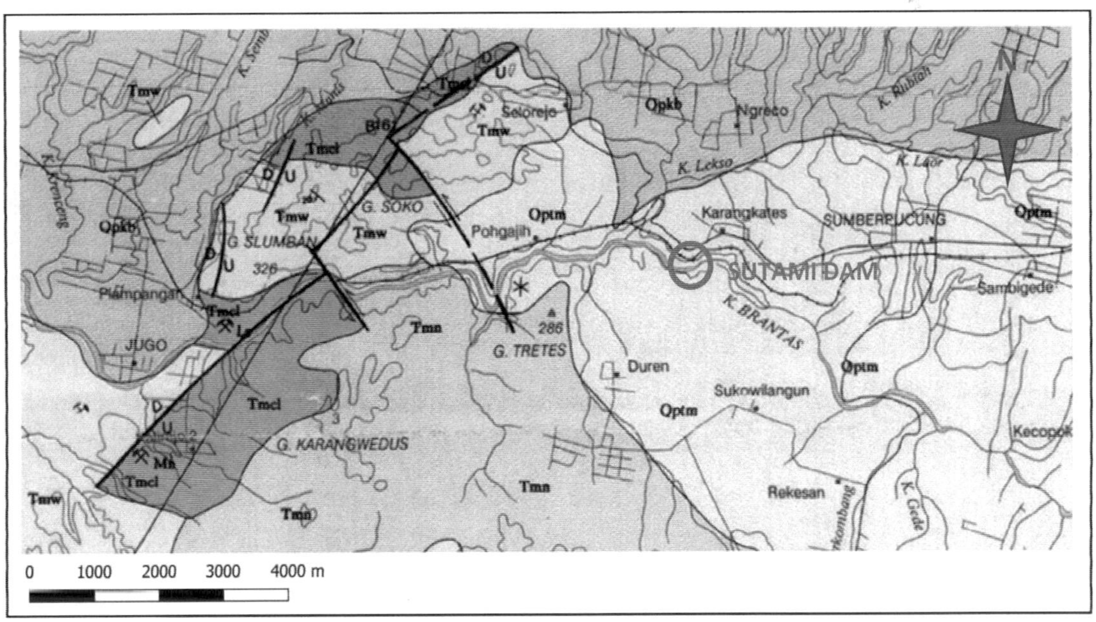

Figure 1. Geological Map of Research Area [8].

3. Methods

The data acquisition area was 225 km^2. The magnetic method data acquisition was carried out with a space between points of 300 meters, and The number of data was 1,372 measurement points. Gravity anomaly data from satellite imagery was downloaded from Gravity Model Plus (GGM plus) with a spacing of 220 meters and obtained 3,000 measurement points. While the gravity anomaly data from satellite imagery, the values obtained are the position of latitude and longitude, altitude, and the value of Free Air Anomaly (FAA).

In the magnetic data, a pseudogravity transformation is carried out first to change the magnetic anomaly that was previously a dipole into a monopole as if it were a gravitational anomaly [9]. This transformation is used to interpret magnetic anomalies because the results are analogous to gravity data from a magnetic object with the density ratio. (Figure 2). the proportionality in this transformation is 100 kg/m^3 every A/m, with 1 A/m = 10^2 gamma then 1 gamma $\approx$ 1 kg/m^3. According to Blakely [10].

For a magnetized body, the magnetic potential V due to proportional density, and the gravitational potential U can be related by the following Poisson' relation:

$$V = -\frac{C_m}{\gamma\rho} M\hat{m} \cdot \nabla_p U \tag{1}$$

$$V = \frac{-C_m M g_m}{\gamma\rho} \tag{2}$$

Where V is magnetic potential, U is gravitational potential (m^2/s^2), g_m is the component of gravity in the direction of magnetization, γ is Newton's constant, ρ is rock density (g/cm^3), C_m is magnetic constant, M is magnetization (Ampere/meter). The assumption in this transformation is that M and ρ are constant, so the Fourier transform equation is:

$$F[g_m] = \frac{\gamma}{C_m}\frac{\rho}{M} F[V] \tag{3}$$

Where,

$$F[V] = -\frac{1}{\theta_f |k|} F[\Delta T] \tag{4}$$

So that,

$$F[g_m] = -\frac{\gamma}{C_m \theta_f |k|}\frac{\rho}{M} F[\Delta T] \tag{5}$$

$$F_{(psg)} = -\frac{\gamma}{C_m \theta_f |k|}\frac{\rho}{M} \tag{6}$$

g_m in this case is pseudogravity. From the equation above, it can be seen that there are three stages in this transformation; perform Fourier transform of total magnetic anomaly data, multiply by $F_{(psg)}$, and do an inversion of the multiplication result. The results of the pseudogravity transformation will strengthen the analysis of the subsurface structure by calculating the ratio of the magnetization value to the pseudo density value.

After the magnetic anomaly data becomes a monopole and is equivalent to a complete Bouguer anomaly, then the two data are separated into regional and residual anomalies with upward continuation. Furthermore, the residual anomaly of the two methods is sliced in the same position, so that a 2D cross-section is obtained for comparison. Data processing is carried out based on Figure 2.

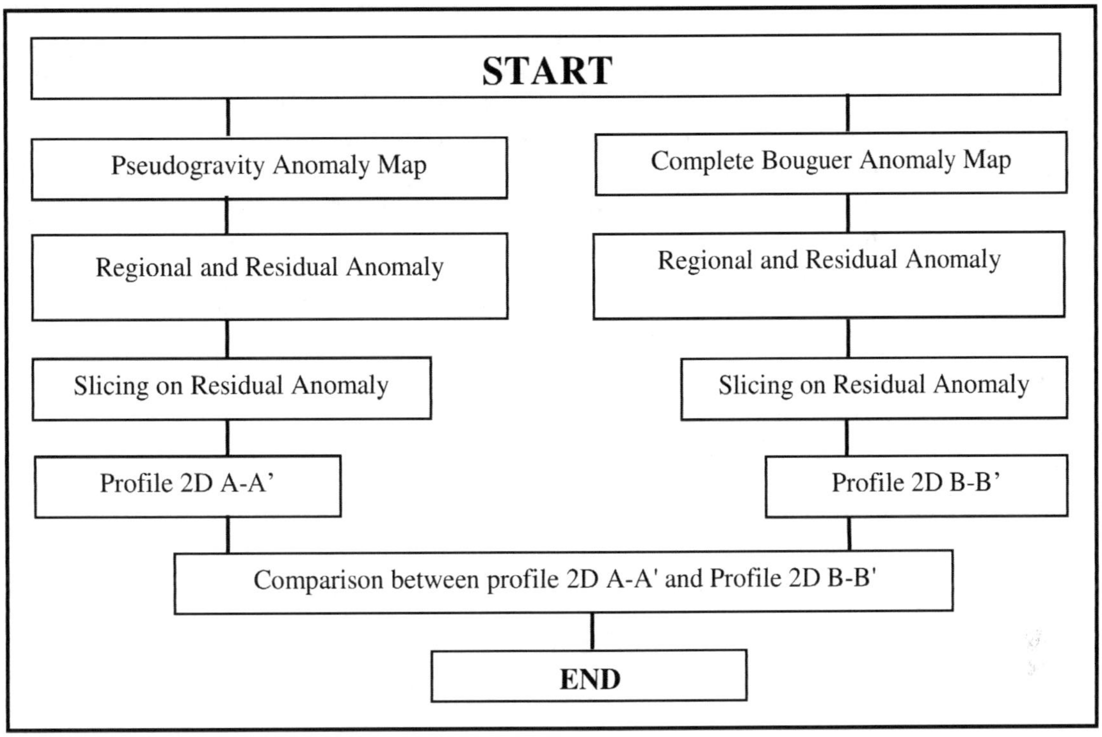

Figure 2. Study Flow Chart.

Data interpretation uses qualitative and quantitative methods. The qualitative method was used to determine the type of anomaly by reading the contours of the pseudogravity anomaly and the Complete Bouguer Anomaly. The quantitative method was used to determine the type of subsurface layer obtained by modeling and adapted to the geological map of the research. The results of data interpretation of the two methods were compared.

4. Result and Discussion

4.1. Result

Pseudogravity transformation is used to convert magnetic anomalies that were previously dipoles into monopoles such as gravitational anomalies. The results of the pseudogravity transformation will strengthen the analysis of the subsurface structure by calculating the proportionality of the magnetization value to the pseudo density value.

The pseudogravity contour map (Figure 3) has a minimum value of -0.2 mGal and a maximum value of 0.5 mGal. High anomaly values in the northwest and north of the study area are due to volcanic deposits originating from Mount Selorejo, Mount Sedayu, and Mount Golo. While the southern part is volcanic deposits from Mount Tretes. And the high anomaly in the northeastern area is still related to the sediment of Mount Kawi.

Based on the range of anomaly values, it can be grouped into 2 groups, namely low anomaly, and high anomaly. The high anomaly is located in the southern part with a value range of 0.0 mGal to 0.5 mGal and is thought to be andesitic lava to basalt and porphyry lattice of the Mandalika Formation (Tomm). Tuff members of the Mandalika Formation (Tomt) in the form of tuff are composed of dacite and rhyolite (liparite). The Mandalika Formation stretches approximately 15 km from the south coast of Blitar. The low anomaly in the north with a value range of -0.2 mGal to -0.1 mGal is a combination of several formations, namely the Campurdarat Formation, Wuni Formation, Nampol Formation, and Wonosari Formation. The Campurdarat formation (Tmcl) consists of limestone and claystone.

The complete Bouguer anomaly contour map (Figure 4) from satellite imagery data has a minimum value of 73.633335 mGal and a maximum value of 89.32008 mGal. Based on the range of anomaly values, it can be grouped into 2 groups, namely low anomaly, and high anomaly. The high anomaly is in the southern part with a range of 83.06951 mGal to 89.32008 mGal which is thought to be andesitic lava to basalt and porphyry latit of the Mandalika Formation (Tomm), with Tuff members of the Mandalika Formation (Tomt) in the form of tuff composed of dacite and rhyolite (liparite). The Mandalika Formation stretches approximately 15 km from the south coast of Blitar. The low anomaly in the north with a value range of 73,63335 mGal to 83,06951 mGal is a combination of several formations, namely the Campurdarat Formation, Wuni Formation, Nampol Formation, and Wonosari Formation.

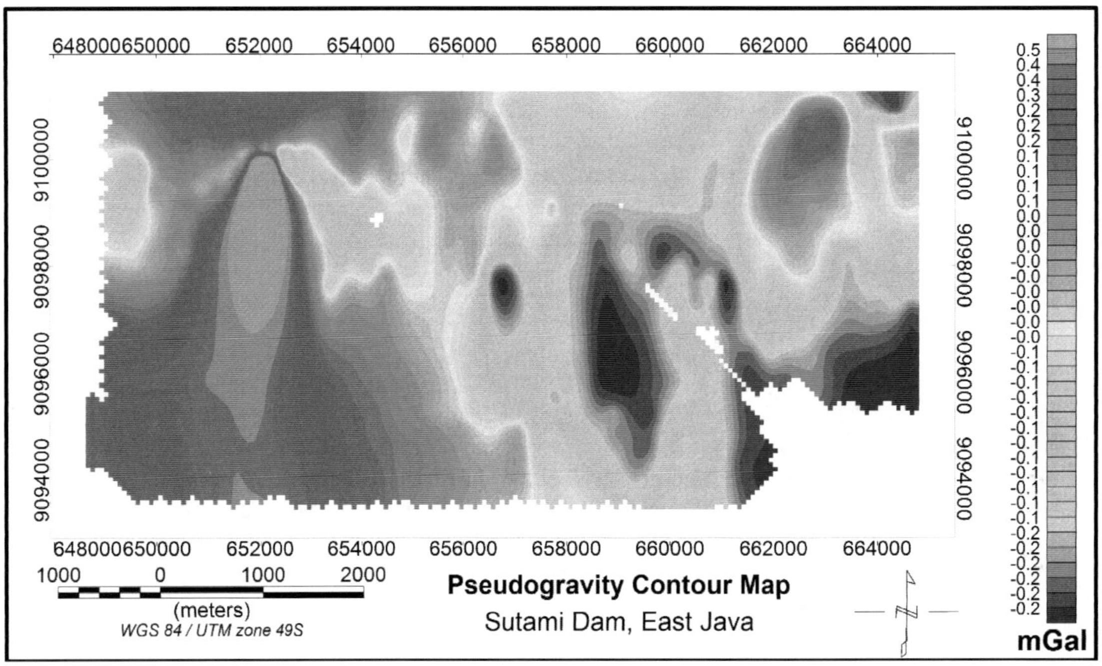

Figure 3. Pseudogravity Contour Map.

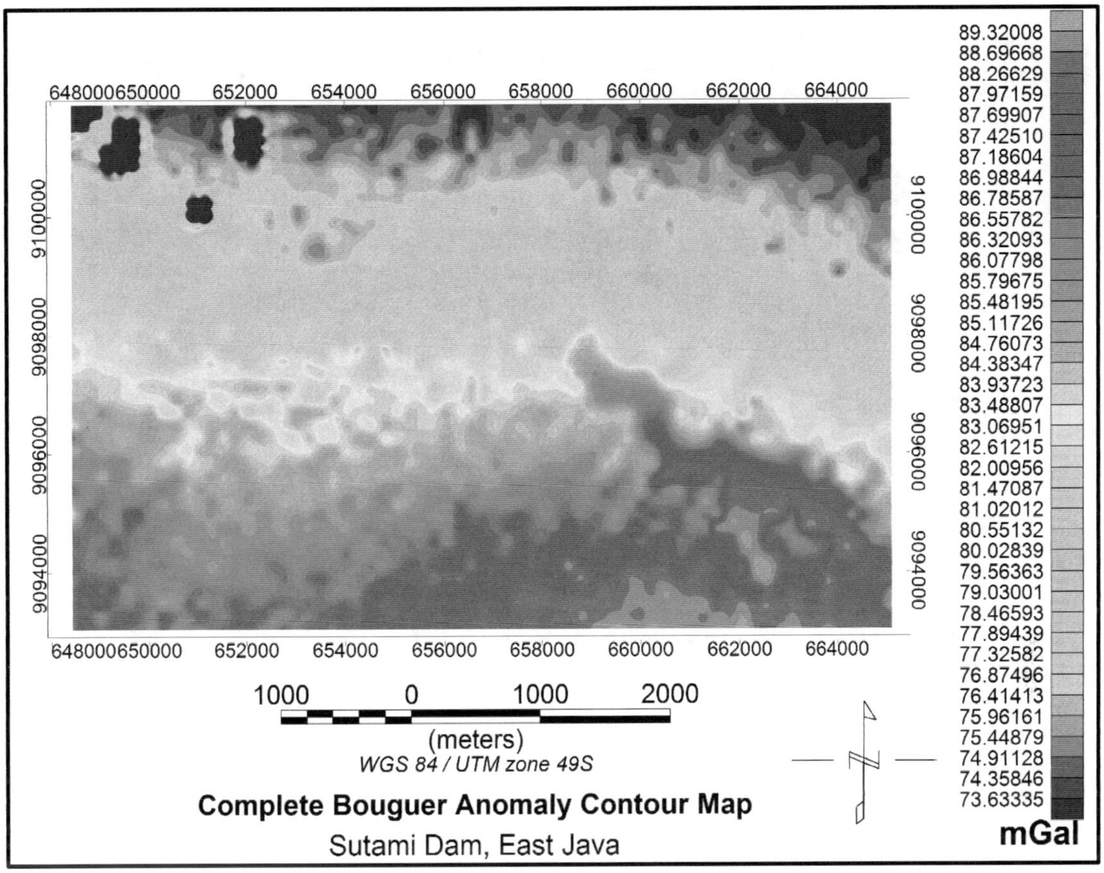

Figure 4. Complete Bouguer Anomaly Contour Map.

The residual anomaly map of the pseudogravity transformation (Figure 5) and the residual anomaly map from the satellite imagery gravity data (Figure 6) are in alignment. Based on Figure 5 and Figure 6, it can be seen that the value of the magnetic field in the south is higher than the value of the magnetic field in the north. The north side of Figure 5 and Figure 6 is thought to be Kawi Butak sediment and this sediment only covers the upper part, while the southern side is marine sediment which has a higher magnetic field value because on this side it was estimated that in ancient times it was an ancient volcano. Ancient volcanoes located in the south include Mount Golo, Mount Selorejo, Mount Selumban, Mount Tretes, and Mount Krowedul.

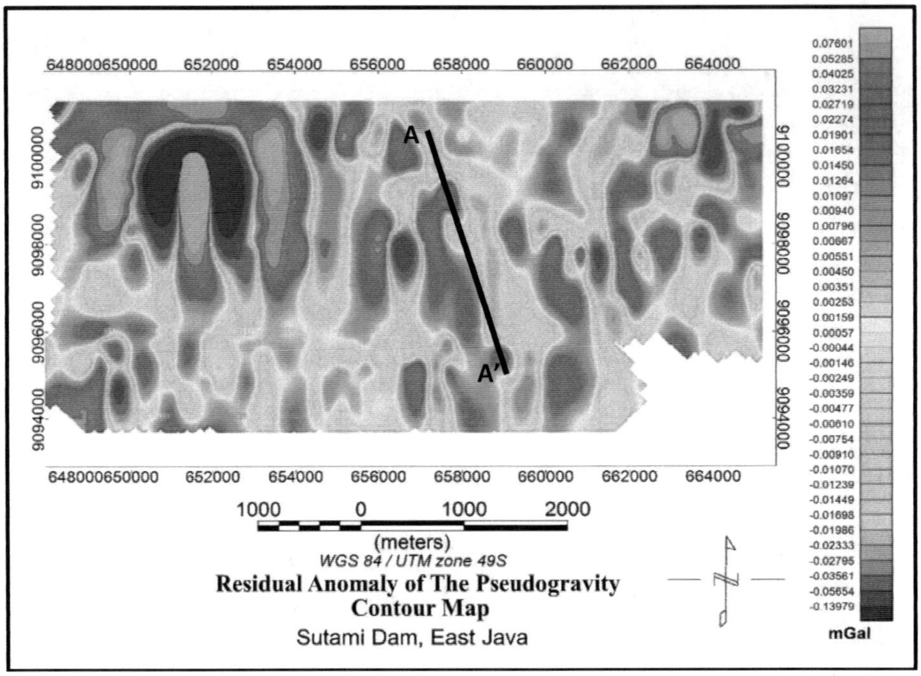

Figure 5. Pseudogravity Residual Contour Map and location of slicing profile 2D A-A'.

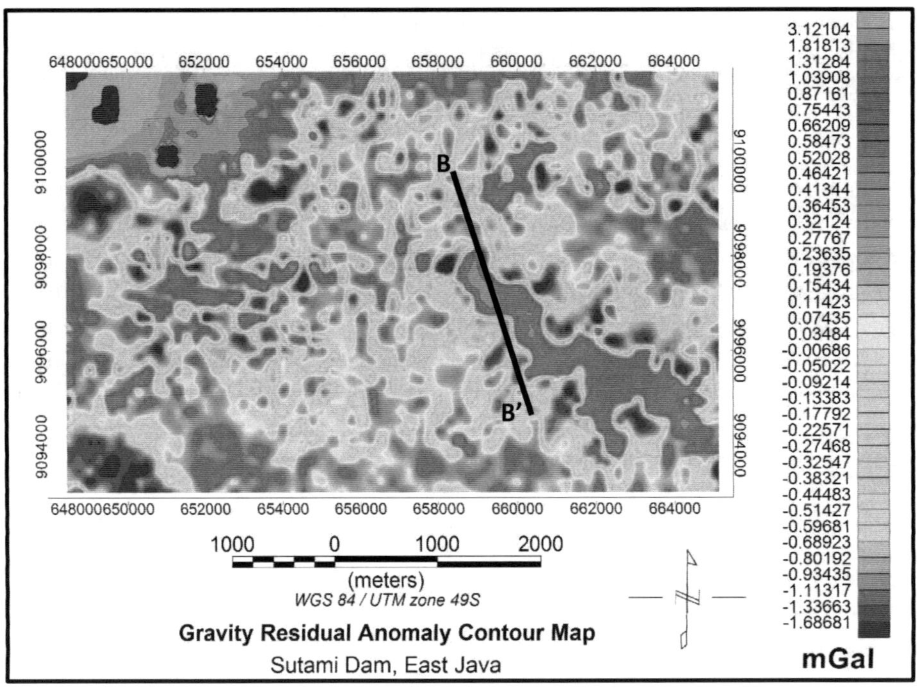

Figure 6. Residual Contour Map of gravity anomaly data from satellite imagery location of slicing profile 2D B-B'.

The results of subsurface modeling on the slice A-A' dan the slice B-B'can be seen in Figure 7 and Figure 8. There are four formations namely soil, Buthak Volcanic Products, Tuff Deposits and Campurdarat Formations. The contact layer of limestone sediments with volcanic sediments is located at the Tuff Deposits and the Campurdarat Formation. Tuff deposits formed in the late Pleistocene to early Holocene during the Quaternary period have rock types that have not yet been consolidated. Meanwhile, the ground mixed layers formed in the early Miocene of the Tertiary era have perfectly condensed rock types.

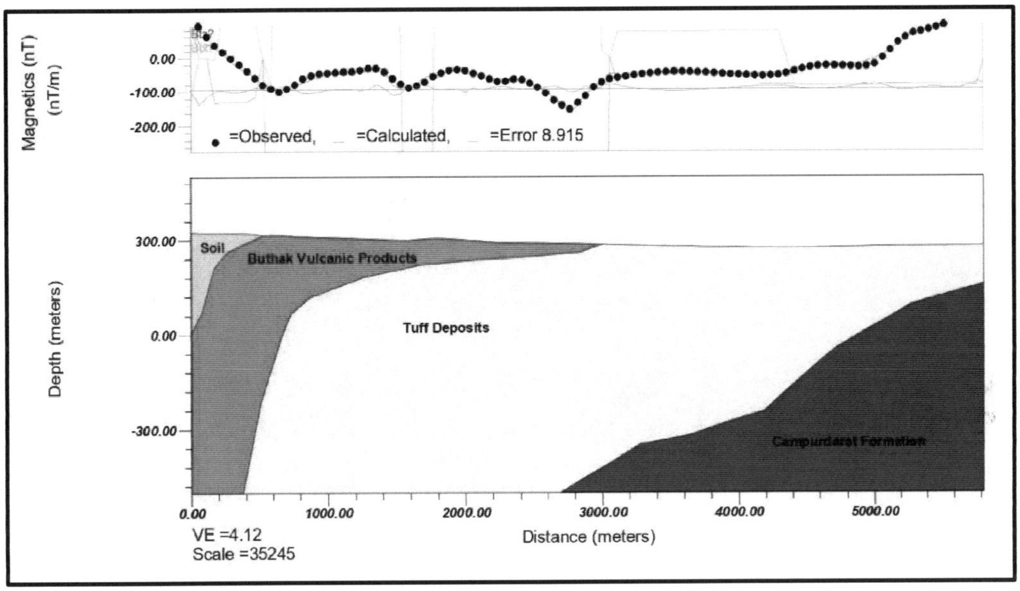

Figure 7. Profile 2D A-A' on Magnetic data of Pseudogravity Transformation.

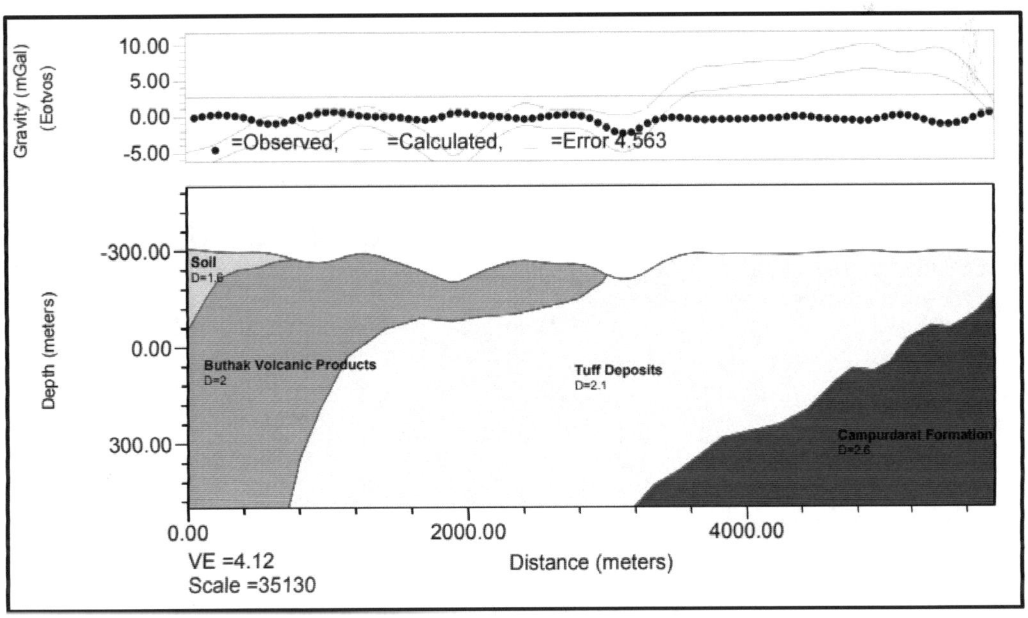

Figure 8. Profile 2D slicing B-B' on Gravity anomaly data from satellite imagery.

4.2 Discussion

There is a difference in the value of the pseudogravity contour with the value of the Complex Bouguer Anomaly map. This is because in the transformation of the total magnetic anomaly into pseudogravity data, the value of M and is ρ considered constant.

There is a contour alignment between the residual map of the magnetic data of pseudogravity transformation and the residual map of the anomaly gravity data of satellite imagery, which is high in the south and low in the north. This is following the geological map of research area. Likewise, with the type of formation resulting from 2D slicing, both methods have the same type of formation, namely soil ($\Delta\rho$ = 1.6 g/cm^3), Butak Volcanic Product ($\Delta\rho$ = 2 g/cm^3), Tuff Deposit ($\Delta\rho$ = 2.1 g/cm^3) and Campurdarat Formation ($\Delta\rho$ = 2.6 g/cm^3).

5. Conclusion

Based on the results of the research that has been done, it can be concluded that Profile 2D slicing A-A' and profile 2D slicing B-B' show that there are similarities in subsurface formations between the magnetic method of pseudogravity transformation and satellite imagery gravity data. There are 4 types of formations in the research area, both by the magnetic method and the gravity method of satellite imagery, namely soil ($\Delta\rho$ = 1.6 g/cm^3), Butak Volcanic Product ($\Delta\rho$ = 2 g/cm^3), Tuff Deposit ($\Delta\rho$ = 2.1 g/cm^3) and Campurdarat Formation ($\Delta\rho$ = 2.6 g/cm^3).

References

[1] Purwana YM, D.H.I RH, Setiawan B and Aulawi N 2019 Seismic hazard analysis for Sutami Dam using probabilistic method *MATEC Web of Conferences* **276** p. 1–9
[2] Sunaryo and Susilo A 2015 Vulnerability of Karangkates Dams Area by Means of Zero Crossing Analysis of Data Magnetic *AIP Conference Proceedings* **1658** p. 1–8
[3] Sunaryo and Susilo A 2017 Seepage Zone Identification at Sutami Dam by Means of Geoelectrical Resistivity Data *IOP science* **012011(75)** p.1–6
[4] Sunaryo and Susilo A 2015 Vulnerability of Karangkates Dams Area by Means of Density Contrast Parameter to Anticipate Energy Sustainability *Basic Science* **5** p. 1689–99
[5] Fitriah F, Sunaryo and Susilo A 2015 Identification of Cracked Zone in Sutami dam Using Geoelectrical Method *Indones J Appl Phys* **5(02) p**:65–72
[6] Sutasoma M, Susilo A, Sunaryo, Sarjiyana, Cahyo RHD and Suryo EA 2021 Identification of Rock Layer Contacts in the Surrounding of the Sutami Dam Using Geomagnetic Methods *Int J GEOMATE.* **21(84)** p.188–93
[7] http://murray-lab.caltech.edu/GGMplus/index.html downloaded in april 29th 2019 at 8.35 PM
[8] Sjarifudin and Hamidi S 1992 *Peta Geologi Lembar Blitar, Jawa*. Pusat Penelitian dan Pengembangan Geologi: Bandung
[9] Subarsyah S and Priohandono YA 2016 Metoda Pseudo-Gravity Dalam Analisis Kelurusan Dan Patahan Di Sekitar Tinggian Asahan, Perairan Selat Malaka. *J Geol Kelaut* **7(2)** p.65–71
[10] Blakely RJ. 1995 Potential *Theory in Gravity and Magnetic* Cambridge University Press: New York p. 313–56

Development of physics module based high order thinking skill (HOTS) to improve student's critical thinking

Y Marnah[1], Suharno[1] and Sukarmin[1]

[1]Master of Physics Education, FKIP, Sebelas Maret University, Jl. Ir. Sutami NO. 36A, Kentingan, Surakarta, Central Java, 57126, Indonesia

E-Mail: ayumarnah@gmail.com, Suharno_71@staff.uns.ac.id, sukarmin67@staff.uns.ac.id

Abstract. Physics learning activities implemented in high school have not directed students to think critically and still tend to be teacher-centered. This results in students not being able to develop critical thinking skills to solve problems that involve reasoning in learning activities. The purpose of this study was to develop a product in the form of a physics module based on High Order Thinking Skills to improve critical thinking skills in student learning outcomes. The research and development method used is Research And Development (R&D). The research and development model uses a 4D Thiagarajan model with the research stages of defining, designing, developing, and distributing. The subjects in this study were students of class XI SMA. The technique of collecting research data is using a questionnaire and test questions Pre-Test-Post-Test. The results showed that the development of physics modules based on High Order Thinking Skills which was developed in class XI students increased by 71%, so that the development of physics modules had a significant effect on students' critical thinking skills in the learning process on student learning outcomes.

1. Introduction

Education is an activity that is carried out consciously or systematically in the form of creating and influencing students to adapt to their environment so that students are expected to function and be useful in community life. the government seeks to improve the quality of education through the development of the education system, including through the 2013 curriculum and the law on the education system [1]. The 2013 curriculum in principle is one of the curricula in education that pioneers the development of 21st century skills. The process of providing education itself can be realized through conducive learning activities so that students can increase their confidence and character which are carried out in a directed and planned manner [2].

The curriculum is something that can arouse students in observing, asking questions, reasoning, increasing creativity and communication skills for learning activities. One of the subjects in the 2013 curriculum taught in high schools (SMA) is physics. The 2013 curriculum can awaken students in conducting observation activities, asking questions, reasoning, increasing creativity and communication skills for learning activities. One of the subjects in the 2013 curriculum that is taught in high schools (SMA) is physics [3].

Success in a learning activity can be seen from the activities carried out by students in carrying out their reasoning process activities. In this case, students are not only required to just listen or take notes, but students can improve their higher-order thinking skills, one of which is critical thinking skills. The learning process that occurs in schools today is very diverse, ranging from models to learning methods, which is an effort to make students easy to understand and learn the material taught by the teacher [4].

The module is one of the learning tools that help students succeed in their learning [5]. Modules are teaching materials that have the advantage of being able to stand alone, so that they can increase student learning motivation because both teachers and students can measure abilities that have not been successful so that learning becomes more focused, students can also easily achieve results with their abilities [6]. High Order Thinking Skill quoted from Bloom's taxonomy is one of students' critical thinking abilities in responding to cognitive abilities, from the arrangement of levels in thinking (cognitive), from the initial level of low thinking to higher levels in the cognitive domain [7]. Higher order thinking is where the process involves mental operations, namely classification, induction, deduction and reasoning [8]. High Order Thinking Skill is a thinking process that involves mental activity to explore complex and reflective experiences that are carried out consciously to get the goal, which is to acquire knowledge which includes the levels of analytical, synthetic, and evaluative thinking [9].

Based on the results of the descriptions of the experts' statements above, it can be concluded that higher order thinking is a process where critical and creative thinking skills can solve problems and are able to increase knowledge, skills in thinking, analyzing, evaluating, achieving goals, and gaining knowledge.

The module is one of the learning tools that help students succeed in their learning. The module is a teaching material that has the advantages of being able to stand alone, so that it can increase student learning motivation because both teachers and students can measure abilities that have not been successful so that learning becomes more focused, students can also easily achieve results with their abilities [6]. High-order thinking (High Order Thinking Skill) quoted from Bloom's taxonomy is an arrangement of levels in thinking (cognitive), from the initial level of low thinking to a higher level in the cognitive domain. Higher order thinking is where the process involves mental operations. Namely classification, induction, deduction and reasoning [10]. High Order Thinking Skill is a thinking process that involves mental activity to explore complex and reflective experiences that are carried out consciously to get the goal, namely to acquire knowledge which includes the levels of analytical, synthetic and evaluative thinking [11].

Based on the results of the descriptions of the statements of the experts above, it can be concluded that higher order thinking is a process where critical and creative thinking skills can solve problems and are able to increase knowledge, skills in thinking, analyzing, evaluating, achieving goals, and gaining knowledge.

The results of initial observations indicate that critical thinking skills on interest in learning physics, especially wave material, are still relatively low, thus affecting the reasoning process and learning outcomes. This is because the learning process carried out by the teacher is still very monotonous, causing boredom and making learning physics less interesting for students. The learning process should be directed at efforts to teach students to be able to overcome every problem and obstacle in a rapidly changing life, through every competency that must be possessed[12]. The learning model applied by the teacher in learning is still less varied, thus making the interaction between the teacher and students passive in the learning process.

On the other hand, there are several obstacles in the teacher's efforts to create interactive conditions in learning, namely the lack of facilities and infrastructure that support learning activities. So that the average value obtained in physics subjects is still relatively low. Based on the data found, from the results of the odd semester daily test scores, it was found 23 students who scored below the KKM, while the remaining 12 students got an average score above the KKM. This shows that only 35% of students experience complete learning outcomes. Based on these problems, the researchers tried to develop a module that can help students develop the process of critical thinking skills to improve their learning outcomes, namely by developing a physics module based on higher order thinking skills to improve students' critical thinking skills.

2. Methods

This research is a Research And Development (R&D) research. [13] Stating that research and development is a step in developing a new product or perfecting an existing product and can be accounted for. research methods used to produce module products, and test the effectiveness of module products in the field of education. The research carried out is the development of teaching materials in the form of physics learning modules for SMA/MA class IX on wave material. The learning device system development model used is the Thiagarajan model. Thiagarajan's model consists of 4 stages of research which include the stages of research, namely definition, design, development, and dissemination. The subjects in this study were 20 students of class XI.

The data collection instrument in this study was the test and questionnaire method. [14] states that the test method is a way of collecting data that confronts students with a number of questions. Data collection techniques were carried out through tests consisting of Pre-Test and Post-Test using paired sample t-test and structured interviews. The test used to measure critical thinking skills on student learning outcomes uses Pre-Experimental Design with the type of One-Group Pretest-Posttest Design.

3. Result and Discussion

Based on the results of the analysis carried out, it shows that the ability to think critically on student learning outcomes before and after the physics module on the mechanical wave material in class XI IPA using the experimental class and control class is shown in the Table 1.

Table 1. Data on cognitive learning outcomes of experimental class students.

Class XI	Test type	Total students	Mean	Std dev	Min	Max
Experiment	*Pre-test*	20	65.00	9.70	50.00	80.00
Control	*Post-test*	20	90.00	5.30	80.00	100.00

Based on the table above, it shows that the average pre-test & post-test scores in both the experimental and control classes were found to be different between student learning outcomes. [12] stated that the application of Higher Order Thinking Skills-based modules in learning was able to improve students' critical thinking skills. In the experimental class before applying learning using modules and after applying learning using modules there was a very significant difference.

The results of the analysis using the paired sample $T -_{\text{Test}}$ showed that the significance value of $(2 -_{\text{Tailed}})$ was $0{,}000 < 0{,}05$, which was smaller than the significance value, where H_0 was rejected. So it can be concluded that the results of students' critical thinking skills on learning outcomes before using the Higher Order Thinking Skills-based module have different results from the results of students' critical thinking abilities on student learning outcomes after using the physics module in their learning activities.

States that learning is indicated by a change in behavior as a result of experience, which means that the best learning is by experiencing, because by experiencing students it will be easier to remember an event [12] Learning uses module-based. Based on the results of structured interviews using questionnaires, it shows that students' critical thinking skills towards the reasoning process are in the high category. The results of the calculation of these values are shown in the following table:

Table 2. Results of critical thinking ability.

Observer	Learning Activities		Average (%)	Category
	I	63.93	7.10	High
XI IPA I	II	64.30	71.00	High

Based on the table above, it shows that the results of the analysis of students' critical thinking skills in class XI science in the pre-test obtained an average value of 63.93 with an average percentage of 7.10 % and is in the high category. While the post-test obtained an average score of 64.30 students with an average percentage of 71.00 % and are in the same category, namely high. It can be concluded that the physics module developed is very effective in improving students' critical thinking skills in the physics learning process, especially on learning outcomes.

4. Conclusion

In general, it can be concluded that students' critical thinking skills show that before and after the use of Higher Order Thinking Skills-based physics modules have different results on student learning outcomes in their learning activities. This also shows that the physics module developed by the researcher as a support in the learning process, especially on the material of mechanical waves in class XI high school students is very effectively used to improve students' critical thinking skills in the physics learning process, especially on their learning outcomes.

References

[1] F. A. Perdana, "Keterampilan Proses Sains Untuk Meningkatkan Kemampuan Berpikir Kritis Dan Motivasi Belajar Siswa Sma / Ma Kelas X Pada Materi Dinamika Gerak," vol. 6, no. 3, hal. 61–76, 2017

[2] O. Nurwahida, "Pengaruh pendekatan higher order thinking skills (hots) terhadap hasil belajar ilmu pengetahuan sosial murid kelas IV SD inpres bontomanai kecamatan tamalate kota makassar," 2018

[3] G. H. Desi Fitriani, Yusuf Suryana, *Indonesian Journal of Primary Education Pengembangan Instrumen Tes Higher-Order Thinking Skill pada Pembelajaran Tematik Berbasis Outdoor Learning*, vol. 2, no. 1. 2018

[4] I. Kurniawan, *Penggunaan Model Pembelajaran Discovery Learning Berorientasi Hots (Higher Order Thinking Skill) Sebagai Upaya Meningkatkan Hasil Belajar Siswa*, vol. 5, no. 1. 2020

[5] Nurkholis, *Pengukur Kecepatan Gelombang Bunyi di Udara*, vol. 53, no. 9. 2014

[6] D. Oktaviana, S. Hartini, dan M. Misbah, *Pengembangan Modul Fisika Berintegrasi Kearifan Lokal Membuat Minyak Lala Untuk Melatih Karakter Sanggam*, vol. 5, no. 3. 2017

[7] E. Mariana, Sukarmin, dan Cari, "Untuk Meningkatkan Kemampuan Berpikir Kritis Siswa Sma / Ma Kelas X Pada Materi Listrik Dinamis," *J. Inkuiri*, vol. 6, no. 2, hal. 137–146, 2017

[8] S. Zubaidah, "Berfikir Kritis : Kemampuan Berpikir Tingkat Tinggi Yang dapat Dikembangkan Melalui Pembelajaran Sains," *Semin. Nas. Sains 2010 dengan Tema "Optimalisasi Sains untuk Memberdayakan Manusia,"* no. January 2010, hal. 11, 2010

[9] E. Rofiah, N. S. Aminah, dan W. Sunarno, *Pengembangan Modul Pembelajaran IPA Berbasis High Order Thinking Skill (HOTS) Untuk Meningkatkan Kemampuan Berpikir Kritis Siswa Kelas VIII SMP/MTs*, vol. 7, no. 2. 2018

[10] K. V. Febrianti, F. Bakri, dan H. Nasbey, *Pengembangan Modul Digital Fisika Berbasis Discovery Learning Pada Pokok Bahasan Kinematika Gerak Lurus*, vol. 2, no. 2. 2017

[11] K. Y. K. Yusuf, "Penerapan Model Discovery Learning Untuk Meningkatkan Keterampilan HOTS Dan Prestasi Belajar IPA Siswa Kelas VII SMP Negeri 1 Garung Semester 1 Tahun

Pelajaran 2017/2018," *SPEKTRA J. Kaji. Pendidik. Sains*, vol. 4, no. 1, hal. 41–48, 2018

[12] F. Yanti, S. Sukarmin, dan S. Suparmi, "Pengembangan Modul Pembelajaran Fisika Sma/Ma Berbasis Masalah Untuk Meningkatkan Keterampilan Berpikir Kritis Siswa," *Inkuiri*, vol. 4, no. 3, hal. 96–103, 2016

[13] M. Auliya dan K. Kosim, "Pengembangan Modul Fisika Materi Optik Dengan Pendekatan Saintifik Berbasis Fenomena Alam Untuk Meningkatkan Efektivitas Belajar Siswa Sma," *J. Pijar Mipa*, vol. 12, no. 2, hal. 71–80, 2017, doi: 10.29303/jpm.v12i2.344

[14] A. K. Pustaka dan M. Pedesaan, "BAB II Kajian Teori A. Kajian Pustaka 1.," hal. 18–45, 2002

Non-contact technique for CO_2 gas monitoring using thermal imaging camera

U Salamah[1,2], S P Sakti[2], H Soetedjo[1] and A Naba[2]

[1]Departement of Physics, Universitas Ahmad Dahlan, Jl. Ring Road Selatan, Tamanan, Banguntapan, Bantul, Yogyakarta 55191, Indonesia
[2]Departement of Physics, Universitas Brawijaya, Jl. Veteran, Malang, Jawa Timur 65145, Indonesia

E-mail: umi.salamah@fisika.uad.ac.id

Abstract. CO_2 gas is a poisonous gas that is invisible and odorless. Now days, the CO_2 gas detection system uses the direct contact (in situ) method with the gas to identify its existance and quanities. The further determination is being carried out in the laboratories. This method is ineffective and inefficient because of some unpractical procedures as in the real situation collecting the gas could be dangerous. Therefore the development of a non-contact detection system will be important and challenging wor. One method of developing a non-contact gas detection system is using the Optical Gas Imaging (OGI) method. In this paper, a non-contact CO_2 gas measurement system has been designed based on infrared absorption using an infrared thermal camera. The experiment is carried out on a laboratory scale with three main components of an infrared light source, a gas source and a detector. From the results, unique pattern of gas of CO_2 coudl be distinguised from the open air and also mixed gas through the equalized histogram.

1. Introduction

CO_2 is a poisonous gas that is colorless and odorless, it can interfere the metabolism of calcium phosphorus, resulting in an increase in calcium deposits in soft tissues. This gas also causes the weakening of the contractile force on the heart which is at high risk of death. The limit on the average adult exposure to this gas is 5,000 ppm within 8 hours per day [1]. The colorless and odorless properties of CO_2 make this gas more dangerous than other toxic gases. CO_2 gas comes from the respiration of living things, emissions of fossil fuels, geothermal and also volcanoes. The presence of CO_2 can be used as a parameter and approach in studying the fields of health, environmental and natural phenomena. In the medical field CO_2 is used as an approach in measuring the spread of Covid-19 [2]–[4], in the environmental field CO_2 is measured to determine air pollution levels, in the nature, CO_2 is one of the parameters indicate the volcanic activity. Therefore, it is very important to develop a CO_2 detection system thatcould show the early warning phenomenon of some disasters. Currently CO_2 gas measurement systems involve some components of infrared spectrophotometers and electrochemical [5]–[8].

The measurement using those components introduce a direct contact with the sample/CO_2 gas that will be analyzed further method of using a spectrometer, chromatography, and volumetric [9]. These methods have limitations including time consumption, risk of gas loss, and controllers the risk of being exposed to invisible and potentially hazardous chemicals. Meanwhile, measurements based on the contact system have a high risk of the accuracy of the measurement results because the sampel of CO_2 gas are reactive and corrosive. Therefore, the development of a non-contact CO_2 detection system is very important to overcome these problem and improve the limitations. Some experiments have been done to monitor CO_2 gas by using IR imaging camera [10]. They used two kinds of camera with spectral of MW band (3 - 5 μm) and LW (8 - 14 μm), respectively for detecting CO_2. The results show that image of CO_2 gas signature could be obtained from that camera. Moreover, by using the bi-spectral imaging camera (3 - 4 μm and 4 - 5 μm channels), it can be used to visualize the CO_2 content in the exhalation air [11]. In this paper, we report the investigation of CO_2 gas flown in the chamber using commercial EasIR-4 in long wave IR based on non-contact technique. The use of this technique could be encouraging as for wide applications it can avoid any hazardous effects and more convenient.

2. Methods

CO_2 gas can absorb any infrared radiation when they make the interaction. Infrared radiation energy causes vibrations in molecules particularly for CO_2 gas thorugh four vibrational modes, three of which are active against infrared radiation energy (asymmetric stretching vibrations and two bending vibrations). The infrared absorption band is very distinctive and specific for each type of chemical bond or functional group so that it can identify compounds. In the IR spectrum, the peak absorption of CO_2 gas is established to occur in the wavelengths spectrum from 2.5 - 2.6 μm, 4.3 - 4.5 μm, also for 9.5 μm, 10.5 μm, and 14 μm, respectively with some weak peaks of absorption as shown in Figure 1. Use of the thermal imaging camera that covers the spectrum from 8 - 14 μm (as this spec is relatively less expensive in the in the market) has been done for non contact gas detection such of CO_2. This unique peak absorption will be an advantage for CO_2 gas determination from other gases in the atmosphere.

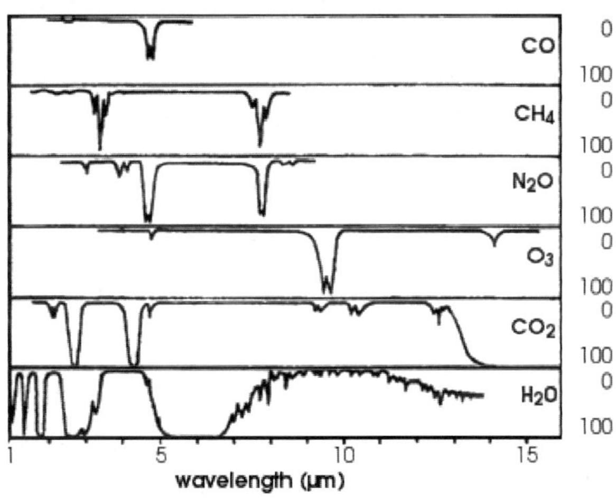

Figure 1. The percentage of radiation that gases found in the earth's atmosphere absorption [12].

In this experiment, by using a vacuum chamber, CO_2 gas (99.7 %), CO_2 mixed environment air were flown for various times into the chamber to investigate the characteristic based on IR image taken from the camera (The camera of EasIR-4). The gas flow was set for a rate of 5 L/min. Single gas of CO_2 and ambient air were tested as reference for each gas prior to the mixing of both gases. The experimental set up for CO_2 gas detection using thermal imaging camera is carried out by using a vacuum tube as shown in Figure 2 (a). During the experiment, the camera was positioned directly at the right side after to the chamber filled with gas as shown in Figure 2 (b). The distance between the outer chamber and the

Journal of Physics: Conference Series **2165** (2022) 012019 doi:10.1088/1742-6596/2165/1/012019

camera is 2 cm. Meanwhile, the bulb lamp (Tungsten) of 3 Watt radiates IR spectrum is installed in the chamber at the left side as shown in Figure 3. With this set up the absorption phenomenon by the CO_2 gas will occur and the change of image will be recorder by the camera. Experiment was carried out by varying the flowing time of CO_2 gas into the vacuum chamber, at room temperature (27 °C).

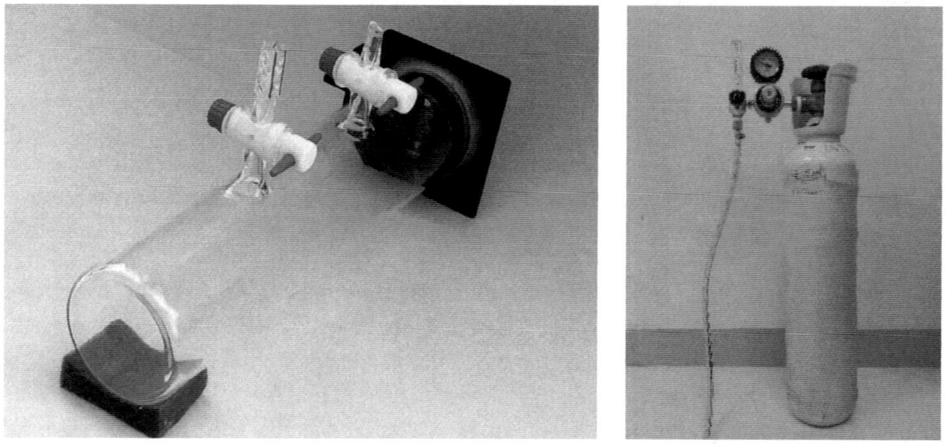

(a) (b)

Figure 2. Vacuum chamber (a) and pure CO_2 gas container used for the experiment (b).

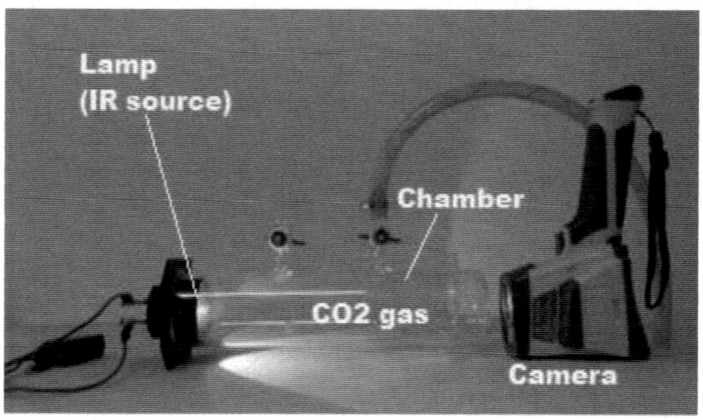

Figure 3. Experiment set up of camera used for CO_2 gas detection during the experiment.

Image taken by the camera will be converted into the grayscale histogram and histogram of equalized image for further investigation.The gray histogram is carried out to reduce noise of the image obtained. Meanwhile, the histogram equalization is done to improve the quality of the digital images [13]–[18].

3. Result and Discussion

From the experiment, by varying the flow time of CO_2 gas into the chamber, the image could be obtained by the Thermal Imaging camera. The camera covers the range of 8 - 14 μm that make possible the signature of CO_2 gas could be obtained. This results show that by using Thermal Imaging camera covers the range of 8 - 14 μm, the signature of CO_2 gas could be obtained. The absorption occurs dominantly in the wavelength from 8 - 13 μm as could be referred to Figure 1. Although in that spectrum H_2O peak of absorption is exist, that is negligible because in the chamber, it is believed no significant amount of water molecules. By obtaining some results from the camera, that is very interesting as that camera could be used for non-contact detection of CO_2 gas. The images obtained from the camera for the following condition in the chamber with CO_2 gas flown for (a) 0 min, (b) 1 min, (c) 1.5 min, (d) 2.0 min, and (e) 2.5 min, respectively, as given in Table 1. In that table, the images were then converted into the gray scale histogram and histogram of equalized image.

Table 1. Various images for CO_2 gas filled in the chamber obtained from the experiments and their conversion to histograms.

Image	Grayscale Histogram	Histogram of Equalized Image
 Vacuum (no CO_2 or Air)	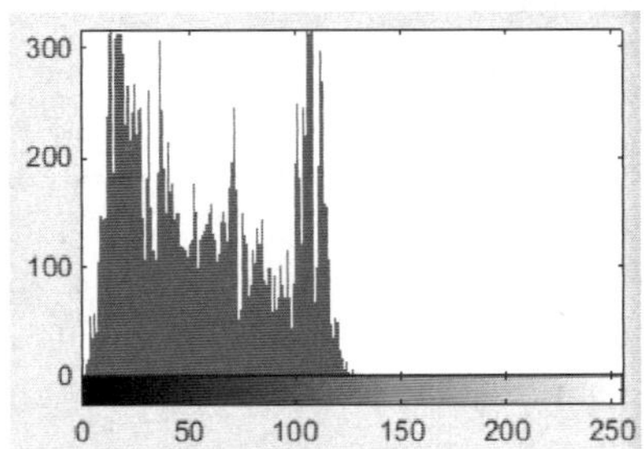	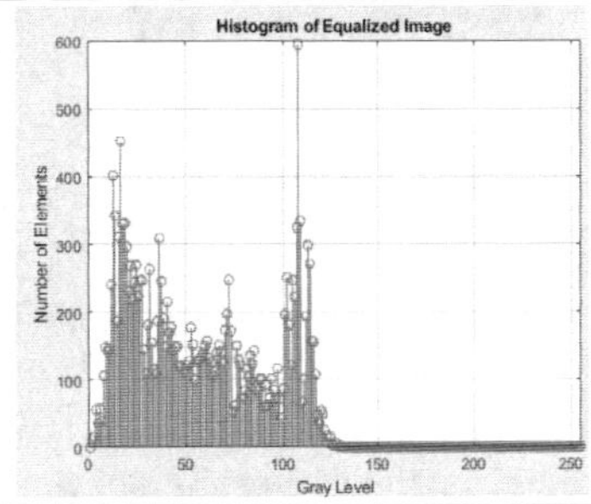
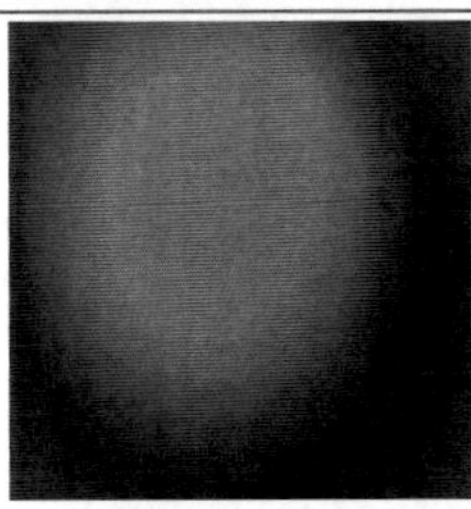 CO_2 (1 min)	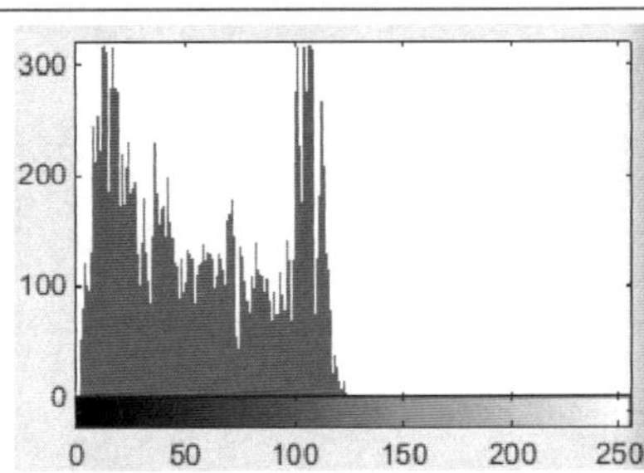	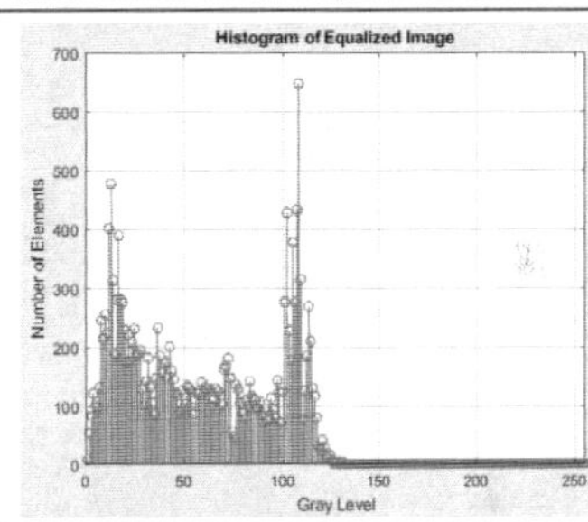
 CO_2 (1.5 min)	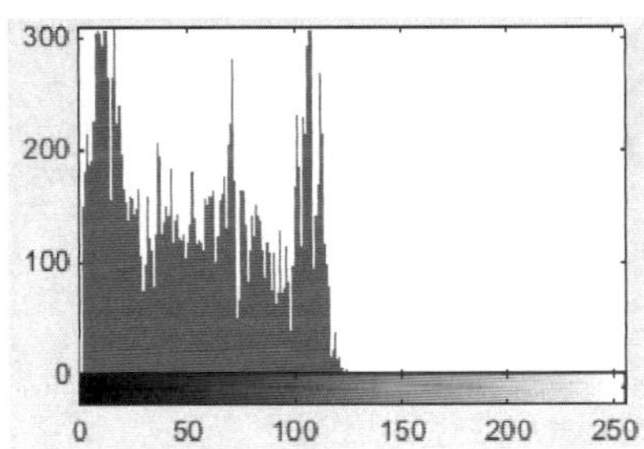	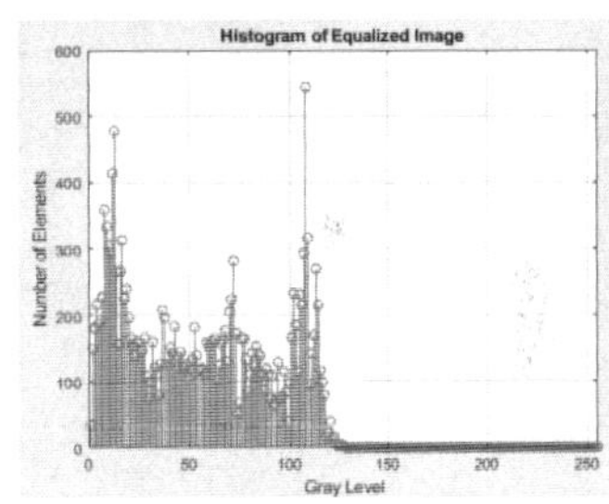
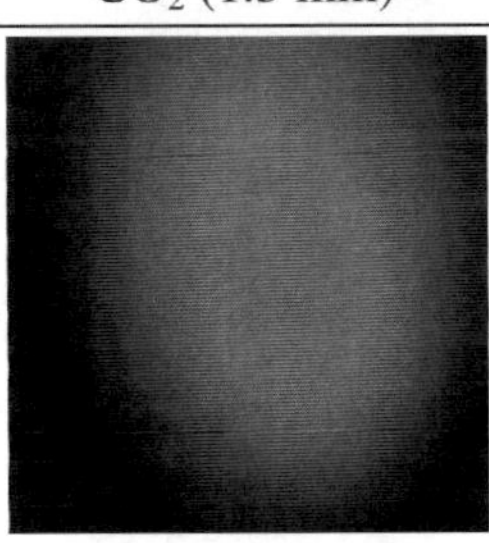 CO_2 (2 min)	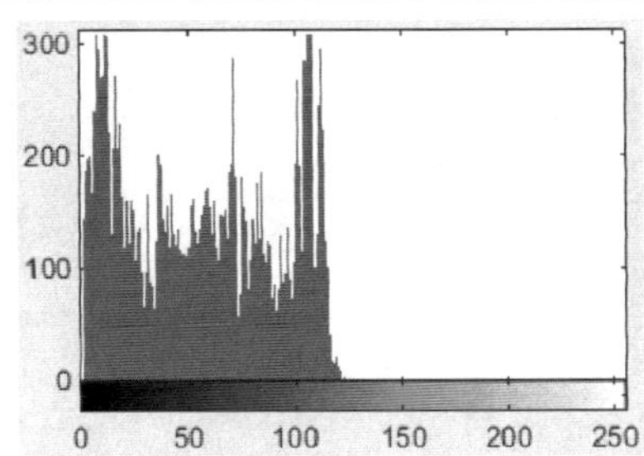	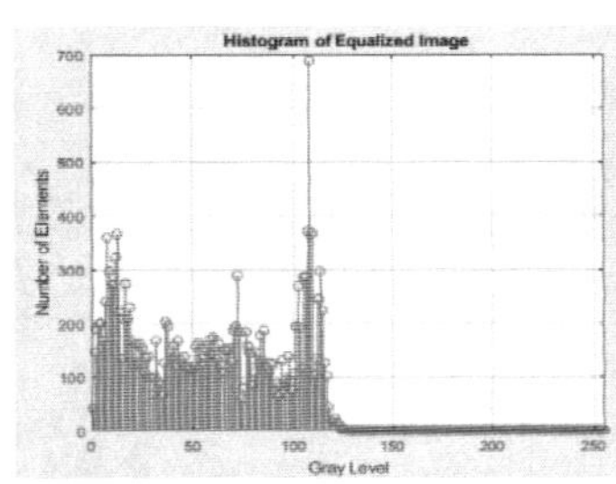

 CO_2 (2.5 min)	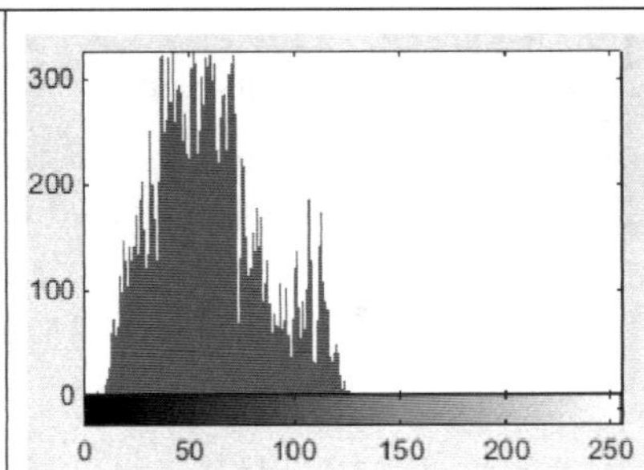	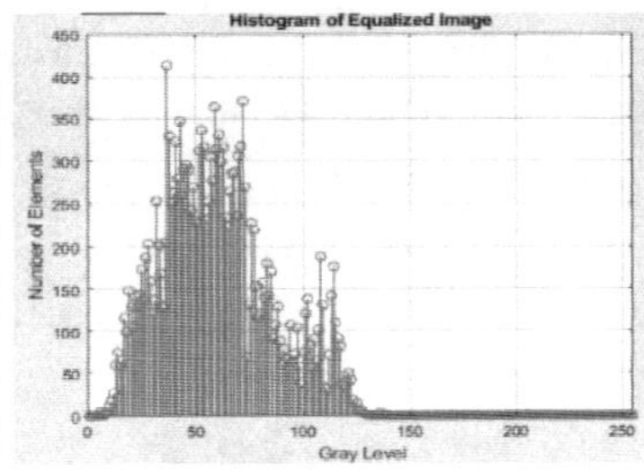

The images was then converted into the greyscale histogram and histogram of equalized image as given in Table 1. From that figures, the resulting histogram is a dark image where the histogram graph donates on the left side. The peaks of the histogram, which show the prominent pixel intensity, on CO_2 filling 1, 1.5 and 2 minutes show almost the same pattern. The significant different found for the exposure time of 2.5 minutes. By using the direct human eyes observation, the CO_2 gas in the chamber is invisible, the interesting one is by using the commercial long wave EasIR-4 we could catch the images that identity the CO_2 gas at a small quantities, as shown in Figure 4.

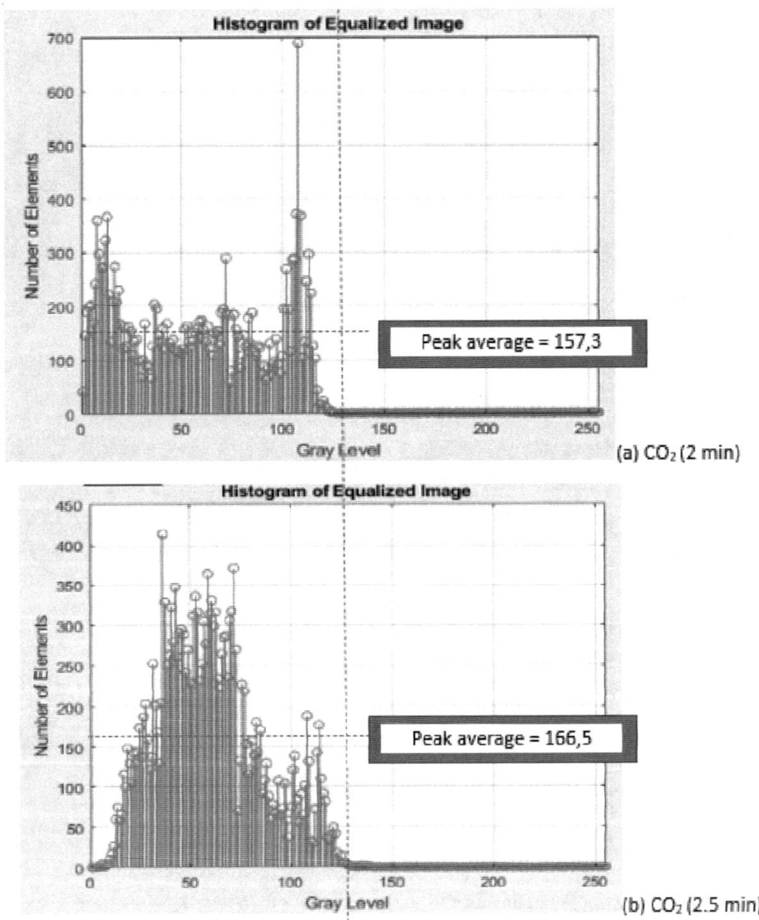

Figure 4. The histogram equalized changes to the filling of (a) CO_2 2 minutes and (b) CO_2 2.5 minutes.

Journal of Physics: Conference Series **2165** (2022) 012019 doi:10.1088/1742-6596/2165/1/012019

The pattern of CO_2 histograms for 2.5 minutes of filling also has a significant difference with the histogram pattern of the ambient air. Table 2 shows the differences in the histogram of CO_2, ambient air and a mixture of CO_2 with ambient air. The width of the CO_2 graph on the 2.5 minutes filling ranges from 0 to 110, while the ambient air is 0 to 50. In the CO_2 mixture with the ambient air, an interesting pattern is obtained where the width of the histogram graph returns in the range 0 to 110.

Table 2. Grayscale histogram and histogram of equalized image obtained for various exposure time of CO2 gas and ambient air for each image data.

<table>
<tr><th>Image obtained from the camera</th><th>Grayscale Histogram</th><th>Histogram of Equalized Image</th></tr>
<tr><td>
Vacuum (no CO_2 or air flown)</td><td>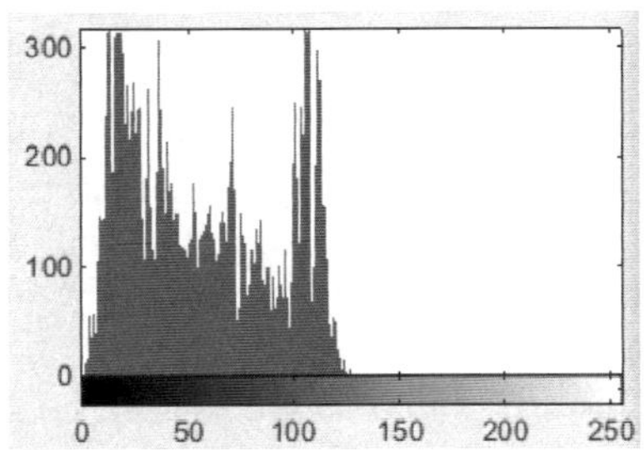
</td><td>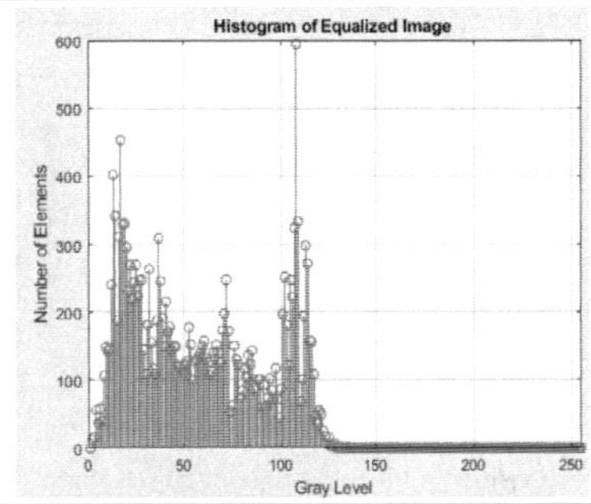
</td></tr>
<tr><td>
Ambient Air (2.5 min.)</td><td>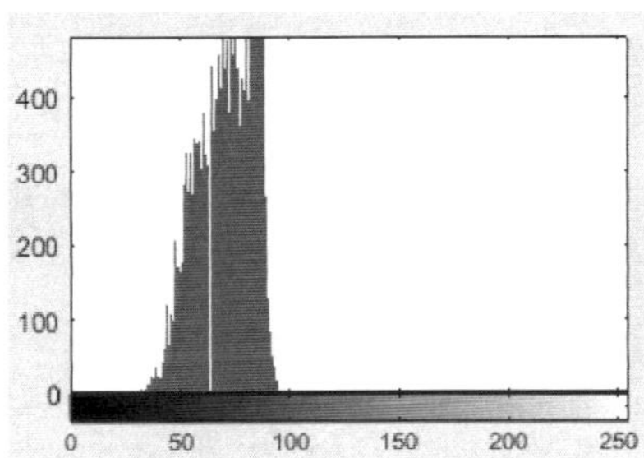
</td><td>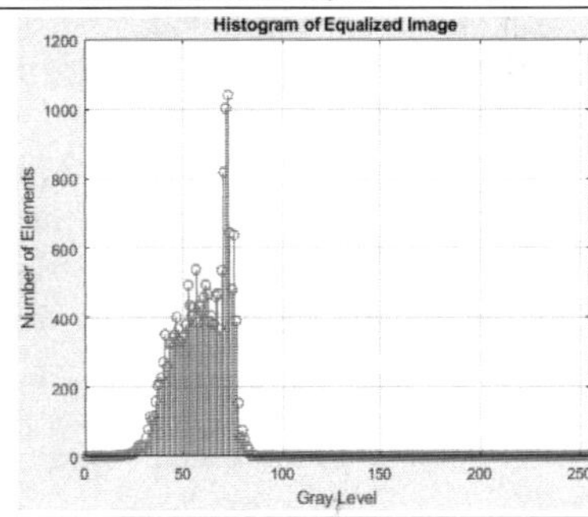
</td></tr>
<tr><td>
Pure CO_2 flown (2.5 min.)</td><td>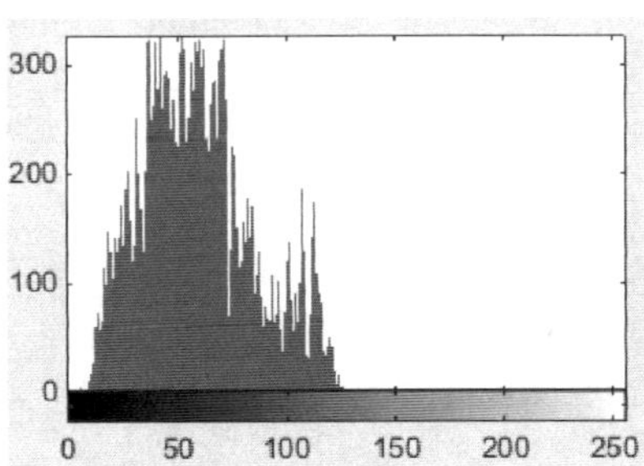
</td><td>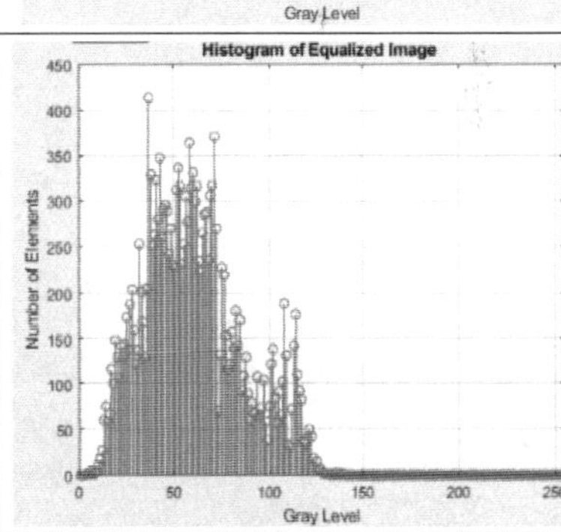
</td></tr>
<tr><td>
Pure CO_2 + Air flown (2.5 min.)</td><td>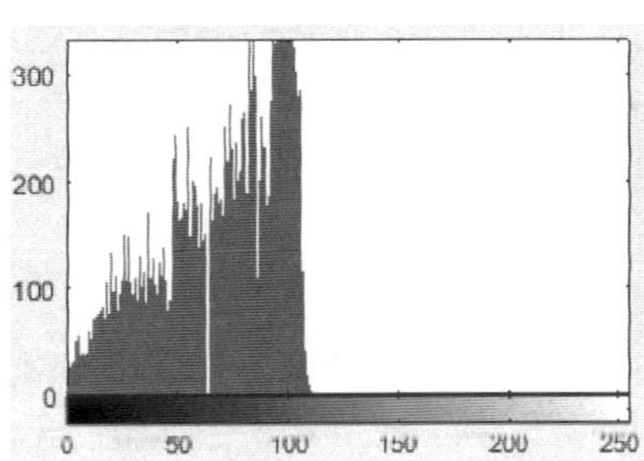
</td><td>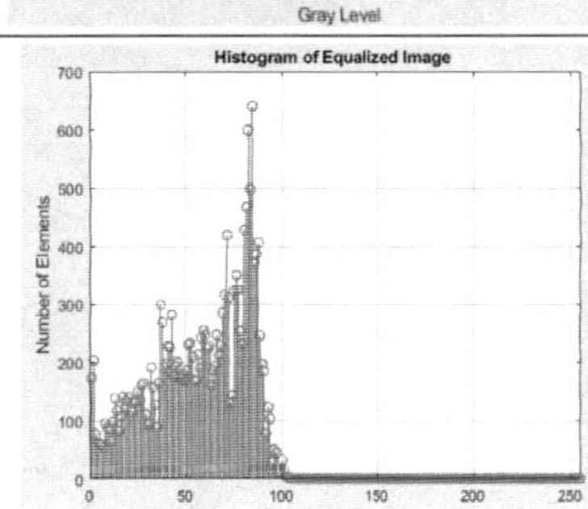
</td></tr>
</table>

Table 2 shows the comparison of optical imaging, grayscale histogram and histogram equalized from the chamber with three conditions, namely without gas (in a vacuum), with ambient air filled for 2.5 minutes, with CO_2 gas filled for 2.5 minutes and with a mixture of CO_2 and CO_2 filled. ambient air for 2.5 minutes. The gray level obtained under vacuum is 0 to 125, while those containing ambient air are 40 to 100, the gray level containing CO_2 is 10 to 125 and those containing a mixture of ambient air gas with CO_2 are 0 to 110. Differences in patterns of equalized histogram also looks significant where the shape of the equalized histogram pattern in vacuum optical imaging shows an even peak at every gray level, while the optical imaging pattern in the air has one prominent peak. Equalized histogram on CO_2 optical imaging resembles a normal distribution and the equalized histogram pattern on a mixture of CO_2 gas and ambient air has a prominent peak on the right edge. The unique pattern of each optical imaging proves that the non-contact detection system in this study can distinguish CO_2 gas from ambient air gas.

4. Conclusion

From the results and discussion above, it can be concluded that there is a linear relationship between the CO_2 levels that enter the vacuum tube and the intensity value and the frequency of appearance of the resulting pixels. The more CO_2 gas that filled the vacuum tube, the resulting histogram pattern formed a normal distribution with the peak of the gray level area tending to be in the middle. This pattern of changes proves that CO_2 gas can be detected non-contact using a thermal imaging camera. Thus, the result of this research can be used as a basis for non-contact CO_2 detection that can be developed into remote CO_2 detection systems.

References

[1] S. a Rice, "Health Effects of Acute and Prolonged CO_2 Exposure in Normal and Sensitive Populations *," Third Anu. Conf. Carbon Sequestration, pp. 5–8, 2003

[2] R. Sussmann and M. Rettinger, "Can we measure a COVID-19-related slowdown in atmospheric CO_2 growth? Sensitivity of total carbon column observations," *Remote Sens.*, vol. 12, no. 15, 2020, doi: 10.3390/RS12152387

[3] H. F. Dacre, L. M. Western, D. Say, S. O'Doherty, T. Arnold, C. Rennick and E. Hawkins5, " Detectability of COVID-19 global emissions reductions in local CO_2 concentration measurements ," *Environ. Res. Lett.*, vol. 16, no. 9, p. 094043, 2021, doi: 10.1088/1748-9326/ac1eda

[4] H. Sugawara, S. Ishidoya, Y. Terao, Y. Takane, Y. Kikegawa, and K. Nakajima, "Anthropogenic CO_2 Emissions Changes in an Urban Area of Tokyo, Japan, Due to the COVID-19 Pandemic: A Case Study During the State of Emergency in April–May 2020," *Geophys. Res. Lett.*, vol. 48, no. 15, pp. 1–10, 2021, doi: 10.1029/2021GL092600

[5] S. Gersen *et al.*, "Detection of H2S, SO_2 and NO_2 in CO_2 at pressures ranging from 1- 40 bar by using broadband absorption spectroscopy in the UV/VIS range," *Energy Procedia*, vol. 63, no. 2, pp. 2570–2582, 2014, doi: 10.1016/j.egypro.2014.11.279

[6] S. Inguaggiato, I. S. Diliberto, C. Federico, A. Paonita, and F. Vita, "Review of the evolution of geochemical monitoring, networks and methodologies applied to the volcanoes of the Aeolian Arc (Italy)," *Earth-Science Rev.*, vol. 176, no. March 2017, pp. 241–276, 2018, doi: 10.1016/j.earscirev.2017.09.006

[7] M. Shofar and S. Suryono, "Sistem Telemetri Pemantau Gas Karbon Dioksida (CO_2) Menggunakan Jaringan Wifi," *Youngster Phys. J.*, vol. 3, no. 3, pp. 243–248, 2014

[8] T. J. Roberts, C.F. Braban, C. Oppenheimer, R.S. Martin, R.A. Freshwater, D.H. Dawson, P.T. Griffiths, R.A. Cox, J.R. Saffell and R.L. Jones., "Electrochemical sensing of volcanic gases," *Chem. Geol.*, vol. 332–333, pp. 74–91, 2012, doi: 10.1016/j.chemgeo.2012.08.027

[9] J. Fei and I. Pavlidis, "Analysis of Breathing Air Flow Patterns in Thermal Imaging," *Annu. Int. Conf. IEEE Eng. Med. Biol. - Proc.*, pp. 946–952, 2006, doi: 10.1109/IEMBS.2006.260117

[10] M. Vollmer and K.-P. Möllmann, "3.2 - IR Imaging of CO_2: Basics, Experiments, and Potential Industrial Applications," pp. 59–64, 2020, doi: 10.5162/irs11/i3.2

[11] F. Rutz, R. Rehm, A. Wörl, J. Schmitz, M. Wauro, J. Niemasz, M. Masur, M. Walther, R. Scheibner and J. Ziegler, "Imaging detection of CO_2 using a bispectral type-II superlattice infrared camera," no. 1, pp. 1–7, 2012, doi: 10.21611/qirt.2012.233

[12] S. L. Valley, "CORPUSCULAR RADIATION: A Revision of Chapter 17, Handbook of Geophysics and Space Environments," 1968

[13] J. E. Johnson, J. A. Shaw, R. Lawrence, P. W. Nugent, L. M. Dobeck, and L. H. Spangler, "Long-wave infrared imaging of vegetation for detecting leaking CO_2 gas," *J. Appl. Remote Sens.*, vol. 6, no. 1, p. 063612, 2012, doi: 10.1117/1.jrs.6.063612

[14] P. Kandhway, A. K. Bhandari, and A. Singh, "A novel reformed histogram equalization based medical image contrast enhancement using krill herd optimization," *Biomed. Signal Process. Control*, vol. 56, p. 101677, 2020, doi: 10.1016/j.bspc.2019.101677

[15] B. S. Rao, "Dynamic Histogram Equalization for contrast enhancement for digital images," *Appl. Soft Comput. J.*, vol. 89, p. 106114, 2020, doi: 10.1016/j.asoc.2020.106114

[16] I. G. Torre, J. J. Martín-Sotoca, J. C. Losada, P. López, and A. M. Tarquis, "Scaling properties of binary and greyscale images in the context of X-ray soil tomography," *Geoderma*, vol. 365, no. June 2019, p. 114205, 2020, doi: 10.1016/j.geoderma.2020.114205

[17] Y. Li, Y. Zhang, A. Geng, L. Cao, and J. Chen, "Optics & Laser Technology Infrared image enhancement based on atmospheric scattering model and histogram equalization," vol. 83, pp. 99–107, 2016, doi: 10.1016/j.optlastec.2016.03.017

[18] M. Wan, G. Gu, X. Maldague, W. Qian, K. Ren, and Q. Chen, "Particle swarm optimization-based local entropy weighted histogram equalization for infrared image enhancement," *Infrared Phys. Technol.*, 2018, doi: 10.1016/j.infrared.2018.04.003

Effect of additional magnetic material Ni on structure and magnetic properties $Eu_{2-x}Ce_xCu_{0.99}Ni_{0.01}O_{4+\alpha-\delta}$

L P Rochman[1], H D Anggia[1], R Pratama[1], T Maulana[1], S Winarsih[1], Y Maryati[1], M A Syakuur[1], U Widyaiswari[2,3], D P Sari[2,4], T Saragi[1*] and Risdiana[1]

[1]Department of Physics, Padjadjaran University, Jl. Raya Bandung-Sumedang Km. 21 Jatinangor, Sumedang, 45363, Indonesia
[2]Meson Science Laboratory, RIKEN Nishina Center, 2-1 Hirosawa, Wako, Saitama 351-0198, Japan
[3]Department of Physics, Faculty of Mathematics and Natural Sciences, Universitas Indonesia, Depok 16424, Indonesia
[4]Graduate School of Engineering and Science, Shibaura Institute of Technology, Saitama City, Saitama 337-8570, Japan

E-mail: t.saragi@phys.unpad.ac.id

Abstract. Superconductors are materials with the unique characteristics when the temperature of the material is below its critical temperature, namely zero resistivity. There are many superconducting phenomena that need to be explained theoretically. One of them is the effect of addition of impurities on the physical properties of superconducting materials. In this study, the addition of 0.01 Ni magnetic impurities in the $Eu_{2-x}Ce_xCu_{O4+\alpha-\delta}$ material has been carried out to form the structure $Eu_{2-x}Ce_xCu_{0.99}Ni_{0.01}O_{4+\alpha-\delta}$, in an under-doped area with a concentration of Ce (x) = 0.09 and 0.10. Material synthesis was carried out by the solid reaction method with prefire at 900 °C for 20 hours, sintering at 1,000 °C for 16 hours, and annealing at 900 °C for 10 hours by flowing argon (Ar) gas. The XRD measurements were carried out to investigate the crystal structure, while the magnetic properties were carried out by SQUID measurement at temperature of 2 K to 30 K. From the XRD measurements, it was found that the crystals formed had a tetragonal crystal structure T' which was indicated by the appearance of two main peaks with Miller indices (103) and (110), lattice parameter a is 3,904 Å in both samples, lattice c is 11.8752 Å and 11.9738 Å for the concentrations of Ce (x) = 0.09 and 0.10. The volume of unit cells obtained were 181.0178 $Å^3$ and 181.0244 $Å^3$ with the bond distance between Cu-O of 1.9521 Å and 1.9523 Å. It was also obtained that the crystal size were 86.2604 nm and 91.6735 nm. From the SQUID measurement, it was found that the material $Eu_{2-x}Ce_xCu_{0.99}Ni_{0.01}O_{4+\alpha-\delta}$ has paramagnetic properties. The value of the effective magnetic moment was decreased by reducing the addition of Ni impurities of 0.0057 and 0.0051.

1. Introduction

Superconductor is a material that has special characteristics when the fabric temperature is underneath its basic temperature (T_c). The special property of superconductors is zero resistivity. High-T_c superconducting cuprates (HTSC) pull in to inquire about intrigued since of their different physical properties such as high temperature, high critical current esteem, and crtitical magnetism. There are two cuprate superconducting system, be specific cuprate superconductors with gap doping and cuprate superconductors with electron doping. Both have comparative stage graphs to each other [1–4]. Some properties in both systems have been found to be similar. From the point of view of the effect of impurities on superconductivity, it has been found that, for hole-doped cuprates, superconductivity is suppressed by non-magnetic impurities such as Zn more markedly than by magnetic impurities such as Ni [5], which contradicts the results in electron-doped cuprates. Nuclear magnetic resonance/quadrupole measurements have revealed that Ni operates to scatter holes weakly, while Zn operates to conduct strong ones. Ni has a weak influence on the electronic and magnetic states in the CuO_2 field [6]. Ni^{2+} ions with spin quantum number $S = 1$ tend to trap holes, forming Ni^{2+} ions with ligand holes, i.e., the so-called Zhang-Rice doublet state (the effective value of S is 1/2) so as not to disturb the antiferromagnetic correlation (AF) between Cu^{2+} spins with $S = 1/2$. This indicates that Ni is no longer considered a common magnetic impurity at high T_c cuprates [7]. A report given by Sahoo *et al.* showed that the AFM and SC regimes were shortened while the gap showing the spin-glass state become widens in the under-doped region of the hole-doped cuprates phase diagram. Similarly, in electron-doped couplings, there is less overlap of the two phases [8].

Electron doping superconductors are known to be single layer CuO_2 materials with two atoms of rare earth elements such as Nd, Pr, Sm, or Eu doped with Ce^{4+} atoms to produce compounds (Nd, Pr, Sm, Eu)$_{2-x}$Ce$_x$CuO$_4$. The element of Eu does not have a magnetic moment in the ground state, so that the effect of impurities on the superconductivity can be clearly observed [9]. On the other hand, when a Eu^{3+} is replaced by Ce^{4+}, the CuO_2 plane gets an excess electron, leading to the formation of an electron-doped system of $Eu_{2-x}Ce_xCuO_4$ [10]. However, for electron-doped system, the research report determining the effect of magnetic impurity on the physical properties is still very limited.

In the superconducting phase, there are three regions based on the doping concentration, namely the underdoped, optimum doped, and overdoped region [11,12]. Recently, Risdiana, *et.al.*, reported the effect of the addition of magnetic impurities Ni in $Eu_{2-x}Ce_xCu_{1-y}Ni_yO_4$ material with $x = 0.12$. It was found that the onset T_c of superconducting properties decreases from about 13 K without magnetic impurity to 4 K with the addition of magnetic impurity Ni of 0.01. The lattice parameters for this material did not so change much even though the impurity was applied up to $y = 0.05$ [9]. A study of on the effect of Ni impurity in the underdoped region with a smaller concentration of x needs to be done to obtain more comprehensive information. In this research, we report the effect of the addition of magnetic Ni on the structure and magnetic properties of the electron-doped superconducting material $Eu_{2-x}Ce_xCu_{0.99}Ni_{0.01}O_4$ in the underdoped region with $x = 0.09$ and 0.10.

2. Methods

Electron doping superconducting materials from $Eu_{2-x}Ce_xCu_{0.99}Ni_{0.01}O_{4+\alpha-\delta}$ were prepared using variations of $x = 0.09$ and 0.10 with variations of $y = 0.01$ made by the solid-state reaction method. Raw materials were mixed in a stoichiometric ratio and preheated in the air at 900°C for 20 h. The preheated materials were reground and pressed into pellets of 10 mm in diameter, and sintered in air at 1,000 °C for 16 h with repeated regrinding. The sample after sintering is called as-grown sample. As-grown samples of ECCO were annealed in flowing Ar gas of high purity at various temperatures 900 °C for several hours in between 8 h and 20 h in order to remove the excess oxygen at the apical site. The value of δ is observed in between 0.011 and 0.005 for the first sintering and for the second sintering, the value of δ is 0.057 and 0.046 [2,12,13].

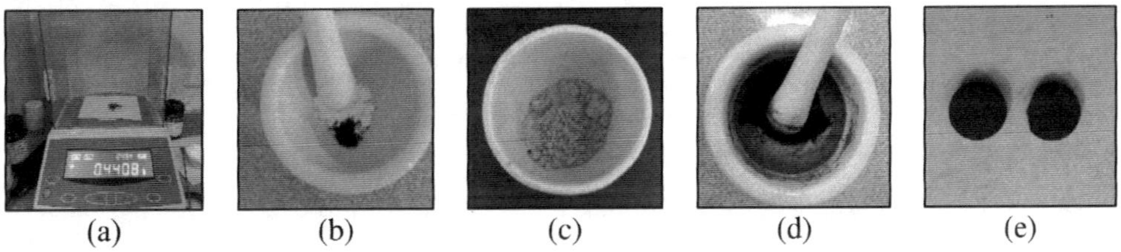
(a) (b) (c) (d) (e)

Figure 1. Weighing the material Eu_2O_3, CeO_2, CuO_2, NiO (a), First ground (b), Homogeneous sample and ready for prefire (c), Sample after prefired and sintered (d), Annealed sample (e).

The characterization carried out in this study included a sample characteristic test to determine the crystal structure carried out by XRD measurements and susceptibility measurements by using Magnetic Properties Measurement System (MPMS) with Superconducting Quantum Interference Devices (SQUID). The magnetic susceptibility measurement was performed from 2 K to 30 K at the Physical and Chemical Research Institute (RIKEN), Japan.

3. Result and Discussion

The XRD pattern is shown in Figure 1 on the ECCNO samples $x = 0.09$ and 0.10 with a concentration of Ni $y = 0.01$. Miller index and sample purity were identified using software by matching with reference data using High Score Plus Software. All samples showed sharp peaks indicating good sample crystallinity. All the main peaks can be observed and matched with tetragonal diffraction peaks with a planar configuration in the CuO_2 plane or called the T' structure. Specific peaks indicating the T' structure were clearly observed with Miller indices (110) and (103) and there was a small impurity peak at $2\theta = 28^o$ which indicated that oxygen reduction was still not optimal. The sample purity is 99.1 % and 98.2 %, for each sample with Ni $y = 0.01$ impurities.

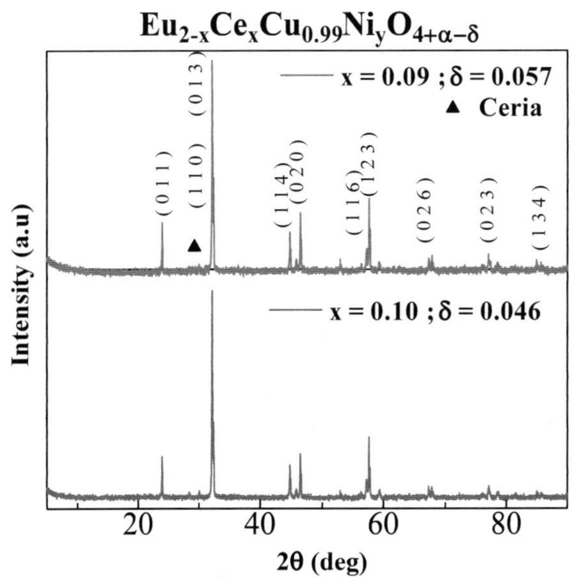

Figure 2. The results of the XRD ECCNO characterization with $x = 0.09$ and 0.10 and $y = 0.01$.

Table 1. Comparison of lattice parameter values, unit cell volume, Cu-O bond distance, and crystal size at a concentration of Ce (x) = 0.09; 0.10 and the addition of impurities Ni with $y = 0$ [14] and 0.01.

Ce (x)	Ni (y)	a (Å) = b (Å)	c (Å)	V (Å^3)	Cu-O bond (Å)	$\overline{D}$ (nm)
0.09	0	3.9050	11.8682	180.9820	1.9525	87.0669
	0.01	3.9042	11.8752	181.0178	1.9521	86.2604
0.10	0	3.9056	11.8656	180.9965	1.9528	88.6125
	0.01	3.9045	11.9738	181.0244	1.9523	91.6735

Tabel 1 shows Comparison of lattice parameter values, unit cell volume, Cu-O bond distance, and crystal size at a concentration of Ce (x) = 0.09; 0.10 with $y = 0.01$. For comparrison, structure information of ECCO with $x = 0.09$; 0.10 and $y = 0$ are also displayed [14]. The crystal size for ECCNO with $x = 0.09$ and 0.10 and Ni concentration $y = 0.01$ which was successfully calculated by the Debye-Scherrer calculation was 86.2604 nm and 91.6735 nm, respectively. On the other hand, the crystal size of ECCNO with $y = 0$ was 91.4683 nm for $x = 0.09$ and 94.4484 nm for $x = 0.10$ [10]. It can be seen that the crystal size decreases by addition of Ni.

Temperature dependence of susceptibility of ECCNO samples with concentrations of $x = 0.09$ and 0.10 and $y = 0.01$ and 0 measured at $H = 5$ Oe are shown in Figure 2. For all samples of ECCO with $y = 0$ [14] and $y = 0.01$, the paramagnetic characteristic was observed. The values of susceptibility are in between 10^{-3} and 10^{-5} emu/gr.Oe. These values indicate that the sample did not show a superconducting phase. Compared to the results obtained by Risdiana *et al.*, the samples with $x = 0.12$ and concentrations of $y = 0$ and 0.01 are reported to have diamagnetic behavior with the onset of T_c at about 11 K and 4 K for $y = 0$ and 0.01, respectively [9]. The different results between $x = 0.09$; 0.10 and $x = 0.12$ are probably due to the presence of excess oxygen content in the conduction layer which is still not completely reduced. The reduced oxygen contents is very small. On the other hand, the purity value of samples were 99.1% and 98.2% for $x = 0.09$; 0.10, respectively, which indicates the presence of impurities.

$Eu_{2-x}Ce_xCu_{1-y}Ni_yO_{4+\alpha-\delta}$

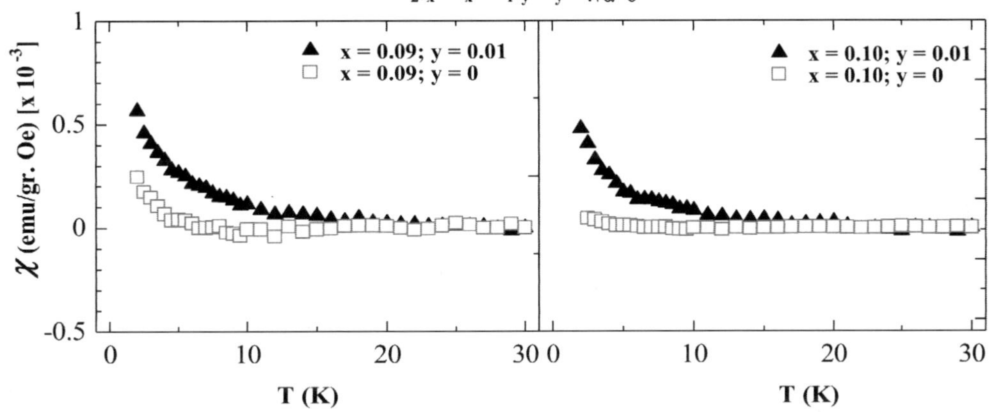

Figure 3. Temperature dependence of magnetic susceptibility of ECCNO with $y = 0.01$ and 0 [14] and (a) $x = 0.09$ and (b) $x = 0.10$ at magnetic field of 5 Oe.

Figure 3 shows a graph of susceptibility against 1/T of the ECCNO sample in the temperature range of 13 K – 30 K.

$Eu_{2-x}Ce_xCu_{0.99}Ni_{0.01}O_{4+\alpha-\delta}$

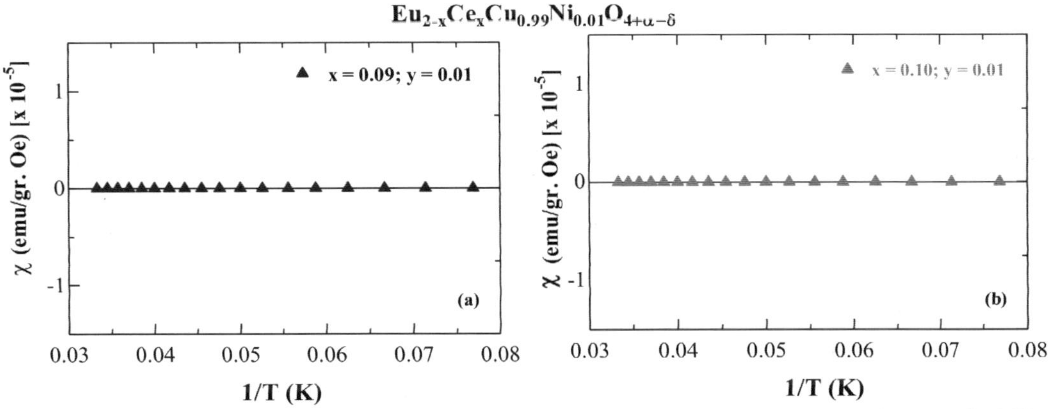

Figure 4. The graph of the relationship between suceptibilty and 1/T in the normal state of 13 K – 30 K for the ECCNO sample with $x = 0.09$ (a); $x = 0.10$ (b) and $y = 0.01$.

Table 2 shows the value of the curie constant, magnetic moment per unit volume ($n\mu^2$) and effective magnetic moment (μ_{eff}) with the addition of Ni impurities for x = 0.09; 0.10 and y = 0.01. It can be seen that the value of the Curie constant (C) increased with increasing Ce concentration from 1.7897 x 10^{-3} to 1.4045 x 10^{-3} emu.K.Oe/mol. The magnetic moment per unit volume and effective magnetic moment also increased with increasing Ce concentration. Compared to the ECCNO sample for x = 0.12, the values of C and n.μ^2 decreased due to the concentration of Ce which contributed to the smaller magnetic moment per unit volume.

Table 2. Comparison of Curie constant (C), magnetic moment per unit volume ($n\mu^2$), and effective magnetic moment (μ_{eff}) in the normal state of the ECCNO sample x = 0.09 and y = 0.01.

Doping Ce (x)	C (x 10^{-3}emu.K.Oe/mol)	n.μ^2 (x 10^{-20} emu.J.m/mol)	μ_{eff} (μ_B)
0.09	1.7897	1.9664	0.1196
0.10	1.4045	1.5431	0.1059

4. Conclusion

To investigate the effect of Ni impurities on the structure and magnetic characteristics of the electron-doping system in the underdoped region, electron-doped superconductors of $Eu_{2-x}Ce_xCu_{0.99}Ni_{0.01}O_{4+\delta}$ with $x = 0.09$ and 0.10 with $y = 0.01$ have been successfully produced. All of the principal peaks of the ECCO system were seen with a typical peak of the T' tetragonal structure, indicating the presence of CeO_2 impurities, according to XRD measurements. Paramagnetic behavior was detected in samples with $x = 0.09$; 0.10; and $y = 0.01$ based on magnetic susceptibility data. The value of C, magnetic moment per unit volume (n.μ^2), and effective moment μ_{eff} decrease with increasing x. Compared to the ECCNO sample for $x = 0.12$, the values of C and n.μ^2 decreased due to the concentration of Ce which contributed to the smaller magnetic moment per unit volume. And for the results of research that has been carried out aimed at studying the phenomenon of superconductivity in the normal state in the electron doped High-Tc superconducting cuprates ECCO which is influenced by magnetic impurities Ni.

Acknowledgement

We would like to thank Dr. Isao Watanabe from The Physical and Chemical Research Institute (RIKEN, Japan) for providing SQUID and hosting for Sakura Program. The authors would also like to thanks Kemenristek of Indonesia for financial support for the research in the scheme of Fundamental Research Funding (PDUPT) 2021, contract no. 1207/UN6.3.1/PT.00/2021 and ALG of Universitas Padjadjaran 2021, no. 1959/UN6.3.1/PT.00/2021.

References

[1] Risdiana, S. Pratiwi, D. Suhendar, W.A. Somantri, L. Safriani, T. Saragi, Dependence of oxygen content with physical properties in electron-doped superconducting cuprates Eu1.85Ce0.15CuO4, AIP Conf. Proc. 1712 (2016) 0–4. https://doi.org/10.1063/1.4941903

[2] M.A. Syakuur, Y. Maryati, U. Widyaiswari, D.P. Sari, T. Saragi, Risdiana, The dependence of magnetic moments on magnetic impurities of ni in eu1.86ce0.14cu1-yniyo4+αδ, Key Eng. Mater. 860 KEM (2020) 148–153. https://doi.org/10.4028/www.scientific.net/KEM.860.148

[3] Y. Tanabe, K. Suzuki, T. Adachi, Y. Koike, T. Kawamata, Risdiana, T. Suzuki, I. Watanabe, Change of the ground state upon hole doping unveiled by Ni impurity in high- T_c cuprates, J. Phys. Soc. Japan. (2010). https://doi.org/10.1143/JPSJ.79.023706

[4] Risdiana, T. Adachi, Y. Koike, I. Watanabe, μSR study of the Cu-spin dynamics in the electron-doped high-T c cuprate of Pr0.86LaCe0.14Cu 1-y(Zn,Ni)yO4, Phys. B Condens. Matter. 374–375 (2006) 218–220. https://doi.org/10.1016/j.physb.2005.11.059

[5] G. Xiao, M.Z. Cieplak, J.Q. Xiao, C.L. Chien, Magnetic pair-breaking effects: Moment

formation and critical doping level in superconducting La1.85Sr0.15Cu1-xAxO4 systems (A=Fe,Co,Ni,Zn,Ga,Al), Phys. Rev. B. (1990). https://doi.org/10.1103/PhysRevB.42.8752

[6] Risdiana, L. Safriani, W.A. Somantri, T. Saragi, T. Adachi, I. Kawasaki, I. Watanaber, Y. Koike, Possible existence of the stripe correlations in electron-doped superconducting cuprates Eu1.85Ce0.15Cu1-yNiyO4+ α-δσ studied by muon-spin-relaxation, Adv. Mater. Res. 896 (2014) 354–357. https://doi.org/10.4028/www.scientific.net/AMR.896.354

[7] K. Suzuki, T. Adachi, Y. Tanabe, Y. Koike, T. Kawamata, Risdiana, T. Suzuki, I. Watanabe, Hole trapping by Ni, Kondo effect, and electronic phase diagram in nonsuperconducting Ni-substituted La2-xSrxCu 1-yNiyO4, Phys. Rev. B - Condens. Matter Mater. Phys. 82 (2010). https://doi.org/10.1103/PhysRevB.82.054519

[8] B.K. Sahoo, B.N. Panda, Theoretical study of phase diagram of electron-doped cuprates, Solid State Commun. 152 (2012) 2151–2155. https://doi.org/10.1016/j.ssc.2012.09.012

[9] Risdiana, M.A. Syakuur, Y. Maryati, M. Manawan, T. Saragi, Structure and magnetic properties of eu2-xcexcu1-yniyo4+α-δ, Key Eng. Mater. 855 KEM (2020) 123–127. https://doi.org/10.4028/www.scientific.net/KEM.855.123

[10] Risdiana, M. Saputri, M.F. Sobari, A.I. Hanifah, W.A. Somantri, T. Saragi, Structural and Magnetic Properties of n-type Superconductor Eu2-xCexCuO4+α-δ, IOP Conf. Ser. Mater. Sci. Eng. 196 (2017) 2–6. https://doi.org/10.1088/1757-899X/196/1/012012

[11] Risdiana, M. Manawan, L. Safriani, T. Saragi, A. Aprilia, N. Syakir, S. Hidayat, A. Bahtiar, F. Fitlilawati, R.E. Siregar, Study of structure and resistivity of electron-doped superconducting cuprates Eu1.85Ce0.15Cu1-yZnyO4+α-δ, J. Phys. Conf. Ser. 1091 (2018) 1–6. https://doi.org/10.1088/1742-6596/1091/1/012019

[12] Risdiana, M. Saputri, M.F. Sobari, A.I. Hanifah, W.A. Somantri, T. Saragi, Structural and Magnetic Properties of n-type Superconductor Eu2-xCexCuO4+α-δ, IOP Conf. Ser. Mater. Sci. Eng. 196 (2017) 1–5. https://doi.org/10.1088/1757-899X/196/1/012012

[13] Y. Maryati, A.I. Hanifah, M.A.B. Subardhi, E.A. Rahayu, Y.R. Tayubi, M. Manawan, T. Saragi, Risdiana, The effect of heating treatment in electron doped superconductor Eu1.85Ce0.15CuO4+α-δ, J. Phys. Conf. Ser. 1080 (2018) 0–5. https://doi.org/10.1088/1742-6596/1080/1/012022

[14] T. Maulana, Studi Ukuran Kristal Dan Pengaruh Doping Atom Cerium Terhadap Sifat Magnetik Pada Superkonduktor Eu, (2021)

Initial evaluation of land use potential using geospatial analysis at Sendang Biru Beach, Malang Regency, East Java Province, Indonesia

A Wahid[1], Sunaryo[2], A Susilo[2] and Wiyono[2]

[1]Doctor Program, Department of Physics, University of Brawijaya, Indonesia Veteran Street, Malang, East Java, Indonesia
[2]Geophysical Engineering, Department of Physics, University of Brawijaya, Indonesia Veteran Street, Malang, East Java, Indonesia

Email: awundana@gmail.com, awahid@staf.undana.ac.id

Abstract. The use of ArcGIS for geospatial analysis is vital as the first step in earth research. The research focused on Sendang Biru Beach, Tambak Rejo Village, Malang Regency, East Java Province, Indonesia. Based on the analysis of topographic, geological, and land use data, Sendang Biru Beach is located in a basin to a hill with an average height of 125 masl. Geologically, it is located in the Wonosari Formation, flanked by the Qas (swamp and river deposits) and the Tomm (Mandalika Formation). Furthermore, land use is dominated by dry forests. The basin is used for plantations, settlements, and scattered fields and has a karst aquifer flowing all year round, including springs that have diffuse and fissure channel types. The water quality in the area is still within normal limits, has not experienced contamination of seawater intrusion. Sendang Biru Beach Land in Tambak Rejo Village, the ideal development is directed at Fisheries and Coastal Tourism, supported by the presence of extensive dry forest land, karst aquifer resources, and the existence of Sempu Island as a nature reserve.

1. Introduction

Geospatial data and information are needed to conduct further research related to natural resource exploration, especially geological data, topography, land use. Sendang Biru Beach is located in Tambak Rejo Village, Sumber Manjing Wetan District, Malang Regency, has the potential for fisheries resources and the Sempu Island nature reserve as well as it has karst water sources originating from springs, surface flows in the form of rivers and lakes [1]. Furthermore, underground rivers and the increasing population, consequently there is pressure on coastal areas, especially the opening of new residential areas, excessive use of resources, and land conversion resulting in changing the function of the ecosystem [2].

ArcGIS applications have been widely used in research related to geosciences, the use of ArcGIS for mapping the fields of geohazard disaster mitigation, natural resources, energy, the environment and land use, including: Dual Mitigation System is a system combining Indonesia Tsunami Early Warning System (INA TEWS) and Automatic Packet Reporting System (APRS) South Coast of Malang [3], evaluation of regional geohazard investigation vulnerability in risk assessment, disaster loss assessment and regional disaster prevention Yaozhou District China [4], identification and mapping of landslide susceptibility zones in rivers by integrating all indicators in ArcGIS [5] and [6]. Environmental Impact Assessment of Air Pollution Dispersion Based on ArcGIS application of GIS based software to represent atmospheric pollutant concentrations with and visualization of contours on maps [7]. Mapping of wind resource land using ArcGIS multi criteria decision analysis technique [8], influence of slope, elevation, and Weibull wind characteristics on wind power capacity in wind farms using ArcGIS software [9]. Simulating water and pesticide transport in the soil with the development of maps of pesticide accumulation and risk in a GIS in the form of a python tool in ArcGIS [10], real time ArcGIS applied to show distributions and yields changes in residual chloramine concentrations in water distribution pipeline systems [11].

As an invasive alien species alternanthera philoxeroides by analyzing spatial and temporal distribution patterns and estimating suitable areas for this species in China using MaxEnt and ArcGIS models [12], application of ArcGIS to predict rainfall run-off in the Kuronagi River [13], development of Marine Geospatial Ecology Tools (MGET) from ArcGIS a collection of open source geoprocessing tools that is easy to use, and can be extended to suit ecologists [14], application of ArcGIS to construct and prioritize ecological corridors maximize the social and ecological functions of urban green spaces in the urban landscape in Tongzhou District, Beijing, China [15]. Design of wetlands in urban construction based on the ArcGIS system to encourage residents to participate in urban park development, urban planning, and urban development [16], classify land cover and assess the impact of agricultural expansion on land surface temperature (LST) in El Province Beheira, Egypt. four categories of agricultural land cover, urban, desert and waterland were distinguished and mapped using ArcGIS Desktop 10.6 [17], integration of thematic maps was actualized with overlay analysis ArcGIS 10.3.1 resulting in groundwater recharge zone maps GIS based recharge maps for managers water and urban planners [18].

Combined geological, geomorphological and geotechnical data with ARCGIS to provide a framework for the capital city of Antananarivo in Madagascar in the form of lithology, slope, land use and groundwater level combined to build a geotechnical map using mapping soil carrying capacity combined with land use constraints and geohydrology [19], Nigeria digital soil map (DSM) database and soil map with digital cartography and GIS operation [20] , ArcGIS overlay analysis to get a groundwater potential map With geology, lineament density, slope and geomorphology found as dominant and sensitive factors for groundwater zoning of Lake Tana watershed in Blue Nile, Ethiopia [21], hydrogeological environmental impacts on groundwater availability, groundwater potential zone maps were created using ArcGIS software [22]. Making a map of the coastal area of Sempu Island, Sendang Biru Village, with ArcGIS and Global Mapper as map making software [23,24], Banyu Cave tunnel level in Donomulyo District, Malang Regency by using GIS. It has a clustered appearance and a pattern that develops in the form of a curvilinear passage dominated by underground rivers and bedding planes [25].

Based on the above. It is necessary to analyze geospatial information and data in the form of topographic data, geology, land use as well as physical and chemical properties (Conductivity, Density, pH, Salinity) Sendang Biru Beach, Tambak Rejo Village, Sumber Manjing Wetan District, Malang Regency as initial research to explore natural resources and land use.

2. Methods

2.1. ArcGIS geographic information system software

ArcGIS is a commercial geographic information system software, and spatial information and its characteristics can wholly and accurately be output to users in the form of text or images according to the specified needs [1,20], generating generalized maps according to user expectations with ArcGIS procedures applied sequentially [26], disaster studies are an essential part of disaster prevention and mitigation, providing a reference for local infrastructure development, tourism development, planning city and eco development regional nominations using the ArcGIS software [6]. The method used in this research is to make geological maps, topographic maps, and land use maps by utilizing the ArcGIS 9.3 software application, with a flow chart as shown in Figure 1.

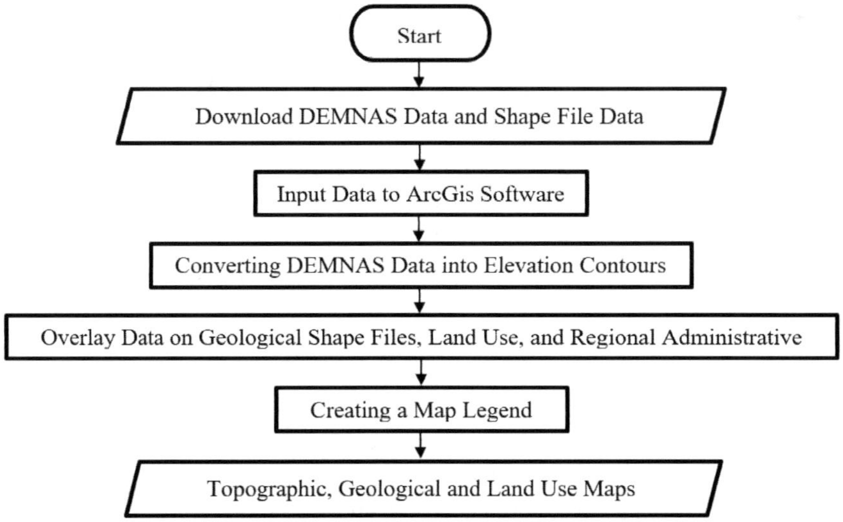

Figure 1. Flowchart of map making with ArcGis 9.3.

2.2. *Physical and chemical properties of springs in the Sendang Biru Coastal Area*
While taking data on the physical and chemical properties of springs in the Sendang Biru Coastal Area. The use of a density, conductivity, salinity and pH meter for sampling methods from two springs line 1 and Line 6 and one sample from a drilled well Line 5 is conducted in this research. Sample are analyzed at the Analysis and Measurement Unit, Department of Chemistry, Faculty of Mathematics and Natural Sciences University of Brawijaya Malang. Sample submitted is on April 21, 2021, and the results of the analysis were received on May 25, 2021.

3. Result and Discussion

3.1. *Geological Maps*
The geological map obtained by using ArcGIS 9.3 in Figure 2, it can be seen that the location of Sendang Biru Beach, Tambak Rejo Village, and Sempu Island is in Tmwl (Wonosari Formation), which is dominated by limestone, sandy marl, claystone inserts [27,28]. The Wonosari Formation is flanked by Qas (swamp and river deposits) which are dominated by gravel, sand, clay, plant remains in the west and north. Tomm (Mandalika Formation), which is dominated by andesite lava, basalt, trachyte, dacite, propitiate andesite breccia in the upper part. The upper north is the Mandalika Formation and the Wuni Formation. There is a river flowing along the Qas Formation.

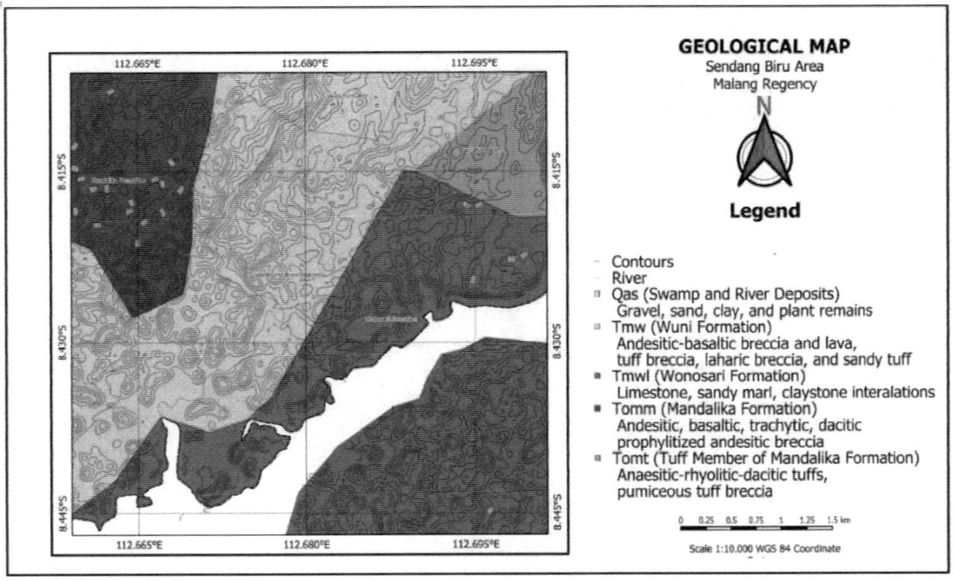

Figure 2. Geological Map with ArcGis 9.3.

3.2. Topographic maps

The topographic map generated by using ArcGIS 9.3 in Figure 3, it can be seen that the location of Sendang Biru Beach in Tambak Rejo Village is in a basin. Most of them are at an altitude of 0-45 masl mainly in coastal locations. At some points there are hills that spread at an altitude of 45-90 masl and flanked by a topographic surface with an altitude of 40-125 masl. The further north, the topography reaches 180 masl. There are hills spread on the surface of the island at an altitude of 45-90 masl. The topography of Sempu Island on the coast is at an altitude of 0-45 masl.

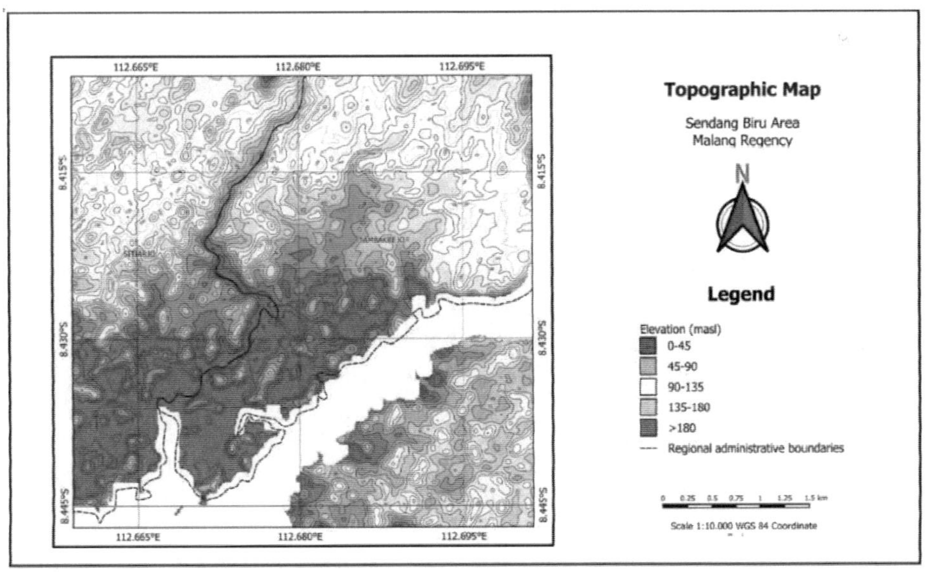

Figure 3. Topographic Map with ArcGis 9.3.

Journal of Physics: Conference Series **2165** (2022) 012021 doi:10.1088/1742-6596/2165/1/012021

3.3. Land Use Map
Utilization map generated by using ArcGIS 9.3, in Figure 4, it can be seen that the location of Sendang Biru Beach in Tambak Rejo Village is most of the land is dominated by dry forest surrounding it and spreads to the coast. Part of the land use basin in the form of plantations, residential use and utilization as agricultural fields spread in groups on plantations. In the north, east, and west, and the southern part to the coast, it is dominated by dry forest; in the southern part of the bottom, there is a small part of alkaline forest and swamp. Sempu Island is a protected nature reserve opposite the beach of Sendang Biru.

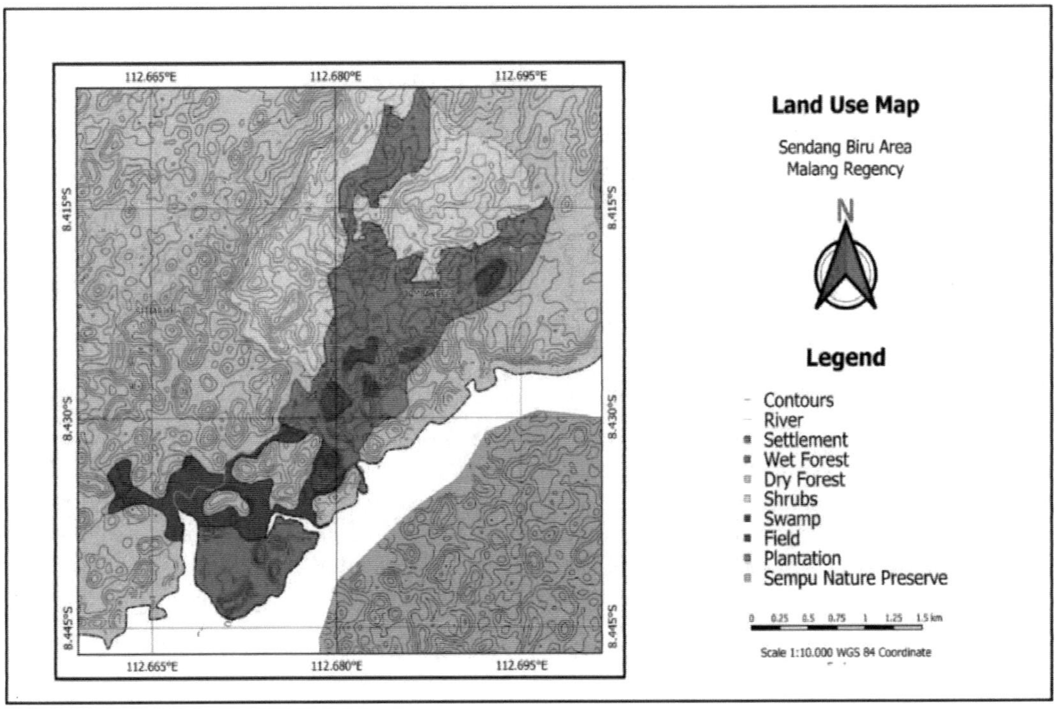

Figure 4. Land use map with ArcGis 9.3.

3.4. The nature of the Spring Blue Spring
There are two spring lines (line 1 and line 6) and several water sources used as bore water, varying between 14 – 30 m. Sampling was carried out at springs line 1 and line 6 as well as samples from drilled wells line 5, obtained the density value of the springs and boreholes of 1,000 g/liter, while the conductivity, pH, and salinity values with the results in Table 1. The conductivity value of the springs has a different value; line 1 is more significant than line 6, where line 1 has been used by PDAM while line 6 is used by the community freely for MCK, so it has much contamination. The pH and salinity values for line 1 and line 6 springs have values close to the same, while for boreholes, they have large values due to the sampling obtained through the pump results. In general, the spring water quality of Sendang Biru is still within normal limits and has not experienced contamination or seawater intrusion based on data on density, conductivity, salinity, and pH.

Table 1. Results of analysis of the properties of spring in Sendang Biru.

CODE	Parameter	Analysis Result		Analysis Method	
		Level	Unit	Reactor	Method
Line 1	Conductivity	297.50 ± 0.71	$\mu S/_{cm}$	-	Conductometer
Line 5	Conductivity	356.50 ± 0.71	$\mu S/_{cm}$	-	Conductometer
Line 6	Conductivity	342.00 ± 1.41	$\mu S/_{cm}$	-	Conductometer
Line 1	pH	7.24 ± 0.01	-	$AgNO_3$	pH meter
Line 5	pH	2,45 ± 0.01	-		pH meter
Line 6	pH	7.25 ± 0.00	-		pH meter
Line 1	Salinity	28.24 ± 0.00	%	$AgNO_3$	Argentometer
Line 5	Salinity	47.06 ± 0.00	%	$AgNO_3$	Argentometer
Line 6	Salinity	25.10 ± 0.00	%	$AgNO_3$	Argentometer

Geographic Information Systems is an interdisciplinary field developed in recent years, including geography, computer science, remote sensing, management science, and other disciplines [6]. ArcGIS is a commercial geographic information system software, and spatial information and its characteristics can wholly and accurately be output to users in the form of text or images according to the specified needs [4,20], generating generalized maps according to user expectations with ArcGIS procedures applied sequentially [26], disaster studies are an essential part of disaster prevention and mitigation, providing a reference for local infrastructure development, tourism development, planning city and eco development regional nominations using the ArcGIS software [6].

The Sendang Biru Beach, at a geographical position of 08o37'- 08o41'LS and 112o35'-112o43'E with an altitude of 0-100 masl and has a significant potential for fishery resources in East Java, the attraction in the form of a fish auction place and Sempu island located opposite the coast is designated as a nature reserve with an area of 877 ha. Sumbermanjing Wetan sub district is an area located in the Wonosari Formation area, this area is dominated by limestone, sandy marl, and claystone inserts. In addition, this area is also in a sedimentary rock area with rock ages in the late Miocene Tertiary type [27,28], having karst water sources from springs, surface runoff in the form of rivers and lakes and underground rivers. The distribution of water resources; 36 springs are found in the southern part, springs in the Sendang Biru karst area are in valleys, basins, and slopes [2]. Sendang Biru spring has a flow rate fluctuation during the rainy and dry seasons that is not too large, including perennial flow (flow throughout the year), the increase in flow rate, which is not too large, indicates that it is a spring that has diffuse and fissure types. Land use in the Sendang Biru spring catchment area is heterogeneous, dominated by shrubs and forests that grow irregularly. The use of unspoiled land and the presence of quite a lot of Karren formations indicate that karst development is quite massive [29]; having karst springs looks like a natural reservoir that holds rainwater, releasing water on the ground indicates a large enough water source in residential areas, water output from natural reservoirs, excess karst springs, the length of time delay between rainwater coming out to springs so that some karst springs will issue large discharges in the dry season [2].

Based on the analysis using ArcGis 3.9 software through Geological Maps, Topographic Maps, and Land Use Maps. Sendang Biru Beach, Tambak Rejo Village, has the potential for natural resources in the form of groundwater. This is indicated by the presence of 2 natural springs and is located in a basin with an average height of 0-45 masl, and the presence of drilled wells that spread at several points and have a specific pattern, based on the geological map is in the Wonosari Formation which is dominated by limestone, sandy marl, claystone inserts. Water resources in the form of karst aquifers which are still undergoing a fairly massive development process, are marked by the presence of several hills that spread on the surface (Karren). The presence of 2 springs that appears in limestone layers with the year-round flow, diffuse channel type and water discharge fissure that does not fluctuate with seasonal changes [29]. Recharge comes from the overflow of rainfall and flow and seepage from the Qas formation (swamp and river deposits), dominated by gravel, sand, clay, plant sides in the west and north, and Tomm (Mandalika Formation), which is dominated by andesite lava, Basalt, trachyte, dacite, propitiate, andesite breccia in the east. Sendang Biru has water resources in the form of a karst aquifer [2], with a gravity system flow that can be explored and exploited to meet the need for clean water with water quality still within normal limits.

Plantations dominate land use and agricultural land that spreads along the surface of the karst aquifer land and residential land; over time, the increasing number of people will require more and more land for settlement. Land use as a livelihood will cause problems in the form of exploitation of aquifer resources, and Excessive karst land will cause damage to the ecosystem and cause contamination, especially in terms of seawater intrusion and pollutants. The land use of Sendang Biru Beach is desirable for the development of fisheries and tourism resources because most of the land is dry and dominated by dry forest, the existence of Sempu Island as a protected nature reserve, and the presence of groundwater resources in the form of karst aquifers.

4. Conclusion

ArcGis application is beneficial in initial research mainly related to geospatial information (geology, topography and land use). Sendang Biru Beach, Tambak Rejo Village, and Sempu Island are located in Tmwl (Wonosari Formation), dominated by limestone, sandy marl, claystone inserts, and located in basins. Most of which is at an altitude of 0-90 masl, mainly in coastal locations increasingly to the north, the topography reaches 180 masl. Very attractive for the development of fishery resources and tourism because most of the land is dry and dominated by dry forests. The existence of water resources in the form of a karst aquifer which is still undergoing a fairly massive development process is characterized by the presence of several hills that spread on the surface (Karren). The presence of 2 springs that appear in the limestone layer with the flow throughout the year diffuse channel type and water discharge fissure. It does not experience fluctuations in seasonal changes; recharge comes from the overflow of rainfall and flows and seepage from the Qas Tomm formation (Mandalika Formation). Aquifer karst can be explored and exploited to meet the need for clean water with water quality still within normal limits.

Acknowledgment

On this occasion, the author would like to thank the Head of the Geophysics Laboratory and the Head of the Chemistry Laboratory, Faculty of Mathematics and Natural Sciences, University of Brawijaya and the staff for the use of the laboratory and analysis sample.

References

[1] Widiana F. and Wikantiyoso R, 2018, "Implementation of local wisdom in the Sendang Biru Beach Tourism Development strategy for the Conservation of Sempu Island," *Local Wisdom J. Ilm. Kaji. Kearifan Lokal*, vol. 10, no. 1, doi: 10.26905/lw.v10i1.2397

[2] Sahrina A., Fitrianti D., Suprianto A. and Labib M. A., 2020, "Potential and Challenges of Karst Water Resources in Sumbermanjing Wetan District of Malang Regency," *IOP Conf. Ser. Earth Environ. Sci.*, vol. 412, no. 1, doi: 10.1088/1755-1315/412/1/012032

[3] Asri A.K., Elya H., Duantari N., Suryaningsih E. and Victoria L. D. D. D., 2016, "Dual Mitigation System: Database System Combination of EWS and APRS for Disaster Management (Case Study: Malang Southern Coast)," *Procedia - Soc. Behav. Sci.*, vol. 227, no. November 2015, pp. 435–441, doi: 10.1016/j.sbspro.2016.06.098

[4] Cheng Y, Zhang J. and Peng J., 2013, "ArcGIS-based evaluation of geo-hazards at Yaozhou County, Shaanxi, China," *J. Rock Mech. Geotech. Eng.*, vol. 5, no. 4, pp. 330–334, doi:

10.1016/j.jrmge.2012.11.002

[5] Basu T. and Pal S., 2019, "RS-GIS based morphometrical and geological multi-criteria approach to the landslide susceptibility mapping in Gish River Basin, West Bengal, India," *Adv. Sp. Res.*, vol. 63, no. 3, pp. 1253–1269, doi: 10.1016/j.asr.2018.10.033

[6] Tan Q., Bai M., Zhou P., Hu J. and Qin X., 2021, "Geological hazard risk assessment of line landslide based on remotely sensed data and GIS," *Meas. J. Int. Meas. Confed.*, vol. 169, no. August 2020, p. 108370, doi: 10.1016/j.measurement.2020.108370

[7] Zhong S., Zhou L., and Wang Z., 2011, "Software for environmental impact assessment of air pollution dispersion based on ArcGIS," *Procedia Environ. Sci.*, vol. 10, no. PART C, pp. 2792–2797, doi: 10.1016/j.proenv.2011.09.433

[8] Mathew S. A. and Mariappan V. E. N., 2014, "Wind resource land mapping using arcgis, wasp and multi criteria decision analysis (MCDA)," *Energy Procedia*, vol. 52, pp. 666–675, doi: 10.1016/j.egypro.2014.07.123

[9] Sharma P. K., Gautam A., Baredar P., Warudkar V., Bhagoria J. L. and Ahmed S., 2020, "Analysis of terrain of site Mamatkheda Ratlam through wind modeling tool ArcGIS and WAsP," *Mater. Today Proc.*,no. xxxx, pp. 3–7, doi: 10.1016/j.matpr.2020.09.638

[10] Anlauf R,, Schaefer J. and Kajitvichyanukul P., 2018, "Coupling HYDRUS-1D with ArcGIS to estimate pesticide accumulation and leaching risk on a regional basis," *J. Environ. Manage.*, vol. 217, pp. 980–990, doi: 10.1016/j.jenvman.2018.03.099

[11] Bai X., Zhi X., Zhu H., Meng M. and Zhang M., 2015, "Real-time ArcGIS and heterotrophic plate count based chloramine disinfectant control in water distribution system," *Water Res.*, vol. 68, pp. 812–820, doi: 10.1016/j.watres.2014.10.041

[12] Wang Z., Torres M., Paudel P., Hu L., Yang G., and Chu X., 2020, "Assessing the karst groundwater quality and hydrogeochemical characteristics of a prominent dolomite aquifer in Guizhou, China," *Water (Switzerland)*, vol. 12, no. 9, pp. 9–13, doi: 10.3390/W12092584

[13] Baiyinbaoligao, Wang D., and Li X. Y., 2011, "Application of ArcGIS in the calculation of basins rainfall-runoff," *Procedia Environ. Sci.*, vol. 10, no. PART C, pp. 1980–1984, doi: 10.1016/j.proenv.2011.09.310

[14] Roberts J.J., Best B. D., Dunn D. C., E. A. Treml E. A., and Halpin P. N., 2010, "Marine Geospatial Ecology Tools: An integrated framework for ecological geoprocessing with ArcGIS, Python, R, MATLAB, and C++," *Environ. Model. Softw.*, vol. 25, no. 10, pp. 1197–1207, doi: 10.1016/j.envsoft.2010.03.029

[15] Wanghe K. *et al.*, 2020, "Gravity model toolbox: An automated and open-source ArcGIS tool to build and prioritize ecological corridors in urban landscapes," *Glob. Ecol. Conserv.*, vol. 22, p. e01012, doi: 10.1016/j.gecco.2020.e01012

[16] Liu, N., 2020, "Wetland design in sponge city construction based on ArcGIS and Internet of Things system," *Microprocess. Microsyst.*, no. October, p. 103435, doi: 10.1016/j.micpro.2020.103435

[17] Dashti Barmaki M., Rezaei M. and Madadi S., 2020, "Use of fractal dimensions analysis in geographic information system and remote-sensing techniques to identify groundwater prospective zones in the Anar-Dashtegol anticline, Iran," *Carbonates and Evaporites*, vol. 35, no. 1, 2020, doi: 10.1007/s13146-019-00535-x

[18] Abdullateef L., Tijani M. N., Nuru N. A., John S. and Mustapha A., 2020, "Heliyon Assessment of groundwater recharge potential in a typical geological transition zone in Bauchi , NE-Nigeria using remote sensing / GIS and MCDA approaches," *Heliyon*, vol. 7, no. August 2020, p. e06762, 2021, doi: 10.1016/j.heliyon.2021.e06762

[19] Andriamamonjisoa S. N. and Hubert-Ferrari A., 2019, "Combining geology, geomorphology and geotechnical data for a safer urban extension: Application to the Antananarivo capital city (Madagascar),"*J.African Earth Sci.*,vol.151,pp.417–437,2019, doi: 10.1016/j.jafrearsci.12.003

[20] Nkwunonwo U. C., Okeke F. I., Ebinne E. S. and Chiemelu N. E., 2020, "Free, open,

quantitative and adaptable digital soil map data and database for Nigeria," *Data Br.*, vol. 31, p. 105941, doi: 10.1016/j.dib.2020.105941

[21] Berhanu K. G. and Hatiye S. D., 2020, "Identification of Groundwater Potential Zones Using Proxy Data: Case study of Megech Watershed, Ethiopia," *J. Hydrol. Reg. Stud.*, vol. 28, no. January, p. 100676, doi: 10.1016/j.ejrh.2020.100676

[22] Das N., Mondal P., Sutradhar S., and Ghosh R., 2020, "The Egyptian Journal of Remote Sensing and Space Sciences Assessment of variation of land use / land cover and its impact on land surface temperature of Asansol subdivision," *Egypt. J. Remote Sens. Sp. Sci.*, vol. 24, no. 1, pp. 131–149, doi: 10.1016/j.ejrs.2021.03.003

[23] Cahyadi. A and Prabawa B., 2017, "Study Of Genesis And Dynamics of Coastal Areas of Island Karst Area," no. December 2018, pp. 0–9, doi: 10.31227/osf.io/wru5k

[24] Wibawa I. G. N. A. and Luthfi, O. M., 2016 "Water Quality In Coral Reef Ecosystems In Sempu Strait, Sendang Biru, Malang," vol. 12, no. 3, pp. 121–127

[25] Suprianto A. and Labib M. A., 2019, "Estimation of Determination of Levels and Alley Patterns of Banyu Cave and Surrounding Areas Using GIS," *JPIG (Jurnal Pendidik. dan Ilmu Geogr.*, vol. 4, no. 1, pp. 1–8, doi: 10.21067/jpig.v4i1.3086

[26] Smirnoff A., Paradis S. J., and Boivin R., 2008, "Generalizing surficial geological maps for scale change: ArcGIS tools vs. cellular automata model," *Comput. Geosci.*, vol. 34, no. 11, pp. 1550–1568, doi: 10.1016/j.cageo.2007.10.013

[27] Ramadhani W., Wasis D., and Ab M., 2010,"Identification of Subsurface Structures Using the Gravity Method in Sumbermanjingwetan Village and Druju Village – South Malang," *University of Brawijaya*

[28] Boimau. Y, Sunaryo, and Susilo A., 2018,"Identification of Underground River Flow in Karst Area of Sumber Bening-Malang, Indonesia Based on Geoelectrical Self-Potential and Resistivity Data," *Int. J. Appl. Phys.*, vol. 5, no. 3, pp. 1–7, doi: 10.14445/23500301/ijap-v5i3p101

[29] Randhiki Gusti Perdana, 2017, "Temporal Variations Of HCO_3 - Dissolved Content In Sendang Blue Spring and Beji Spring In Sumbermanjing Wetan and Gedangan District," pp. 16–21

Preparation and characterization of CTAB surfactant modified TiO_2 nanoparticles as antibacterial fabric coating material

B F Bukit[1], E Frida[2*], S Humaidi[2] and P Sinuhaji[2]

[1]Doctoral Program Department of Physics, Universitas Sumatera Utara, 20155, Medan, Indonesia
[2]Department of Physics Universitas Sumatera Utara, 20155, Medan, Indonesia

E-mail: ernafridatarigan@gmail.com

Abstract. TiO_2 preparation has been carried out with the addition of cetyl trimethyl ammonium bromide (CTAB) surfactant. The sol-gel method is used in the preparation of TiO_2, where $TiCl_4$ is used as a precursor. The resulting TiO_2 was characterized using XRD, FTIR, XRF and SEM. XRD results show the crystal size of TiO_2 17.75 nm with the rutile phase. The FTIR results show a broad absorption band between 800 and 400 cm^{-1} by Ti–O vibrations in the crystal lattice. The XRF results showed that the TiO_2 content was 80.80%. Morphological results showed an irregular ball-like structure that was less aggregated. The characterization results show that TiO_2 can be used as a coating on fabrics with antibacterial properties.

1. Introduction

In the last decade, nanostructured materials have attracted significant attention. Size and phase control of nanomaterials can be helpful in a wide variety of applications. For example, titanium dioxide (TiO_2) is one of the most attractive materials widely used in various applications [1]. In addition, TiO_2 accounts for 70% of the total volume of pigment production worldwide and is included in the top five nanoparticles used [2].

Titania is naturally formed in four main phases: rutile, anatase, brookite, and TiO_2(B). The stability of the four phases of titania depends on the particle size. The rutile phase is thermodynamically stable, but anatase is the most stable phase at sizes below 14 nm. Brookite and TiO_2(B) are metastable forms that are not commonly observed in minerals and are challenging to synthesize in pure form [3]. In photocatalytic activity, the anatase and rutile phases have an essential role. Thermodynamically, the anatase phase is less stable than the rutile. The structure of anatase and rutile can be described as an octahedral chain of TiO_6 in which six O_2 surrounds each Ti^{4+} ion- ions; both crystal structures are distinguished by their octahedral distortion octahedral chain arrangement pattern. The photocatalytic properties of TiO_2 can be utilized in the treatment of water contaminated with sewage, as a self-cleaning and anti-bacterial material on fabrics and anti-fogging on glass [4], [5]. The photocatalytic properties of TiO_2 can be used when TiO_2 is irradiated with light, and the bandgap will produce electron and hole pairs which cause a redox reaction on the surface of TiO_2. As a result, electrons will move from the valence band to the conduction band so that electron and hole pairs will form on the photocatalyst surface [6], coats the polyester fabric TiO_2, which has high sedimentation stability. As a result, a very thin layer is formed on the surface of each polyester fabric fibre. The layer formation is not accompanied by the deposition of TiO_2 between the fibres, allowing the fabric to maintain its softness well.

Many methods have been developed to make TiO_2 nanoparticles, such as sol-gel, hydrothermal, emulsion, anodization, vapour deposition, sonication, and microwave [7]. The sol-gel method is the most widely used method for synthesizing TiO_2 because it can be carried out at room temperature and is cost-effective [8]. On the other hand, the wet phase or sol method is the best way to control particle size, shape and composition [9]. This control can be achieved by using molecular precursors in combination with organic molecules. Organic molecules act to direct the structure in solution to control and limit the growth of the structure. To obtain small-sized crystals, about < 50 nm, a surface stabilizer can be used to prevent agglomeration and increase solubility. The surface stabilizer is because nanostructures are usually unstable in solution and tend to agglomerate to minimize their surface energy. Stabilization of this system can be achieved by using steric or electrostatic interactions to keep the particles apart. One of the steric interactions carried out is by adding surfactants. These surfactants can slow down the formation of the titania lattice [10].

The use of surfactants has been investigated to produce porous materials [11]. The sol-gel method uses a non-ionic surfactant as a template, such as the sol-gel method's synthesis of TiO_2 thin films. Among the surfactants studied, Tween 80 is the most promising surfactant in film homogeneity, UV absorbance, methylene blue adsorption, and can control the volume and pore size [12]. Mesoporous TiO_2 material was synthesized using PVA as a surfactant agent. Synthesis was carried out by sol-gel, ultrasonic, and hydrothermal methods. Mesoporous TiO_2 can remove inorganic and organic pollutants because of the large pore volume of TiO_2 synthesized using surfactants [13]. The surfactant cetyl trimethyl ammonium bromide (CTAB) has been commonly used to control particle size in methods such as the sol-gel and hydrothermal methods, and non-ionic surfactants such as P123 have also been used to obtain nanorods and nanoneedle structures. However, it should be emphasized that due to the nature of surfactants (anionic, cationic or non-ionic), their behaviour may differ depending on the solvent used for synthesis and the pH level of the medium, leading to differences in morphology and other properties [14], [15].

In this study, TiO_2 was synthesized using the sol-gel method where $TiCl_4$ was used as a precursor, and by adding CTAB surfactant, the results of this study were later used as a coating material on fabrics with antibacterial properties.

2. Methods

TiO_2 synthesis was carried out using the sol-gel method using $TiCl_4$ as a precursor. First, $TiCl_4$ was mixed with distilled water and stirred for 2 hours using a magnetic stirrer. While stirring, the NH_4OH solution is added to the solution little by little until the solution is white for 4 hours at 70 ^oC. Next, the reaction product in the form of a gel, separated and washed with deionized water to remove chlorine ions. Then CTAB surfactant was added and stirred again. Finally, the gel was dispersed into 300 ml of ethanol solution. The precipitate was dried at 70 ^oC for 5 hours. From the results of this study, a sample code (TiO_2 CTAB) was made. The schematic of the synthesis of TiO_2 nanoparticles is shown in Figure 1.

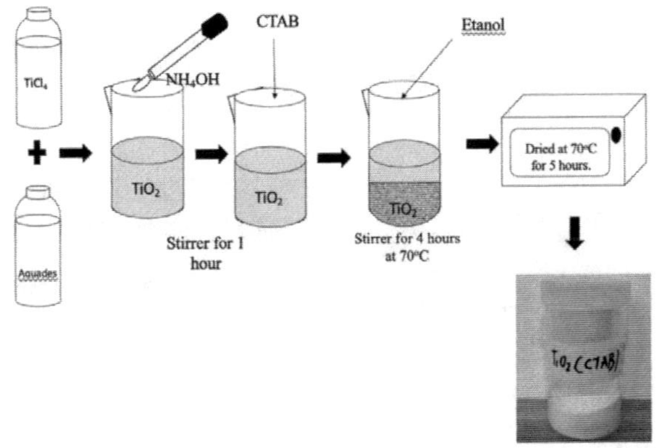

Figure 1. Schematic of TiO_2 Nanoparticle Synthesis.

3. Result and Discussion

3.1. X-Ray Diffraction (XRD) Characterization

X-ray diffraction (XRD) analysis was used to measure particle size and structural analysis of commercial TiO_2, and TiO_2 sol-gel process and CTAB surfactant used a Goniometer type Shimadzu 6000 instrument with a Cu/Kα1 X-ray source with a wavelength, = 1, 54056. The analysis was carried out with an angle of 2θ ranging from 7° to 70°. In addition, the crystal size (D) of the material is carried out using the Debye Scherrer equation.

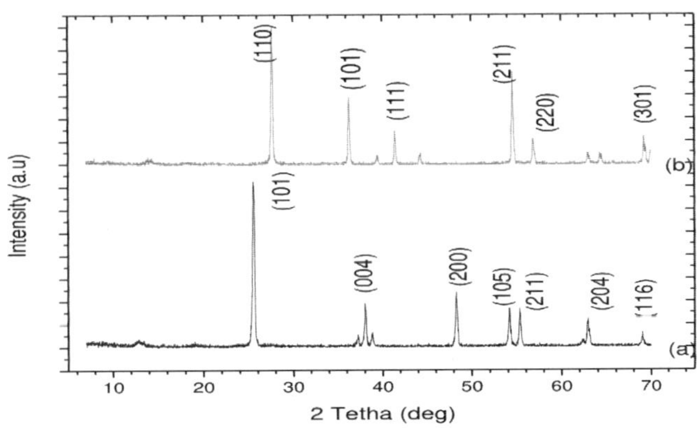

Figure 2. Diffraction pattern of a). Commercial TiO_2, b). TiO_2 CTAB.

From Figure 2. The diffraction pattern on commercial TiO_2 is obtained; the maximum intensity is found at an angle of 2θ = 25.525° with a distance of 3.4897, has an anatase phase (TiO_2). Room group I41/amd(141). has a tetragonal system with lattice parameters a = 3.777 c = 9.501 and has a mass of 3.915 g/cm³ with dhkl (101). The particle size of commercial TiO_2 obtained from XRD using the Debye Scherrer method is 32.16 nm. Meanwhile, for TiO_2 nanoparticles synthesized with CTAB surfactant, the maximum intensity is found at an angle of 2θ = 27.65° with a distance of 3.2414 having a rutile phase (TiO_2). The results of the X-ray diffraction pattern have a tetragonal system of space group P42/mnm (136) with lattice parameters a = 4.584 c = 2.953 and have a mass of 4.248 g/cm³ with dhkl (110). The particle size of TiO_2 with CTAB surfactant obtained from XRD using the Debye Scherrer method is 17.75 nm. The decrease in crystal size in TiO_2 synthesized with CTAB surfactants was caused by CTAB, preventing agglomeration and limiting crystal growth [16], [17].

3.2. Scanning Electron Microscope (SEM)Characterization.

SEM analysis was carried out to describe the surface of TiO_2 Comerscial and TiO_2 CTAB the composition of the compounds contained in it. The analysis was carried out using the ZEISS VO MA 10 SEM tool.

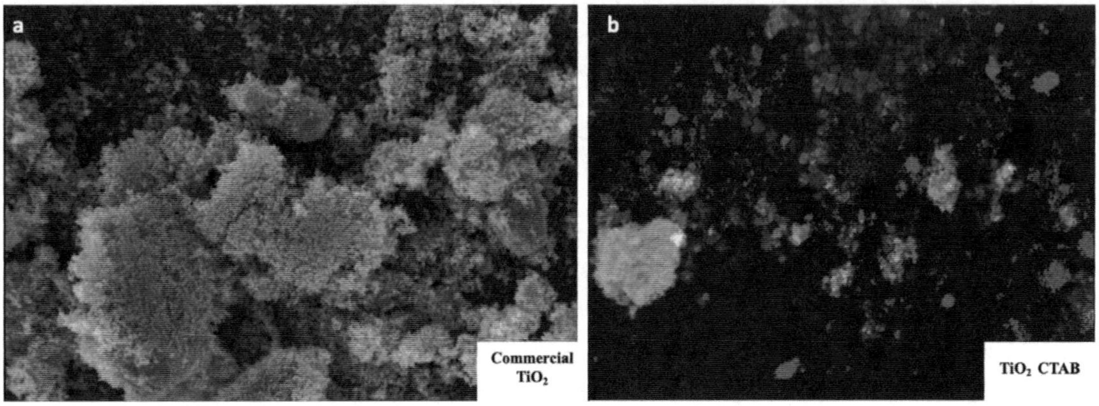

Figure 3. SEM a). Commercial TiO_2, b). TiO_2 CTAB.

The SEM results in Figure 3b of the nanoparticles become less agglomerated than commercial TiO_2 due to the presence of a long CTAB hydrocarbon chain that prevents agglomeration. As a result, a less aggregated irregular ball-like structure is seen in TiO_2 CTAB [18], [19].

3.3. Fourier-transform Infrared Spectroscopy (FTIR) Characterization.

Characteristics of the compounds that compose it. One chemical property is brought from the functional groups of the compounds that make up the material. Each functional group of the compound has specific energy to vibrate in several modes. The energy absorbed by each functional group is detected to vibrate in various modes. The amount of energy absorbed (frequency) can be determined by what compounds are contained in the material. Several other aspects of the measurement can be used as material to analyze the material being tested. It can be characterized using FTIR.

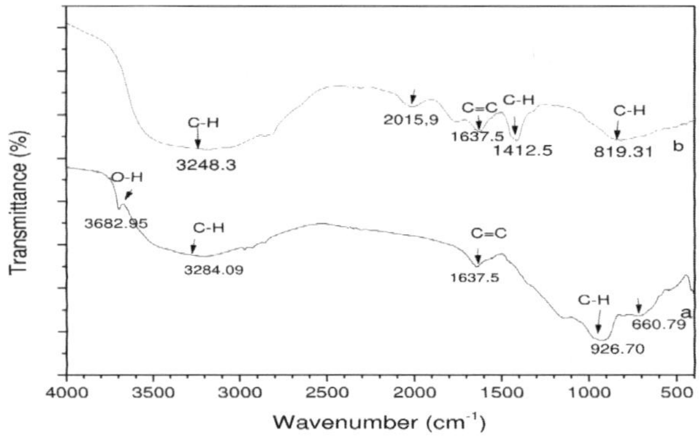

Figure 4. FTIR Nanoparticles a). Commercial TiO_2, b). TiO_2 CTAB.

Figure 4 shows a broad absorption band between 800 and 400 cm^{-1} correspondings to Ti–O vibrations in the crystal lattice. The broad absorption bands between 3600 and 3000 cm^{-1} and bands at 1637 cm^{-1} are due to the deformation and strain vibration of the OH group, respectively, and these bands indicate the presence of weakly bound water molecules on TiO_2. This characteristic is needed in photocatalysis because the OH group can react with holes and prevent the recombination of electron-hole pairs. The 2,015.9 absorption band in the FTIR spectrum of the CTAB samples can be attributed to the CH bond vibrations of the methyl and methylene groups, respectively, of the surfactant residues [20]–[22].

3.4. X-Ray Fluorescence (XRF) Characterization

XRF characterization of TiO_2 samples was carried out using XRF PANalytical Minipal 4. This tool can be used to test the elemental content of a material ranging from Sodium-Uranium. Samples can be solid, powder and liquid. The X-ray tube of this instrument allows the use of different target materials to avoid the possibility of masking the sample spectrum with the tube spectrum. The rhodium target is standard, with chromium, molybdenum and tungsten target materials available as options. Table 1 shows the compound content of commercial TiO_2 and TiO_2 CTAB obtained from characterization using XRF.

Table 1. Compound content of commercial TiO_2 and TiO_2 CTAB.

Compound	TiO_2 Comersil (%)	TiO_2 CTAB (%)
TiO_2	99.01	80.80
K_2O	0.14	-
CaO	0.10	0.18
P_2O_5	0.31	-
Fe_2O_3	-	0.75
Cl	-	16.5
V_2O_5	-	0.64
WO3	0.26	-
PtO2	0.18	0.089
BrO	-	0.15
Yb_2O_3	-	0.17
ZnO	-	0.13

The content of commercial TiO_2 compounds is 99.01%, while TiO_2 CTAB is 80.80%, followed by 16.5 Cl. Thus, the Cl content is most likely obtained from the $TiCl_4$ residue that is not entirely lost. Likewise, the presence of BrO content most likely comes from CTAB [23], [24]. TiO_2 can be used as a coating because it has antimicrobial properties on photocatalyst activity. TiO_2 coating can prevent the emergence of gram-positive bacteria, gram-negative bacteria, viruses, and fungi. When TiO_2 is irradiated with light, the bandgap will produce electron-hole pairs, which cause a redox reaction on the TiO_2 surface. Therefore, electrons will move from the valence band to the conduction band so that a pair of electrons and holes will be formed on the surface of the photocatalyst. Negatively charged electrons and oxygen will combine to form O^{2-} while positively charged holes and water will produce hydroxyl radicals. Therefore, various radical oxygen species (ROS) can be oxidized to carbon dioxide (CO_2) and water (H_2O) [25].

4. Conclusion

From the results of preparation and characterization of TiO_2 nanoparticles modified by CTAB surfactants, it was found that these materials can be used as coatings on fabrics. After the modification, there was a decrease in particle size and a reduction in agglomeration of TiO_2 CTAB compared to commercial TiO_2. The rutile phase obtained from TiO_2 CTAB allows photocatalyst activity to prevent bacterial contamination. In addition, with the modification, the TiO_2 content is sufficient. TiO_2 as a coating on fabrics can be used in hospital facilities, including bed sheets, pillowcases, blankets, curtains, towels, and patient clothes, which can be contaminated by bacteria, viruses, and fungi from blood, skin, feces, urine, and vomit.

Acknowledgement

Universitas Sumatera Utara funded this research for the TALENTA Doctoral Research Grant. The fiscal Year 2021 Number: 35/UN5.2.3.1/PPM/SPP-TALENTA USU/2021.

References

[1] J. Yuenyongsuwan, N. Nithiyakorn, P. Sabkird, E. A. O'Rear, and T. Pongprayoon, "Surfactant effect on phase-controlled synthesis and photocatalyst property of TiO2 nanoparticles," Materials Chemistry and Physics, vol. 214, pp. 330–336, 2018, doi: 10.1016/j.matchemphys.2018.04.111

[2] M. Montazer and S. Seifollahzadeh, "Enhanced self-cleaning, antibacterial and UV protection properties of nano TiO2 treated textile through enzymatic pretreatment," Photochemistry and Photobiology, 2011, doi: 10.1111/j.1751-1097.2011.00917.x

[3] H. Zhang and J. F. Banfield, "Understanding Polymorphic Phase Transformation Behavior during Growth of Nanocrystalline Aggregates: Insights from TiO 2," no. 120, pp. 3481–3487, 2000

[4] B. F. Bukit and S. H. Sirait, "Preparasi dispersi antibakteri berbahan dasar titanium dioksida (TiO_2) serta aplikasinya pada kain dengan metoda dip coating," JUITECH (Jurnal Ilmiah Fakultas Teknik Universitas Quality), 2019, doi: 10.36764/ju.v3i2.259

[5] Y. Yuan, J. Ding, J. Xu, J. Deng, and J. Guo, "TiO 2 nanoparticles co-doped with silver and nitrogen for antibacterial application," 2010. doi: 10.1166/jnn.2010.2225

[6] N. Prorokova, T. Kumeeva, and I. Kholodkov, "Formation of coatings based on titanium dioxide nanosolson polyester fibre materials," Coatings, vol. 10, no. 1, 2020, doi: 10.3390/coatings10010082

[7] M. Malekshahi Byranvand, A. Nemati Kharat, L. Fatholahi, and Z. Malekshahi Beiranvand, "A Review on Synthesis of Nano-TiO2 via Different Methods," Journal of Nanostructures, vol. 3, no. 1, pp. 1–9, 2013, doi: 10.7508/jns.2013.01.001

[8] D. P. MacWan, P. N. Dave, and S. Chaturvedi, "A review on nano-TiO2 sol-gel type syntheses and its applications," Journal of Materials Science, vol. 46, no. 11, pp. 3669–3686, 2011, doi: 10.1007/s10853-011-5378-y

[9] M. Cargnello, T. R. Gordon, and C. B. Murray, "Solution-phase synthesis of titanium dioxide nanoparticles and nanocrystals," Chemical Reviews, vol. 114, no. 19, pp. 9319–9345, 2014, doi: 10.1021/cr500170p

[10] I. E. Fernández and J. E. Rodríguez-Páez, "Wet-chemical preparation of TiO2-nanostructures using different solvents: Effect of CTAB concentration and tentative mechanism of particle formation," Journal of Alloys and Compounds, vol. 780, pp. 756–771, 2019, doi: 10.1016/j.jallcom.2018.12.007

[11] D. S. Kim, S. J. Han, and S. Y. Kwak, "Synthesis and photocatalytic activity of mesoporous TiO2 with the surface area, crystallite size, and pore size," Journal of Colloid and Interface Science, vol. 316, no. 1, pp. 85–91, 2007, doi: 10.1016/j.jcis.2007.07.037

[12] H. Choi, E. Stathatos, and D. D. Dionysiou, "Synthesis of nanocrystalline photocatalytic TiO2 thin films and particles using sol-gel method modified with nonionic surfactants," Thin Solid Films, vol. 510, no. 1–2, pp. 107–114, 2006, doi: 10.1016/j.tsf.2005.12.217

[13] R. Thahir, H. Bangngalino, A. Wahid Wahab, N. la Nafie, and I. Raya, "Direct synthesis of mesoporous TiO2 using PVA as surfactant template and assessment of their photocatalytic activities," IOP Conference Series: Materials Science and Engineering, vol. 509, no. 1, 2019, doi: 10.1088/1757-899X/509/1/012124

[14] J. Payormhorm, S. Chuangchote, and N. Laosiripojana, "CTAB-assisted sol-microwave method for fast synthesis of mesoporous TiO2 photocatalysts for photocatalytic conversion of glucose to value-added sugars," Materials Research Bulletin, vol. 95, pp. 546–555, 2017, doi: 10.1016/j.materresbull.2017.08.016

[15] C. Xie, P. Li, Y. Liu, F. Luo, and X. Xiao, "Preparation of TiO2 nanotubes/mesoporous calcium

silicate composites with controllable drug release," Materials Science and Engineering C, vol. 67, pp. 433–439, 2016, doi: 10.1016/j.msec.2016.05.041

[16] H. F. Lu, R. Y. Hong, and H. Z. Li, "Influence of surfactants on co-precipitation synthesis of strontium ferrite," Journal of Alloys and Compounds, 2011, doi: 10.1016/j.jallcom.2011.08.058

[17] B. P. Zeng, Y. Jia, G. G. Xu, M. Li, and R. Feng, "Preparation of TiO2/g-C3N4 by CTAB-assisted and photocatalytic degradation of unsymmetrical dimethylhydrazine waste water," Cailiao Gongcheng/Journal of Materials Engineering, vol. 47, no. 9, 2019, doi: 10.11868/j.issn.1001-4381.2017.000933

[18] P. Gogoi, D. Dutta, and T. K. Maji, "Equilibrium and kinetics study on removal of arsenate ions from aqueous solution by CTAB/TiO2 and starch/CTAB/TiO2 nanoparticles: A comparative study," Journal of Water and Health, vol. 15, no. 1, pp. 58–71, 2017, doi: 10.2166/wh.2016.127

[19] Z. Jiang, J. Xie, D. Jiang, X. Wei, and M. Chen, "Modifiers-assisted formation of nickel nanoparticles and their catalytic application to p-nitrophenol reduction," CrystEngComm, vol. 15, no. 3, pp. 560–569, 2013, doi: 10.1039/c2ce26398j

[20] E. M. Ginting, N. Bukit, E. Frida, and B. F. Bukit, "Microstructure and thermal properties of natural rubber compound with palm oil boilers ash for nanoparticle filler," Case Studies in Thermal Engineering, 2020, doi: 10.1016/j.csite.2019.100575

[21] Z. W. Li, Y. B. Gan, X. Xin, and G. Q. Lin, "Characteristics of effective permeability and resonance frequency for barium-ferriteepoxy composites," Journal of Applied Physics, 2008, doi: 10.1063/1.2899958

[22] W. Zhuang et al., "Facile synthesis of amino-functionalized mesoporous TiO2 microparticles for adenosine deaminase immobilization," Microporous and Mesoporous Materials, vol. 239, pp. 158–166, 2017, doi: 10.1016/j.micromeso.2016.09.006

[23] T. Huang, S. Mao, J. Yu, Z. Wen, G. Lu, and J. Chen, "Effects of N and F doping on structure and photocatalytic properties of anatase TiO2 nanoparticles," RSC Advances, vol. 3, no. 37, pp. 16657–16664, 2013, doi: 10.1039/c3ra42600a

[24] D. Huang, Y. J. Wang, Y. C. Cui, and G. S. Luo, "Direct synthesis of mesoporous TiO2 and its catalytic performance in DBT oxidative desulfurization," Microporous and Mesoporous Materials, vol. 116, no. 1–3, pp. 378–385, 2008, doi: 10.1016/j.micromeso.2008.04.031

[25] R. Dastjerdi and M. Montazer, "A review on the application of inorganic nano-structured materials in the modification of textiles: Focus on anti-microbial properties," Colloids and Surfaces B: Biointerfaces, vol. 79, no. 1, pp. 5–18, 2010, doi: 10.1016/j.colsurfb.2010.03.029

Journal of Physics: Conference Series **2165** (2022) 012023 doi:10.1088/1742-6596/2165/1/012023

Preparation and characterization of Bentonite-OPBA nanocomposite as filler

E Frida[1], N Bukit[2], F R A Bukit[3*] and B F Bukit[4]

[1]Department of physics Universitas Sumatera Utara, 20155, Medan, Indonesia
[2]Department of physics, Universitas Negeri Medan, 2022, Medan, Indonesia
[3]Department of Electrical Engineering, Universitas Sumatera Utara, 20155, Medan, Indonesia
[4]Universitas Quality Berastagi, 22153, Berastagi, Sumatera Utara, Indonesia

E-mail: ferryrahmatastiantabukit@gmail.com

Abstract. Bentonite-OPBA nanocomposite has been made. The preparation of OPBA and bentonite was carried out by calcination process at a temperature of 700°C for bentonite and 500 °C for OPBA and a ball mill process for 10 hours and coprecipitation method with a solution 6 M HCl. Then nano bentonite was synthesized with surfactant Cetyl Trimethyl Ammonium Bromide (CTAB). The XRD analysis obtained a Bentonite-OPBA particle size of 14.46 nm, with a Tetragonal crystal system and density of 3.01200 g/cm³. XRF analysis obtained SiO_2 content of 55.1 % FTIR analysis obtained absorption band detected at 981.33 cm^{-1} indicates that the increase in silica strengthening has in the polymerization process.

1. Introduction

In recent years, many researchers have produced environmentally friendly products such as geopolymers. The production of geopolymers is an alumina-silicate based material rich in silicon (Si) and Aluminium (Al) activated by alkaline solutions. Previous studies have shown that geopolymers have excellent mechanical properties, high chemical resistance and the ability to bind hazardous wastes (Javed et al., 2020). One of the alumina silicate originating from geopolymer comes from oil palm boiler ash (OPBA). In Indonesia in 2018, the total area of oil palm plantations was recorded at 14,326,350 hectares (Statistics, 2018), so it is likely to generate OPBA waste. OPBA comes from the burning residue of palm kernel shells and fruit fibres containing the chemical element silica (SiO_2) as much as 31.45% and lime (CaO) as much as 15.2%. This material is formed in the exhaust gases and collected using an electrostatic precipitator. Since these particles solidify while suspended in the exhaust gases, they are generally spherical in shape [1, 2]. Based on the research conducted, it turns out that OPBA contains a lot of silica (SiO_2) which is a pozzolan and filler material for composites [3], [4].

Bentonite is an abundant natural resource in Indonesia, but its utilization has not been optimal. One application of bentonite currently being studied by many international and national research institutions is its use as a nano-sized filler, better known as a nanofiller—modification of bentonite into organoclay material with surfactants, as organic layer silica (OLS). In the OLS structure, the distance between the basalts in bentonite (d-spacing) is enlarged, wherein processing with polymer materials in the melting phase, the d-spacing will increase in intercalate and finally, the layer structure contained in bentonite is separated from each other . exfoliation so that bentonite is dispersed into the polymer system with a size smaller than 100 nm. Thermal and acid treatment processes can carry out the synthesis of silica nanoparticles from bentonite to reduce alumina and increase silica content [5]. Bentonite modification by combining other materials can improve the performance of bentonite. For example, combining bentonite/rice husk in the HDPE matrix can improve the mechanical properties of the composite [6]. In addition, combining bentonite with carbon black, which is used as a natural rubber filler, can affect tensile strength, elongation at break, modulus of elasticity, and composite hardness [7, 8]. In addition, the addition of nano SiO_2 and bentonite was able to increase the compressive strength of the mortar [9].

In this study, bentonite nanoparticles were prepared by calcining bentonite and ball mill, coprecipitation, and CTAB surfactant. At the same time, the manufacture of OPBA nanoparticles is carried out by calcining OPBA followed by the ball mill and coprecipitation method. The addition of CTAB surfactant is expected to prevent agglomeration and control pore volume and size [10–12].

2. Methods

Synthesis of Nano Bentonite

Bentonite calcination at a temperature of 700 ^oC for 5 hours and milling with a ball mill type Planetary Ball Mill for 10 hours with a rotation of 250 rpm. Furthermore, the synthesis of bentonite by coprecipitation method using 6 M HCl using a stirrer at 70 ^oC for 4 hours. Next, mix bentonite with 7 M NaOH using a stirrer at 70 ^oC for 4 hours. Wash bentonite with distilled water until it produces a neutral pH. It is drying bentonite at 250 ^oC for 4 hours. CTAB was dissolved in distilled water and then heated and stirred at 80 ^oC for 1 hour. Besides that, mix bentonite with deionized water at 70 ^oC for 2 hours. Mixing the CTAB solution and bentonite dispersion at 70 ^oC for 3 hours and then filtered. Adding $AgNO_3$ with distilled water. The mixture was then dried in an oven at 100 ^0C for 10 hours. The schematic for the manufacture of nano bentonite is shown in Figure 1.

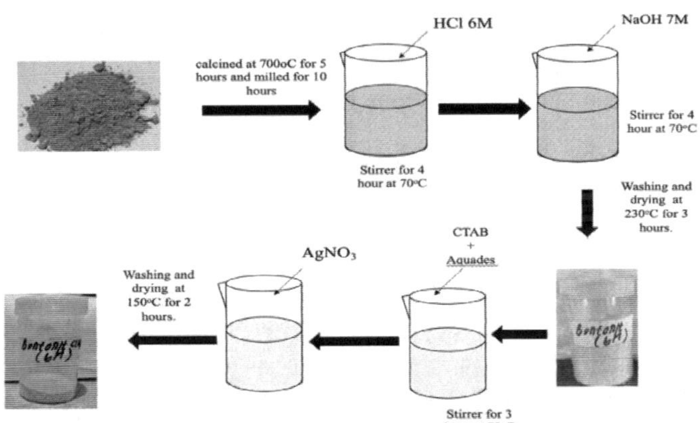

Figure 1. Schematic of nano bentonite preparation.

Synthesis of OPBA nanoparticle

Synthesis of OPBA nanoparticles was carried out by ball mill and coprecipitation methods. OPBA obtained was calcined at a temperature of 500 ^oC for 5 hours; the calcined results were milled with a ball mill type Planetary Ball Mill type Retsch PM200 for 10 hours with a rotation of 250 rpm. The OPBA ball mill process was mixed with 6 M HCl solution at a temperature of 70 ^0C for 4 hours. It was mixing OPBA with 7 M NH_4OH for 4 hours at 70 ^0C. The OPBA was washed with distilled water until it produced a neutral pH on the OPBA and then precipitated. The precipitate was dried at 200 ^oC for 2 hours. The schematic for the preparation of OPBA nanoparticles is shown in Figure 2.

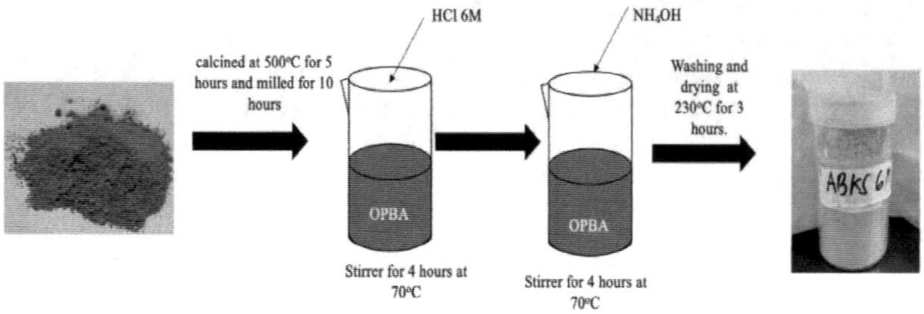

Figure 2. Schematic of making OPBA nano nanoparticles.

Bentonite/OPBA nanocomposite preparation
Nano bentonite and OPBA were mixed by adding 100 ml of ethanol using a stirrer at a temperature of 70 °C for 2 hours, then washed with distilled water to produce a neutral pH on the nanocomposite precipitated. The precipitate was dried at 150 °C for 2 hours. The schematic for the manufacture of bentonite/OPBA nanocomposites is presented in Figure 3.

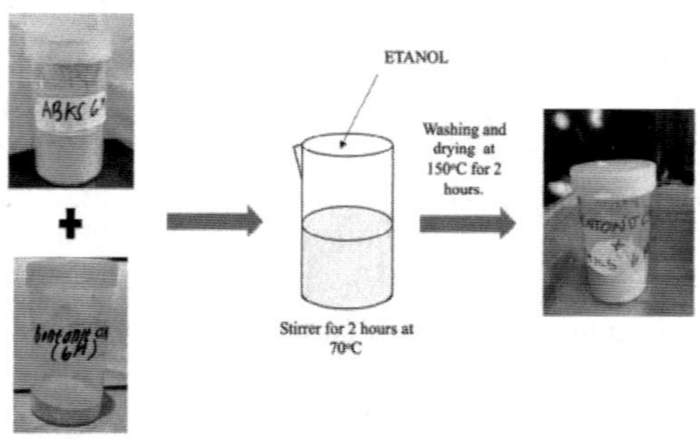

Figure 3. Schematic of making bentonite-OPBA nano nanocomposite.

3. Result and Discussion

3.1. XRD analysis
X-ray diffraction (XRD) analysis to measure particle size and structure analysis of bentonite and OPBA used a Goniometer type Shimadzu 6,000 instrument with a Cu/Kα1 X-ray source with a wavelength, = 1.54056. The analysis was carried out with an angle of 2θ starting from 7° to 70°. To identify the diffraction pattern of the synthesized material, the resulting diffractogram is compared with the JCPDS (Joint Committee on Powder Diffraction Standard) database. The crystal size (D) of the material is carried out using the equation:

$$D = \frac{K\lambda}{\beta \cos\theta} \quad (1)$$

Where D is the crystal size (Å), is the X-ray wavelength used, namely CuKα radiation (λ = 1.54056), is the Bragg angle, = Scherrer's constant has a general value of 0.9 and is the half-peak width of Full Width at Half Maximum (FWHM) in radians.

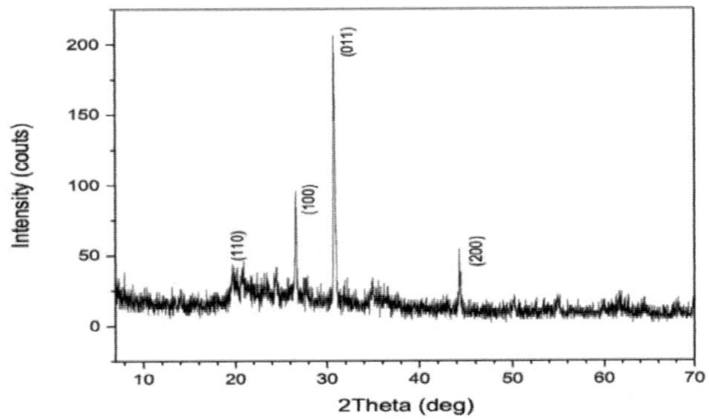

Figure 4. X-ray diffraction analysis for bentonite and OPBA.

Bentonite-OPBA XRD diffraction pattern is presented in Figure 4. The peak angles occur at 2 theta angles of 20.25°, 26.50°, 30.61° 44.43°, which indicate a geopolymer material containing silica, anorthite/albite, and calcite minerals. Anorthite and Albite are hydrothermal products in nature (in marshlands etc.) as well as in geopolymers. This zeolitic product exhibits a polymerization process in which geopolymers are made from aluminosilicates such as fly ash, slag, bottom ash, etc. This proves the success of combining bentonite with OPBA [13–15]. XRF and FTIR results where the dominant silica content is obtained [16]. Meanwhile, the data on the physical properties and nanostructures produced by the match program is shown in Table 1. From equation 1 using an analysis with the origin graph, the Bentonite-OPBA particle size obtained is 14.46 nm.

Table 1. XRD Analysis of OPBA and Bentonite nanoparticles with ball mill processes.

Date	Bentonite-OPBA
Crystal system	Trigonal (hexagonal axes)
Space group	P 31 2 1 (152)
Unit cell	a = 4.6764 Å c = 5.2475 Å
Density	3.01200 g/cm³
2 theta angle	30.61
Maximum d_{hkl}	011
Intensity I/I_0	1000
Lattice distance d (Å)	3.2061
2 theta angle	26.50
Maximum d_{hkl}	100
Intensity I/I_0	858.2
Lattice distance d (Å)	4.0499
2 theta angle	44.43
Maximum d_{hkl}	200
Intensity I/I_0	527.3
Lattice distance d (Å)	2.0249
2 theta angle	20.25
Maximum d_{hkl}	110
Intensity I/I_0	375.8
Lattice distance d (Å)	2.3382

3.2. SEM Analysis
To describe the surface morphology of Bentonite-OPBA and the composition of the compounds contained in it, SEM analysis was carried out using the EVO MA 10 Zeiss instrument. SEM of Bentonit-OPBA is presented in Figure 5.

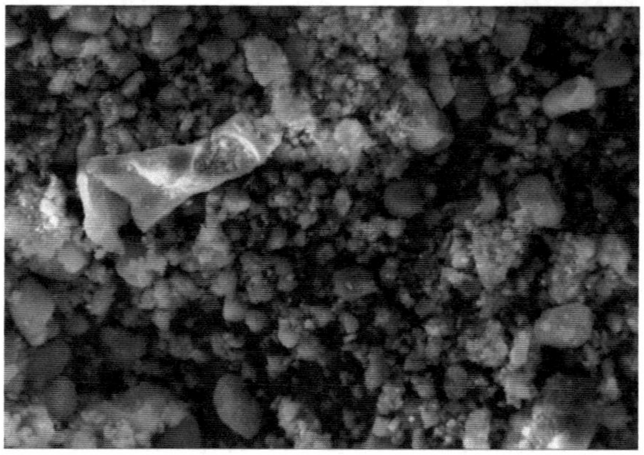

Figure 5. Scanning Electron Microscope of Bentonit-OPBA.

SEM analysis showed that structure is less aggregated this is probably caused by modifications with the use of CTAB surfactants on bentonite [17–19].

3.3. FTIR Analysis
Each functional group of the compound has specific energy to vibrate in several modes. The energy absorbed by each functional group is detected to vibrate in various modes. The amount of energy absorbed as a function of frequency can be determined by what compounds are contained in the material. Several other aspects of the measurement can be used as material to analyze the material being tested. It can be characterized using FTIR. To see the functional groups of organic bentonite and OPBA, characterization was carried out using the FT-IR Shimadzu IR Prestige 21. FTIR spectrum of bentonit-OPBA showed in Figure 6.

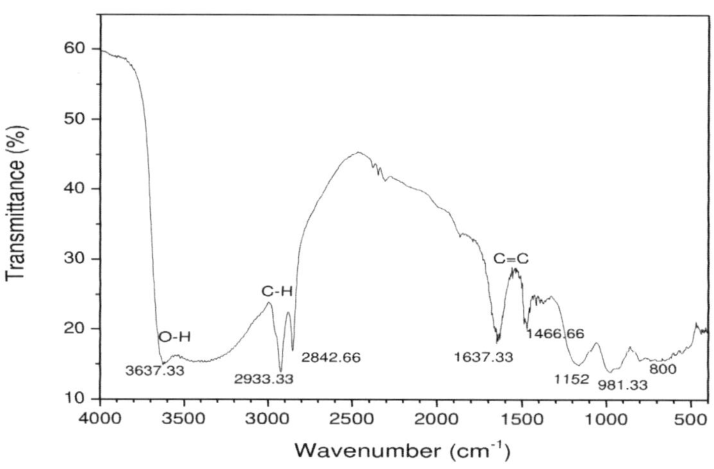

Figure 6. Fourier transform-infrared analysis for bentonite and OPBA.

The degree of polymerization can be expressed quantitatively from the Si–O–Si stretching peaks in the FTIR spectrum. The absorption band detected at 981.33 cm^{-1} indicates that the increase in silica strengthening has increased in the polymerization process. The stretching deformation (SiO–Al/Fe/Mg) of 800 cm^{-1} is related to the vibrational characteristics of the Si–O quartz. The deformation of Al–O 981.33 cm^{-1} is associated with Mg-Al–OH. The signal at 1637.33 cm^{-1} corresponds to the deformation of H_2O (adsorbed between the aluminum-silicate layers). The band at 3630 cm^{-1} is associated with Al-OH stretching [20].

3.4. XRF Analysis

XRF characterization on Bentonit-OPBA samples was carried out using XRF PANalytical Manipal 4. This tool can be used to test the elemental content of a material ranging from Sodium-Uranium. Table 2 showed the content of Bentonit – OPBA.

Table 2. Elements obtained from XRF analysis.

Compound	Bentonit-OPBA
SiO_2	55.10
K_2O	3.22
CaO	9.19
Fe_2O_3	12.10
Al_2O_3	12.00
Ti	1.60
MnO	0.20
Cr_2O_3	0.06
ZnO	0.03
V_2O_5	0.03
P_2O_5	3.20
Br	3.53
Re_2O_7	0.10
Eu_2O_3	0.20

The main content of Bentonite-OPBA is SiO_2, Fe_2O_3, Al_2O_3. This is by the main constituents of bentonite and OPBA, respectively (Aljlil, 2020; Bukit et al., 2019; Javed et al., 2020; Kueaket & Tonnayopas, 2021). The SiO_2 content obtained is 51.1 %, this is greater than the results of previous studies because OPBA is calcined to a temperature of 500 °C [21, 22].

4. Conclusion

From the results of the study, it was found that the peak angle occurred at an angle of 2 thetas 20.25°, 26.50°, 30.61°, 44.43° which indicated a geopolymer material containing silica, anorthite/albite, and calcite minerals. In addition, the particle size obtained is 14.46 nm. SEM results show a structure like an irregular ball that spreads evenly. The absorption band detected at 981.33 cm^{-1} indicates that the increase in silica strengthening has led to an increase in the polymerization process. This is by the XRF results obtained, where SiO_2 is the dominant element. Therefore, Bentonite-OPBA Nanocomposite can be used as a filler in thermoplastics and elastomers.

Acknowledgement

Thank you to the Directory of Research and Community Service Deputy for Research Strengthening and Development of the Ministry of Research and Technology/National Research and Innovation Agency for Financing Penelitian Dasar Unggulan Perguruan Tinggi (PDUPT). The fiscal Year 2021 No: 12 / E1/KP.PTNBH/ 2021, dated 08 March 2021.

References

[1] E. M. Ginting, N. Bukit, E. Frida, and B. F. Bukit, "Microstructure and thermal properties of natural rubber compound with palm oil boilers ash for nanoparticle filler," Case Studies in Thermal Engineering, 2020, doi: 10.1016/j.csite.2019.100575

[2] M. Yang et al., "Facile Preparation of Robust Superhydrophobic Cotton Textile for Self-Cleaning and Oil-Water Separation," Industrial and Engineering Chemistry Research, vol. 58, no. 1, 2019, doi: 10.1021/acs.iecr.8b04433

[3] C. Jaturapitakkul, J. Tangpagasit, S. Songmue, and K. Kiattikomol, "Filler effect and pozzolanic reaction of ground palm oil fuel ash," Construction and Building Materials, 2011, doi: 10.1016/j.conbuildmat.2011.04.073

[4] N. H. A. Khalid et al., "Palm oil fuel ash as potential green micro-filler in polymer concrete," Construction and Building Materials, 2016, doi: 10.1016/j.conbuildmat.2015.11.038

[5] A. Ramadhan, M. I. Fathurrohman, and B. Soegijono, "Chemical Modification of Upgraded Bentonite as Filler of Natural Rubber/Organobentonite Composite," Procedia Chemistry, 2015, doi: 10.1016/j.proche.2015.12.028

[6] F. Mulana, "Study on formation and mechanical properties of rice husk / natural bentonite prepared composites," pp. 83–89, 2013

[7] N. Bukit and E. Frida, "The Effect Zeolite Addition in Natural Rubber Polypropylene Composite on Mechanical, Structure, and Thermal Characteristics," MAKARA Journal of Technology Series, 2014, doi: 10.7454/mst.v17i3.2926

[8] E. M. Ginting, N. Bukit, Muliani, and E. Frida, "Mechanical properties and mophology natural rubber blend with bentonit and carbon black," 2017, doi: 10.1088/1757-899X/223/1/012003

[9] V. K. Karunarathne, S. C. Paul, and B. Šavija, "Development of nano-SiO2 and bentonite-based mortars for corrosion protection of reinforcing steel," Materials, vol. 12, no. 16, pp. 1–14, 2019, doi: 10.3390/ma12162622

[10] F. F. Cai, Z. H. Yang, J. Huang, G. M. Zeng, L. ke Wang, and J. Yang, "Application of cetyltrimethylammonium bromide bentonite-titanium dioxide photocatalysis technology for pretreatment of aging leachate," Journal of Hazardous Materials, 2014, doi: 10.1016/j.jhazmat.2014.04.050

[11] H. Choi, E. Stathatos, and D. D. Dionysiou, "Synthesis of nanocrystalline photocatalytic TiO2 thin films and particles using sol-gel method modified with nonionic surfactants," Thin Solid Films, vol. 510, no. 1–2, pp. 107–114, 2006, doi: 10.1016/j.tsf.2005.12.217

[12] B. P. Zeng, Y. Jia, G. G. Xu, M. Li, and R. Feng, "Preparation of TiO2/g-C3N4 by CTAB-assisted and photocatalytic degradation of unsymmetrical dimethylhydrazine waste water," Cailiao Gongcheng/Journal of Materials Engineering, vol. 47, no. 9, 2019, doi: 10.11868/j.issn.1001-4381.2017.000933

[13] E. U. Haq et al., "Reinforcement of Silica Particles in Bentonite Clay Based Porous Geopolymeric Material," Silicon, no. 2, 2020, doi: 10.1007/s12633-020-00633-9

[14] E. Ul Haq, S. Kunjalukkal Padmanabhan, and A. Licciulli, "Synthesis and characteristics of fly ash and bottom ash based geopolymers-A comparative study," Ceramics International, vol. 40, no. 2, pp. 2965–2971, 2014, doi: 10.1016/j.ceramint.2013.10.012

[15] M. Sirait, N. Bukit, and N. Siregar, "Preparation and characterization of natural bentonite in to nanoparticles by co-precipitation method," AIP Conference Proceedings, vol. 1801, no. January, 2017, doi: 10.1063/1.4973084

[16] M. Sirait, S. Gea, N. Bukit, N. Siregar, and C. Sitorus, "Synthesis of nanobentonite as heavy metal adsorbent with various solvents," Oriental Journal of Chemistry, vol. 34, no. 4, pp. 1854–1857, 2018, doi: 10.13005/ojc/3404020

[17] P. Gogoi, D. Dutta, and T. K. Maji, "Equilibrium and kinetics study on removal of arsenate ions from aqueous solution by CTAB/TiO2 and starch/CTAB/TiO2 nanoparticles: A comparative study," Journal of Water and Health, vol. 15, no. 1, pp. 58–71, 2017, doi: 10.2166/wh.2016.127

[18] Z. Jiang, J. Xie, D. Jiang, X. Wei, and M. Chen, "Modifiers-assisted formation of nickel nanoparticles and their catalytic application to p-nitrophenol reduction," CrystEngComm, vol. 15, no. 3, pp. 560–569, 2013, doi: 10.1039/c2ce26398j
[19] N. Bukit, E. M. Ginting, E. A. Hutagalung, E. Sidebang, E. Frida, and B. F. Bukit, "Preparation and characterization of oil palm ash from boiler to nanoparticle," Reviews on Advanced Materials Science, vol. 58, no. 1, pp. 195–200, 2019, doi: 10.1515/rams-2019-0023
[20] C. M. Legarto, A. Scian, and M. B. Lombardi, "Preparation and characterization of bentonite nanocomposites via sol–gel process," SN Applied Sciences, vol. 1, no. 7, pp. 1–7, 2019, doi: 10.1007/s42452-019-0801-0
[21] N. Bukit, E. M. Ginting, I. S. Pardede, E. Frida, and B. F. Bukit, "Mechanical properties of composite thermoplastic hdpe / natural rubber and palm oil boiler ash as a filler," Journal of Physics: Conference Series, vol. 1120, no. 1, 2018, doi: 10.1088/1742-6596/1120/1/012003
[22] E. M. Ginting, Motlan, N. Bukit, M. T. Saragih, A. H. Sinaga, and E. Frida, "Preparation and Characterization of Oil Palm Empty Bunches Powder," Journal of Physics: Conference Series, vol. 1120, no. 1, 2018, doi: 10.1088/1742-6596/1120/1/012004

Journal of Physics: Conference Series **2165** (2022) 012024 doi:10.1088/1742-6596/2165/1/012024

Simulation of photomultiplier tube malfunction and its effect on SPECT imaging system

A T Oktaviana[1,2], M Meireni[1] and K Fahmi[1]

[1]Study Program of Physics, Faculty of Mathematics and Sciences Military, Indonesia Defense University, Bogor, 16810, Indonesia
[2]Department of Physics, Faculty of Mathematics and Sciences, Universitas Indonesia, Depok, 16424, Indonesia

E-mail: aditya.oktaviana@idu.ac.id

Abstract. Ring artifact appears on the image from SPECT (Single Photon Emission Computed Tomography) imaging system is related to the uniformity. Generally, ring artifacts arose due to the poor performance of the photomultiplier tubes (PMT) in the gamma camera. If one of the PMT cannot perform optimally, defects known as ring artifacts on the image will occur. Detecting the position of the PMT malfunction is a challenge. To study the detection of ring artifacts due to PMT malfunction, artifact reconstruction was conducted by masking one of the PMT on the gamma camera using a copper sheet sizes 4×4 cm^2 and thickness 1 mm. The position of the copper sheet was varied on the horizontal (x) and vertical (y) axes from the center of the detector. Jaszczak phantom without inserts (cold rods and sector cold rods) was used as an imaging object during acquisition. The image results obtained are ring artifact if the copper sheet is positioned on the (x,0), and half-ring artifact on the (0,y).

1. Introduction

SPECT commonly use as a diagnostic imaging modality in nuclear medicine. It requires routine quality control (QC) before being used to obtain a high-quality image. QC of the SPECT system consists of three parameters; homogeneity, contrast, and spatial resolution [1]. Artifact, which causes the poor image quality, that appeared on the image from SPECT imaging system must be considered.

Artifact is a defect that appears on digital image processing. Those artifacts arose from several cases; such as from radiopharmaceutical sources [2], gamma camera [3], patient [4], and image acquisition to reconstruction procedures [5]. Some methods conducted to detect and reduce those artifacts. Analytical methods for identifying the circular and noncircular artifacts were introduced by Gullberg [6]. O'Connor and Vermeersch tried to quantify the magnitude of ring artifact and its relation with image noise [7]. The annular ring sampling technique was used for quantifying ring artifact on the image by Madsen [8]. Nonetheless, the best way to minimize those artifacts that appear in the image is by doing a better QC [9].

Ring artifact is related to the homogeneity parameter, in other words, uniformity of the image. Generally, those artifacts arose from gamma cameras on SPECT imaging systems [10]. The gamma camera consists of the collimator, NaITl crystals and PMT arrays in detecting the emitted radiation from the radiopharmaceutical source used [11]. Ring artifact can be generated from angle-shifted collimator from COR (Center of Rotation), which made mispositioning of radiation count in the image reconstruction [12]. Even using a multi-pinhole collimator design, the artifact still exists [13]. As well hybrid system like SPECT-CT new design will generate artifacts [14]. The performance of PMT must be considered too, consider temperature increased on one of the PMT in the array can generate serious artifacts [15]. Likewise, one more important thing is maintaining the collimator from radioactive contamination, and the motion of the patient cannot be ignored [16].

In this research, uniformity measurement was conducted based on the QC protocol using a double head SPECT system. The Jaszczak phantom was used as an imaging object by removing the inserts (cold rods and sphere rods) and Tc-99m used as a radiopharmaceutical source. Continuing from the previous research [17], a PMT malfunction simulation was made by masking in face of the parallel hole LEHR (Low Energy High Resolution) collimator which convinced as in front of PMT position using the copper sheet. The position of the copper sheet then varied above, below, left, and right from the center of one of the heads, so that the image results with artifact due to PMT malfunctions will be obtained.

2. Methods

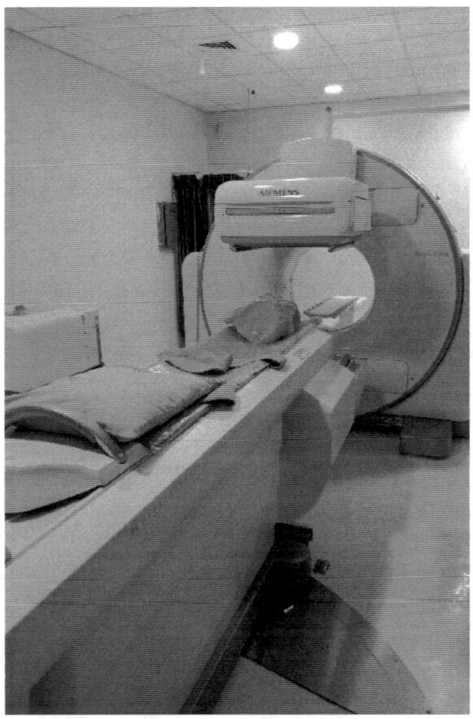

Figure 1. SPECT dual head imaging.

The SPECT variant Siemens Symbia E series intevo 16 with dual-head gamma cameras is used as the imaging system (Figure 1). The measurement of uniformity used a Jaszczak phantom Flangeless Deluxe type with the insert was removed so that it only a cylindrical phantom with a diameter of 20.4 cm, height 18.6 cm, and a volume of 6.4 L. Tc-99m radiopharmaceutical sources of approximately 3 mCi widely used as radioactive sources for imaging were mixed in water on the phantom until uniformly distributed (resulting the activity concentration to be approximately 16 kBq/mL).

Journal of Physics: Conference Series **2165** (2022) 012024 doi:10.1088/1742-6596/2165/1/012024

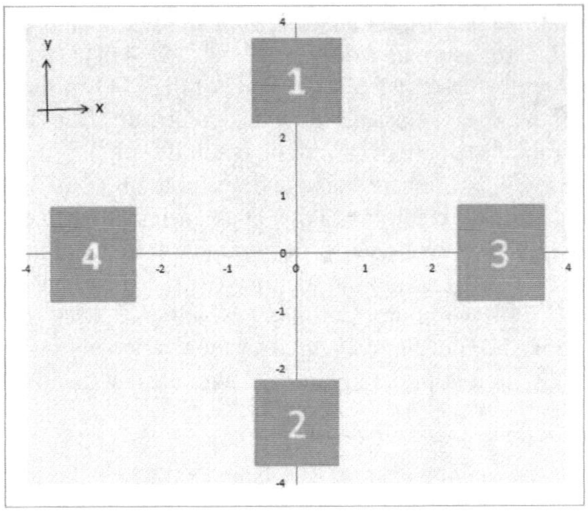

Figure 2. Variation position of copper sheet ondetector; [1] (0,3), [2] (0,-3), [3] (3,0), [4] (-3,0).

The positioning of the copper sheet, size 4 × 4 cm^2 and thickness 1 mm, as shown in the Figure 2 was used to cover the PMT so that the incoming radiation was attenuated. It was done to reconstruct the ring artifact in the resulting image and to simulate the malfunction of one of the PMT in the array on the gamma camera. The phantom is positioned at the center of the FOV (Field-of-View). The rotation of each gamma camera was 180° clockwise. The detector distance was set as close as the phantom and two gamma cameras can be moved. Image acquisition was conducted for 15 minutes for each copper sheet position. The resulting image slice was 128 slices with size 128 × 128 pixels and slice thickness 0.3 cm. Furthermore, image reconstruction generated by using filtered back projection technique and Shepp-Logan filter, including attenuation correction using Chang's method.

3. Result and Discussion

The artifact image was reconstruction was shown in the Figure 3. Artifact that generated from these images divided by three types; upper half-ring artifact, lower half-ring artifact, and possibly ring artifact. The upper half-ring artifact generated from the PMT malfunction at position (-3,0), while for the bottom half-ring artifact from position (3,0). For the full ring artifact, it was the result of PMT malfunction which is at (0,3) and (0,-3). Therefore, it can be concluded that ring artifact will be formed if the PMT malfunction is in position $(0,y)$ with $y \in \mathbb{R}$. The difference between the ring artifact generated at the PMT malfunction positions (0,3) and (0,-3) was the image slice numbers that artifact founded. For position (0,3) ring artifact appeared on slice numbers 45-54, while for position (0,-3) ring artifact appeared on slice numbers 85-94.

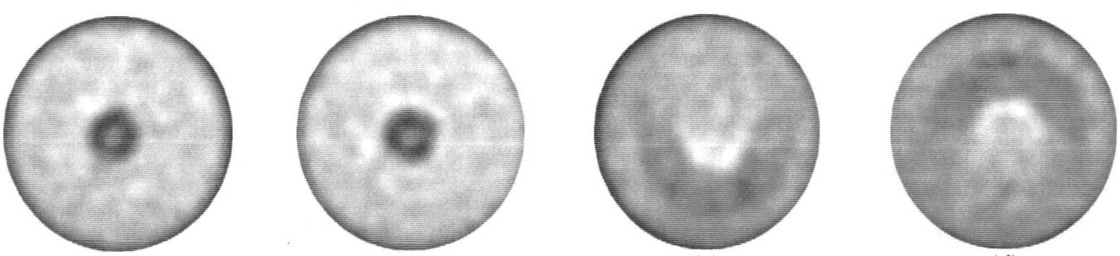

(a) (b) (c) (d)

Figure 3. Artifact image reconstruction from PMT malfunction at position (a) (0,3), (b) (0,-3),(c) (3,0) and (d) (-3,0).

The artifact image generated from the attenuation of copper material when the acquisition was performed. Although it can still receive radiation from radiopharmaceutical sources, the PMT can be considered degradation because the presence of artifact (both half-ring and full-ring) is visible. This can be seen from the distribution of pixel values on image.

For the half-ring artifact, it generated because during acquisition each gamma camera only rotate 180° [18]. For the position of the malfunction $(0,y)$, it will generate artifact on the COR during acquisition. Logically the shape of the artifact with PMT malfunction $(0,y)$ generate full circle (without ring) according to the shape of the collimator used. The possibilities as to why those ring artifact occured was emitted radiation scattering occurs in the copper sheet, thus it will produce secondary electron and new emitted photon radiation that allowed into the PMT [19].

The shape of the ring artifact (both half and full) was not smooth on the sides considering the LEHR collimator used will produce artifact at the edge [20]. To identify more accurate, it was necessary to quantify the ring artifact on image. Plot profile from image was taken from image as example shown in Figure 4, and the resulted plot profile in Figure 5. Thus the width of the ring is measured using imageJ. The width of the half ring (both generated above and below) was approximately 4 cm. However, the full ring artifact has a ring width of approximately 1 cm, with an annulus diameter of 2 cm. If the total measured from edge to another end edge of the ring artifact, the full ring artifact is actually approximately the same size as the copper sheet used. In other words, the reconstruction of ring artifact at the $(0, y)$ position with a copper sheet was still incorrect. Nevertheless, for predict the location of PMT malfunction, it can still be considered. By knowing the generated form of the artifact (half-ring or full-ring), it can identify the problem of which part of the PMT has been degradation or severe damaged.

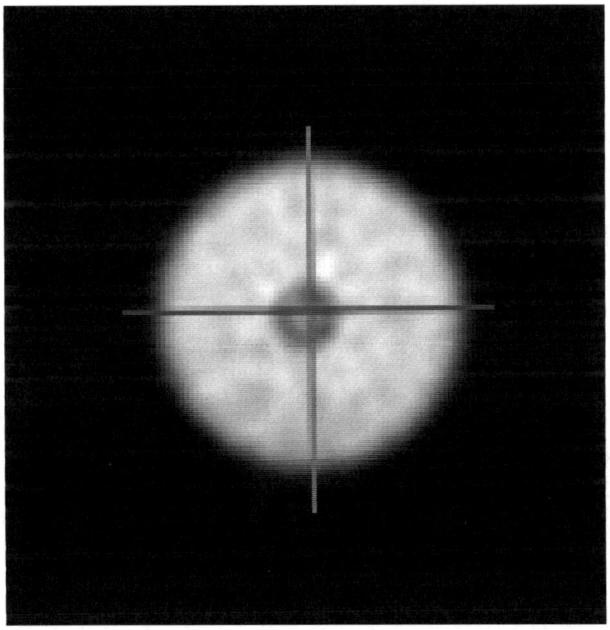

Figure 4. Blue line and red line used for plotting the profile of artifact image.

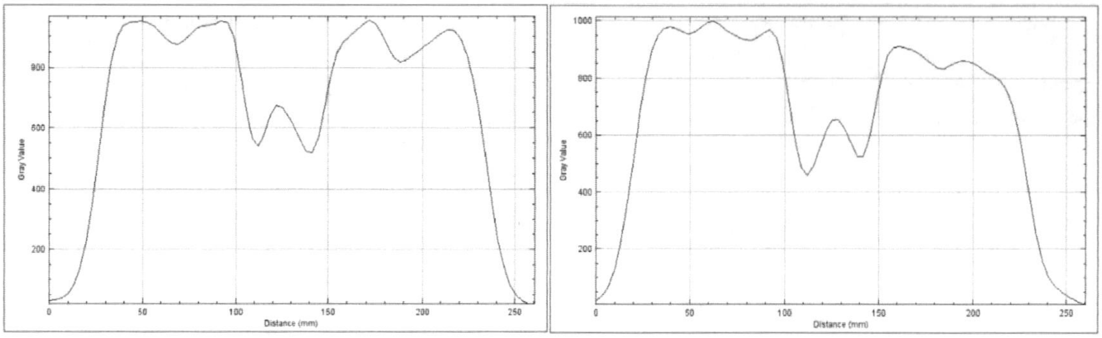

Figure 5. Plot profile distance in mm vs gray value generates from blue line (left) and red line (right).

4. Conclusion

Detecting ring artifact due to PMT malfunction (masking at possible PMT position by copper sheet) can be conducted by made acquisition of the phantom uniformity section. It can be studied to find out the location of where the degraded PMT is. For using dual head gamma cameras which rotate 180° during acquisition, ring artifact generated in half-ring form if PMT malfunction position at $(x,0)$ and full-ring form at $(0, y)$ with $x, y \in \mathbb{R}$. These results probably can make it easier to repair the damage that occurs to the gamma camera detector due to PMT malfunction. For the further investigation, analysis of the position of the PMT malfunction that produces ring artifact can also be done by comparing the image results obtained by performing a Monte Carlo simulation.

Acknowledgement

The authors thanks to Rumah Sakit Kanker Dharmais - National Cancer Center Indonesia, Jakarta, to arrange the phantom measurements and for providing the data, and thanks to Hibah PITTA UI 2018 grant number 2318/UN2.R3.1/HKP.05.00/2018 for funding this research.

References

[1] Graham L S, Fahey F H, Madsen M T, Aswegen A and Yester M V 1994 *Med. Phys* **22** 401-409
[2] Rogers W L, Clinthorne N H, Harkness B A, Koral K F and Keyes Jr. J W 1982 *J. Nucl. Med.* **23** 162-168
[3] O'Connor M K 1996 *Seminar in Nuclear Medicine* **26** 256-277
[4] Howarth D M, Forstram L A, O'Connor M K, Thomas P A and Cardew A P St E 1996 *Seminar in Nuclear Medicine* **26** 295-307
[5] Madsen M T 1994 *J. Nucl. Med. Technol.* **22** 3-11
[6] Gullberg G T 1987 *Med. Phys.* **14** 105-114
[7] O'Connor M K and Vermeersch C 1991 *Med. Phys.* **18** 190-197
[8] Madsen M T 1997 *Med. Phys.* **14** 1696-1700
[9] Groch M W and Erwin W D 2001 *J. Nucl. Med. Technol.* **29** 12-18
[10] Frye S, Frye R, Sterkel B, Muzaffar R and Osman M M 2019 *J. Nucl. Med. Technol.* **47** 223-226
[11] Peterson T and Furenlid L R 2011 *Phys. Med. Biol.* **56(17)** R145-R182
[12] Malmin R E, Stanley P C and Guth W R 1990 *J. Nucl. Med.* **31** 655-659
[13] Vunckx R, Nuyts J, Vanbilloen B, De Scent-Hubert M, Vanderghinste D, Rattat D, Mottaghy F M and Defrie M 2009 *IEEE Transactions on Nuclear Science* **56(5)** 2696-2705
[14] Seret A, Nguyen D and Bernard C 2012 *EJNMMI Research* **2** 45-64
[15] Bashir S, Shahbaz M and Iqbal D 2017 *J. Nucl. Med. Technol.* **45** 119-120
[16] Kumar N, Verma S, Singh R K R, Datta D, Kheruka S C and Gambhir S 2017 *J. Nucl. Med. Technol.* **45** 241-242

[17] Oktaviana A, Pawiro S A, Siswatining T and Soejoko D S 2019 *J. Phys.: Conf. Series* **1248** 012030
[18] Hirtl A, Bergmann H, Knausl B, Beyer T, Figl M and Hummel J 2017 *Med. Phys.* **44** 1638-1645
[19] Synder D L, Miller M I, Thomas Jr L J and Politte D G 1987 *IEEE Transactions on Medical Imaging* **MI-6(3)** 228-238
[20] ALehyani S H A A 2009 *Journal of King Saud University (Science)* **21** 109-117

Electrical impedance analysis of NaCl and $CaCl_2$ solutions based on equivalent electric circuit

E Rahmawati[1,2], D R Santoso[1], J A F Noor[1] and A Nadhir[1]

[1]Physics Department, Brawijaya University, Jl. Veteran 65145 Malang, East Java, Indonesia
[2]Physics Department, State University of Surabaya, Jl. Ketintang 60232 Surabaya, East Java, Indonesia

E-mail: endahrahmawati@unesa.ac.id, dieks@ub.ac.id

Abstract. In this paper, we report the behaviour of electrical impedance of sodium chloride (NaCl) and Calcium chloride ($CaCl_2$) with the concentration of 1.25 to 100 mM/L in frequencies ranging from 10 mHz to 1 MHz. The complex electrical impedance (real and imaginary parts) is obtained using the electrical impedance spectroscopy (EIS) technique. The experimental data had been fitted with three models of equivalent electrical circuit (EEC), model-a is Randles circuit, model-b consist of R solution (R_s) in series with constant phase element (CPE) and model-c consist of Randles circuit in series with CPE. Best fit data is obtained from model-c with goodness of fit less than 0.005. The CPE in this circuit associated with the electric double-layer. As expected from the experimental and fitting spectra, even for a low concentration for both salts, an inverse relationship was observed between impedance and salt concentration. The number of ions contains in $CaCl_2$, which are more than NaCl is estimated to be the cause that $CaCl_2$ impedance is lower than NaCl. The total impedance influenced by the CPE component increased significantly at frequencies about 100 Hz, indicating that the equivalent electric circuit used can estimate the effect of double-layer on the measurement.

1. Introduction

Ionic molecules that dissolve in a solvent produce an ionic solution that can conduct electricity. The presence of ions in this solution produces an electrostatic field that affects the surrounding ions or molecules. Many researchers examine the electrical properties of complex fluid and ionic solutions [1], various electrolyte and ionic solutions[2-4], oils [5], and beverages [6]. Most experiments are conducted at relatively high frequencies or investigated by computer-based simulation [7-8].

A powerful and non-invasive measurement technique to investigate the electrical properties of samples is Electrochemical Impedance Spectroscopy (EIS). This technique consists of applying a sinusoidal current (galvanostatic mode) or voltage (potentiostatic mode) with a certain amplitude and frequency into the sample and then measuring the amplitude and phase shift of the output voltage or current. This procedure is repeated in a specific frequency range and produces an impedance spectra. The spectra are like black boxes that must be extracted to obtain the electrical properties and behaviour of the sample. To extract information from the impedance spectrum is by fitting it using the equivalent electrical circuit model. EIS is used in a widely used in many applications as a quick and easy automated technique to characterize solid, liquid, semiliquid, organic, and inorganic materials [6, 9].

We can analyze or predict the processes that occur during measurement in three frequency ranges. High frequency corresponds to the internal ohmic resistance, including the solutions or electrolyte, active material and electrode contact. Intermediate frequencies are commonly due to electrochemical processes and electron transport at electrode/sample interfaces. The low frequency refers to surface effects due to the particles dynamic where the diffusion processes and effects on the surface due to interaction electrode-electrolyte may has a significant role [9,10].

By the applying of Ohm law, the complex electric impedance Z can be obtained from expressed by Equation (1)

$$Z = \frac{V_0\, exp(j\omega t)}{I(t)} \qquad (1)$$

Since we used galvanostatic mode, $I(t)$ is the amplitude of applied current, $j = \sqrt{-1}$, $\omega = 2\pi f$, ω is the angular frequency and V_0 is the voltage. The complex impedance $Z(\omega)$ consist of a real part $Z'(\omega)$ and an imaginary part $Z"(\omega)$ as presented in equation (2).

$$Z(\omega) = Z'(\omega) + jZ"(\omega) \qquad (2)$$

Considering applying general Ohm's law, the complex impedance can be used to describe the frequency (f), the concentration of the sample, the charge of the electric ions involved, and the temperature of the medium [2-4]. Furthermore, in measuring the impedance of ionic liquid NaCl and CaCl2, an electrical double layer phenomenon will appear at the electrode-sample interface, characterized as the ratio between changes in surface charge density to changes in electric surface potential [1].

This investigation aims to examine the spectra using three models of EEC. The first one is model-a, the Randles circuit, and the second is model-b consisting of a series of R solution (Rs) and a constant phase element (CPE) and the last one is model-c consisting of the Randles circuit in series with a CPE. The proposed EEC is used for evaluating; (1) the effect of salt concentration (from 1.25 mM/L to 100 mM/L) on electrical properties of the ionic solution of $CaCl_2$ and NaCl in ranging frequency of 10 mHz to 1 MHz, (2) the influence of the ions of NaCl and $CaCl_2$ in the electrical behaviour of salt solutions, (3) the behaviour of the double layer in interface of electrode-salt solution.

2. Methods

2. 1. Equivalent electric circuit model with CPE impedance

The EEC has been a valuable tool to evaluate the electrical properties of samples and define physical models that fit the experimental results. As shown in figure 1(a), a Randles circuit is an EEC that coomposed of an active solution resistance R_s in series with the parallel of the double-layer capacitance C_{dl} and the charge transfer resistance R_{ct}. The Randles circuit is a simple model that illustrating processes at the electrochemical interface. In actual electrochemical systems, impedance spectra are commonly more complex, and this circuit may not give suitable results. In modelling the impedance, we must consider that the electrical double layer appears at the interfaces of two-electrodes [4].

In model-b, the double layer impedance is modelled by a CPE which is in series with R_s. The CPE is a component that delineates double layer on interfacial electrode-solution. Generally, the CPE behavior has been assigned to the surface of electrodes including non-uniformity, roughness and surface reactivity [4,10,11]. We found that at 1 kHZ total impedance start rising which is not fit with experimental data that start rising from 100 Hz.

The model-c composed of a parallel R_{ct} and C in series with R_s and CPE. The parallel R_{ct} and C is assumed as charge transfer resistance and bulk capacitance closed to electrode-solution interfacial. It assumes the polarization impedance is caused by both electrodes interface and combined into a CPE, with the CPE impedance described by equation 3, where the constant Y_o is equal to capacitance when

Journal of Physics: Conference Series **2165** (2022) 012025 doi:10.1088/1742-6596/2165/1/012025

the exponent $\alpha = 1$. Commonly a constant phase element, the exponent $\alpha < 1$. The double-layer capacitor on real cells often behaves like a CPE instead of like a capacitor.

$$Z_{CPE} = \frac{1}{Y_o(j\omega)^{\alpha}} \quad (3)$$

The total complex impedance of the model and the conductivity are described in equation (4) and (5) respectively. In equation (5), d is distance between electrodes and A is area of electrode.

$$Z = R_s + Z_{CPE} + \frac{1}{1/R_{ct} + i\omega C} \quad (4)$$

$$\sigma = \frac{1}{Z}\frac{d}{A} \quad (5)$$

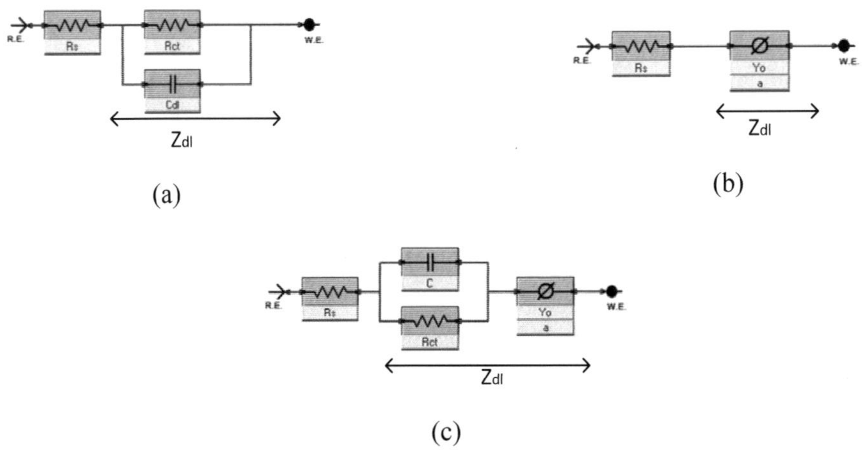

Figure 1. Equivalent electrical circuit model: (a) model-a, (b) model-b, (c) model-c.

2. 2. Experimental method

We set up several saline solutions of NaCl and $CaCl_2$ with variations in concentration from 1.25 mM to 100 mM. The complex impedance is measured with a *Gamry 3,000 potentiostat* connected to a holder with a pair of parallel copper electrodes, where d = 6 mm and A = 2 cm. The copper electrodes were cleaned using nitric acid and water mixed with 1-part nitric acid and 3-parts water for 30 seconds to let the outer part of the electrode dissolve the dirt. Then it was washed using ultrasonic bath for two minutes. In this experiment, we used galvanostat mode by injecting 5 mA current through the electrodes and then measured the voltage. Experimental data then is fitted by *Gamry Echem Analyst* using an electrical circuit model such as in Figure 1.

3. Result and Discussion

3.1. Measurement result of NaCl and $CaCl_2$ solution

The real and imaginary impedance data of NaCl and $CaCl_2$ at various concentrations is shown in Figure 2. However, at low and medium frequencies, starting from 10 mHz to 100 Hz, both types of solutions' impedance declines significantly and forms plains in the curve at high frequency from 1 kHz to 1 MHz, corresponding to solution resistance. The impedance is relatively steady in this frequency range, and we can easily differentiate the solution concentration based on its impedance. The higher the concentration, the lower the impedance. The imaginary part of impedance in this measurement has a similar pattern with the report of Sanabria for conductive solution [4]. Impedance declines to negative values and rising forming a bell-shaped curve at around 10 kHz to 1 MHz for high concentration solution (more than 5mM). This form of phenomena is discovered in over-damped oscillators.

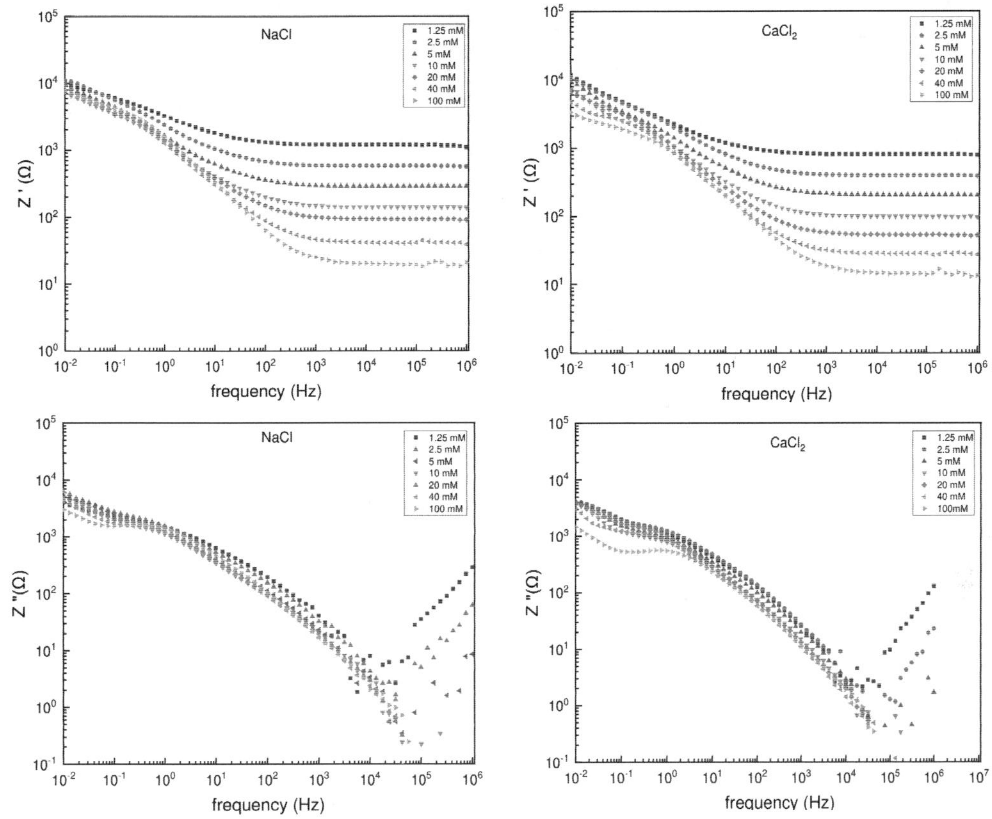

Figure 2. Experiment data of real impedance (Z') and imaginary impedance (Z'') spectra of various concentration NaCl and $CaCl_2$.

3.2. *Impedance spectrum of NaCl and $CaCl_2$ solution based on the proposed equivalent electrical model*

Figure 3 shows the comparison fit obtained with three models of EEC. Model-c provides the best fit with the goodness of fit less than 0.005 for NaCl and $CaCl_2$. The goodness of fit of model-a and b is 0.05 to 0.01 and 0.01 to 0.005, respectively. The best-fitting parameter of model-c in table 1 is used to calculate complex impedance and complex conductivity (equations 4 and 5).

Figure 4 shows the fit spectra of total impedance and its components for 10 mM NaCl and $CaCl_2$. The total impedance of model-c is including R_s, the impedance of parallel R_{ct} and C, and Z_{CPE}. Rs is remain constant over the range of measurement frequency. The impedance of parallel R_{ct} and C is rising until about 1Hz and stay constant for low frequency, while Z_{CPE} kept increasing at low frequency. It is clear that CPE causing the total impedance to rise at low frequency. It shows that CPE can predict the double layer phenomenon causes a significant increase in total impedance, especially at low frequencies.

The parameter of α has been correlated to the fractal of the electrode surface [12], which value is $0< \alpha <1$. In this experiment, α has slight differences between solutions, around 0.49 to 0.51. However, this range is below the previous study using stainless steel electrodes with $\alpha =0.78$ [2-4]. Since the electrodes used in this study differed from those reported, we can assign those discrepancies to the electrode material and surface roughness. Y_o are not constant but vary as a solution concentration, especially for concentration more than 5 mM. It indicates that the relaxation will probably also be influenced by the solution that contacts the electrode, not only on the physical properties of the electrodes. Y_o, with the unit of $S{\times}s\alpha$ (S corresponds to Siemens, s to seconds, and α to the power-law exponent). If the CPE element $\alpha = 1$ presents perfectly smooth and flat electrodes, then CPE will behave as a pure capacitor[9-12].

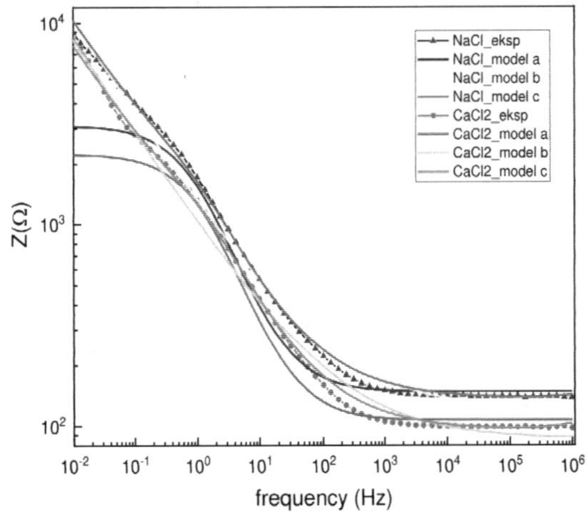

Figure 3. Fitting comparison three model EEC for 10 mM NaCl and $CaCl_2$.

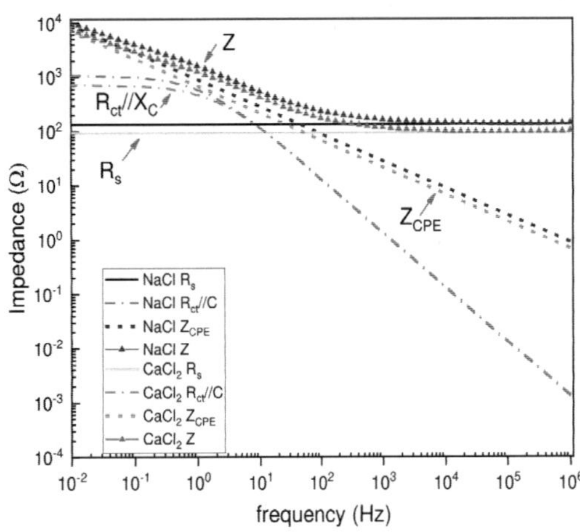

Figure 4. Comparison of complex impedance Z, R_s, Z_{CPE} and the impedance of parallel C and R_{ct} for 10 mM NaCl and $CaCl_2$.

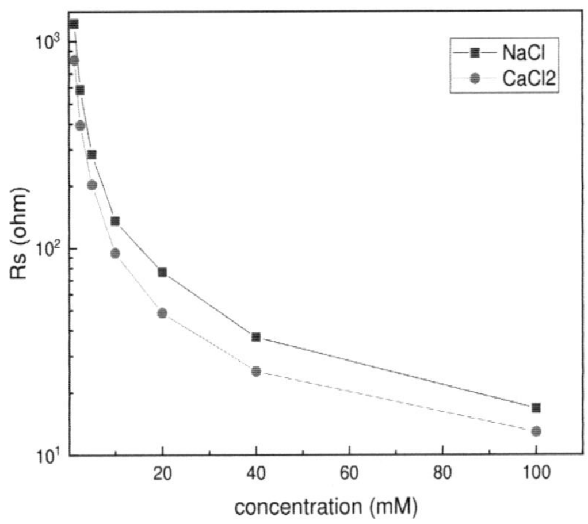

Figure 5. Solution impedance (R_s) of NaCl and $CaCl_2$ with concentration range of 1.25 mM to 100 mM.

From Figure 5, it can be confirmed the inverse relation of R_s and the concentration of solution for both NaCl and $CaCl_2$. R_s of NaCl tend to be higher about one and a half times than $CaCl_2$. It decreases sharply from 1.25 mM to 10 mM and manages to reach an exponential decay for higher salt concentration. The conductance of $CaCl_2$ is higher than NaCl; this is because the number of $CaCl_2$ ions is more than NaCl.

The conductivity spectra in Figure 6 show the complex conductivity spectrum for the various salt solutions in units of mS/cm which is similar with the result of Sanabria [4]. These graphs show that as the concentration of the ions increased, independent of the type of ions. There was a relaxation process at frequencies less than 1 kHz. This relaxation varies related to ion concentration, and it shows the same pattern for the two solutions.

Table 1. Best-fitting parameters obtained by model-c.

Solution	Concentration (mM)	Rs (Ω)	C (μF)	R_{ct} (Ω)	Yo ($S{\times}s^a$)	α	σ at 10KHz (mS/cm)
NaCl	1.25	1,222.0	36.6	2,250	$4.4{\times}10^{-4}$	0.49	0.024
	2.5	589.9	65.1	1,860	$3.3{\times}10^{-4}$	0.50	0.049
	5	285.3	114.0	1,340	$3.7{\times}10^{-4}$	0.50	0.101
	10	135.4	122.0	1,070	$4.4{\times}10^{-4}$	0.50	0.130
	20	85.53	107.0	1,390	$4.4{\times}10^{-4}$	0.50	0.318
	40	36.9	88.3	1,480	$5.0{\times}10^{-4}$	0.51	0.665
	100	16.7	67.7	1,980	$6.3{\times}10^{-4}$	0.51	1.280
$CaCl_2$	1.25	813.0	83.0	1,330	$4.0{\times}10^{-4}$	0.50	0.036
	2.5	393.0	63.5	1,360	$3.8{\times}10^{-4}$	0.49	0.074
	5	202.4	95.7	880.3	$4.5{\times}10^{-4}$	0.50	0.141
	10	80.7	116.0	722.0	$5.8{\times}10^{-4}$	0.50	0.326
	20	48.7	1,330.0	1,220	$5.7{\times}10^{-4}$	0.51	0.532
	40	25.3	129	1,090	$8.1{\times}10^{-4}$	0.49	0.984
	100	12.87	73.4	740	$1.15{\times}10^{-3}$	0.50	1.804

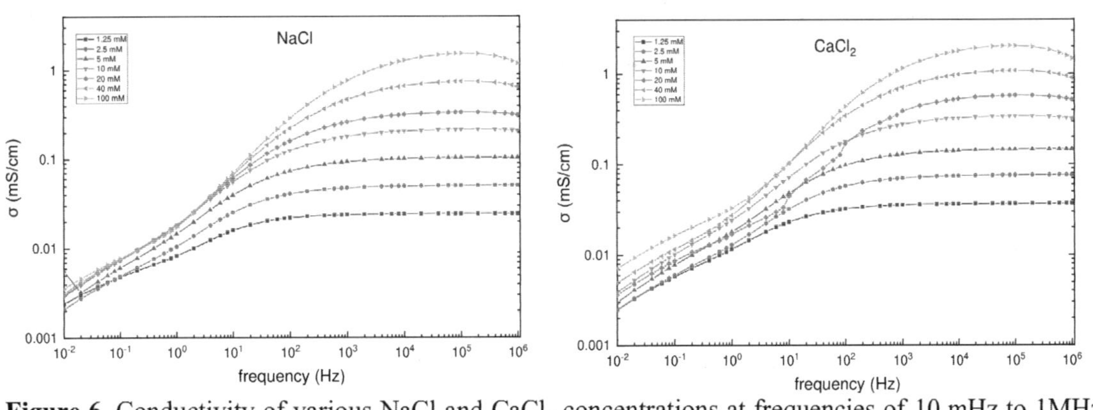

Figure 6. Conductivity of various NaCl and $CaCl_2$ concentrations at frequencies of 10 mHz to 1MHz based on model-c.

4. Conclusion

The spectra of impedance measurement spectrum for all solutions form a plateau from 100 Hz to 1 MHz, rising at low frequencies and starting to decrease and steady at high frequencies. The number of ions contains in $CaCl_2$, which are more than NaCl, was estimated to be the cause that $CaCl_2$ impedance is greater than NaCl. Experimental data was fitted using three EEC models. Best fitting was obtained from the model-c. The impedance generated by the CPE component increases significantly at frequencies less than 100 Hz and significantly influence the total impedance. It is indicating that the equivalent electric circuit used can estimate the effect of double-layer properties on the measurement. As expected from the best fit parameter, the fitting spectrum of conductance shows that the increase in the concentration of the solution causes an increase in conductance.

Acknowledgement
This research is partially funded by the Hibah Penelitian Unggulan LPPM Universitas Brawijaya 2021 with contract number 537.35.2/UN10.C10/PN/2021.

References
[1] Becchi M, Avendaño C, Strigazzi A and Barbero G 2005 J. Phys. Chem. B 109(49) 23444
[2] Lima L F, Vieira A L, Mukai H, Andrade C M G and Fernandes P R G 2017 J. Mol. Liq. 241 530
[3] Grossi M, Parolin C, Vitali B and Riccò B 2019 Eng. Sci. Technol. an Int. J. 22(1) 102
[4] Sanabria H and Miller J H 2006 Phys. Rev. E - Stat. Nonlinear, Soft Matter Phys. 74(5) 1
[5] Grossi M, Di Lecce G, Toschi T G and Riccò B 2014 IEEE Sens. J. 14(9) 2947
[6] Grossi M and Riccò B 2017 J. Sensors Sens. Syst. 6(2) 303
[7] Sala J, Gùrdia E and Martí J 2010 J. Chem. Phys. 132(21) 214505-1
[8] S. Cruz Manzo S and Chen R 2015 Insights Anal. Electrochem. 1(1) 1
[9] Lasia A Electrochemical impedance spectroscopy and its applications2014 (London: Springer)
[10] Ben Ishai P, Talary M S, Caduff A, Levy E and Feldman Y 2013 Meas. Sci. Technol. 24(10)1
[11] Nyikos L and Pajkossy T 1990 Electrochim. Acta. 35(10) 1567–1572
[12] Geddes L A 1997 Ann. Biomed. Eng. 25(1) 1

Synthesis of Polydimethylsiloxane with hydrolysis and condensation methods using monomer of Dichlorodimethylsilane as vitreous humour substitute

U Fauziah[1], D Sandi[1], V F Arini[1], D G Aulia[1], S Setiadji[2,3], Fitrilawati[1] and Risdiana[1]

[1]Departement of Physics, Faculty of Mathematics and Natural Sciences, Universitas Padjadjaran, Jl. Raya Bandung-Sumedang km 21 Jatinangor, Sumedang, West Java, Indonesia
[2]Departement of Chemistry, Faculty of Science and Technology, UIN Sunan Gunung Djati Bandung, Jl. A. H. Nasution No. 105 Cibiru, Bandung, West Java, Indonesia
[3]Doctoral Program of Chemistry, Faculty of Mathematics and Natural Sciences, Universitas Padjadjaran, Jl. Raya Bandung-Sumedang km 21 Jatinangor, Sumedang, West Java, Indonesia

E-mail: fitrilawati@phys.unpad.ac.id

Abstract. Vitreous humour is a fluid in the eye socket, that can be disturbed due to changes in its chemical or physical structure. Polydimethylsiloxane (PDMS) that also known as a silicone oil is a liquid that is commonly used as a substitute for vitreous humour in vitreoretinal surgery. PDMS can be synthesized from the monomer of Octamethylcyclotetrasiloxane (D4). However, in Indonesia, it is still difficult to obtain D4 monomer, so an alternative is needed. In this study, a synthesis of PDMS was carried out using the monomer precursor Dichlorodimethylsilane (DCMS) through a hydrolysis method under alkaline conditions using potassium hydroxide (KOH). Then the monomer obtained from the hydrolysis process was self-polymerized by aging to form a polymer gel. The concentration of KOH was varied at 0.5, 0.6 and 1 molar. The synthesized PDMS gel have viscosity values of 0.57, 1.53 and 4.49 Pa.s, density values of 0.986, 1.004, and 1.005 g/mL, refractive index in the range of 1.4001-1.4012, and surface tension in the range 19-21 mN/m. Those values of synthesized PDMS are similar to properties of commercial PDMS. We obtained that PDMS sample that synthesized using 1 M KOH shows the fastest polymerization time.

1. Introduction

The normal human eye contains vitreous humour which fills about 80% of the eye's volume, this fluid fills the space between the lens and the retina [1]. Vitreous humour serves as a buffer for the retina to keep it in the position [2]. If the vitreous humour is disturbed due to changes in its physical and chemical structure, it can cause retinal detachment.

Retinal detachment occurs when the neurosensory retinal segment detaches from the retinal pigment epithelium cells, it can be caused by acute posterior vitreous detachment (PVD), myopia, aphakia, and trauma [3]. Retinal detachment can cause a partial loss of vision in the eye. One of the ways to treat retinal detachment is vitreoretinal surgery using the vitrectomy method. The vitrectomy method is an eye surgical procedure to replace the vitreous humour using an artificial replacement fluid that has similar characteristics with vitreous humour, such as high surface tension, optically clear, and chemically and biologically inert [3]. One fluid that is often used as a substitute for vitreous humour in vitreoretinal surgery is Polydimethylsiloxane (PDMS) [4].

PDMS or silicone oil is also known as fluid tamponade in vitreoretinal surgery [5]. PDMS has several characteristics, including biologically and chemically inert, optically clear, and the same viscosity value as vitreous humour [6]. However, in Indonesia, obtaining commercial PDMS is quite difficult due to its limited supplies and its prices that relatively expensive. One way to overcome this is by producing PDMS. There are several polymerization methods in the synthesis of PDMS such as an addition polymerization, a condensation polymerization, and a ring-opening polymerization [7]. Ring-opening polymerization is commonly used for synthesis of the PDMS. This polymerization method requires a monomer type of cyclic siloxane group of Octamethylcyclotetrasiloxane (D4) [8,9]. But in Indonesia, D4 is also very difficult to obtain and its price is relatively expensive, while the number of patients undergoing vitreoretinal surgery is increasing every year [10]. For that reason, it is necessary to find a solution in order to produce PDMS from materials that are cheaper and easier to obtain. In this work, we study a synthesis of PDMS using a monomer precursor of Dichlorodimethylsilane (DCMS). The DCMS material was chosen because it has a structure similar to the structure of PDMS, which has silicon (Si) and methyl (CH3) side chain. In addition, DCMS is also relatively cheap and easy to obtain.

The method used in this study is to obtain monomer is hydrolysis method. Since the material used is not a cyclic siloxane so that the ring-opening polymerization method can not be applied. In the hydrolysis method, the first step was done by breaking down elements of Cl from the DCMS structure and forming an OH functional groups. In this study, DCMS was hydrolysis by adding KOH solution as a catalyst and heating treatment to accelerate the hydrolysis process. KOH can react with HCl which is an unnecessary side product. In addition, the presence of KOH during hydrolysis also affects the aging time of monomer to self-polymerize and become a gel [10]. KOH was chosen because KOH is easy wash, easy to obtain, and relatively inexpensive [11]. After the hydrolysis, the monomer sample was aged to self-polymerize and form a PDMS gel. The aging process was carried out at a temperature of 15° - 20° C. Here, we reported the study of the synthesis of PDMS with hydrolysis and condensation methods using monomer of DCMS as a substitute for vitreous humour with specific characteristics and good quality using materials that easily to obtain.

2. Methods

The synthesis of PDMS consisted of several steps. The first step is the hydrolysis process to obtain a monomer sample, then the monomer aging process at a temperature of 15° - 20° C to form a PDMS gel, and finally the purification process to purify the PDMS sample.

The monomer sample was obtained by the hydrolysis method, that was carried out by mixing 100 ml of Dichloromethane as a solvent with 25 ml of DCMS. 12.5 mL of KOH solution in Milli-Q water was added slowly into the solution. The hydrolysis reaction was carried out using the reflux method with stirring the solution using magnetic stirrer at 200 rpm at a temperature of 40° - 45° C for 2 hours. Then, a precipitate and HCl were separated from the clear solution using a separating funnel. Further, the monomer solution was separated from the solvent using a rotatory evaporator at 40 °C for 30 minutes. The monomer sample resulting from the hydrolysis will then be aged naturally at a temperature of 15° - 20° C. During the self-polymerization process, the propagation was taken place trough hydroxyl group (OH) and formed dimers, trimers and soon, therefore of the polymer chains was getting longer.

During the aging process, the monomer samples undergo self-condensation to form polymer chain. The condensation process occurs naturally and form a PDMS gel with a fairly high viscosity. The PDMS gel was then purified to remove residual monomer and KOH. The purification method carried out follows previous research [7,13]. The process of monomer synthesis is described by the scheme in Figure 1.

Then after the PDMS gel synthesis, it is continued with a characterization process. The hydrolysed PDMS monomer was characterized before the aging process. The yield and density of the hydrolysed monomer was determined, then characterized using the SEKONIC VISCOMATE viscometer model VM-10A-MH to measure viscosity, the AS ONE I-500 refractometer (Brix 0 ~ 90%) to measure the refractive index and surfgauge to measure surface tension.

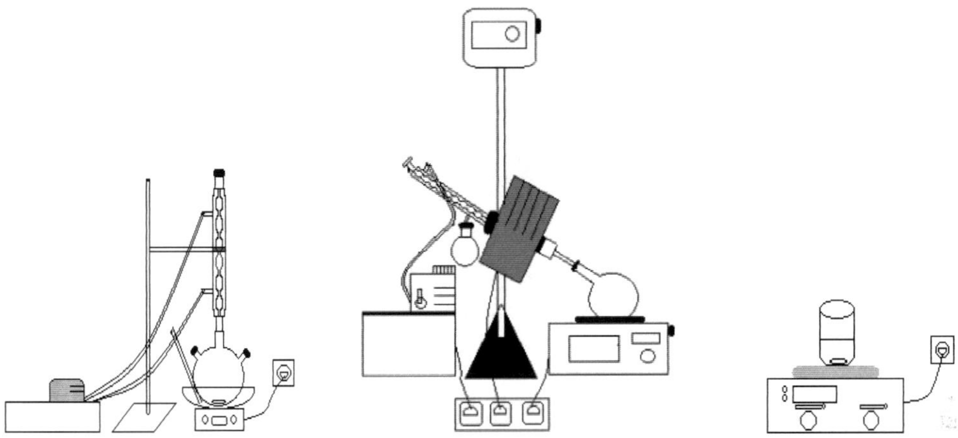

Figure 1. Schematic of the PDMS monomer synthesis process; (a) hydrolysis process, (b) evaporation process and (c) purification process after condensation.

Table 1. The synthesis parameter of PDMS and its sample code.

Sample code	Concentration of KOH (Molar)	Reaction Temperature (°C)	Reaction Time (Minutes)
PDMS-0.5	0.5	40-45	120
PDMS-0.6	0.6	40-45	120
PDMS-1.0	1.0	40-45	120

Then for purified PDMS samples, all PDMS samples were characterized by using a SEKONIC VISCOMATE viscometer model VM-10A-MH to measure viscosity, AS ONE I-500 refractometer (Brix 0 ~ 90%) to measure refractive index and surfgauge to measure surface tension, and measured the density value.

3. Result and Discussion

3.1. Synthesis result of Polydimethylsiloxane (PDMS)

The hydrolysis process under alkaline conditions produces a two-phase reaction mixture, namely a clear solution and a yellow HCl solution mixed with a white precipitate of KCl salt. After being separated using a separation funnel, the monomer still contains a Dichloromethane solvent and DCMS precursor which has not reacted completely. Then after the evaporation process, the monomer sample became clearer and odorless, which indicated that the solvent had disappeared. The monomer product of hydrolysis process that have been evaporated is clear and transparent as shown in Figure 2. After aging at a temperature of 15° - 20° C, a polymer propagation process was occurred by condensation. As a result of the condensation process, a PDMS monomer turn into a gel with a fairly high viscosity.

Figure 2. The PDMS monomer as a result of hydrolysis after evaporated is clear and transparent.

3.2. Purification result of the Synthesize Polydimethylsiloxane (PDMS)

The purification of PDMS gel as result of aging process was carried out by diluting the gel using chloroform in order to dissolve the remaining precursor and follow then by addition of Milli-Q water. The water was added into the gel solution in order to dissolve and separate KOH from the gel sample. The mixture was allowed to stand until the gel and water separated into two phases. Then water was separated from the polymer gel. The step was repeated until the pH is 7 that indicates the KOH washed out from the polymer sample. After the purification process, the PDMS gel appears a clearer and more transparent as shown in Figure 3. The result shows that the synthesized PDMS is pure and it is expected that the synthesized gel can fulfil a requirements of vitreous humour substitutions.

Figure 3. Of the appearance of PDMS gel after purification process.

3.3. Characteristics of monomer and gel PDMS

The results of characterization of the hydrolysis gel are shown Table 2. The values of density are in the range between 0.83 and 0.99 g/mL. If it is compared with D4, this value is close to a density value of D4. The yield of the monomer synthesis is about 40%, this might relate to hydrolysis temperature. When the hydrolysis temperature exceeds the boiling point of the Dichloromethane solvent, the solvent will be evaporated faster and the yield can be low. The table shows that the viscosity of the monomer is proportional to its density and KOH concentration. The viscosity value is related with a molecular weight of the sample. The synthesized monomer has a higher viscosity compare to the commercial D4. This might indicate that during the hydrolysis process a propagation of the Si-O bond taken place. Further, it is obtained that values of refractive index and surface tension of synthesized monomer is slightly higher compared to the commercial D4.

Table 2. Density, yield, viscosity, refractive index, and surface tension of hydrolysis monomer samples synthesized at various KOH concentrations.

Sample Name	Density (g/mL)	Yield (%)	Viscosity (Pa.s)	Refractive Index	Surface Tension (mN/m)
PDMS-0.5	0.83	40	0.0117	1.4012	19
PDMS-0.6	0.96	41	0.0206	1.4003	19
PDMS-1.0	0.99	40	0.0239	1.4001	20
D4	0.96	-	0.0023	1.3950	18

Table 3. Condensation time, density, viscosity, refractive index, and surface tension of synthesized PDMS gel.

Sample code	Condensation Time (Days)	Density (g/mL)	Viscosity (Pa.s)	Refractive Index	Surface Tension (mN/m)
PDMS-0.5	48	0.986	0.57	1.4012	19
PDMS-0.6	63	1.004	4.49	1.4008	21
PDMS-1.0	18	1.005	1.53	1.4001	20
PDMS of 1300 cSt [13]	-	-	1.08	1.4044	20
PDMS of 5500 cSt [13]	-	-	3.55	1.4040	19

Table 3 shows characteristics of synthesized PDMS gel with varied condensation time. The density of PDMS gel are around 0.986 to 1.005 g/mL. When compared to human vitreous humour, this value is close to the density value of vitreous humour of 1.0053-1.0089 g/mL [12]. The concentration of KOH affects the rate of self-polymerization process. The density of the sample related to intermolecular bonds and chains length, higher density cause a higher viscosity of the sample. It can be seen that sample of PDMS-0.6, has high viscosity of 4.49 Pa.s, and sample of PDMS-1.0 has low viscosity of 1.53 Pa.s. Sample of PDMS-0.5 has a low viscosity that is not sufficient for the needs as a substitute for vitreous humour (0.57 Pa.s). Table 3 also shows values of the refractive index of the purified PDMS gel. It can be seen in the table that the refractive index of the synthesized PDMS in the similar range of the commercial PDMS refractive index (1.400-1.404). By examine the changes of the refractive index of the synthesized PDMS, it shows quality of PDMS to changes in the accommodation ability of the eye that occur. These data show that the surface tension is directly proportional to the viscosity. The increasing of surface tension and viscosity indicates the strength of intermolecular bonds and formation of long-chain bonds in the sample. The surface tension on PDMS to determine the formation and stability of the emulsion, and the possibility of particle dispersion in PDMS when applied to the eye. The surface tension of PDMS affects the tension between the interface and the sub-retinal fluid in the eye socket [13].

Figure 4 shows a relationship between viscosity and the concentration of KOH used in the synthesis process. The data show that the concentration of KOH affects the rate of self-polymerization. KOH as a catalyst works by lowering the activation energy in the reaction process so that the reaction can run faster. The red curve shows the viscosity value of the hydrolysed gel before condensation process. The blue curve shows the viscosity value of the PDMS gel after the condensation and purification process. In the sample with 0.5 molar KOH concentration, the sample experienced an increase in viscosity from 0.0117 Pa.s to 0.57 Pa.s within 48 days. In the sample with a concentration of 0.6 molar KOH, the sample experienced an increase in viscosity to reach medium viscosity from 0.0206 Pa.s to 4.49 Pa.s within 63 days. In the sample with 1 molar KOH concentration, the sample experienced an increase in viscosity from 0.0239 Pa.s to 1.53 Pa.s within 15 days. It can be seen that samples containing KOH with a concentration of 1 molar can produce a faster condensation time than samples containing KOH with a concentration of 0.5 and 0.6 molar.

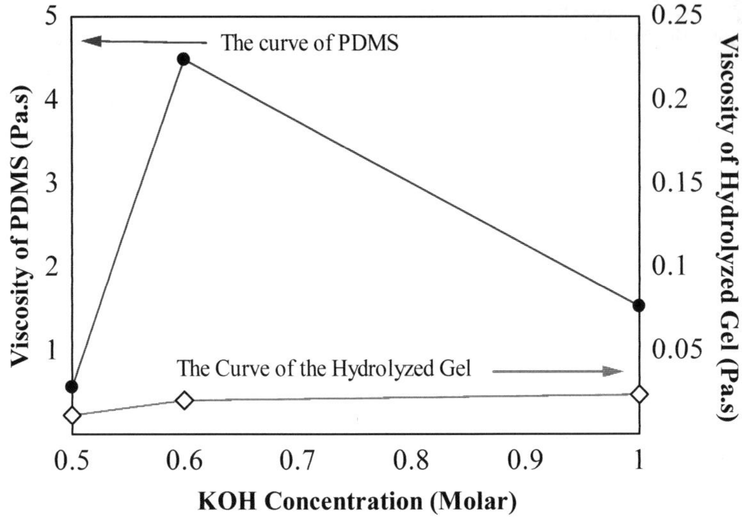

Figure 4. Correlation between viscosity and the concentration of KOH in PDMS synthesis.

4. Conclusion

PDMS has been successfully synthesized by hydrolysis and condensation method using DCMS as the precursor monomer. Two samples were successfully synthesized with viscosity, refractive index, and surface tension characteristics similar to those of commercial PDMS. The viscosity values of PDMS-0.6 and PDMS-1.0 were 4.49 and 1.53 Pa.s. The effect of KOH concentration showed that the higher the concentration of KOH, the faster the condensation time. Samples with 1 M KOH content had the fastest polymerization aging time. The successful synthesis of PDMS is expected to overcome the need for PDMS as a substitute for vitreous humor in vitreoretinal surgery.

Acknowledgment

The authors would like to thanks Kemenristek-DIKTI of Indonesia for financial support in the scheme of Fundamental Research (Penelitian Dasar) 2021, contract no. 1207/UN6.3.1/PT.00/2021. This work was also partially supported by the Academic Leadership Grant of Universitas Padjadjaran 2021, contract no. 1959/UN6.3.1/PT.00/2021.

References

[1] Alovisi C, Panico C, De Sanctis U and Eandi C M 2017 Vitreous substitutes: old and new materials in vitreoretinal surgery *J. Ophthalmol* doi: 10.1155/2017/3172138

[2] Snell R S 2012 *Clinical Anatomy* (Philadelphia: Lippincott Williams & Wilkins) chapter 11 pp 556–62

[3] Khurana A K 2015 *Comprehensive Ophtalmology by Region* (New Delhi: Jaypee Brothers Medical Publishers (P) Ltd) chapter 11-12 pp 257–309

[4] Giordano G G and Refojo M F 1998 Silicone oils as vitreous substitutes *Prog. Polym. Sci.* **23** 509–32

[5] Caramoy A, Hagedorn N, Fauser S, Kugler W, Groß T and Kirchhof B 2011 *Investig. Ophthalmol. Vis. Sci.* **52** 5432–36

[6] Soman N and Banerjee R 2003 Artificial vitreous replacements *Biomed. Mater. Eng.* **13**

[7] Fitrilawati *et al* 2018 Effect of KOH concentration on characteristics of polydimethylsiloxane synthesized by ring opening polymerization method *J. Phys. Conf. Ser.* **1080**

[8] Setiadji S *et al* 2020 Synthesis of polydimethylsiloxane and its monomer from hydrolysis of

dichlorodimethylsilane *Key Eng. Mater.* **860** 234-38
[9] Fauza A N, Ardi A and Novianti R M 2019 penentuan kurva kalibrasi untuk estimasi viskositas pdms **03** 33-9
[10] Setiadji S *et al* 2019 Uji stabilitas bahan polydimethylsiloxane *J. Material dan Energi Indonesia* **09** 8-17
[11] Odian G 2004 *Ionic chain polymeryzation; in principles of polymerization* (New York: Wiley-Interscience)
[12] Swindle K E and Ravi N 2007 Recent advances in polymeric vitreous substitutes Expert Rev. Ophthalmol. **2** 255-65
[13] Setiadji S *et al* 2019 Optimization of Polydimethylsiloxane synthesized parameters as vitreous humour substitutes *Mater. Sci. Forum* **966** 189-93

Analysis of physics learning needs due to the covid-19 pandemic

W Sudarsana[1], Sarwanto[1] and A Marzuki[1]

[1]Master of Physics Education Program, University of Sebelas Maret, Surakarta, Central Java, Indonesia

E-mail: wayansudarsana@student.uns.ac.id

Abstract. This study aims to determine the need for learning physics due to the covid-19 pandemic. This type of research is a preliminary study with a descriptive approach and the instrument used is a questionnaire. The research subjects were 30 high school teachers consisting of 30 schools in two provinces of East Java and Central Sulawesi. The data analysis of this research is descriptive qualitative obtained based on the teacher's answers. This study shows that several learning components influence overcoming learning problems during the COVID-19 pandemic, namely technological developments, learning models, and learning media. This study concludes that it is necessary to develop physics learning devices as an alternative to learning during the covid-19 pandemic.

1. Introduction

Learning is the behavior of a person's behavior to be better based on the experience that has been passed where there are influences from internal and external aspects that are interconnected so that they gain new experiences in the form of knowledge and skills [1,2]. By learning a person does not only experience changes in knowledge but is accompanied by changes in attitudes, behavior, and patterns of thinking and skills from before [3,4].

Learning is an activity that involves educators and students who have a link between the steps of learning activities that have been planned to achieve a field of study [5]. Learning can be interpreted as an effort to condition or organize students to learn effectively in the classroom [6].

The knowledge that is arranged systematically and organically obtained based on experiences such as observation and experimentation so that it can define a concept being studied [7,8] and can be linked to everyday life [9] and can understand and understand these concepts and connected with other concepts [10].

Due to the COVID-19 pandemic, face-to-face learning is dangerous to do [11,12] This is under Circular Letter Number 4 of 2020 of the Minister of Education and Culture regarding distance learning, studying, and working from home [13]. During this bold learning period, students' mastery of concepts decreases. Based on the results of observations that have been made, one of the problems that occur in education in Indonesia is the emergence of the Covid-19 Virus which has spread throughout Indonesia. This causes disruption of learning in schools, learning is usually only done face-to-face. But with the emergence of the Covid-19 Virus so learning is done online. Therefore, an alternative is needed in the form of interactive learning media that can help the learning process run amid the Covid-19 Virus. This is evidenced by research conducted [14], where physical lessons have problems.

Then from the results of interviews with teachers and students related to the process of learning physics in class from the beginning to the end of learning. Interviews conducted with teachers are more directed to the learning tools used such as approaches to learning, learning models, learning methods, learning media, learning resources, the use of the internet in learning, the obstacles faced during the learning process, and student learning outcomes. Meanwhile, interviews conducted with students led to students' responses to the teacher's way of teaching, the needs needed in learning physics, and suggestions for making physics learning more interesting for students to actively studyMany difficulties faced by students in receiving physics learning materials online [15,16] This is due to mass learning with unlimited participants and has also been proven by previous research where physical learning during the covid 19 period was not appropriate.

The COVID-19 pandemic has had a major impact on the physics learning process that must be done online. This causes educators to be creative in planning Physics learning activities. Therefore, it is necessary to conduct preliminary research to determine the need for learning physics during the COVID-19 pandemic.

2. Methods

This research is a preliminary study that aims to determine the need for learning physics due to the covid-19 pandemic. The method used in this study is a descriptive research method with a qualitative approach. Qualitative descriptive research aims to describe, describe, explain, explain and answer in more detail the problems to be studied by studying as much as possible an individual, a group, or an event [17]. This research was conducted in two provinces, namely East Java and Central Sulawesi. the object of research under study was the learning process during the covid-19 pandemic. Meanwhile, 30 high school teachers were used as subjects in this study.

The instrument was used in the form of a questionnaire consisting of 25 statements with 4 categories, then distributed to respondents and made a descriptive analysis of the results of the existing questionnaire. The analytical questionnaire used is the initial stage of the 4D development model that was developed [18], which is define with the main objective being to find out which products are suitable to be developed for the problems encountered. Furthermore, the needs analysis questionnaire was distributed to 30 teachers via the google form link. The data analysis technique used in this research is descriptive qualitative from the answers given by the teacher. Data analysis using the interactive model of [19] can be described as follows:

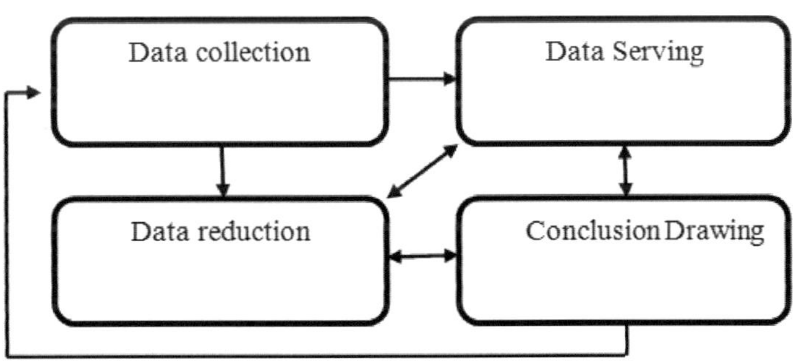

Figure 1. Miles & Huberman's interactive analytics model.

3. Result and Discussion

This research was carried out by giving 25 statements to 30 teachers who teach at each grade level, which can be seen in Figure 2 below.

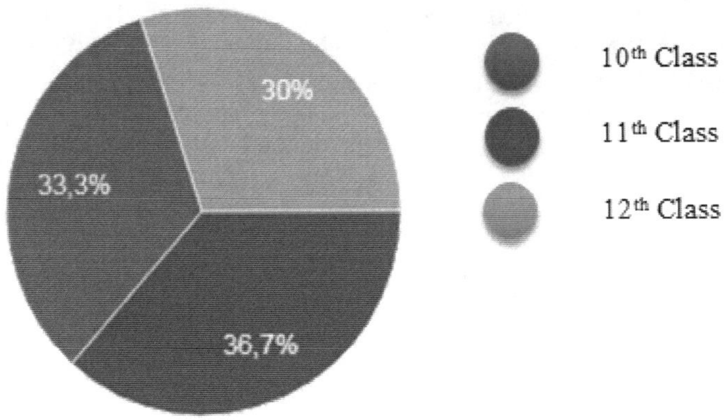

Figure 2. Teacher teaching in class.

Based on Figure 2 ,it can be seen that there are 30 teachers as respondents who teach in 10th class by 36.7%, teachers teaching in 11th class by 33.3%, and teachers teaching in 12th class by 30%.
The needs analysis for teachers was carried out using a questionnaire containing statements related to learning physics. The results of the analysis of needs for teachers are in accordance with the state of physics learning during covid-19.

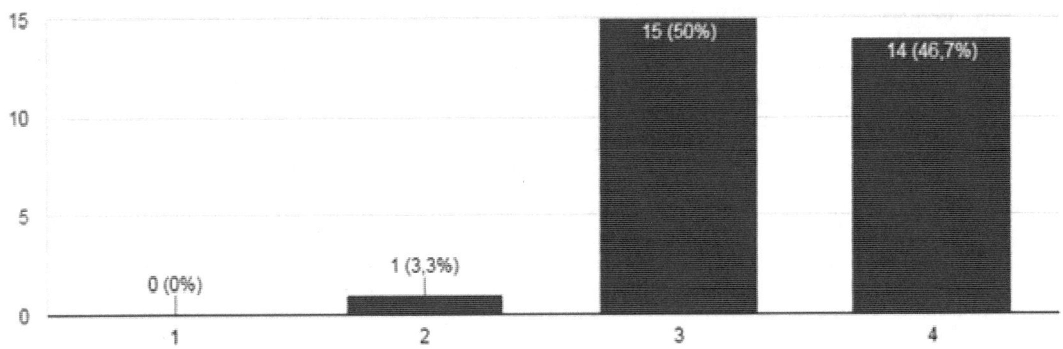

Figure 3. Answers to media use.

Based on Figure 3, the teachers who use media in learning physics are 50% with 15 teachers who answered strongly agree, 46.7% with 14 answered agree, and 3.3% with 1 teacher who answered disagree. This proves that almost all teachers teach by utilizing learning media.

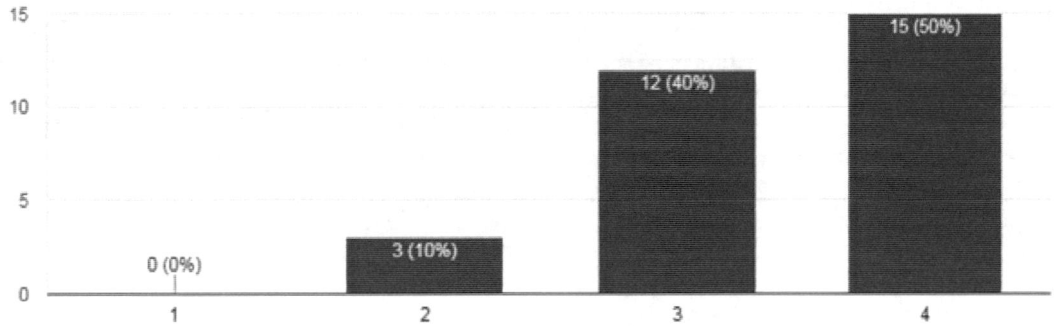

Figure 4. The answer to the solution to the lack of face-to-face time.

Based on Figure 4, the teacher believes that one solution to overcome the shortage of face-to-face hours due to covid-19 is to use e-modules during online learning. This can be seen from the teacher's answers by 50% with 15 teachers who answered strongly agree, 40% with 12 teachers who answered agree, and 10% with 3 teachers who answered disagree.

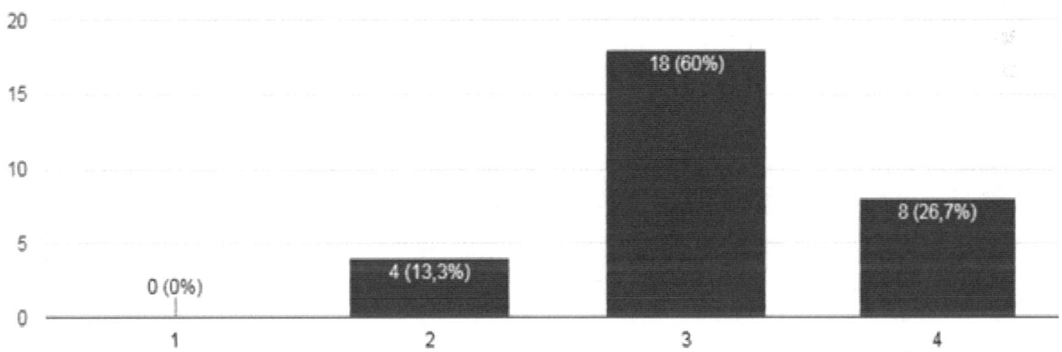

Figure 5. Answers to the learning model.

Based on Figure 5, the teacher believes that using the learning model in the electronic module will make learning activities more effective. This can be seen from the teacher's answer of 26.7% with 8 teachers who answered strongly agree, 60% with 18 teachers who answered agree, and 13.3% with 4 teachers who answered disagree.

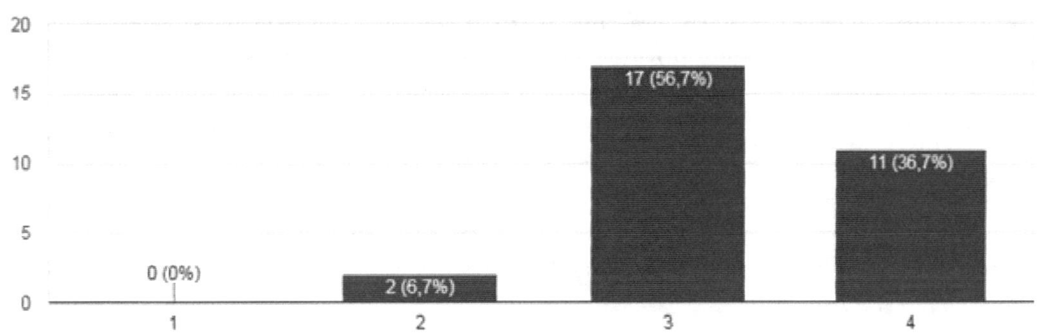

Figure 6. The answer to the COVID-19 pandemic.

Based on Figure 6, the teachers believe that the website can be used as a medium for learning physics and can be a learning solution because of the covid-19 pandemics. This is evidenced by the teacher's answer of 36.7% with 11 teachers who answered strongly agree, 56.7% with 17 teachers who answered agree, and 6.7% with 2 teachers who answered disagree.

The initial stage carried out in this preliminary research is to make observations, distribute pre-research questions to teachers and conduct interviews to teachers and students about the learning media used. After completing the initial stage, then conduct a needs analysis from observations and interviews that have been carried out. Based on the results of the needs analysis in schools, it is known that teachers have not optimized learning media and the learning media used are still media in general. Given the Covid-19 situation which requires online learning processes. Therefore, students need learning media that can be used online which includes materials, pictures, videos, and simulations so that the online learning process does not affect students' understanding of absorbing material as well as face-to-face learning. Based on some of the teacher's answers that have been presented above, it can be seen that teachers need learning tools that can be used to teach online due to the covid 19 pandemic. Research [20] also revealed that distance learning during the covid 19 pandemics was not by with the characteristics of e-learning so that the development of learning tools is needed as an alternative in learning.

This research is supported by research conducted by [21]regarding the development of an electronic module based on problem-based learning flipbooks on electronic induction materials for 12th class Senior High School that the flipbook module developed has the characteristics of being able to be used remotely. , students do not need to come to class to use the flipbook module media so that students can easily access the flipbook module anytime and anywhere without attending class. This is certainly an advantage of the flipbook module as a learning medium that will support distance learning of physics during the COVID-19 pandemic. Likewise, what was done by [22] that with the characteristics of materials made in electronic form using professional flip pdf software, teaching materials were developed in an attractive and clear manner making it easier for students to learn physical teaching electronically. learning videos that are able to help clarify material that cannot be conveyed through writing so that through videos it can be clearly visualized and there are interactive quizzes that can hone students' skills directly so that it creates student interest in physics.

4. Conclusion

This preliminary research concludes that ongoing learning requires that it be done online due to covid-19. Therefore, teachers are required to plan effective learning activities so that they do not interfere with ongoing educational programs. In this case, several components influence physics learning during the COVID-19 pandemic, namely technological developments, learning models, and learning media. So it is necessary to develop physics learning devices as an alternative due to covid-19.

Acknowledgement

Thank you to the Journal of Research in Science Education Mataram University for providing a forum to share and exchange knowledge, thank you to Dr. Sarwanto, S. Pd., M.Si. and Ahmad Marzuki, S, Si., Ph.D. who had guided and provided direction to me so that I could complete this research, as well as thanks to the principal and respondents who have been willing to be the subject of this research and all those who support this research.

References

[1] Dahar R W 1989 *Study and Learning Theories* (Jakarta: Erlangga) p 21
[2] Wiradintana R 2018 The Cognitive Revolution Through The Application Of Bruner's Theory Of Learning in Refining The Behavioral Approach *Journal Study Of Economics Education and Economics* **2** 1
[3] Khodijah N 2014 *Educational Psychology* (Jakarta: Raja Grafindo Persada) p 34
[4] Sardiman 2009 *Interaction and Motivation Teaching Learning* (Jakarta: PT. Rajawali Pers) p

20
[5] Widoyoko E P 2009 *Learning Program Evalution* (Yogyakarta: Pustaka Belajar) p 46
[6] Hamalik O 2012 *Curriculum Development Management* (Bandung: Remaja Rosdakarya) p 12
[7] Desmawati R 2018 Design Stage Of Senior High School Physics Teaching Materials Integrated Faith and Piety Based on Discovery Learning on Tempeature and Heat Materials *Natural Science Journal* **4** 516-526
[8] Hanna D Sutarto and Harijanto A 2016 Learning Model with Theme Consepts Accompanied by Picture Media in Physics Learning in High School *Journal Of Physics Education* **5** 23-29
[9] Neizhela A and Mosik 2015 Improving Learning Outcomes Through Contextual Approches with the Think Pair Share Method for Middle School Students Calorie Materials *Unnes Physics Education Journal* **4** 50229
[10] Lubis I L 2009 Level of Understanding of Students' Physics Consepts *Infotama Media* **4** 14-22
[11] Suseno *et* 2021 Learning Model Combining Online System and Practicum to Improve Cognitive and Psychomotor Ability *Journal Of Physics Education* **9** 42-54
[12] Zainudin and Wijayanti R 2021 Development of Physics Learning Devices During the Covid-19 Pandemic Based on Critical Thingking Skills Using the Moodle Application *Eduproxima Journal* **3** 43-49
[13] Eva Y and Nadi S 2021 Feasibility Analysis of Youtube Platform Based Physics Learning Videos on Work and Energy Materials Physics Education Innovation **10** 21-31
[14] Kharisma N N Roesminingsih M V and Suhanadji 2020 An Overview of the Online Learning Needs of PKBM Budi Utama Surabaya During the Covid-19 Pandemic *Journal of Non-Formal Education* **15** 38-44
[15] Saeri M 2020 Physics Learning with Video Scren Recording in the Covid-19 Pandemic Period to Optimize the Learning Ability of Class XII-MIPA 1 Students with subject of Dynamic Electricity at State High School 2 Bangkalan *Journal of Education, Learning, and Development* **02** 26-32
[16] Mulyadi E 2020 Physics Online Learning Through Whatsapp, Google Form, and Email in Active Attendance Achievements and Student Learning Outcomes *Journal of Teacher Scientific Works* **5** 1
[17] Craswell J W 2013 *Research Design Pendekatan Kualitatif, Kuantitatif dan Mixed* (Yogyakarta: Pustaka Pelajar) p 127
[18] Thiagaranjan S Semmel D S & Semmel M I 1974 *Instructional Development for Training Teacher of Expectional Children* (Minneapolis, Minnesota: Leadership Training Institute/ Special Education, University of Minnesota) p 43
[19] Moleong L J 2011 *Qualitative Survey Method* (Bandung: PT Remaja Rosdakarya) p 38
[20] Suryandari W A and Burhendi A C F 2020 Preliminary Study of the Characteristics of Physics Online Learning During the Covid-19 Pandemic Period *Journal of Basic Education (Proceedings of the National Seminar and Discussion on Basic Esucation)* E-ISSN 2549-5801
[21] Hermawati M H Sunaryo and Rustana E C 2020 Development of Problem Based Learning-Based Flipbook Electronic Module on Electronic Induction Materials for 12th Class Senior High School. *Proceedings of the National Physics Seminar (E-Journal) SNF2020.* P-ISSN: 2339-0654 E ISSN: 2476-9398. DOI: doi.org/10.21009/03.SNF2020.02.PF.04
[22] Sriwahyuni I Risdianto E and Johan H 2019 Development of Electronic Teaching Materials Using Professional Pdf Flip on Optical Equipment Materials In Senior High School *Journal of Coil Physics* **3** 145-152

Journal of Physics: Conference Series **2165** (2022) 012028 doi:10.1088/1742-6596/2165/1/012028

Study of crystal structure and magnetic properties of $Eu_{2-x+0.01}Ce_{x-0.01}Cu_{0.99}Fe_{0.01}O_{4+\alpha-\delta}$

H D Anggia[1], L P Rochman[1], R Pratama[1], T Maulana[1], S Winarsih[1], Y Maryati[1], M A Syakuur[1], U Widyaswari[2,3], D P Sari[2,4], T Saragi[1*] and Risdiana[1]

[1]Department of Physics, Padjadjaran University, Jl. Raya Bandung-Sumedang km. 21 Jatinangor, Sumedang, 45363, Indonesia
[2]Meson Science Laboratory, RIKEN Nishina Center, 2-1 Hirosawa, Wako, Saitama 351-0198, Japan
[3]Department of Physics, Faculty of Mathematics and Natural Sciences, Universitas Indonesia, Depok 16424, Indonesia
[4]Graduate School of Engineering and Science, Shibaura Institute of Technology, Saitama City, Saitama 337-8570, Japan

E-mail: t.saragi@phys.unpad.ac.id

Abstract. Superconductor is material which has ability to perform zero resistivity in certain temperature. There are some phenomena in superconductor that still cannot be describe well. One of the phenomena is the effect of impurities to the physical properties of superconductor. In this research, the partially substitution of magnetic impurity Fe for Cu in electron-doped superconducting cuprates of $Eu_{2-x+0.01}Ce_{x-0.01}Cu_{0.99}Fe_{0.01}O_{4+\alpha-\delta}$ in over doped regime x = 0.16 and 0.20 with 1% of Fe content was prepared by solid state method in order to investigate the structure and magnetic properties of ECCFO. The results of the characterization on crystal structure using XRD show that the material has a tetragonal T' crystal structure which is indicated by the appearance of two main peaks with miller index (103) and (110). Magnetic-susceptibility measurements were carried out at low temperatures from 10 to 30 K. The values of curie constant, magnetic moment per unit volume and effective magnetic moment decreased with increasing of Cerium concentrations.

1. Introduction

The study of Impurity effect has played a key part in the elucidation of the mechanism of high-T_c superconductivity (HTSC) in recent study. Magnetic Ni impurity suppressed superconductivity more effectively than non-magnetic Zn impurity in electron-doped materials. In a hole-doped system, however, the substitution of nonmagnetic Zn for Cu suppressed superconductivity more than the substitution of magnetic impurity Ni [1–3].

Another impurity substitution for Cu in hole-doped system is magnetic impurity Fe. This element is trivalent and known to have a stronger magnetic moment than Ni yet still having relatively close atomic number with Cu which 3d metals. In a wide range of hole concentrations where superconductivity arises in $La_{2-x}Sr_xCuO_4$, it has been discovered that a 1% Fe substitution increases the magnetic transition temperature and magnetic correlation. The substitution of magnetic impurity Fe also suppresses superconductivity, most rapidly among other impurity such as Zn and Ni [1,4].

In electron-doped, this study still unresolved yet because the superconductivity is also depending on the oxygen content. The oxygen content is often having excess and can be reduced by annealing process yet it still hard to control [5,6]. In order to study the magnetic properties in Cu substitution, other element of the superconductor such as the rare earth element must have zero magnetic moment. Among Nd, Pr, Sm and Eu rare earth elements, Eu^{3+} has no magnetic moment in its ground state [6]. In previous study of $Eu_{2-x+y}Ce_{x-y}Cu_{1-y}Fe_yO_{4+\alpha-\delta}$ (ECCFO) in optimum doped regime, Fe substitution successfully disturbing spin-spin correlation in Cu layer shown by the superconductivity completely disappear for Fe concentration larger than 2% and 5% [7]. However, a comprehensive report on the effects of substituting magnetic impurity Fe for Cu on the structure in electron-doped cuprates has yet to be published in a wide Ce concentration range, making it difficult to draw a clear conclusion for comparing the effects of impurity in electron- and hole-doped systems.

Higher doping concentration research should be conducted on $Eu_{2-x+y}Ce_{x-y}Cu_{1-y}Fe_yO_{4+\alpha-\delta}$ (ECCFO), because this area is still in the superconducting phase, and there is a possibility of a quantum critical point, which occurs when a continuous transition between competing phases occurs at zero temperature, as seen in the hole-doped system [8]. In this paper, we investigate effects of substitution of magnetic impurity Fe for Cu to the structure and magnetic properties in electron-doped superconducting cuprates of $Eu_{2-x+0.01}Ce_{x-0.01}Cu_{0.99}Fe_{0.01}O_{4+\alpha-\delta}$ (ECCFO) in over doped regime $x = 0.16$ and 0.20 with 1% of Fe content.

2. Methods

A solid-state reaction method was used to synthesize samples of $Eu_{2-x+0.01}Ce_{x-0.01}Cu_{0.99}Fe_{0.01}O_{4+\alpha-\delta}$ the over doped regimes of x = 0.16 and 0.20 with 1% Fe content. Dried Eu_2O_3, CeO, CuO, and Fe_2O_3 powders were mixed in a stoichiometric ratio, ground and prefired in air for 20 hours at 900°C, then pulverized and sintered in air for 16 hours at 1,000 °C. The sintered materials were reground and compacted into 10 mm diameter pellets, which were sintered in air for 16 hours at 1,000 °C. Samples were annealed at 900 °C for 10 hours in an Argon flow. The weight changes before and after annealing was used to determine the reduced oxygen content (δ). Samples of ECCFO were checked by powder x-ray diffraction to be single phase. The crystal structure was also studied using X-ray diffraction measurements. Using a standard SQUID magnetometer in a magnetic field of 5 Oe on field cooling, DC magnetic susceptibility (χ) measurements were carried out at low temperatures from 10 to 30 K.

(a) (b) (c) (d)

Figure 1. (a) Eu_2O_3, CeO, CuO, and Fe_2O_3 powders were mixed together and ground, (b) After grinding sample turns grey (homogeneous) and then prefired, (c) Sample after prefired, reground and sintered (d) Compated sample after sintered and reground, which were sintered for the second time and annealed.

3. Result and Discussion

The $Eu_{2-x+0.01}Ce_{x-0.01}Cu_{0.99}Fe_{0.01}O_{4+\alpha-\delta}$ sample in over doped regime $x = 0.16$ and 0.20 with 1% of Fe content are produced. The value of δ is almost equivalent to the value of α. However, we have no control over the value of δ well. For sample $x = 0.16$, the oxygen reduction was 0.041, and for sample x = 0.20, it was 0.102.

Figure 2 shows XRD pattern of $Eu_{2-x+0.01}Ce_{x-0.01}Cu_{0.99}Fe_{0.01}O_{4+\alpha-\delta}$ in over doped regime $x = 0.16$ and 0.20 with 1% of Fe content. All main peaks of tetragonal crystal structure with a planar (T') configuration of CuO_2 was observed in all samples. In sample with $x = 0.16$ and 0.20 were found peak at $2\theta = 28°$ and 33° caused by impurity of CeO_2. The peak was then identified as CeO_2 by using Rietveld Refinement Analysis. The volume fraction from ECCFO is 99.5% for sample $x = 0.16$ and 92.3% for sample $x = 0.20$. The presence of CeO_2 material is suspected as a result of the precursors that do not fully react in the solid reaction synthesis.

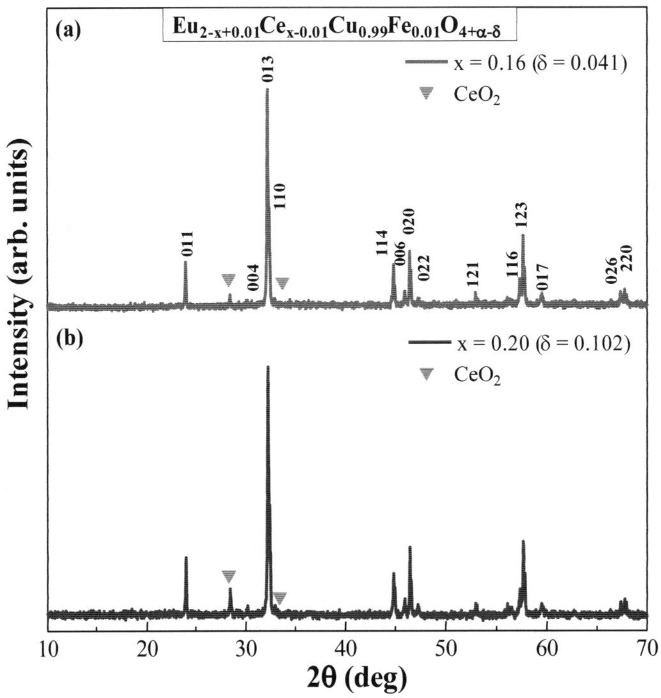

Figure 2. XRD Pattern of ECCFO with (a) $x = 0.16$ and (b) $x = 0.20$.

Table 1 summarizes crystal structure, lattice parameters (*a*-axis and *c*-axis), unit cell volume, Cu-O bond length, and crystallite size. All lattice parameter values are calculated from d_{hkl} values and Miller index. It was found that the value of the lattice parameter on the *a*-axis did not change with a value of 3.9068 Å for samples $x = 0.16$ and 0.20. This could be due to the substitution of Fe^{3+} with an ionic radius of ~0.55 which is relatively smaller than other elements in the ECCFO material, especially Cu^{2+} in the planar conduction layer, reducing the effect of increasing the *a*-axis lattice parameters due to an increase in Ce concentration. It is also shown that the Cu-O bond distance in the planar conduction layer corresponding to the *a*-axis lattice parameter is constant with increasing Ce. The *c*-axis are 11.8489 Å for the sample $x = 0.16$ and 11.8513 Å for the sample $x = 0.20$. The unit cell volumes are 180.8535 Å^3 for sample $x = 0.16$ and 180.8834 Å^3 for sample $x = 0.20$. The *c*-axis lattice parameters and unit cell volume increase can be indicated as an increase in the distance between the conduction layer and the charge reservoir.

The crystallite size obtained from by using Debye-Scherrer equation for nano-crystallite as shown in Equation 1 [9]. The value of the Full Width at Half Maximum (*FWHM*) from main peak and its Bragg angle are used.

$$D = k\lambda/\beta_{hkl}\cos\theta \tag{1}$$

where D is the crystallite size (nm), λ is the X-ray wavelength (nm), is the Bragg angle (rad) β_{hkl} is the FWHM (Bragg peak width at half maximum height) (rad) and k is the shape factor constant (rad) with value 0.9 rad.

The crystallite size values are 86.53 nm for the sample x = 0.16 and 84.12 nm for the sample x = 0.20. The size decreased with the addition of Ce is related to the crystal heating process during synthesis where the larger the crystallite size, the more effective the sintering process [10].

Table 1. The lattice parameter, Cu-O bond length, and crystalline size of of ECCFO with 1% Fe impurity at x = 0.16 and 0.20.

x	$a = b$ (Å)	c (Å)	V (Å^3)	*Cu-O* (Å)	Crystallite Size (nm)
0.16	3.9068	11.8489	180.8535	1.9534	86.53
0.20	3.9068	11.8513	180.8834	1.9534	84.12

Magnetic susceptibility to temperature at Field Cooling (FC) 5 Oe for ECCFO samples with x = 0.16 and 0.20 at 1% Fe concentration is shown in Figure 3. Temperature dependence of dc magnetic-susceptibility (χ) for ECCFO x = 0.16 and x = 0.20 with 1% Fe impurity in 10 to 30 K. For the sample with x = 0.16 and x = 0.20 shows paramagnetic behavior where the susceptibility is positive with a value of around 10^{-5} to 10^{-3} emu/gr. Oe. Sample with x = 0.16 has higher magnetic susceptibility than that of sample with x = 0.20.

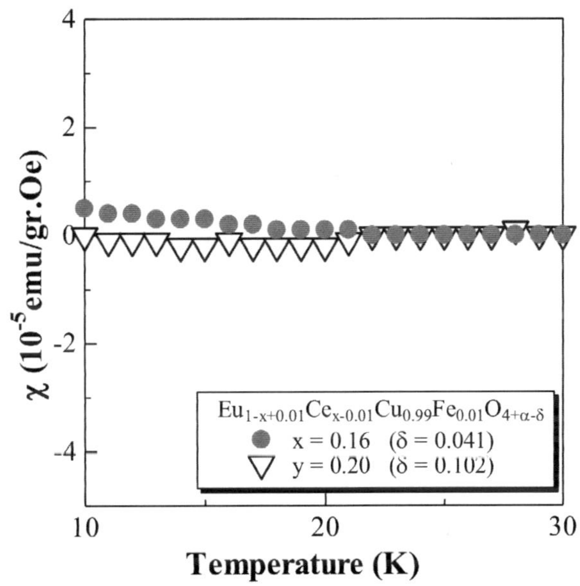

Figure 3. Temperature dependence of dc magnetic-susceptibility (χ) on field cooling at 5 Oe for ECCFO at x = 0.16 and x = 0.20 with 1% Fe impurity.

Figure 4 illustrates the relationship between dc magnetic susceptibility (χ) and 1/T of ECCFO x = 0.16 and 0.20 at 1% Fe impurity in the temperature range of 10 to 30 K. The Curie Constant (C) calculated using gradient of the linear combination that well fitted for all data. And magnetic moment per unit volume, and effective magnetic moment can be calculated from the curie constant as shown in Equation 2 and 3 [11].

$$n\mu^2 = \frac{3k_B C}{\mu_0} \quad (2)$$

$$\mu_{eff} = \sqrt{\frac{3k_B C}{N_A \mu_B^2}} = 2.828\sqrt{C}\,\mu_B \tag{3}$$

where n is the number of atoms per unit volume, μ is magnetic moment per atom, N_A is Avogadro's constant, k_B is Boltzmann's constant, μ_{eff} is effective magnetic moment and μ_B is magneton Bohr unit.

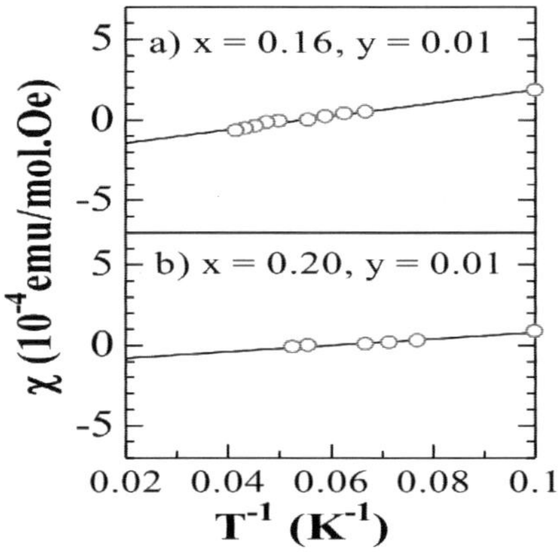

Figure 4. Magnetic-susceptibility (χ) versus *1/T* in the normal state of ECCFO (a) $x = 0.16$ and (b) $x = 0.20$ at 1% of Fe content.

The Curie Constant (C), magnetic moment per atom per unit volume ($n\mu^2$), and effective magnetic moment are all shown in Table 2. The value of the Curie constant, magnetic moment per unit volume ($n\mu$), and effective magnetic moment (μ_{eff}) all decrease as Ce concentration (x) increases. The Curie constant of the sample with $x = 0.16$ is 4.1372 x 10^{-3} emu. K. Oe/mol and the Curie constant of the sample with $x = 0$ is 4.1372 x 10^{-3} emu. K. Oe/mol.

Table 2. Curie constant, magnetic moment per unit volume and effective magnetic moment in the normal state at x = 0.16 and 0.20 with 1% of Fe impurity.

x	δ	C (10^{-3} emu. K. Oe/mol)	$n\mu^2$ (10^{-19} emu. J. Oe/mol)	μ_{eff} (μ_B)
0.16	0.0410	4.1372	1.3636	0.1819
0.20	0.1020	1.9916	0.6564	0.1262

For comparison, ECCFO superconductor at optimum doped with $x = 0.15$ and $y = 0.01$, 0.02 and 0.05 reported by Risdiana, et al shows that there was a decrease in the Curie constant and T_c with increasing Fe concentration. The Curie constant at 1% Fe was 3.64 x 10^{-5} emu. K. Oe/gr [7]. Compared with the two samples, this result has a larger Curie constant value than sample with x = 0.16 and 0.20 which has value of 0.828 x 10^{-5} emu. K. Oe/gr and 0.3223 x 10^{-5} emu. K. Oe/gr. Which have the same Curie Constant tendency to decrease with increasing Ce concentrations.

The magnetic moment per unit volume and the effective magnetic moment decreases with the addition of Ce concentration. The magnetic moment per unit volume for sample with $x = 0.16$ is 1.3636 x 10^{-19} emu. J. Oe/mol and for sample with $x = 0.20$ is 0.6564 x 10^{-19} emu. J. Oe/mol. And the effective magnetic moment for sample with $x = 0.16$ is 0.1819 μ_B and sample with $x = 0.20$ is 0.1262 μ_B. The effective magnetic moment is proportional to the square root of the curie constant, and the magnetic moment per unit volume is proportional to the curie constant. The effective magnetic moment shows the total magnitude of the magnetic dipole moment of the system where this moment comes from the contribution of the movement of spin, orbital and nuclear charges.

From the previous work, the ECCFO sample at optimum doped regime with $x = 0.15$ and $y = 0.01$, 0.02 and 0.05 [7] shows that the addition of Fe concentration will decrease the value of magnetic moment per volume, probably because the correlation of Cu spins in the antiferromagnetically arranged in conduction layer weakens due to the contribution of high-spin Fe orbital moments to the total magnetic moments. In the ECCFO with $x = 0.16$ and 0.20 the increasing of Ce concentrations will decrease the value of effective magnetic moment. This may be due to the presence of electrons as charge carriers from the addition of Ce which is also disturbed due to the presence of Fe which has a strong magnetic moment.

4. Conclusion

$Eu_{1-x+0.01}Ce_{x-0.01}Cu_{0.99}Fe_{0.01}O_{4+\alpha-\delta}$ (ECCFO) material has been produced with concentrations of Ce x = 0.16 and 0.20 and 1% Fe concentration using the solid-state reaction method and also annealed to reduce excess oxygen content. The resulting oxygen reduction of 0.041 for sample $x = 0.16$ and 0.102 for sample $x = 0.20$. XRD characterization results show that for ECCFO samples with $x = 0.16$ and 0.20 with 1% Fe, that all the main peak is tetragonal with a planar configuration of CuO_2 plane (T') with volume fraction of ECCFO 99.5% for sample $x = 0.16$ and 92.3% for sample $x = 0.20$. The addition of Ce concentrations has no effect on the a-axis lattice parameters and Cu-O. The addition of Ce concentrations increases the c-axis lattice parameters and unit cell volume, due to the substitution of Ce with an ionic radius greater than Eu. The c-axis lattice parameters and unit cell volume increase can be indicated as an increase in the distance between the conduction layer and the charge reservoir. The results of SQUID characterization of dc magnetic susceptibility to temperature dependency showed that all sample has paramagnetic behavior at temperature 10 to 30 K. The value of Curie constant, magnetic moment per unit volume and effective magnetic moment decreases with the addition of Ce which may occur due to the more charge carrier from Ce that induced by the substitution of Fe for Cu that also weaken the correlation of Cu spins in the antiferromagnetically arranged in conduction layer.

Acknowledgement

We would like to thank Dr. Isao Watanabe from The Physical and Chemical Research Institute (RIKEN, Japan) for providing SQUID and hosting Sakura Program. Authors would also like to thanks Kemenristek-DIKTI of Indonesia for financial support for the research in the scheme of Fundamental Research (Penelitian Dasar Unggulan Perguruan Tinggi) 2021, Contract No. 1207/UN6.3.1/PT.00/2021 and Academic Leadership Grant (ALG) of Universitas Padjadjaran 2021, No. 1959/UN6.3.1/PT.00/2021.

References

[1] Xiao G, Cieplak M Z, Xiao J Q and Chien C L 1990 Magnetic Pair-Breaking Effects: Moment Formation and Critical Doping Level in Superconducting $La_{1.85}Sr_{0.15}Cu_{1-x}A_xO_4$ systems (A=Fe,Co,Ni,Zn,Ga,Al) *Physical Review B* 42 8752–5

[2] Risdiana, Adachi T, Oki N, Yairi S, Tanabe Y, Omori K, Suzuki T, Watanabe I, Koda A, Higemoto W and Koike Y 2007 μSR Study Of Impurity Effects on The Cu-Spin Fluctuations in The Overdoped Regime of $La_{2-x}Sr_xCu_{1-y}Zn_yO_4$ *Physica C: Superconductivity and its Applications* 460-462 II 874–5

[3] Risdiana, Syakuur M A, Maryati Y, Manawan M and Saragi T 2020 Structure and magnetic Properties of $Eu_{2-x}Ce_xCu_{1-y}Ni_yO_{4+\alpha-\delta}$ *Key Engineering Materials* 855 KEM 123–7

[4] Suzuki K, Adachi T, Tanabe Y, Sato H, Risdiana, Ishii Y, Suzuki T, Watanabe I and Koike Y 2012 Fe-substitution Effects on the Cu-spin Correlation in $La_{2-x}Sr_xCu_{1-y}Fe_yO_4$ Studied by Muon Spin Relaxation *Physics Procedia* 30 275–8

[5] Armitage N P, Fournier P and Greene R L 2010 Progress and Perspectives on Electron-doped Cuprates *Reviews of Modern Physics* 82 2421–87

[6] Risdiana, Adachi T, Oki N, Koike Y, Suzuki T and Watanabe I 2010 Muon Spin Relaxation

Study of The Cu Spin Dynamics in Electron-doped High- T_c Superconductor $Pr_{0.86}LaCe_{0.14}Cu_{1-y}Zn_yO_4$ *Physical Review B - Condensed Matter and Materials Physics* 82 014506

[7] Risdiana, Suhendar D, Pratiwi S, Somantri W A and Saragi T 2015 The Study of Magnetic and Transport Properties in Normal State $Eu_{1.85+y}Ce_{0.15-y}Cu_{1-y}Fe_yO_{4+\alpha-\delta}$ *International Journal of Chemical, Molecular, Nuclear, Materials and Metallurgical Engineering* 9 600–4

[8] Mandal P R, Sarkar T and Greene R L 2019 Anomalous Quantum Criticality In The Electron-doped Cuprates *Proceedings of the National Academy of Sciences of the United States of America* 116 5991–4

[9] Holzwarth U and Gibson N 2011 The Scherrer Equation versus the "Debye-Scherrer Equation" *Nature Publishing Group* 6 534

[10] Ruys A 2019 4 - Processing, Structure, and Properties of Alumina Ceramics *Alumina Ceramics* ed A Ruys (Duxford: Woodhead Publishing) pp 71–121

[11] Hira U and Sher F 2018 Structural, Magnetic And High-Temperature Thermoelectric Properties of $La_{0.4}Bi_{0.4}Ca_{0.2}Mn_{1-x}Co_xO_3$ ($0 \leq x \leq 0.3$) Perovskites *Journal of Magnetism and Magnetic Materials* 452 64–72

Development of ethnoscience module on pond ecosystem, in Serawai Tribe, Seluma Regency

R G T Kusumah[1], M Andaria[2], A Misriani[3], Purdiyanto[4], Samsilayurni[5]

[1] Universitas Islam Negeri Fatmawati Sukarno Bengkulu, Indonesia
[2] PKBM Cakrawala Kepahiang, Bengkulu, Indonesia
[3] Institut Agama Islam Negeri Curup, Indonesia
[4] Stikes Garuda Putih, Jambi, Indonesia
[5] Universitas Muhammadiyah Palembang, Indonesia

E-mail: raden@iainbengkulu.ac.id

Abstract. This study aims to develop an ethnoscience module as a teaching material for pond ecosystems in science learning at Public Elementary School 36 Seluma Regency. The Research and Development (R & D) method is used in this study. The stages are potential and problem, data collection, product design, design validation, design revision, trial, product revision, usage trial, product revision, mass production. The information came from the village of Tanah Abang, the Serawai tribe and Public Elementary School 36 Seluma. The results showed that the final product in the form of an ethnoscience-based science module with an Artificial Ecosystem in the pond was suitable for use. This feasibility refers to the National Education Standards Agency (BSNP) in the category of having passed stage I and stage II by obtaining an average design feasibility score of 87%, material feasibility 86%, and presentation feasibility 83%. Thus obtained an average feasibility of 85%.

1. Introduction

Education is a learning process either through formal, informal or formal activities whose purpose is none other than individual development, to master various aspects of both cognitive, affective and psychomotor [1–4]. Educational activities are not only carried out and facilitated by teachers in schools but also by parents, families and the environment. The teaching and learning process is the core of educational activities in schools [5]. In order for the purpose of education and teaching to run properly, it is necessary to administer teaching and learning activities, which is commonly called curriculum administration [6–8]. The teacher's duties in the teaching and learning process include teaching and administrative tasks [9,10]. In a teaching situation, the teacher takes the lead and is fully responsible for leadership in the classroom [11–15]. The problem of education that is felt today is the low learning outcomes of students. This can be seen from the violation of school rules by students, as well as the difficulty of students to achieve learning outcomes in accordance with the minimum completeness criteria that are set [16,17]. All elements of compulsory education so that students can improve learning outcomes cannot be separated from government responsibilities and the role of teachers in learning [18–21].

The results of student learning while in the field indicate that the author's initial observations note that the science learning that has been carried out at Public Elementary School 36 Seluma in fifth grade turns the average value of science subjects tends to be low compared to other subjects. Based on the results of the even semester scores of 20 students, students who get a C score are 75%, students who get a B grade are 15%, and those who get an A grade are 10%. The provisional diagnosis was taken that students as passive learning subjects, were unable to remember the concepts that had been studied so they were also unable to answer questions correctly. The low level of mastery of the material can be seen from the poor semester results of students in science subjects [22]. This is very dependent on many things, in general it can be said that difficulties in understanding a teaching can be viewed in terms of materials and methods or teaching approaches [23]. Visual aids and guide books students' creativity in learning is lacking [24]. This learning difficulty needs to be overcome so that student learning outcomes can increase. Science learning that is integrated with local wisdom has a high value of contextual learning. Learning is ethnoscience. Ethnoscience has the goal of enriching the planning of science programs, teacher education, research, and in schools, various ideas are offered with a cultural background on learning. There are five important things that were found from the research, namely; (1) culture, cognition, and learning science as learning collaborations that improve student learning outcomes; (2) theories, objectives, and teaching strategies based on ethnoscience as teaching strategies that focus on observation, meta-analysis of problem solving, and discovery of new things by students; (3) ethnoscience can be integrated into science in elementary schools, ecology, marine fisheries, and general education; (4) useful for teachers because they can get to know various cultures from students; (5) students can work on projects in groups in the learning process [25].

Utilization of nature as a contextual learning media is an duty for teachers. The creativity of teachers in utilizing environmental-based learning media will help students learn science contextually. Tanah Abang Village where 36 Public Elementary School is located has an artificial pond with a river flowing into it. Irrigation canals are made to drain water from rivers to ponds. This site becomes a pond of swift water for the development of fish for local consumption. This pond site has the potential to be a learning media. The physical nature of the water owned by the swift water pool is that it has a fairly swift current when compared to the calm water pool. This is due to the current from the river with a fairly high routine. The water in the pool is changed every day. The speed of this water flow is up to 0.75-1.2 m/second. When compared to a pond without a stream from the river, the current generated only comes from fish that swim only as fast as 0.10 m/s .

The importance of learning science with a scientific approach based on ethnoscience, because Indonesia as a country consisting of various ethnicities and cultures, education in Indonesia should be based on ethnoscience which refers to values and culture, local wisdom that exists in the archipelago [26–28] . Scientific approach in the context and content of the 2013 curriculum based on ethnoscience, this learning will provide great and broad opportunities for science teachers to raise local potential, local mindsets, and indigenous local wisdom in the development of science learning with the theme, topic of science. in the 2013 science curriculum [29].

The ethnoscience-based module was developed because it was found that there was an increase in learning outcomes with the activeness of students, so they have a better understanding of students who study conventionally [26]. A module is a teaching material that is systematically arranged using language that is easily understood by students according to their level of knowledge and age so that users can learn with or without a teacher [30,31]. The module characterized by ethnoscience aims to improve students' scientific literacy on pond ecosystem material.

2. Methods

The research method used in this research is Research and Development (R & D). It is a research method used to produce certain products, and test the effectiveness of these products.

The development model used in this study refers to the development model according to Borg & Gall [32]. The process develop and validate educational products. The steps in this process are generally known as the R & D cycle, which consists of: reviewing the results of previous research relating to the validation of components in the product to be developed, developing it into a product, testing the designed product, and reviewing and correcting the product based on the test results.

The module that was developed first carried out 3 tests, namely; 1) expert judgement of linguist, 2) expert judgement of material, and 3) expert judgement of media. Research syntax can be seen in the Table 1.

Table 1. Syntaxis of research.

Number	Syntax of Research	Results
1	Potention and Problem	Low learning outcome of science learning in elementary school
2	Collecting Data	By research instrument (questionnaire, interview)
3	Product Designing	Design by problem that have research before based on ethnoscience approach
4	Validation of Design	Using 3 expert judgment
5	Revision of Design	Revised product based of note from expert
6	Try out in Small Group	Practicality of Module in small group
7	Revision of Product	Revision based on user notes and practicality
8	Final Product	Revision have been done

3. Result and Discussion

3.1. *Results*

3.1.1. Analysis of Planning Results. Outcome Planning is the process of defining organizational goals, creating strategies to achieve those goals, and developing organizational work activity plans. Planning is the most important process of all management functions because without planning the other functions of organizing, directing, and controlling will not work. Planning consists of two important elements, namely goals and the plan itself.

Based on the results of the preliminary study stage, the next step is to plan the modules that are needed by teachers and students. This planning stage includes determining the module components, such as learning objectives, competency standards, basic competencies, achievement indicators, materials, and media used in the modules to be developed in an Outline of Module Content.

The planned modules have the content specifications listed in the outline. The appearance of the module is made colorful with the placement according to the text components and paying attention to the page layout. Outlines are arranged in more detail for each component of the model. The outline becomes an overview of each page contained in the module. The process of making this outline includes determining the image or illustration that will be on a module page, details of the text, along with information that clarifies the arrangement of the module.

The initial design of the module is the module writing stage. The writing of the module refers to the Outline of Module Content and the outline which is always through guidance and mentoring to produce a good initial product. Various components that have been determined in the outline are collected through several sources, books and the internet. This step is the first stage of writing the module, namely the stage of collecting materials. The thing that needs to be considered in the collection of materials is the source. Each source taken from books or the internet is a detailed source so that students or module users can access the material accurately.

The addition of material about the characteristics of rushing water pools in the school environment is the main component. Hopefully students are able to learn about the natural science of pond physics in a contextual way.

3.1.2. Analysis of Product Validation Stage Results. The components that become the material for the validator's assessment of the module are the content component, language component, and presentation component. development that is considered to have been oriented towards the achievement of effective and efficient learning objectives.

At the module validation stage, there are several suggestions for improvement given by the validator, namely the use of illustrations in the module must really contain character education, improve writing because there are still typing errors, and use the original site address on the image source used in the module to avoid plagiarism from the original image source.

3.1.3. Analysis of the Results of the Limited Trial Phase. The limited trial was carried out by giving modules to 10 students in grade 5 to read and study. The implementation of this limited trial aims to find out an overview of the quality of the module as well as the strengths and weaknesses of the temporary module before being tested more broadly in classroom learning. Students are given 1 week to study the given module, with the implementation of discussions for 3 times to involve student assessment responses.

This module is equipped with data from observations made by researchers. This data is presented in an Ethno science based module. The pool color observation data can be seen in the Table 2.

Table 2. Observation of artificial pond physical colour.

Week of Observation	Physical Pond Colour Observation	
	Morning	Daylight
1	Light Brown	Greenish Chocolate
2	Dark Brown	Light Brown
3	Light Brown	Greenish Chocolate

With this data, students can better interpret science learning in real life. And find out the cause of the color change in the pool. So that the results of learning science are more meaningful. Based on the student response questionnaire to the developed module, it can be concluded that the illustration developed has a very interesting category. In the second revision stage, the module was only corrected for typing errors, so the developed module is ready to be tested more widely in an expanded test.

3.1.4. Extended Trial Phase Results Analysis. The expanded trial of the module product was carried out by applying an ethnoscience-based module in the learning process of class 5 as many as 20 students who were divided into 3 groups with 6-7 remaining in each group. Every student gets the same treatment at every meeting. During implementation, students use the module in learning so that students learn from the provided module.

3.1.5. Analysis of Ethnoscience-Based Module Science Learning Elementary School Materials Artificial Ecosystem (Pond Ecosystem) Serawai Seluma Tribe. The development of the module from the research results of identifying pond ecosystems in Tanah Abang Village, Seluma Regency can be used as teaching material for science subjects for class V because it has passed the feasibility validation and student response test. This shows that the module that has been developed has met the criteria and is feasible to be used as teaching material in the learning process. The results of the validation by three validators with an average percentage of 91%. The module was validated before being declared eligible and had passed several improvements. The improvements were made based on suggestions and input from the validator. exposure to material regarding preservation, table layout and sources in the image. Improvements to the Module are made and then it is declared feasible by the validator.

This module has been developed according to the criteria for a good module according to the Ministry of National Education, which includes the title, topic of activities according to the Basic Competencies to be achieved, there is a completion time, tools and materials needed to complete the task, brief information, work steps, table of observations, questions and conclusion.

Furthermore, the response test was carried out by the fifth grade students of Public Elementary School 36 Seluma. The results of student responses to the module also fall into the "Very Good" category where the average percentage score is 90.6%. This is in accordance with the criteria table for the ideal percentage of the response test if a score of 80% is obtained then it is included in the "Very Good" category. In the student response questionnaire there are 13 aspects, of these 13 aspects there are two aspects that get a low percentage score, but are still in a good category. The aspect, namely, the use of the language used in the module got a score of 79.2% and the second aspect, namely the questions contained in the module, got a score of 66.7%.

However, the student response questionnaire did not indicate which parts of the language and questions were difficult to understand. The low score for this aspect was thought to be due to differences in the level of understanding of the students themselves. In the aspect of language use and questions, even though they received a low score, they were still in a good category where the number of students who understood the use of language and also questions in the module were still more than this module received a very good response with an average percentage of 90.6% although there are two aspects that score low. This is supported by other aspects, where in other aspects it scores above 80%.

3.2. Discussion

3.2.1. Discussion of the Results of the Preliminary Study Stage. Preliminary study given to science teachers and 5th grade elementary school students to find out their need for alternative science learning modules. Based on the results of the analysis of teacher needs, it can be seen that teachers have difficulty getting teaching materials that can attract students' enthusiasm to learn science. Based on the results of the analysis of student needs, it can be seen that some students have textbooks as a guide, but they find it difficult to study the book because of its monotonous and verbalistic presentation. Finally, students feel bored when they have to read and study the book. Government Regulation No.19 of 2005 Article 21 paragraph 2 states that the implementation of the learning process is carried out by developing a culture of reading and writing. This is contrary to the reality of the findings in the field, namely that there are still many students who judge their textbooks to be less interesting to read and study.

Inappropriate teacher methods and teaching methods that are too fast make students' enthusiasm for learning science decrease [15,33–35]. The selection of good learning models and methods needs to be chosen so that students feel motivated and enthusiastic about learning science, models and learning methods that can involve students directly to find the concept of a material.

The lack of linkage of Biology that students learn with applications in everyday life reduces the usefulness of studying science, especially the material for Artificial Ecosystems (Pond Ecosystems). Therefore, it is hoped that the development of a module that presents a science material that is fun and increases students' enthusiasm to learn it. students and existing ethnoscience values.

3.2.2. Discussion of the results of the planning stage. Based on the results of the preliminary study stage, the next step is planning the appropriate module needed by teachers and students. This planning stage includes determining the module components, such as learning objectives, basic competencies, and material achievement indicators, and the media used in the science learning module which will be developed in an Outline of Module Contents.

The planned science learning module has the content specifications listed in the Outline of Module Contents. The module display is made colorful with the placement of illustrations according to the text components and pay attention to the page layout. The module consists of 3 sub-subjects. The first topic studied the meaning of artificial ecosystems and pond ecosystems, the second topic studied about biotic components and abiotic components, and the third topic studied the interactions of biotic components and abiotic components. This Outline of Module Contents is used as a reference at the stage of making the outline and writing the initial product.

The initial design of the 1st draft module is the stage of writing the science learning module. The writing of the science learning module refers to the Outline of Module Contents and the outline which is always through the guidance of the supervisor in order to produce a good initial product. This step is the first stage of writing an ethnoscience-based science learning module, namely the stage of collecting materials. The thing that needs to be considered in collecting materials is the source of the material taken. Every picture/illustration taken from books, the internet or other sources that are not the result of one's own work must be accompanied by the source The source in question is a detailed source so that students or users of ethnoscience-based science learning modules can access the material accurately. Pictures/illustrations are very important in the module. This is because pictures are a simple form of visual communication.

The material collection stage also includes the selection of articles and supporting materials, such as the use in daily life and ethno-science values of the Serawai Tribe related to artificial ecosystem materials. Various sources of images and materials are included in the module to make it easier to browse related matters more easily.

The mixing process is carried out after the layout is completed. All the materials that have been collected for a page are entered into a worksheet and adjusted to the layout design that has been made. At this mixing stage, the text position settings are in front and below the image or do not touch the image at all.

After the materials on a page have been placed in accordance with the layout, the next step is the finishing stage. This stage is one of the stages that takes the creativity of the module writer. This stage requires the author to determine the appropriate coloring for each component, the size of each component, image cutting, image effects, text arrangement, text color and size, as well as convenience in enjoying a module page. Writing the source of the article, the source of the image, and the address of the image, write the original address under the component by shortening the address using the Google URL Shortener application which can be accessed google. This is done because of the limited place to type the source to type the URL address which is usually very long.

3.2.3. Discussion of the Results of the Product Validation Stage. The components that become the validator's assessment material for the ethmoscience-based science learning module are the content component, the language component, and the media/design component.

Based on the results of the assessment by 3 experts, 1 science teacher, and 2 IAIN Bengkulu lecturers, it can be seen that in general the modules developed have a very good category. This score is obtained because all development components refer to the principles of development planning which are considered to have been oriented towards achieving effective and efficient learning objectives.

At the module validation stage, there are several suggestions for improvement given by the validator, namely the use of illustrations in the module must really contain ethnoscience values, improve writing because there is still typing, improve module coloring to make it brighter so students don't get tired quickly when reading it, and the use of the original site address on the image source used in the module to avoid plagiarism of the image source. In line with Ria's research, the modules developed were also validated by several experts relevant to their knowledge. Suggestions for improvement have been added to improve the quality of the content of the developed module [27].

3.2.4. Discussion of the Results of the Limited Trial Phase. The limited trial was carried out by giving modules to 15 fifth grade students to read and study. The implementation of this limited trial aims to find out an overview of the quality of the module as well as the advantages and disadvantages of the module.

The enthusiasm of students to learn the module can be seen from the very beginning. Students immediately opened the module by paying a glance at each page in the module, occasionally students stopped at one page and it was seen that he was reading it. A lot of knowledge about science info and ethnoscience values of a material in students' daily lives that is obtained from the module and is not found in textbooks. The provision of illustrations/pictures for each component and the continuity of each component in the module makes it easier for students to better understand the information provided. be delivered. Based on the student response questionnaire to the developed module, it can be concluded that the developed science learning module has a very interesting category.

3.2.5. Discussion of the Results of the Limited Trial Phase. The results of the final product stage in the form of a valid and effective ethnoscience-based science learning module using ethnoscience as the basis for development in this module resulted in an increase in students' scientific literacy. In general terms, the developed science learning module has several advantages and disadvantages.
The advantages of the developed science learning module are:

a. The ethnoscience-based science learning module developed is made in full color
b. Ethnoscience-based science learning modules are developed with a good layout arrangement, so as to provide convenience and comfort for students
c. The ethnoscience-based science learning module is equipped with illustrations on each material to make it easier for students to understand the material.
d. There are ethnoscience values that exist in the Serawai Tribe, Seluma Regency related to the material in the module to increase students' insight.
e. Increase students' interest in reading and interest in learning about pond ecosystem materials.

The disadvantages of the ethnoscience-based science learning module developed are:

a. The price is relatively expensive because it is full color
b. It takes a tool in the form of a computer/laptop as well as an internet connection so that the use of ethnoscience-based science learning modules can be maximized.

4. Conclusion

The result of product development is in the form of an ethnoscience-based science module as a science teaching material with an Artificial Ecosystem (Pond Ecosystem) and from collecting information from Tanah Abang Village, Serawai Tribe and Public Elementary School 36 Seluma Regency. The effectiveness of the ethnoscience-based science learning module on Artificial Ecosystem Materials (Pond Ecosystem) for fifth grade elementary school students in the second semester is equipped with ethnoscience values related to Artificial Ecosystem Materials (Pond Ecosystem). The developed module is validated to see its feasibility to the validator, the overall module design is considered very valid by the validator which means the module is suitable for use in learning. The feasibility level of the module based on validation by the validator obtained results with a percentage of 91% with a very valid category and the response test by students showed an overall percentage of 90.6% which means the module is considered very good.

References

[1] Cuc M C 2014 The Influence of Media on Formal and Informal Education Procedia - Soc. Behav. Sci. 143 68–72
[2] Tripon A 2014 Innovative Technology for Sustainable Development of Human Resource Using Non-formal and Informal Education Procedia Technol. 12 598–603
[3] Beddie F and Halliday-Wynes S 2010 Informal and Non-Formal Learning in Vocational Education and Training ed P Peterson, E Baker and B B T-I E of E (Third E McGaw (Oxford: Elsevier) pp 240–6
[4] Anders S, Fischer-Bruegge D, Fabian M, Raupach T, Petersen-Ewert C and Harendza S 2011 Teaching post-mortem external examination in undergraduate medical education—The formal and the informal curriculum Forensic Sci. Int. 210 87–90
[5] Djamarah S B 2008 Psikologi Belajar (Jakarta: Rineka Cipta)
[6] Abdullah Idi 2014 Pengembangan Kurikulum (Jakarta: Grasindo)
[7] Kuswanto E 2015 Peranan Guru PAI dalam Pendidikan Akhlak di Sekolah Mudarrisa J. Islam. Educ
[8] Adang S 2012 Kurikulum dan Pembelajaran Bandung: Yayasan Kesuma Karya

[9] Downs N J, Parisi A V and Igoe D 2014 Measurements of occupational ultraviolet exposure and the implications of timetabled yard duty for school teachers in Queensland, Australia: Preliminary results J. Photochem. Photobiol. B Biol. 131 84–9
[10] Kumar P and Wiseman A W 2021 Teacher quality discourse in India: A national reform agenda content analysis Teach. Teach. Educ. 107 103504
[11] Tiswarni T 2019 Usaha Kepala Sekolah dalam Meningkatkan Disiplin Kerja Guru (Studi Deskriptif Kualitatif di SMP N 1 Argamakmur Kabupaten Bengkulu utara) At-Ta'lim Media Inf. Pendidik. Islam 18 261
[12] Anugrahana A 2020 Hambatan, Solusi dan Harapan: Pembelajaran Daring Selama Masa Pandemi Covid-19 Oleh Guru Sekolah Dasar Sch. J. Pendidik. dan Kebud. 10 282–9
[13] Iriyani D 2008 Pengembangan Supervisi Klinis Untuk Meningkatkan Keterampilan Dasar Didaktika 2 1–10
[14] Sumantri B A 2019 Pengembangan Kurikulum di Indonesia Menghadapi Tuntutan Kompetensi Abad 21 At-Ta'lim Media Inf. Pendidik. Islam 18 27
[15] Indarti S 2019 Investigasi Implementasi Model Discovery Learning Berbasis Pendekatan Saintifik Dalam Meningkatkan Aktivitas dan Hasil Belajar IPA IJIS Edu Indones. J. Integr. Sci. Educ. 1 100
[16] Amri A and Tharihk A J 2018 Pengembangan Perangkat Asesmen Pembelajaran Proyek pada Materi Pencemaran dan Kerusakan Lingkungan Didakt. Biol. J. Penelit. Pendidik. Biol. 2 103–12
[17] Rerung N, Sinon I L . and Widyaningsih S W 2017 Penerapan Model Pembelajaran Problem Based Learning (PBL) untuk Meningkatkan Hasil Belajar Peserta Didik SMA pada Materi Usaha dan Energi J. Ilm. Pendidik. Fis. Al-Biruni 6 47
[18] Yusrizal Y, Suliyanah S and Basri T H 2017 Analysis of Knowledge, Understanding and Skills of Physics Teachers of State Senior High Schools in Developing and Analyzing Test Items J. Pendidik. IPA Indones. 6 335–40
[19] Gao S, Wang Y, Jiang B and Fu Y 2018 Application of problem-based learning in instrumental analysis teaching at Northeast Agricultural University Anal. Bioanal. Chem. 410 3621–7
[20] Pakarinen E and Kikas E 2019 Child-centered and teacher-directed practices in relation to calculation and word problem solving skills Learn. Individ. Differ. 70 76–85
[21] Green M, Mcneese M N and Mississippi S 2007 Using edutainment software to enhance online learning Int. J. E-learning 6 5–16
[22] Sudjana N 2009 Penilaian Hasil Proses Belajar Mengajar (Bandung: PT. Remaja Rosdakarya)
[23] Abu Taleb T F 2013 Requirements of high-quality kindergarten programs according to jordanian parents: An exploratory study J. Res. Child. Educ. 27 472–84
[24] Timmons K 2018 Educator Expectations in Full-Day Kindergarten: Comparing the Factors That Contribute to the Formation of Early Childhood Educator and Teacher Expectations Early Child. Educ. J. 46 613–28
[25] Wahyu Y 2017 Pembelajaran berbasis etnosains di Sekolah Dasar JIPD (Jurnal Inov. Pendidik. Dasar) 1 140–7
[26] Khoiri A and Sunarno W 2018 Pendekatan Etnosains dalam Tinjauan Fisafat Spektra J. Kaji. Pendidik. Sains
[27] Ria Febu Khoerunnisa, N. Murbangun S 2012 Pengembangan Modeul IPA Terpadu Etnosains untuk Menumbuhkan Minat Kewirausahaan J. Innov. Sci. Educ.
[28] Yuliana I 2017 Pembelajaran berbasis etnosains dalam mewujudkan pendidikan karakter siswa sekolah dasar J. Pendidik. dan Pembelajaran Sekol. Dasar 1 98–106
[29] Atmojo S E 2012 Profil keterampilan proses sains dan apresiasi siswa terhadap profesi pengrajin tempe dalam pembelajaran ipa berpendekatan etnosains J. Pendidik. IPA Indonesia
[30] Kristanto A, Mariono A and Nuryati D W 2018 Developing Media Module Proposed to Editor in Editorial Division Journal of Physics: Conference Series vol 947 (IOP Publishing) p 12054

[31] Lestari D, Komarudin K, Mujib M and Mardiyah M 2020 Mathematical module based on islamic values as a development of contextual teaching and learning (CTL) Math Didact. J. Pendidik. Mat

[32] Borg W R and Gall M D 1983 Education Research: An Introducation (New York & London: Longman Inc)

[33] Kawuri M Y R T, Ishafit I and Fayanto S 2019 Efforts To Improve The Learning Activity And Learning Outcomes Of Physics Students With Using A Problem-Based Learning Model IJIS Edu Indones. J. Integr. Sci. Educ. 1

[34] Wolf S 2018 Impacts of Pre-Service Training and Coaching on Kindergarten Quality and Student Learning Outcomes in Ghana Stud. Educ. Eval. 59 112–23

[35] Kane R G and Chimwayange C 2014 Teacher action research and student voice: Making sense of learning in secondary school Action Res

Development of an ethnophysic-based module in the Mandailing Tribe, West Pasaman Regency as science teaching materials

A Walid[1], S D Oktaria[2], R E Putra[3], Nipriansyah[2] and E Susanto[4]

[1]UIN Fatmawati Sukarno Bengkulu, Indonesia
[2]Program Doktor Pendidikan, Universitas Bengkulu, Indonesia
[3]Universitas Negeri Padang, Indonesia
[4]Universitas Dehasen Bengkulu, Indonesia

E-mail: dongawalid19@gmail.com

Abstract. The purpose of this study is to develop ethnosains learning modules that are feasible and practical for use after being validated by several experts and responded by several teachers and students. The research method uses the Borg & Gall development model which consists of 8 stages, namely a preliminary study stage, a research planning stage, a design development stage, an initial trial phase, an initial product revision, a limited field test, a revised stage of the results of a limited field test, and a product stage end. This research is limited to the limited field trial stage namely the feasibility and practicality of the module. The instrument used was a questionnaire for module eligibility and student and teacher response questionnaires to test the practicality of the module. Data from the analysis of the questionnaire by calculating the percentage of achievement in each component of the percentage of achievement criteria of eligibility and practicality of science learning modules based on ethnics that was developed. Ethnographic-based science learning modules developed were declared feasible to be used with a percentage of 90.00% (linguists), 88.00% (material experts), 96.00% (design / media experts) with the very feasible, feasible, and very feasible and the percentage of eligibility of 3 colleagues is 91.46%. Practical to use with the response percentage of 10 students 84.83% (very practical category), and the response of 2 science teachers with a percentage of 93.40% (very practical category). Practical to use with the response percentage a total average of 10 students namely 84.83% are in the interval of $80\% \leq PRM \leq 100\%$. And based on the teacher's questionnaire response that is 93.40% which is included in the category of very practical From the results of these data it can be concluded that the module has been feasible and practical to be used as teaching material in supporting the teaching and learning process on the material interaction of living things with their environment.

1. Introduction

Teachers have an important role in preparing and motivating anintellectual generation that can build and advance the country through education. But to give birth to a quality and superior generation, competent teachers are needed. Teachers are also involved in improving learning by integrating technology. Thus,preparing future teachers to integrate technology in the practice of science learning [1]. In addition to teaching and educating, another task that educators must do is to be able to manage the classroom effectively for the sake of comfort and calmness in the teaching and learning process.

Class management actions are actions taken by educators to provide optimal conditions so that the teaching and learning process takes place effectively [2,3]. So it can be concluded that classroom management is one of the supporters in facilitating the learning process so that teaching and learning activities take place effectively. Direct participation and teacher guidance is essential in this student's work, although each student must work individually, linking theoretical knowledge to the realities of life. Because modern education requires a change in the system, first of all, new approaches to activity, critical analysis and increased responsibility [4]. Learning in schools is a formal civilizing process / alkulturation process. Education serves to preserve positive cultures, and education also functions to create changes in a more innovative direction, so that the original education system in an area has an important role in the development of education and culture. Education can be interpreted as one form of the embodiment of human culture that is dynamic and full of development.

Physics is one of the fields of ethnoscience and a branch of science that studies about natural phenomena including material, human, and the interaction between human and material other. Physics is closely related to other sciences [5]. This is because of physics considered as a trunk of knowledge that is useful for the development of technology, inventions, and other sciences. Physics learning is a learning that implementing everyday life with the principles of physics [6].

Physics learning in schools in general is still centralized on the material available in the book. It's still rare to learn physics that really explains and implement cultural realities around students. The content of the material being taught is not yet many have integrated with the culture. Reflecting on this condition, it is necessary development of learning methods, one of which is the approach used. Application learning physics with an ethnophysics approach requires the teacher's ability to combining original knowledge with scientific knowledge [7] Culture-based physics learning or what can be called Ethnophysics is a cultural relationship with the concept of physics. One of the prospective aspects for studied as content material for learning physics is culture. This is in accordance with the nature of culture as a social heritage that is only owned by citizens by studying it [8]. Given that culture is a reflection of people's lives in the form of belief in science that is trial and error such as the findings of trial and error.

Based on initial observations, it is known that the results of science learning that have been done in the third grade of MIM (Madrasah Ibtidaiyah Muhammadiyah) in Ujung Gading turned out that the average value of science subjects tended to be low compared to other subjects. Based on the results of UTS scores of 24 students, students who got A grades as much as 75%, students who got A grades as much as 15%, and those who got A grades were 10%. Therefore, it makes science subjects a less interesting, boring subject and makes students passive to remember the concepts that have been learned so that they also cannot answer questions properly. The low level of mastery of the material can be seen from the poor results of the student semester on science subjects. According to [5] in his research found some science learning on the most difficult biological topics. It is very dependent on many things, in general it can be said that the difficulty in understanding a teaching can be seen in terms of materials and methods or teaching approaches. While the low-test results are caused by several factors including lack of facilities and infrastructure (props and guidebooks) of student creativity in learning is lacking. Media is one form of tool used to improve and facilitate performance. The demands of technological advancement require development [6-11]. Media innovation is always done to get better quality states that motivation is one of the important elements for science education because it is strongly correlated with students' success in learning science [9-13]. If a student has a high motivation in science learning, a positive attitude towards science will also have an impact on improving learning outcomes in the field of science. The use of teaching materials such as innovative modules creates interesting learning, fosters interest, motivation, reduces dependency and gets ease in learning every indicator contained in the learning tools compiled by teachers [14,15]. This module was chosen because it has self-instruction and self-instruction characteristics, so students are trained independently to manage their study time and understand the subject.

2. Methods

2.1. Retrieval of Ethnoscience Data
The research method is a method used by Research and Development (R & D). The research and development steps according to [16] are; 1. Potential and Problems; 2. Data Collection; 3. Product Design; 4. Design Validation; 5. Design Revision; 6. Product Testing; 7. Product Revision; 8. Usage Trial.

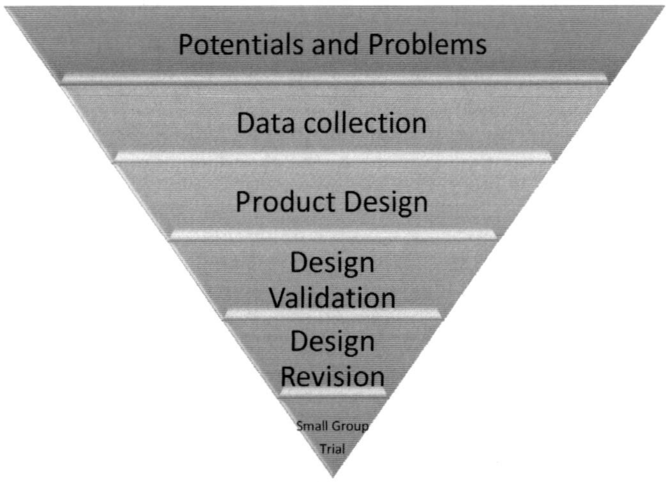

Figure 1. Research steps.

2.2. Research Instruments
The instrument used in this study was a module eligibility questionnaire to get validation data from 3 validators. The language validation questionnaire consisted of 15 statement components with a score of 1-5. The material validation questionnaire consisted of 16 statement components with a score of 1-5. While the media validation questionnaire consisted of 16 statement components with a score of 1-5.

2.3. Data analysis technique
The researcher makes a validation sheet that contains the statements. Then the validator fills out the questionnaire by giving a check mark to the category provided by the researcher based on a Likert scale consisting of 5 rating scores as shown in Table 1.

Table 1. Expert validation rating score.

Explanation	Score
Very good (SB)	5
Good (B)	4
Enough (C)	3
Less (K)	2
Very less (SK)	1

Furthermore, the percentage of Eligibility Criteria [17] obtained is then interpreted into categories based on the Table 2.

Table 2. Eligibility criteria.

Assessment (%)	Interpretation Creteria
$81 \leq P \leq 100$	Very Decent
$61 \leq P < 81$	Worthy
$41 \leq P < 61$	Enough
$21 \leq P < 41$	Not Feasible
$0 \leq P < 21$	Very Unworthy

Module-based teaching materials are declared theoretically feasible if the percentage of eligibility is ≥ 51 % [18].

3. Result and Discussion

3.1. *Results of the Preliminary Study Phase.*

Preliminary study given to third grade science teachers and third grade MI students to find out their need for alternative science learning modules. Based on the results of the analysis of teacher needs, it can be seen that teachers have difficulty getting teaching materials that can attract students' enthusiasm to learn science. This is because the presentation is monotonous and verbalistic. Finally, students feel bored when they have to read and study the book. Government Regulation No.19 of 2005 Article 21 paragraph 2 states that the implementation of the learning process is carried out by developing a culture of reading and writing. This is contrary to the reality of the findings in the field, there are still many students who judge the textbooks they have are less interesting to read and study.

The existence of verbalistic textbooks is the reason that is encountered. The form of verbal communication that has been so dominant in learning in learning creates problems with verbalistic learning experiences in the world of education. Furthermore, he suggested the use of simple, effective, and efficient visual illustrations. Inappropriate teacher methods and teaching methods that are too fast make students' enthusiasm for learning science decrease. Selection of good learning models and methods needs to be chosen so that students feel motivated and enthusiastic about learning science, models and learning methods that can involve students directly to find concepts from a material.

The lack of linking the Biology that students learn with applications in everyday life reduces the usefulness of studying science, especially the material on Science Education. Therefore, it is hoped that the development of a module that presents a science material that is fun and increases students' enthusiasm to learn it. students and existing ethnoscience values.

3.2. *Discussion of the results of the planning stage*

Based on the results of the preliminary study stage, the next step is planning the appropriate module needed by teachers and students. This planning stage includes determining the module components, such as learning objectives, basic competencies, and material achievement indicators, and the media used in the science learning module which will be developed in an Outline of Module Contents (GBIM). The planned science learning module has the content specifications listed in the GBIM. The module display is made colorful with the placement of illustrations according to the text components and pay attention to the page layout. The module consists of 3 sub-subjects. The first topic studied the meaning of artificial ecosystems and rice field ecosystems, the second topic studied about biotic components and abiotic components, and the third topic studied the interactions of biotic components and abiotic components. This GBIM is used as a reference at the stage of making the outline and writing the initial product.

Outlines are arranged in more detail for each component of the model. The outline becomes an overview of each page contained in the module. The process of making this outline includes determining the images/illustrations that will be on the module page, details of the text, along with information that clarifies the arrangement of the module. The selection of images/illustrations is determined at the stage of making the outline, because the selection that is in accordance with the text component will strengthen the impression that will be conveyed by the text component.

The initial design of the 1st draft module is the stage of writing the science learning module. The writing of the science learning module refers to the GBIM and the outline which is always through the guidance of the supervisor in order to produce a good initial product. Various components that have been determined in the outline are collected from various sources, ranging from personal sources, books and the internet. This step is the first stage of writing an ethnoscience-based science learning module, namely the stage of collecting materials. The thing that needs to be considered in collecting materials is the source of the materials taken.

Every picture/illustration taken from books, the internet or other sources that are not your own work must be accompanied by the source. The source in question is a detailed source so that students or users of ethnoscience-based science learning modules can access the material accurately. in the module. This is because the image is a simple form of visual communication. The material collection stage also includes the selection of articles and supporting materials, such as the use in daily life and ethno-science values of the Mandailing Tribe, West Pasaman Regency related to artificial ecosystem materials. Various sources of images and materials are included in the module to make it easier to browse related matters more easily.

The mixing process is carried out after the layout is completed. All the materials that have been collected for a page are entered into a worksheet and adjusted to the layout design that has been made. At this mixing stage, the text position settings are in front and below the image or do not touch the image at all. After the materials on a page have been placed in accordance with the layout, the next step is the finishing stage. This stage is one of the stages that takes the creativity of the module writer. This stage requires the author to determine the appropriate coloring for each component, the size of each component, image cutting, image effects, text arrangement, text color and size, as well as convenience in enjoying a module page. Writing the source of the article, the source of the image, and the address of the image, write the original address under the component by shortening the address using the Google URL Shortener application which can be accessed at http://www.goo.gl. This is done because of the limited place to type the source to type the URL address which is usually very long.

3.3. Discussion of the Results of the Product Validation Stage

The components that become the validator's assessment material for the ethnoscience-based science learning module are the content component, the language component, and the media/design component. Based on the results of the assessment by 3 experts, 1 science teacher, and 2 IAIN Bengkulu lecturers, it can be seen that in general the modules developed have a very good category. This score is obtained because all development components refer to the principles of development planning which are considered to have been oriented towards achieving effective and efficient learning objectives.

Table 3. Validation results of ethnoscience module materials of vegetable natural resources diversity in mandailing tribe, pasaman barat district by validator.

No	Validator	Percentage (%)	Criterion
1	Validator I (Design Lecturer)	96	Highly Valid
2	Validator II (Lecturer of Materials Expert)	88	Highly Valid
3	Validator III (Lecturer of Linguists)	90	Highly Valid
	Average	91	Highly Valid

At the module validation stage, there are several suggestions for improvement given by the validator, namely the use of illustrations in the module must really contain ethnoscience values, improve writing because there is still typing, improve module coloring to make it brighter so students don't get tired quickly when reading it, and the use of the original site address on the image source used in the module to avoid plagiarism of the image source.

3.4. Discussion of the Results of the Limited Trial Phase

The limited trial was carried out by giving modules to 20 third grade students to read and study. The implementation of this limited trial aims to find out an overview of the quality of the module as well as the advantages and disadvantages of the module. The enthusiasm of students to learn the module can be seen from the very beginning. Students immediately opened the module by paying a glance at each page in the module, occasionally students stopped at one page and it was seen that he was reading it. A lot of knowledge about science info and ethnoscience values of a material in students' daily lives that is obtained from the module and is not found in textbooks. Providing illustrations/pictures for each component and the continuity of each component in the module makes it easier for students to better understand the information conveyed.

3.5. Feasibility of the Ethnoscience-Based Module for MI Science Learning Materials on the diversity of plant natural resources of the West Pasaman Mandailing Tribe

The development of the module from the results of the research on the identification of medicinal plants in Ujung Gading Village, West Pasaman Regency, can be used as teaching material for Class III science subjects because it has passed the feasibility validation and student response test. This shows that the module that has been developed has met the criteria and is feasible to be used as teaching material in the learning process. The results of the validation by three validators with an average percentage of 91%. The module was validated before being declared eligible and had gone through several improvements. The improvements were made based on suggestions and input from the validator. exposure of material regarding preservation, table layout and sources in the image. Improvements to the Module are made and then it is declared feasible by the validator.

This module has been developed according to the criteria for a good module according to the Ministry of National Education, which includes the title, topic of activities according to the Basic Competencies to be achieved, there is a completion time, tools and materials needed to complete the task, brief information, work steps, table of observations, questions and conclusion.

Furthermore, a response test was conducted by the students of Ujung Gading Class. The results of student responses to the module also fall into the "Very Good" category where the average percentage score is 90.6%. This is in accordance with the criteria table for the ideal percentage of the response test if a score of 80% is obtained then it is included in the "Very Good" category. In the student response questionnaire there are 13 aspects, of these 13 aspects there are two aspects that get a low percentage score, but are still in a good category. The aspect, namely, the use of the language used in the module got a score of 79.2% and the second aspect, namely the questions contained in the module, got a score of 66.7%.

Based on the results of the analysis of the student practical response sheet assessment on a limited test, the practicality of teaching materials can be seen in that learning materials in the limited group test Ethnographic-based Science Learning Module can be said to be included in the category of very practical with a total average of 10 students namely 84.83% are in the interval of 80% ≤ PRM ≤ 100%. And based on the teacher's questionnaire response that is 93.40% which is included in the category of very practical. This research is reinforced by the results of research show that ethnoscience learning in physics learning has a very significant and effective effect on implementation [19]. Furthermore, the emphasis of this research is strengthened by research which shows the effectiveness of ethnophysics learning in 21st century learning [20].

4. Conclusion

The resulting product development results are in the form of an ethnoscience-based science module as a science teaching material with the sub-theme Diversity of vegetable natural resources and from collecting information from Tanah Abang Village, Mandailing Tribe and MIM Ujung Gading, West

Pasaman Regency. The effectiveness of the ethnoscience-based science learning module on the science education for Grade III students in the second semester of elementary school is equipped with ethnoscience values related to the material diversity of plant natural resources. The developed module is validated to see its feasibility to the validator, the overall module design is considered very valid by the validator which means the module is suitable for use in learning. The feasibility level of the module based on validation by the validator obtained results with a percentage of 91 % with a very valid category and the response test by students showed an overall percentage of 90.6 % which means the module is considered very good. Based on the teacher's questionnaire response that is 93.40% which is included in the category of very practical.

References

[1] Memisoglu PS. The Perception of Teachers About Management Skills of School Principals *. Int J Humanit Soc Sci. 2015;5(2):21–9

[2] Tondeur J, Scherer R, Baran E, Siddiq F, Valtonen T, Sointu E. Teacher educators as gatekeepers: Preparing the next generation of teachers for technology integration in education. Br J Educ Technol. 2019;50(3):1189–209

[3] Sokal L, Trudel LE, Babb J. Canadian teachers' attitudes toward change, efficacy, and burnout during the COVID-19 pandemic. Int J Educ Res Open. 2020;1(October):100016

[4] Saifnazarov I, Abdullahanova G, Alimatova N, Kudratova U. The main trends of increasing the role of the teacher in the innovative development of uzbekistan. Int J Adv Sci Technol. 2020;29(5):1771–3

[5] Ningsih S. Persepsi Mahasiswa Terhadap Pembelajaran Daring Pada Masa Pandemi Covid-19. JINOTEP (Jurnal Inov Dan Teknol Pembelajaran) Kaji Dan Ris Dalam Teknol Pembelajaran. 2020;7(2):124–32

[6] Nyadzi E, Ajayi OC, Ludwig F. Indigenous knowledge and climate change adaptation in Africa: a systematic. 2021

[7] Sudarmin F. R., Nuswowati, M. & Sumarni, W. Development of ethnoscience approach in the module theme substance additives to improves the cognitive learning outcome and student's entrepreneurship. J Phys. 2017;3(824):1–14

[8] Devinta M, Hidayah N, Hendrastomo G. Fenomena Culture Shock (Gegar Budaya) Pada Mahasiswa Perantauan Di Yogyakarta. J Pendidik Sosiol. 2015

[9] Çimer A. What Makes Biology Learning Difficult and Effective: Students' Views. Educ Res Rev. 2012;7(3):61–71

[10] Moolenaar NM, Daly AJ, Sleegers PJC. Occupying the principal position: Examining relationships between transformational leadership, social network position, and schools' innovative climate. Educ Adm Q. 2010;46(5):623–70

[11] Pranitasari D. The Influence of Effective Leadership and Organizational Trust to Teacher's Work Motivation and Organizational Commitment. Media Ekon dan Manaj. 2020;35(1):75

[12] Masters GN, Taylor-Guy P, Fraillon J, Chase A-M. Ministerial Briefing Paper on Evidence of the Likely Impact on Educational Outcomes of Vulnerable Children Learning at Home during COVID-19. Ministerial Briefing Paper on Evidence of the Likely Impact on Educational Outcomes of Vulnerable Children Learning at Home during COVID-19. 2020

[13] Hargett CW, Doty JP, Hauck JN, Webb AMB, Cook SH, Tsipis NE, et al. Developing a model for effective leadership in healthcare: A concept mapping approach. J Healthc Leadersh. 2017;9(March):69–78

[14] Cavas P. Factors Affecting the Motivation of Turkish Primary Students for Science Learning. Sci Educ Int. 2011;22(1):31–42

[15] Putri AS, Aznam N. The Effect of The Science Web Module Integrated on Batik's Local Potential Towards Students' Critical Thinking and Problem Solving (Thinking Skill). J Sci Learn. 2019;2(3):92–6

[16] Sugiyono S. Metode penelitian kuantitatif dan kualitatif dan R&D. Alf Bandung. 2010

[17] Van Spall HGC, Toren A, Kiss A, Fowler RA. Eligibility criteria of randomized controlled trials published in high-impact general medical journals: a systematic sampling review. Jama. 2007;297(11):1233–40
[18] Sapri J, Novitasari N, Saregar A, Topano A, Walid A, Kusumah RGT. Construction Ethnoscience-Based Learning Environment Material in Scientific Knowledge. In: Journal of Physics: Conference Series. IOP Publishing; 2021. p. 12034
[19] Astuti IAD, Bhakti YB. Kajian Etnofisika Pada Tari Piring Sebagai Media Pembelajaran Fisika. In: SINASIS (Seminar Nasional Sains). 2021
[20] Astuti IAD, Bhakti YB. Kajian Etnofisika Pada Tari Piring Sebagai Media Pembelajaran Fisika. SINASIS (Seminar Nas Sains) [Internet]. 2021;2(1):477–82. Available from: http://www.proceeding.unindra.ac.id/index.php/sinasis/article/view/5387

Innovation of simple binoculars as a teaching aid for optical practicum at school

Azhar[1], H Fuadi[2], A Doyan[2,3], Susilawati[2,3], S Ayub[3] and L S Hudha[4]

[1]Physics Education Study Program, FKIP, Riau University, Indonesia
[2]Master of Science Education Study Program, Mataram University, Indonesia
[3]Physics Education Study Program, FKIP, Mataram University, Indonesia
[4]Department of Physics, Faculty of Mathematics and Natural Science, University of Mataram

Email: azhar@lecturer.unri.ac.id, husnul.fuadi@unram.ac.id, aris_doyan@unram.ac.id

Abstract. Learning aids are tools used by teachers in learning and prevent verbalization in students. The purpose of this research is to make simple earth telescopes as visual aids for optical practicum in schools. The method used is the experimental method, namely analyzing, designing, making and testing the props made. The subjects of the research were 17 students of physics education FKIP University of Mataram. The binoculars test saw the object of letters and writing with the font Times New Roman 350 at a distance of 50 meters and 100 meters. The results showed that the object was clearly visible with the satisfaction level of using the teaching aids a score of 3.71 the criteria were satisfied and suitable for use. The use of simple earth telescopes is suitable for use in optical practicums in schools. Teachers are expected to be able to create and develop teaching aids to improve students' understanding of concepts.

1. Introduction

Learning using teaching aids is a series of activities to deliver subject matter that aims to provide opportunities for students to be active in learning, thus enabling students to gain knowledge and develop psychomotor skills as well as foster student creativity to solve problems encountered such as asking questions that have not been understood [1, 2]. The teacher in delivering the material pays less attention to the proportion of the material and the systematics of delivery, and does not emphasize the basic concepts so that it feels difficult for students. The teaching aids available in school laboratories are rarely used as a variety of science learning methods [3, 4, 5]. This makes it difficult for students to understand the real science concepts. Physics as a science that has objects in the form of real objects if delivered only by the lecture method, the material received by students will only be understood as formulas or abstract concepts [6].

Students of the Physics Education Undergraduate Study Program are students who are being prepared to become prospective physics teachers in secondary schools. This means that the implementation of physics learning in schools in the future will be largely determined by them [7]. When viewed from the data of national exam participants, the number of students who choose physics subjects as elective subjects is the least compared to other elective subjects, indicating that the implementation of physics learning in schools is less meaningful [8, 9]. Less meaningful physics learning tends to be unattractive to students, so prospective physics teacher students who are currently still in college are expected to be able to improve this situation [10].

Learning aids are tools used by teachers in learning and prevent verbalization in students [11, 12]. The main function of teaching aids is to reduce the abstraction of the concept and eliminate verbalism in students, so that students are able to capture the true meaning of the concept [13]. Teaching aids are tools that function to accelerate students' understanding of one subject in a particular field of study [14]. Teaching aids can make abstract ideas more concrete to learn, helping students to focus on thoughts and ideas about a problem, which in turn helps them to understand and interpret the information that has been presented. Teaching aids are one of the most important learning media for achieving educational goals [15, 16].

Optics is a part of physics that studies light. In optics studied the properties of light, the nature of light and the use of the properties of light. There are two branches of optics, namely geometric optics and physical optics. In geometrical optics the properties of light reflection and refraction are studied, while in physical optics the properties of interference, diffraction, and polarization of light are studied. The application of this optics has given birth to various tools that are used in certain fields to help human activities. For example, in the field of astronomy, binoculars are used to observe objects in outer space. A telescope or binoculars is an optical instrument used to see objects that are very far away to make them appear closer and clearer. Judging from the course of the rays, binoculars are divided into two, namely refractive binoculars and reflecting binoculars, while from the point of view of the object, binoculars are divided into two, namely telescopes and telescopes. Earth/terrestrial binoculars are one type of binoculars used to see or observe objects on land or at sea, in other words, Earth telescopes are binoculars to observe objects on the earth's surface, including the diversity of the earth's surface. Earth binoculars are often called yojana binoculars or field binoculars. In earth telescopes, the resulting image is clearly visible, close and not upside down, while in star telescopes the resulting image will be inverted. Earth telescopes are divided into two types, namely Galileo binoculars and spyglass binoculars [17]. Galileo binoculars are also called stage binoculars. Named the Galileo binoculars because they were discovered by a scientist named Galileo Galilei. Galileo is a scientist who researches space with a telescope or binoculars [18]. With this research, Galileo eventually made world-shaking discoveries, including the moons of Jupiter, the phases of Venus, the sun's spots, the structure of the moon's surface, and his assertion that the Milky Way galaxy is composed of a large number of individual stars. This binocular has a short size and consists of two lenses, namely a convex lens that functions as an objective lens and a concave lens that doubles as an eyepiece and an inverting lens. The concave lens in Galileo's binoculars serves to produce an upright image [19].

Learning related to binoculars at school is taught in physics subjects in the sub-topics of optics and optical instruments and sub-subjects of astronomy. To better understand the theories and concepts that have been studied, it is important to hold simple practical tools such as binoculars. Research conducted by Sheilia Fitria [20] resulted in the average results of student investigations of the working principle of binoculars by 62.5% in the good category, the average achievement of learning indicators of 84.2% in the very good category, and the average the response of students to the binoculars props was 91.6% with a very good category. Interesting and fun learning is an alternative for teachers to stimulate the creativity of students. To make learning interesting and fun, teachers need to create interesting learning media as well. One of the interesting learning media to use is "simple binoculars" as a tool for students to understand the theories and concepts that have been taught. Based on the description above, the author tries to provide a solution that can be used in physics subjects, namely the manufacture of simple binoculars as a teaching aid for optical learning in schools. This research is important because teachers are taught how to make simple teaching aids from easily found materials. Teachers as facilitators need to plan and design a teaching aid that is easy to make and easy to use using simple materials.

2. Methods

The manufacture of these Galileo binoculars will be carried out at the Physics Education Laboratory, FKIP University of Mataram starting from March to June 2021. To determine the quality and the farthest viewing distance of these binoculars, it is necessary to test the instrument. The use of binoculars will be tested in June 2021 at the Mataram University campus. The test subjects of the tool were 17 students of physics education, FKIP, University of Mataram. The sampling technique was carried out by purposive sampling. Data retrieval using Galileo binoculars by determining the distance of the object to be seen starting from 50 m and 100 m. After using binoculars, students were asked to fill out a questionnaire related to their satisfaction with using the tool.

Table 1. Satisfaction level criteria.

Value Range	Level Criteria
0.00 – 1.79	Very Dissatisfied
1.80 – 2.59	Not satisfied
2.60 – 3.39	Quite satisfied
3.40 – 4.19	Satisfied
4.20 – 5.00	Very satisfied

(Source: Adaptation of Ratumanan and Laurens, 2011)

$$satisfaction = \frac{total\ score}{number\ of\ items} \qquad (1)$$

2.1. *Tools and materials*

The tools and materials that will be used to make Galileo binoculars are as follows; 1). A convex lens as an objective lens with a diameter of 50 mm with a focus of 500 mm. 2). A concave lens as an eyepiece and an inverting lens with a diameter of 40 mm with a focus of – 100 mm. 3). Paralon/ PVC pipe with a diameter of 5 cm. 4). Needle (pipe connection) with a diameter of 5 cm for the objective lens. 5). Needle (pipe connection) with a diameter of 5 cm and 4 cm for the ocular lens. 6). Glue gun. 7). Double-sided. 8). Meter.

Figure 1. Tools and materials for making simple binoculars.

2.2. *How to make binoculars*

How to make simple binoculars follow the following steps; 1). Determine the length of the binoculars body first. With the known physics formula d = fob – fok. 2). Cut the PVC pipe/paralon to the size of the binocular body. 3). Put the objective lens into the pipe connection, then connect the pipe connection that already contains the lens at the very front end of the PVC pipe that has been measured.

Remember that the objective lens is always in front of the eyepiece. 4). Attach the pipe to the far end of the pipe. 5). Place the eyepiece in the eye. 6). Galileo binoculars are ready to use.

Figure 2. The results of making simple binoculars.

2.3. Steps to use binoculars
The steps for using binoculars are as follows: 1). Prepare the Galileo binoculars that have been made. 2). Determine the object to be seen. 3). Determine the distance of the object to be seen starting from 50 meters, 100 meters, and 140 meters. 4). Point the binoculars at a predetermined object by bringing the observer's eye closer to the concave lens as the ocular lens. 5). Record the observations in the prepared table.

3. Results and Discussion

3.1. Object Observation Results
Observations using Galileo binoculars were carried out on Friday June 2021. The object of observation was in the form of letters written on HVS paper with Times New Roman font size 350. Objects were observed at different distances ranging from 50 meters and 100 meters. Observation respondents were 17 physics students at the University of Mataram. Observations were made in the morning around 08.00 – 11.00 WITA in the Mataram University campus environment. Observations showed that Galileo's binoculars functioned well. At a distance of 50 meters all respondents can see clearly and look closer than the original distance. Meanwhile, at a distance of 100 meters from 17 respondents, only one respondent saw the object looked blurry but could still be read. The following are the results of object observations presented in Table 2.

Table 2. Observation results of objects using Galileo's binoculars.

No	Name Student Samples	Object Distance		Body temperature (^{0}C)	Information
		50 m	100 m		
1	M	✓	✓	31.41	Seen
2	N	✓	✓	32.21	Seen
3	P	✓	✓	34.01	Seen
4	M	✓	✓	34.07	Seen
5	A	✓	✓	32.25	Seen
6	J	✓	✓	33.55	Seen
7	N	✓	✓	32.61	Seen
8	D	✓	✓	28.97	Seen
9	A	✓	✓	33.41	Seen
10	I	✓	✓	31.61	Seen
11	N	✓	✓	32.25	Seen
12	N	✓	✓	34.32	Seen
13	F	✓	✓	30.53	Seen
14	D	✓	✓	30.79	Seen
15	A	✓	✓	33.13	Seen
16	L	✓	✓	33.25	Seen
17	L	✓	✓	34.61	Seen

Figure 3. Trial using binoculars on Physics Education Students, FKIP Mataram University.

Table 3. Results of satisfaction using binoculars for physics education students.

No	Respondents	Satisfaction Score	Information
1	Respondent 1	3.60	Satisfied
2	Respondent 2	3.90	Satisfied
3	Respondent 3	3.60	Satisfied
4	Respondent 4	4.00	Satisfied
5	Respondent 5	3.60	Satisfied
6	Respondent 6	3.60	Satisfied
7	Respondent 7	3.70	Satisfied
	Average	3.71	Satisfied

3.2. *Shadow formation by binoculars*

The lens closest to the object is called the objective lens and will form a real image of the object falling on the plane of the focal point fOB (or nearby if the object is not at infinity). Even though the image is smaller than the real object, it forms a larger angle and is very close to the eyepiece, which acts as a magnifier. Thus, the eyepiece magnifies the image produced by the objective lens to produce a much larger second image, which is virtual and upright.

Table 4. Image formation in the objective lens and ocular lens and magnification of the image of the Galileo binoculars (d = 35 cm).

No	*Sob*	*fob*	*S'ob*	*S ok = S'ob – d*	*fok*	$S'ok$	$M = S'ok/Sok$ (Ocular magnification)	$M = \frac{S'ob}{Sok}$ (Binocular magnification)
1	50 m	50 cm	50.50 cm	15.50 cm	-10 cm	-6.07 cm	0.39 x	3.25 x
2	100 m	50 cm	50.25 cm	15.25 cm	-10 cm	-6.03 cm	0.39 x	3.29 x

The results showed that all respondents could see objects that were 50 m and 100 m apart clearly. All respondents know how to use binoculars properly and correctly. This experiment using simple binoculars obtained data that the respondents were very happy and satisfied using it. They hope there will be other props that are easy and cheap to use. Their experience and understanding of concepts related to optical materials is improved. The results of the study [21] showed an increase in the ability of the scientific process skills of the experimental class to increase in the medium category (0.47) and the control class in the medium category (0.34) and almost all of the students' responses to the manufacture of simple binoculars were interested. While the results of the study [22] showed that the understanding of physics lessons using teaching aids with a response of 89%, has the motivation to develop simple teaching aids by 74%. This means that the manufacture and development of teaching aids in science materials in schools is very much needed by teachers and students to improve the process of science process skills and concept understanding.

4. Conclusion

Galileo's binoculars have a short size and consist of two lenses, namely a convex lens that functions as an objective lens and a concave lens that doubles as an eyepiece and an inverting lens. The concave lens in Galileo's binoculars serves to produce an upright image. The lens closest to the object is called the objective lens and will form a real image of the object falling on the plane of the focal point f_{OB} (or nearby if the object is not at infinity). Even though the image is smaller than the real object, it forms a larger angle and is very close to the eyepiece, which acts as a magnifier. Thus, the eyepiece magnifies the image produced by the objective lens to produce a much larger second image, which is virtual and upright. Galileo binoculars are proven to be able to see objects that are far away look closer based on observational data. The magnification of the binoculars using a 500 mm objective lens and an eyepiece – 100 mm is 3.25 times with a binocular length of 35 cm. the farthest distance that can be seen is 100 meters with the object written in alphabet letters with Times New Roman font size 350. Physics education students FKIP Mataram University are satisfied with the use of simple binoculars and hope that there will be simple practical tools that are easy and inexpensive to develop again.

Acknowledgement

The author would like to thank the physics education lecturer, the head of the physics education laboratory and the physics education students at the FKIP University of Mataram.

References

[1] Djahir, et al. (2017). Alat Peraga dalam Geometri Ruang (Sumber Belajar Penunjang PLPG 2017, Mata Pelajaran/Paket Keahlian Matematika). Direktorat Jenderal Guru Dan Tenaga Kependidikan, Kementerian Pendidikan Dan Kebudayaan: Jakarta

[2] Hidayah, I, et al. (2015). Pengembangan dan Pemasaran Media Pembelajaran Matematika Sekolah. Laporan Penelitian RAPID Tahun Pertama

[3] Alfirdausi, abdul Hadi (2011). Pengaruh Penggunaan Alat Peraga Kartu Kotif (Koin Positif Negatif) Terhadap Hasil Belajar Matematika Siswa. Jurnal Pendidikan Universitas Islam Negeri Syarif Hidayatullah. 14 (1): 178-180

[4] Arif F. Sadiman. (2006). Media pendidikan: Pengertian, pengembangan dan pemanfaatannya. Jakarta: Raja Grafindo Persada

[5] Ahmad Rohani (1997). Media instrusional edukatif. Jakarta: Rineka Cipta

[6] Wahyuni, S., Kosim, K., & Gunawan, G. (2018). Pengembangan Perangkat Pembelajaran Fisika Berbasis Inkuiri Terbimbing Berbantuan Eksperimen Untuk Meningkatkan Penguasaan Konsep Siswa. Jurnal Pendidikan Fisika dan Teknologi, 4(2), 240-246

[7] Sutrio, S., Gunawan, G., Harjono, A., & Sahidu, H. (2018). Pengembangan Bahan Ajar Fisika Eksperimen Berbasis Proyek Untuk Meningkatkan Keterampilan Berpikir Kritis Calon Guru Fisika. Jurnal Pendidikan Fisika dan Teknologi, 4(1), 131-140

[8] Hastuti, A., Sahidu, H., & Gunawan, G. (2017). Pengaruh Model PBL Berbantuan Media Virtual Tehadap Kemampuan Pemecahan Masalah Fisika. Jurnal Pendidikan Fisika dan Teknologi, 2(3), 129-135

[9] Muhammad, H. R., Zakaria, D. R., Sandi, A. D. K., Rahayu, P., Putri, R. M., Sari, Y., & Hidayat, M. A. (2019). Ribrium (Alat Peraga Fisika Untuk Siswa SMA)

[10] Prasetyarini, A., Fatmaryanti, S. D., & Akhdinirwanto, R. W. (2013). Pemanfaatan alat peraga IPA untuk peningkatan pemahaman konsep fisika pada siswa SMP Negeri I Buluspesantren Kebumen tahun pelajaran 2012/2013. RADIASI: Jurnal Berkala Pendidikan Fisika, 2(1), 7-10

[11] Aristo Rahardi (2004). Media pembelajaran. Jakarta: Dirjen Dikdasmen

[12] Hidayah, I. (2008). Model Pembelajaran Matematika Sekolah Dasar dengan Optimalisasi Pemanfaatan Lingkungan dan Alat Peraga Manipulatif sebagai Sumber Belajar. Laporan hasil penelitian Hibah Bersaing

[13] Hidayah, I. (2018). Pembelajaran Matematika Berbantuan Alat Peraga Manipulatif Pada Jenjang Pendidikan Dasar dan Gerakan Literasi Sekolah. In PRISMA, Prosiding Seminar Nasional Matematika (Vol. 1, pp. 1-11)

[14] Febri Muhardityana Saputri (2012). Penggunaan Alat Peraga Sederhana Pengukuran Untuk Peningkatan Pemahaman Konsep Fisika Siswa MAN Kutowinangun Tahun Pelajaran 2010/2011. Skripsi, tidak diterbitkan. Universitas Muhammadiyah Purworejo, Purworejo

[15] Yensy, N. A. (2012). Penerapan model pembelajaran kooperatif tipe examples non examples dengan menggunakan alat peraga untuk meningkatkan hasil belajar siswa di kelas VIII SMP N 1 Argamakmur. Exacta, 10(1), 24-35

[16] Oktova, D. T., Ario W, D. H., Lesmana, J. F., Fatrin, M. O. R., Yunitri, N. L. A., Prayogo, D. N., & Sari, Y. (2019). Kazzle Atom (Alat Peraga Mata Pelajaran Kimia Untuk Siswa SMA Kelas X)

[17] Mia Ramdhiani et al. (2013). Instrumen Optik: Disusun Untuk Memenuhi Salah Satu Tugas Kelompok pada Mata Kuliah Optik. Bandung. UIN Sunan Gunung Jati

[18] Surani, S. (2015). Pemanfaatan media pembelajaran Flash pada materi lensa untuk meningkatkan hasil belajar siswa kelas VIII MTs Miftahuth Tholibin Mranggen tahun pelajaran 2014/2015 (Doctoral dissertation, UIN Walisongo)

[19] Amaliyah, H. (2012). Efektivitas Model Pembelajaran Aktif Bermain Jawaban terhadap Hasil Belajar Fisika Peserta Didik Materi Pokok Alat-Alat Optik MAN 1 Semarang Tahun Pelajaran 2011/2012 (Doctoral dissertation, IAIN Walisongo)

[20] Fitriana, S. (2017). Implementasi Alat Peraga Teropong pada Sub Materi Teropong di Kelas XI SMK Bhakti Kencana (Implementation Tools in Sub Materials Binoculars for in-Class XI SMK Bhakti Kencana). PASCAL (Journal of Physics and Science Learning), 1(2), 18-24

[21] Permana, T. (2017). Pengaruh pembuatan teropong bintang sederhana terhadap keterampilan proses sains peserta didik pada Konsep Alat Optik (Bachelor's thesis)

[22] Guswantoro, T., Faradiba, F., Sianturi, M., & Kosnan, T. D. (2019). Persepsi Siswa Sekolah Menengah Pertama Santo Markus 1 Kramatjati Terhadap Alat Peraga Sederhana Teropong Bintang Pada Materi Alat Optik. Jurnal Dinamika Pendidikan, 12(2), 113-124

Designing and constructing mini refrigerator with thermoelectric module

F Tamami[1], A Doyan[1], Susilawati[2], Hikmawati[2], S Ayub[2], Azhar[3] and L S Hudha[4]

[1]Science Education Master Program, University of Mataram, Mataram, Indonesia
[2]Physics Education Study Program, University of Mataram, Mataram, Indonesia
[3]Physics Education Study Program, Riau University, Riau, Indonesia
[4]Department of Physics, Faculty of Mathematics and Natural Science, University of Mataram

E-mail: fuadtamami@unram.ac.id

Abstract. Cooling Engine is one application of the laws of thermodynamics. Cooling machines such as refrigerators are a necessity for everyone. Thermoelectric which has a Peltier effect can be used as an electronic component to cool objects. This research is an experimental research and the object of research is a mini-size refrigerator with thermoelectric. The purpose of this research is to analyse the performance coefficient of the mini refrigerator that has been made. The research method used is a literature study and data are taken by measuring the temperature produced from the refrigerator. The results obtained from this study are that the refrigerator can reduce the system room temperature from 28.5 °C to 13.6 13 within 10 minutes. In the next experiment with a time of one hour, the temperature in the system room was lowered to 7.7 °C. After measuring the outside temperature in the container (reservoir) and inside the refrigerator, the coefficient of performance is 0.3. The coefficient number is relatively low, because the higher the number obtained, the more efficient the refrigerator is made.

1. Introduction

Physics is a branch of science that studies the phenomena that are around us. Physics as a science seeks to arouse human interest so that they want to increase their intelligence and understanding of nature and everything in it which is full of inexhaustible secrets [1]. It is undeniable that the rapid development of science goes hand in hand with the development of technology. Because essentially the product of science is technology itself. Science, especially physics, has several scientific studies, including mechanics, fluids, optics, thermodynamics, and modern physics. This field has many applications in everyday life and not a few have given birth to technology that is useful for human needs. One of the fields of physics that has greatly contributed for today's technology is thermodynamics. Thermodynamics comes from the word 'thermos' which means heat and 'dynamics' which means change. Thermodynamics is a field of physics that studies the characteristics of heat and its transformation into other forms of energy. One of the applications of thermodynamics is cooling machines, especially refrigerators.

The refrigeration engine is a heat engine that works in reverse. So, a heat engine takes heat from a place with a low temperature and dissipates that heat to a place with a high temperature. For this process to occur, the machine must do work. However, heat naturally only wants to flow from places of high temperature to places of low temperature. Heat cannot flow by itself from a place of low temperature to a place of high temperature. So, we need a tool that can do this by giving effort. That's how the cooling machine works where cold air is flowed to a place with a high temperature but with given effort. Various types of cooling machines include refrigerators, air conditioners, PC coolers, and heat pumps. Refrigerator is one application of thermodynamics in which there is isolation of a system in a container with its environment. Refrigerators have many uses such as making ice, cooling foodstuffs so they don't spoil, and so on. Along with the development of the era, the refrigerator underwent changes after changes which used to use freon for cooling to become Freon less and work effectively. This means that innovation from each technology is still carried out to find more effective and easy ways to achieve the same technology. Some of study even develop a refrigerator that can be carried anywhere. Thus, people do not have to bother bringing their large refrigerator to and fro. In addition, the development of a refrigerator that utilizes the Peltier effect of thermoelectric is also one of the innovations made to make a refrigerator.

Thermoelectric is one of the electronic components that is composed of a row of n and p type semiconductors which, when energized by DC electric current, has a hot temperature on one side and a cold temperature on the other side [2–4]. The cold side of the thermoelectric is used as a cooling material for the refrigerator. Several developments have been made using thermoelectric by combining them with heat sinks and fans on the hot side in order to obtain optimal cold temperatures [5–7]. There are also those who use a water block to cool the hot side of the thermoelectric to get optimal results [8], [9]. Of course, this will be more interesting if a combined system of the two is used. Besides using a water block as a thermoelectric hot side cooler, a heatsink and fan are also added to get optimal results.

So, the purpose of this experiment is to make a mini refrigerator by combining the heatsink and fan components to optimize cooling in the hot side of the Peltier. Thus, from the experiment, the good value coefficient of performance (COP) was obtained.

2. Methods

This research is an experimental study with the object of the experiment is a mini refrigerator using a thermoelectric cooler module. The time of the research was carried out from March to June 2021 and took place at the FKIP Physics Laboratory, University of Mataram.

2.1. *Tools and Material*

Table 1. Tools and material provide to made a mini refrigerator.

No	Equipment	Amount
1	Scissor	1
2	Cutter	1
3	Tape	1
4	Water Container	1
5	Thermal Paste	1
6	Styrofoam 40x60 cm	4
7	Water block	1
8	Heat sink	2
9	Fan	3
10	Adaptor	1
11	Pipe	1

Journal of Physics: Conference Series **2165** (2022) 012032 doi:10.1088/1742-6596/2165/1/012032

2.2. Material Specification

2.2.1. Thermoelectric (Peltier). The peltier used in this study is a TEC1-12706 type peltier with the following specifications.

Thermoelectric Cooler

TEC1-12706

Performance Specifications

Hot Side Temperature (°C)	25°C	50°C
Qmax (Watts)	50	57
Delta Tmax (°C)	66	75
Imax (Amps)	6.4	6.4
Vmax (Volts)	14.4	16.4
Module Resistance (Ohms)	1.98	2.30

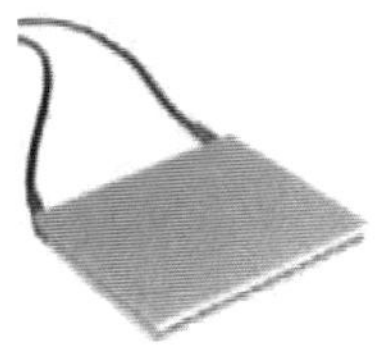

Figure 1. Specifications of peltier.

2.2.2. Heatsink. The heatsink used here is a black anodite heatsink with dimensions of 40 mm x 40 mm with a thickness of 10 mm.

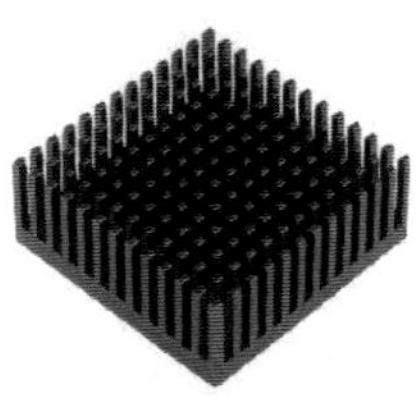

Figure 2. Heatsink.

2.2.3. Adaptor. To support the performance of the Peltier thermoelectric, the adapter used is an adapter with an output of 12 V 5 amperes.

2.2.4. Fan. There are three fans used in this tool, namely, 2 fans measuring 40 mm x 40 mm and 1 fan measuring 100 mm x 100 mm. A 40 mm fan is used on the heatsinks, one of which is used to cool the heatsink and the other is used as a medium for the ice that forms on the cold side of the Peltier. While the 100 mm fan is used to cool the performance of the adapter.

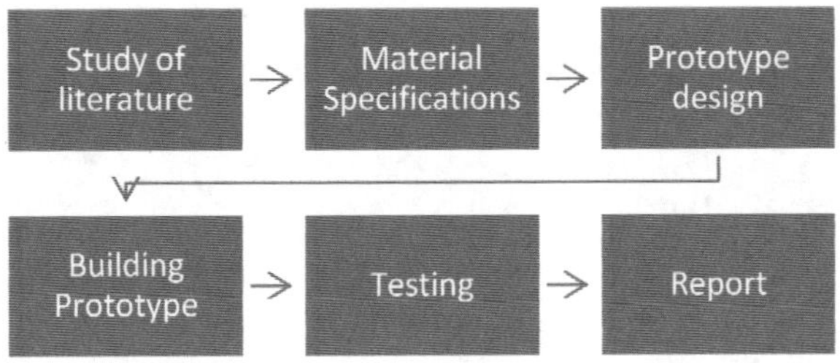

Figure 3. Prototype development diagram.

This research was carried out through several stages, starting from searching for literature related to the manufacture of mini refrigerators with thermoelectric. Then, map the tools and materials used in the manufacture of mini refrigerators. Next, make a design of the tool which is then continued by building the tool. After the mini refrigerator is complete, the next step is to test the tool to find out whether it is functioning properly or not. Furthermore, with the tools that have been tested, data is collected in the form of the temperature produced by the mini refrigerator and finally a report is prepared.

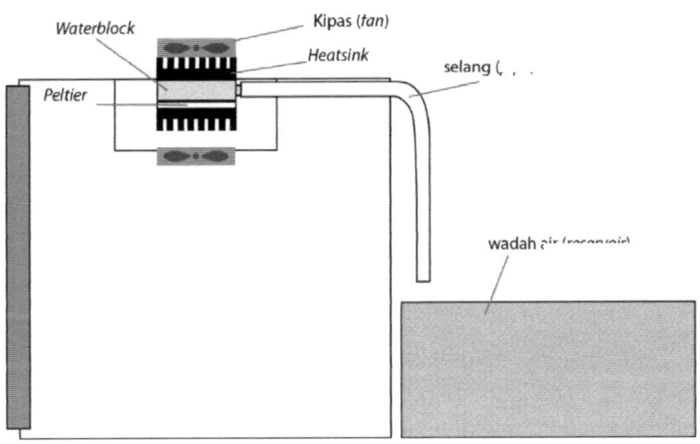

Figure 4. A series of materials for the manufacture of refrigerators with thermoelectric.

3. Result and Discussion

3.1. Results

Table 2. The temperature is generated and measured with a digital thermometer.

No	Time (Minute)	Temperature (Celcius)	
		Trial 1	Trial 2
1	1st	28.5	28.5
2	2nd	22.2	18.3
3	3rd	20.0	16.3
4	4th	18.9	14.9
5	5th	18.4	14.3
6	6th	18.2	13.8
7	7th	18.2	13.7
8	8th	18.3	13.7
9	9th	18.6	13.6
10	10th	18.9	13.8

The data in Table 1 when displayed in a graph will be as shown in Figure 4.

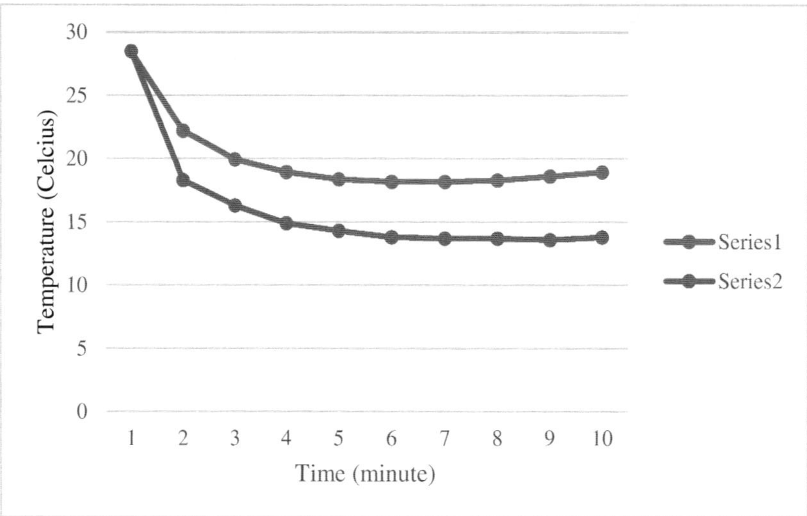

Figure 5. Graph of temperature inside the refrigerator in first 10 minute.

The experiment was carried out again with the intention of knowing how many degrees the temperature was on the heatsink or closer to the cold side of the Peltier. The measurement results obtained are shown in Figure 5.

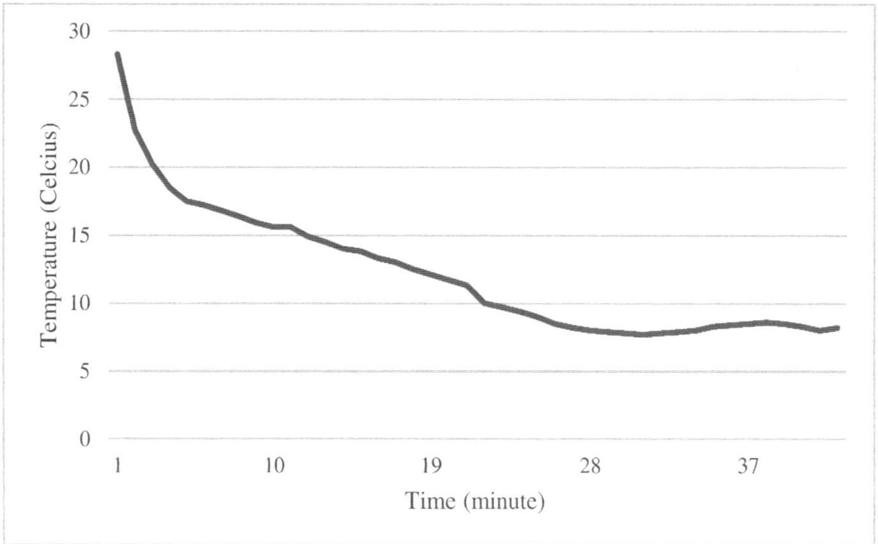

Figure 6. Graph of temperature drop in the area around the heatsink for a few minutes.

3.2. *Discussion*

A good refrigerant is a refrigerator that has a high coefficient of performance. Of course, to get a good coefficient of performance, the refrigerator must waste more heat than the given effort. This has something to do with the amount of electrical power consumed. In this study, to maximize the performance of the thermoelectric (peltier) used in the refrigerator, we use an adapter with an output of 12 V and a current of 5 amperes. If we calculate based on the specifications of the adapter, the electricity consumption for this tool is the output voltage multiplied by the current, which is 12 V multiplied by 5 amperes and produces 60 watts. So, it is known that to turn on the refrigerator, 60 watts of power are used.

As shown in Table 1, the experiment was carried out twice with different thermometers, namely a digital thermometer with a digital thermometer with two sensors, namely a sensor to measure the temperature outside the refrigerator and the temperature inside the refrigerator. The experiment in Table 1 was carried out by placing the thermometer sensor in the refrigerator, more precisely, it was placed to measure the room temperature in the refrigerator. Based on these data, in the first 10 minutes of the first experiment the refrigerator worked, the temperature inside the refrigerator which was initially 28.5 °C dropped to 18.2 °C then rose again to 18.9 °C. Then in the second experiment using a digital thermometer, in the first 10 minutes the refrigerator worked, the temperature was lowered from 28.5 to 13.6 °C and rose again to 13.8 °C. This increase is a natural thing because the refrigerator needs time to reach its stability point [10, 11].

The next experiment was carried out by placing the thermometer sensor on the cooling heatsink. The data obtained can be seen in Figure 10 and based on these data, it is known that the temperature obtained in the heatsink reaches 7.7 °C for a period of 30 minutes. From these data, it can also be seen that there was an increase again after reaching 7.7 °C but the temperature increase decreased again and again and again. This means that this temperature is the stable temperature of the refrigerator.

Thus, we can calculate the coefficient of performance of the refrigerator with the thermoelectric module that has been made using equation 1. In the second experiment, the temperature of the reservoir or water container where the heat was removed was also measured. After being measured, the temperature in the refrigerator is 7.7 °C, the water temperature is 38.3 °C in the water container. Thus, the magnitude of the performance coefficient of this refrigerator is 0.3. This figure is small for the coefficient of performance, because the performance of a good cooling engine is seen from the high coefficient of performance. This is a challenge in itself to make a thermoelectric device with a high coefficient of performance [12].

4. Conclusion

Based on the results and discussions and theoretical studies that have been carried out, it can be concluded that the manufacture of a mini refrigerator using a thermoelectric module has been successfully carried out. The temperature in the refrigerator can be lowered from 28.5 to 13.6 °C. Meanwhile, the temperature of the cold heatsink was successfully lowered from 28.5 to 7.7 °C with a performance coefficient of 0.3. Further research on the development of refrigerators using thermoelectric is interesting to do, especially for those who managed to get a higher coefficient of performance. Studying different thermoelectric materials is very important to improve the coefficient of performance of thermoelectric refrigeration engines.

References

[1] Gunawan, Model Pembelajaran Sains Berbasi ICT. Mataram: FKIP Universitas Mataram, 2015
[2] K. Ivanov, A. Aleksandrov, and I. Belovski, "Synthesis and Study on Waste Heat Thermoelectric Generator," in 2019 International Conference on High Technology for Sustainable Development, HiTech 2019, Oct. 2019, doi: 10.1109/HiTech48507.2019.9128262
[3] B. Praveen, S. Pranay, S. Jhabarmal, R. P. B. Mechanical, A. Group, and O. Institutions, "Fabrication of Portable Solar Thermo Electric Refrigerator by Liquid Cooling," vol. 4, no. 7, pp. 248–255, 2018, Accessed: Jun. 20, 2021. [Online]. Available: https://www.google.com/search?q=Fabrication+of+Portable+Solar+Thermo+Electric+Refrigerator+by+Liquid+Cooling&safe=strict&ei=R_3OYNOSM6TqrgTn3YeQCQ&oq=Fabrication+of+Portable+Solar+Thermo+Electric+Refrigerator+by+Liquid+Cooling&gs_lcp=Cgdnd3Mtd2l6EANQ8rcCW
[4] M. J. Adams, M. Verosky, M. Zebarjadi, and J. P. Heremans, "Active Peltier Coolers Based on

Correlated and Magnon-Drag Metals," Phys. Rev. Appl., vol. 11, no. 5, May 2019, doi: 10.1103/PhysRevApplied.11.054008

[5] M. Fairuz Remeli, N. Ezzah Bakaruddin, S. Shawal, H. Husin, M. Fauzi Othman, and B. Singh, "Experimental study of a mini cooler by using Peltier thermoelectric cell," in IOP Conference Series: Materials Science and Engineering, Jun. 2020, vol. 788, no. 1, doi: 10.1088/1757-899X/788/1/012076

[6] S. Jugsujinda, A. Vora-Ud, and T. Seetawan, "Analyzing of thermoelectric refrigerator performance," in Procedia Engineering, 2011, vol. 8, pp. 154–159, doi: 10.1016/j.proeng.2011.03.028

[7] V. Vaidya, P. Manwatkar, M. Narnaware, U. Gadve, and S. Anvikar, "Design and Experimentation of Thermoelectric Refrigerator for Cold Storage Application," vol. 5, no. 2, pp. 2015–2018, 2017

[8] A. Prashanth, G. Basha, and M. Varaprasad, "Fabrication of Thermoelectric Refrigerator cum Oven," J. Emerg. Technol. Innov. Res., 2020

[9] A. Venugopal, K. Narang, K. Prakash, and M. Joshi, "Cost-effective Refrigerator Using Thermoelectric Effect and Phase Change Materials," Int. J. Sci. Eng. Res., vol. 5, no. 2, pp. 624–627, 2014, Accessed: Jun. 20, 2021. [Online]. Available: https://www.google.com/search?q=Cost-effective+Refrigerator+Using+Thermoelectric+Effect+and+Phase+Change+Material&oq=Cost-effective+Refrigerator+Using+Thermoelectric+Effect+and+Phase+Change+Material&aqs=chrome..69i57.492j0j7&sourceid=chrome&ie=UTF-8

[10] S. M. A. Rahman, A. A. Hachicha, C. Ghenai, R. Saidur, and Z. Said, "Performance and life cycle analysis of a novel portable solar thermoelectric refrigerator," Case Stud. Therm. Eng., vol. 19, Jun. 2020, doi: 10.1016/j.csite.2020.100599

[11] S. P. Gupta et al., "Design and development of thermoelectric refrigerator," Invertis J. Renew. Energy, vol. 8, no. 4, p. 194, 2018, doi: 10.5958/2454-7611.2018.00028.0

[12] H. Moria, M. Ahmed, A. Alghanmi, T. I. Mohamad, and Y. Yaakob, "Experimental study of solar based refrigerator using thermoelectric effect," in Energy Procedia, 2019, vol. 158, pp. 198–203, doi: 10.1016/j.egypro.2019.01.074

Digital-based thermoelectric generator

Azhar[1], M F Lathifah[2], A Doyan[2,3], Susilawati[2,3] and L S Hudha[4]

[1]Physics Education, FKIP, Riau University, Riau, Indonesia
[2]Master of Science Education Program, University of Mataram, Lombok, West Nusa Tenggara, Indonesia
[3]Physics Education, FKIP, University of Mataram, Lombok, West Nusa Tenggara Indonesia
[4]Department of Physics, Faculty of Mathematics and Natural Science, University of Mataram

Email: azhar@lecture.unri.ac.id, fajarilathifah@gmail.com, aris_doyan@unram.ac.id

Abstract. A thermoelectric generator or also known as a *Seebeck* generator is an electrical generator device that converts temperature differences into electrical energy using a phenomenon called the *Seebeck* effect. To find out the magnitude of the voltage generated by the thermoelectric, a series of tests on the thermoelectric must be carried out where one of the test methods is to provide a significant temperature difference on both sides of the thermoelectric generator using hot water media for the hot side and using ice cubes for the cold side. The test results show that the developed thermoelectric generator can generate an electric voltage of 1189.31 mV.

1. Introduction

In this era of globalization, electricity is one of the main human needs in everyday life. Various activities carried out by humans cannot be separated from electricity, such as watching TV, fan/AC fan activities, charging cellphones and laptops, and even new devices that require electricity such as electric water heaters and electric motors appear. In line with this, the development of science and technology (IPTEK) is growing rapidly which is then used by humans to make it easier for them to complete work so that they become more effective and efficient using electricity that utilizes solar power, wind power, and micro-hydro. However, this method of generating electrical energy has limitations, including weather and environmental conditions.These limitations are the reason that an innovation is needed in developing a device that can produce electrical energy that does not depend on the weather and the surrounding environment. By utilizing advances in science and technology, we can design an electric generator using electronic components known as thermoelectric generators.

Thermoelectric has been discovered around the 18th century by a German scientist named Thomas Johann *Seebeck*. Thermoelectricity is a direct conversion process from a temperature difference (heat energy) into an electric voltage directly (thermoelectric generator) or vice versa from electricity to produce cold (coolant thermoelectric) [1,2]. Thermoelectric are influenced by three effects, namely, the *Seebeck*, Thompson Peltier effect [3]. A thermoelectric generator or also known as a *Seebeck* generator is an electrical generator device that converts temperature differences into electrical energy using a phenomenon called the *Seebeck* effect [4]. The working principle of the *Seebeck* effect that works on a thermoelectric is, if two connected metal materials are in an environment with different temperatures, it will produce an electric current or electromotive force [5]. To produce electricity, it is enough to place thermoelectric material in a circuit that connects a hot and cold source. From the circuit, a certain amount of electricity will be generated according to the type of material used.

Thermoelectrics that are generally widely circulated in Indonesia, one of which is type 12705 thermoelectric. This type of thermoelectric has a relatively cheap price and is quite easy to obtain on the market. The electric voltage produced by one thermoelectric is relatively small. Therefore, to find out the voltage generated by the thermoelectric, a series of tests on the thermoelectric must be carried out where one of the test methods is to provide a significant temperature difference on both sides of the thermoelectric generator using hot water media for the hot side and using ice cubes for the cold side [6].

Thermoelectric generators can be made digitally by adding electronic components to the circuit. This component is known as Arduino. Arduino is a microcontroller which can function as the control center of a system as a whole and is displayed on an LCD screen [7]. Arduino is a microcontroller board based on ATmega328. This device has 14 input/output pins consisting of 6 pins that can be used as PWM outputs, 6 analog inputs, a 16 MHz crystal oscillator, a USB connection, a power jack, an ICSP head, and a reset button. Arduino is able to support microcontrollers so that it can be connected to a computer or laptop using a special USB cable. To operate the Arduino, you can use a USB cable that is connected to a computer or other power source such as a battery [8,9].

An Arduino board can be classified into two parts, namely:

1.1. Hardware

The Arduino board consists of several components, while the main components consist of:

1.1.1. *USB plug*. The USB Plug is part of the Arduino which is used to upload programs to the microcontroller and has a power regulator of 5 volts which also serves to power the Arduino board.

1.1.2. External Power Supply. This section serves to power the board and can be connected with a voltage of 9 to 12 volts.

1.1.3. Reset button. This button serves to reset the Arduino, we can press this button when we have uploaded a new program into the Arduino. This button is generally red.

1.1.4. Microcontroller. The microcontroller is part of the Arduino which functions as a receiver of information and forwards it to the programmed part. Before being used, the first thing we do is upload a program or command first so that the microcontroller can only run if it has been filled in with a command or program first.

1.1.5. Analog pins (0-5). This section is an analog input pin consisting of A0 to A5.

1.1.6. Digital I/O pins. This section is a digital input pin, the output pin consists of pins 2 to 13.

1.1.7. *In-circuit programmer*. This is another part that can be used to drive or program Arduino. Consists of 0 (RX) and 1 (TX). Used for receiving (RX) and sending (TX).

1.1.8. Analog and digital *ground pins*.

1.1.9. Power pin. Usually consists of 3.3 and 5 volts.

1.2. Arduino Software

This software is a series of instructions or code that contains information for the hardware about what to do and how to do it. Arduino IDE (Integrated Development Environment) is divided into 3 main parts, namely:

1.2.1. Command Area. This section is an area that consists of menu items such as file, edit, Sketch, Tools, Help and several icons such as verify icon to verify, upload to upload the desired program, new to create a new page, open to open saved files, save to save the file that has been done, and the serial monitor used to send and receive data between the Arduino and its IDE.

1.2.2. Text Area. This section serves to write code that uses a simplified version of the C programming language or also known as a sketch.

1.2.3. Massage *Windows Area*. This section serves to show messages from the IDE, we can find out whether the program we have written is appropriate or not. This section is black and is at the very bottom of the Arduino software display [10].

Journal of Physics: Conference Series **2165** (2022) 012033 doi:10.1088/1742-6596/2165/1/012033

2. Methods

This research was conducted from May to June 2021 at the Physics Laboratory of the FKIP University of Mataram.

2.1. Tool Design Stage

The first step in developing this light intensity measuring instrument is to create a block diagram which can be seen in Figure 1. Next, the hardware design of the developed tool can be seen in Figure 2.

2.2. Tool Trial Stage

Testing the tool is done by giving hot water to one of the cans connected to the temperature sensor, then cold water on the other side.

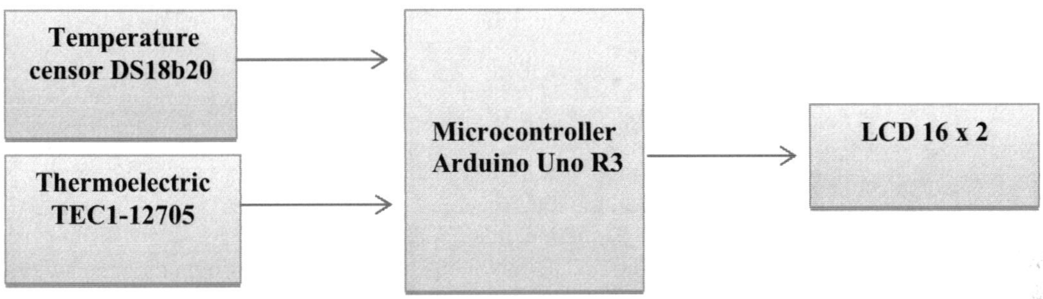

Figure 1. Block Diagram.

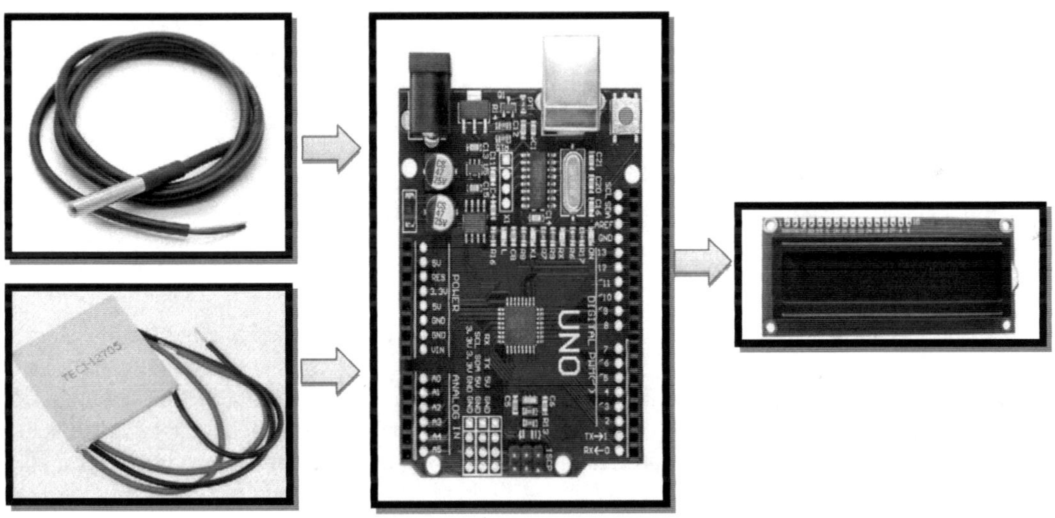

Figure 2. Hardware Design.

3. Result and Discussion

In the activities that have been carried out based on the design and testing of the Digital-based Thermoelectric Generator, the following data are obtained.

Table 1. Temperature and voltage data obtained from a thermoelectric generator.

No	Temperature (°C)		ΔT	Voltage (mV)
	T_1	T_2		
1	21.44	83.00	61.56	1,124.14
2	21.37	83.12	61.75	1,129.03
3	21.37	83.31	61.94	1,129.03
4	21.19	83.62	62.43	1,251.22
5	21.12	83.94	62.82	1,251.22
6	21.12	84.19	63.07	1,251.22
Average	21.27	83.53	62.26	1,189.31

This research was conducted with the aim of knowing how to make a thermoelectric generator, knowing the working principle of thermoelectric so that it can generate electrical energy and to determine the relationship between heat and electrical energy produced in a thermoelectric generator. The first step is to find sources of information from various literatures and scientific journals that support knowledge related to Digital-based Thermoelectric Generators. Furthermore, the design of the tool scheme is carried out to make it easier to design tools in hardware. The tools and materials needed in the manufacture of this tool include: Arduino Uno microcontroller, DS18b20 temperature sensor, thermoelectric, box-shaped can, 16x2 I2C LCD, breadboard, 9 V battery, resistor, jumper cable, styrofoam, transparent box, double-sided tape, cutter, solder and tin. Then as the medium used for testing the tool, hot water and cold water are needed which will later be placed in each can.

After all the tools and materials are ready, the tool starts to be designed in the form of a block diagram, in this case what acts as input is the DSb1820 temperature sensor which has waterproof characteristics and two cans that are connected to a thermoelectric in the middle. Furthermore, the two components are connected to the Arduino and programmed according to the purpose of the device being designed. Then as the output used 16x2 LCD. On the LCD screen will display the temperature value of the cans containing hot water (T1) and cans containing cold water (T2). In accordance with the working principle of thermoelectric which when the two metals are connected then given a different temperature it will produce an electric voltage directly. Therefore, the LCD screen will also display the value of the voltage generated by the thermoelectric.

Testing the tool can be done by preparing hot water and cold water first (inserted into each container) then a temperature sensor is inserted into each can so that the temperature of the test media can be processed by Arduino and read on the LCD. Based on the tool testing that has been carried out, the data obtained are as in table 1. It can be seen that the average temperature value in cold water (T1) is 21.27 °C, hot water (T2) is 83.53 and the average T is 62.26. The temperature difference is able to produce an electric voltage even though it is still in the low category, while the average voltage generated is 1,189.31 mV. If we observe the test results table of the tool, it is known that the value of T is proportional to the resulting voltage. This means that the greater the value of T, the greater the value of the resulting voltage.

4. Conclusion

Based on the description of the test results above, it can be concluded that the designed tool (Thermoelectric Generator) can be used as an alternative energy source directly without having to depend on weather and environmental conditions. We can adjust the temperature in each part of the can according to the expected voltage. However, the weakness of this tool is that its shape is still simple so that the resulting voltage is still quite small.

Acknowledgment
The author would like to thank the Master of Science Education Program at the University of Mataram which has facilitated this research through the Mechanics and Thermostatistics courses. The author also expresses his gratitude to fellow students of the Master of Science Education at the University of Mataram, especially in the concentration of Physics and all those who have helped so that this research can be carried out properly.

References

[1] Khalid, M., Syukri, M., & Gapy, M. (2016). Pemanfaatan energi panas sebagai pembangkit listrik alternatif berskala kecil dengan menggunakan termoelektrik. Jurnal Karya Ilmiah Teknik Elektro, 1(3)
[2] He, W., Zhang, G., Zhang, X., Ji, J., Li, G., & Zhao, X. (2015). Recent development and application of thermoelectric generator and cooler. Applied Energy, 143, 1-25
[3] Sumarjo, J., Santosa, A., & Permana, M. I. (2017). Pemanfaatan sumber panas pada kompor menggunakan 10 termoelektrik generator dirangkai secara seri untuk aplikasi lampu penerangan. SINTEK JURNAL: Jurnal Ilmiah Teknik Mesin, 11(2), 123-128
[4] Rachmanita, R. E., Karimah, C. N., & Azizah, N. (2021, March). Experimental investigations on the performance of thermoelectric generator as energy conversion system. In IOP Conference Series: Earth and Environmental Science (Vol. 672, No. 1, p. 012103). IOP Publishing
[5] Djafar, Z., Putra, N., & Koestoer, R. A. (2010). Kajian Eksperimental Pengembangan Generator Termoelektrik Sebagai Sumber Listrik. SNTTM ke-9, Palembang
[6] Tyas, N. R. (2017). Uji Karakteristik Modul Termoelektrik Generator Seri 12705 (Doctoral dissertation, Fakultas Teknik)
[7] Ramdhani, A. R., & Saripudin, A. (2018, July). Radio Altimeter (Radalt) Simulator on Aircraft using Arduino UNO. In IOP Conference Series: Materials Science and Engineering (Vol. 384, No. 1, p. 012074). IOP Publishing
[8] Kumar, N. S., Vuayalakshmi, B., Prarthana, R. J., & Shankar, A. (2016, November). IOT based smart garbage alert system using Arduino UNO. In 2016 IEEE Region 10 Conference (TENCON) (pp. 1028-1034). IEEE
[9] Ismail, N., Wahid, S., & Ahmad, N. (2020, April). Arduino Based RFID Vehicle Tracking for Home Security. In Journal of Physics: Conference Series (Vol. 1529, No. 2, p. 022060). IOP Publishing
[10] Badamasi, Y. A. (2014, September). The working principle of an Arduino. In 2014 11th international conference on electronics, computer and computation (ICECCO) (pp. 1-4). IEEE

Digital calorimeter for measuring the specific heat of liquids

Azhar[1], A S Melita[2], A Doyan[2, 3], Susilawati,[2, 3], S Ayub[3] and L S Hudha[4]

[1]Physics Education Study Program, Riau University, Riau, Indonesia
[2]Science Education Master Program, University of Mataram, Mataram, Indonesia
[3]Physics Education Study Program, University of Mataram, Mataram, Indonesia
[4]Department of Physics, Faculty of Mathematics and Natural Science, University of Mataram

E-mail: azhar@lecture.unri.ac.id, Ayusafitrimelita10@yahoo.com, aris_doyan@unram.ac.id

Abstract. The purpose of developing this digital calorimeter is to make a temperature measuring device using Arduino Uno that can be used for physics practicum at the Physics Laboratory of the University of Mataram. This tool is designed according to the function of the calorimeter. Calorimeter is a tool used to determine the heat capacity, specific heat capacity, and latent heat capacity of an object or material. In making this digital calorimeter, the ds18b20 sensor is used to measure the temperature of the liquid, when the sensor is inserted into the calorimeter, it will be able to read the temperature rise that occurs in it. The graph is obtained from the relationship between time (t) and heat energy (Q), it can be seen that the longer the liquid is given electrical energy, the higher the heat generated.

1. Introduction

Physics has become the basis for technological progress. One of the fields of study that has rapid development is Electronics and Instrumentation, including physics practicum. New theories are developed through experiments conducted during the practicum. Experiments in the laboratory generally still use analog measuring instruments [1]. Analog measuring instruments cause the practitioner to have difficulty because the data obtained is inaccurate. The errors that cause the data to be inaccurate are errors in tool calibration, errors in reading scales and accuracy in using the tools. One of the experiments that still uses analog measuring instruments is the calorimeter.

Calorimeter is a tool used to determine the heat capacity, specific heat capacity, and latent heat capacity of an object or material [2]. The calorimeter tool that is often used in laboratory experiments is a glass calorimeter or a thermos calorimeter as shown at of the figure 1.

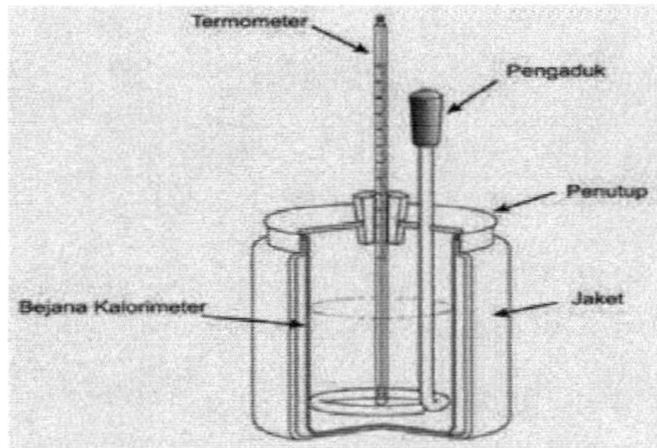

Figure 1. The calorimeter tool

The way to get the value of the calorimeter's heat capacity is to assume that a calorimeter is filled with m1 gram of cold water at a temperature of T_1C. Then, m_2 grams of hot water with a temperature of T_2C are added to the calorimeter. After being stirred and considered to be in equilibrium, it turned out that the final temperature was T_aC. For this situation, Black's Principle applies: The amount of heat released by hot water = a lot of heat received by cold water and the calorimeter.

Hot Q_{water} = Cold Q_{water} + $Q_{calorimeter}$
$M_2C_a (T_2 - T_a) = m_1 C_a (T_a - T_1) + C (T_a - T_1)$

So the heat capacity of the calorimeter is:

$$C = \frac{m^2 C_a(T_a - T_1) - m_1 C_a(T_a - T^1)}{(T_a - T^1)} \quad (1)$$

Information:
m_1 = mass of cold water (kg)
m_2 = mass of hot water (kg)
ca = specific heat of water (J/kg °C)
C = heat capacity of the calorimeter (J/°C)
T_1 = cold water temperature (°C)
T_2 = hot water temperature (°C)
T_a = temperature of the mixture (°C)

The quantitative relationship between heat and other forms of energy is called thermodynamics. Thermodynamics can be defined as the branch of chemistry that deals with the relationship of heat, work, and other forms of energy to equilibrium in chemical reactions and in changes of state [3]. The specific heat of an object is the heat required or released to increase or decrease the temperature of each unit mass of the object by 1°C which satisfies the equation:

$$C = \frac{Q}{m \Delta T} \quad (2)$$

The way to get the specific calorific value of a material is to assume a calorimeter filled with $m_1$grams of cold water at T_1 °C. Then put a metal that has been heated with a mass of m^2 grams and a temperature of T_2 °C into the calorimeter. After being stirred and equilibrated, it turns out that the final temperature is T_a °C. For this situation, the Black Principle applies.

After conducting a field review, problems were found related to the calorimeter experiment, namely in the physics laboratory of the University of Mataram, which still used analog measuring instruments in temperature and mass readings. The importance of modifying an analog calorimeter to a digital calorimeter so that the measurement data has a small error and the data collection time also does not require a long time. With the existing calorimeter, researchers will develop a calorimeter by adding a DS18B20 temperature sensor. The measurement data will be programmed using Arduino Uno to be displayed on the Liquid Crystal Display (LCD).

Figure 2. Temperature sensor ds18b20

A sensor is a device that receives a stimulus and converts it into an electrical signal [4]. The DS18B20 temperature sensor is a temperature sensor that has a digital output. The DS18B20 has a fairly high level of accuracy, which is 0.5°C in a temperature range of -10°C to +85°C. Temperature sensors generally require an ADC and several port pins on the microcontroller, but this DS18B20 does not require an ADC to communicate with the microcontroller and only one requires wire[5].

The microcontroller is an IC with a very high density, where all the parts needed for control are packaged in one chip. Arduino Uno is a microcontroller board that has been integrated with the ATmega328 microcontroller. Microcontroller is a microprocessor system in which there is already a CPU, Read Only Memory (ROM), Random Access Memory (RAM), Input-Output, timers, interrupts, clocks, and other internal equipment that are interconnected and well organized on a single chip which is ready to use [13]. Arduino has advantages such as being equipped with a bootloader so that it does not require a programmer chip, programming is relatively easy because it has many easily accessible libraries and ready-to-use modules.

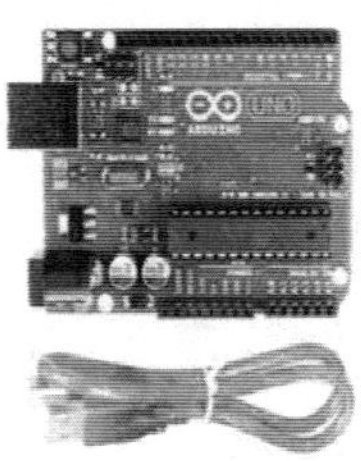

Figure 3. Arduino uno

From figure 3 it can be explained that Arduino Uno is a microcontroller board that has 14 digital input/output pins (6 pins can be used for PWM output), 6 analog inputs and several other devices. This microcontroller can be directly connected to a computer using a USB cable [6].The microcontroller functions as center controlling in the whole system and the result (output) will be displayed on the LCD [7]. LCD (Liquid Crystal Display) is a type of display media where the main viewer uses a liquid crystal. LCDs have been used in various fields, such as electronic devices such as televisions, calculators, or computer screens. LCD provides several advantages compared to other devices for displaying data, including energy saving, light weight and a relatively easier design process and able to display characters as desired. The LCD used in this study is an LCD with a size of 2 x 16 cm.

2. Methods

This research was conducted at the Physics Laboratory of the University of Mataram. This type of research is classified as engineering research. The stages carried out in this research include ideas and clarity of tasks, conceptual design, functional geometry arrangement, detailed design, model making and testing. The implementation of this research requires tools and materials. The tools and materials needed in this study include a multimeter, ds18b20 sensor, Arduino Uno, and LCD.

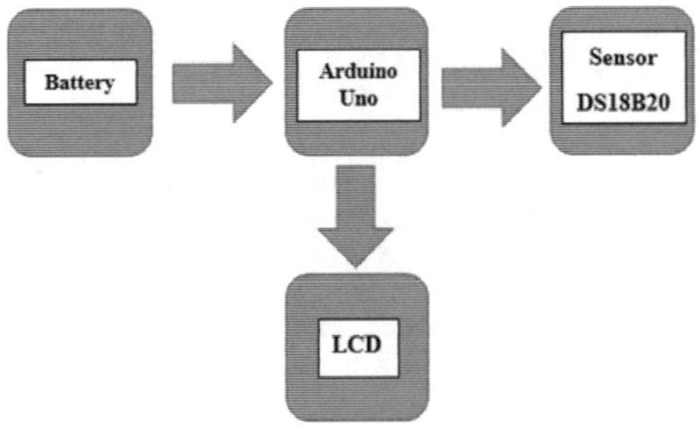

Figure 4. Diagram block

Ideas and clarity of tasks are obtained through literature study searches. Conceptual design is the realization of a system of research to be made. The important thing before doing research is to design a block diagram. Block diagrams contain electronic designs that will aid in tool performance and results. The block diagram for this digital calorimeter experimental set can be seen in Figure 4.

The block diagram in Figure 4 above explains that the tool to be made consists of a power supply as a power supply, the input signal coming from the ds18b20. The signal captured by the sensor will be sent to the Arduino Uno via the signal processing module. Signal from sensor will be programmed in the Arduino Uno to be displayed on the LCD screen. Device design in this study is divided into two, namely hardware design and software design. Hardware design is the mechanical design of the product under study.

3. Result and Discussion

The first paragraph after a heading is not indented (Bodytext style) The results of the research and discussion in this chapter will explain more about the digital calorimeter set system that has been developed with its appropriate use based on the research objectives and also the data obtained when conducting research. Measurement data has an important meaning in a study. Based on the data obtained, it can be seen how the component specifications used and how the relationship between one quantity and other quantities contained in the tool. Analysis of the tool set to the data obtained in the form of performance specifications of measuring instruments and specifications of measuring instruments design. Presentation of data is expressed in the form of tables and graphs. Accuracy data is obtained through comparison of temperature and mass measurements by sensors that are read on the LCD.

3.1. System performance specifications

System performance specification is the identification or description of the functions of each part of the system builder. This needs to be done by testing the tool and analyzing the tool so that it can be seen that the tool is working properly. The components contained in this digital calorimeter tool set are the DS18B20 sensor, Arduino Uno, and LCD. To more easily understand the shape of the digital calorimeter set, see figure 5.

Figure 5. Digital calorimeter experiment set

3.2. The accuracy of temperature, thermal energy and electrical energy readings

The accuracy of the temperature reading on the digital calorimeter experimental set with the ds18b20 sensor that has been programmed using Arduino, the temperature read by the ds18b20 sensor is displayed on the LCD.

Table 1. Data on the relationship between time and temperature.

t (minutes)	T_1 (°C)	T_2 (°C)	E (J)	Q (J)
5	29.87	49.94	5638.17	8389.26
10	29.87	63.94	11305.68	19241.26
15	29.87	76.81	19620.92	19620.92
20	29.87	87.12	22552.68	22552.68
25	29.87	97.19	31949.63	28139,76
30	29.87	100.00	37587.80	29314.34

After knowing the temperature accuracy value by the ds18b20 sensor, it can be analyzed regarding the electrical energy and heat calculated so that it can be calculated. It is known the energy usage required by the digital calorimeter in unit time.

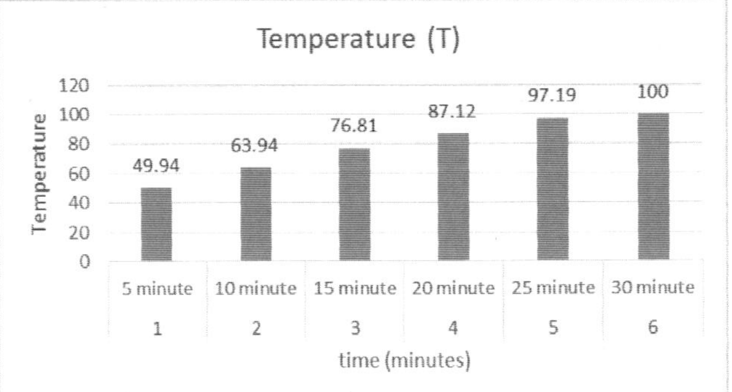

Figure 6. The relationship between time and temperature

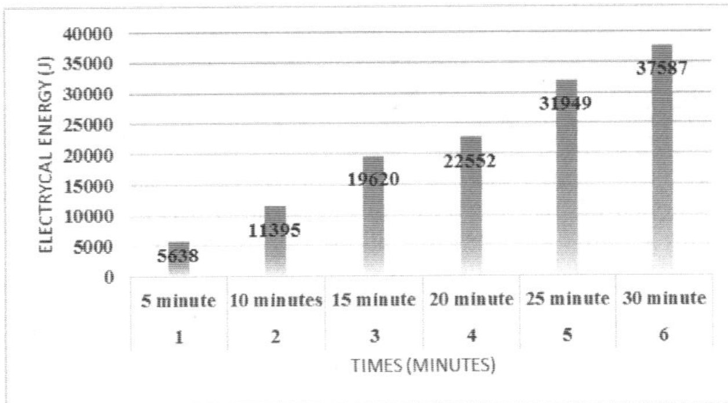

Figure 7. Time relationship with electrical energy

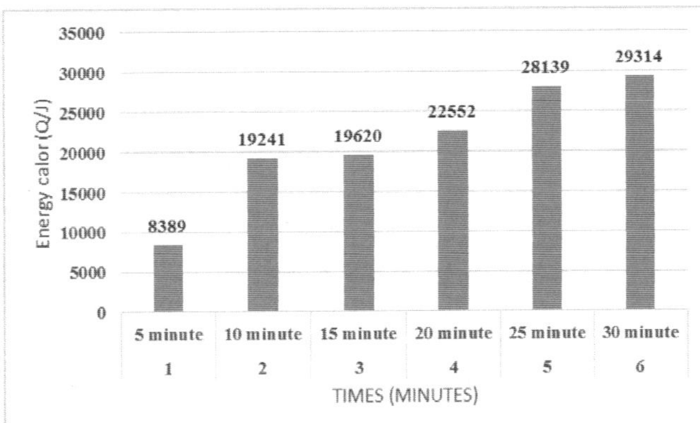

Figure 8. Relationship between times and energy calor.

In the experiment of using a calorimeter practicum tool, researchers used a digital calorimeter which was developed as a temperature and heat measurement tool. In simple terms, a calorimeter is a device for measuring heat.

Researchers observe the heat transfer contained in a digital calorimeter. This arises because there is a change in conserved energy, meaning that the energy contained in electricity comes from a power supply that is will flow into the calorimeter and will turn into heat energy. Then, when stirring, the emergence of an energy, namely mechanical energy which turns into heat energy due to the use of a calorimeter and is not created from electrical and mechanical energy, but the result of changes in the form of the two energies.

The working principle of the calorimeter is known as the name of the mixed method, which is a temperature of a substance heated to a high temperature measured using a calorimeter. The heat lost from the sample will be received by the water and the calorimeter. By calculating the final mixture, the specific heat can be calculated.

Calorimeter is a tool used to determine the accumulated value by measuring the temperature that occurs or changes in other thermal effects. A calorimeter is an instrument used for experiments related to heat. The calorimeter is designed so that the transfer of heat to the surroundings is minimal. Basically the calorimeter consists of two vessels separated by an air chamber. The inner vessel is made of shiny aluminum to reduce heat absorption by the vessel walls. The lid of the vessel is made of plastic which is a poor conductor of heat so that not much heat is lost.

The use of Arduino Uno in the development of the calorimeter is very helpful for researchers to measure the heat produced by the calorimeter. This is because Arduino can directly read the estimated value generated by the calorimeter, and it is directly displayed on the LCD screen used.

In the experiment of using a calorimeter, the aim is to find the heat energy and electrical energy produced through the use of a digital calorimeter. In experiments obtained by researchers, the determination of electrical energy is certainly different from the first experiment to the end.

Researchers conducted the experiment 6 times, respectively, the electrical energy values obtained were 5638.78 J., 11305.68 J., 19620.92 J., 22553.68 J., 31949.63 J., and 37587.8 A. Because the researcher sees the equation of the heat energy which is directly proportional to the mass of water, the specific heat, and the change in time used in the measurement, the heat energy produced will also be different.

At time t = 5 minutes, the heat energy produced is 8389,26 J, the second time at t = 10 minutes the result is 19241.26 J., the third time at t = 15 minutes the result is 19620.92 J, at t = 20 minutes the result is 22552.68 J, at the time t = 25 minutes the result is 28139.76 J, and at t = 30 minutes the result is 29314.34 J.

The final temperature produced during the experiment using a digital calorimeter is also directly proportional to the time used, namely at t = 5 minutes the resulting temperature is 49.94 C, then at t = 10 minutes, the resulting temperature is 63.94 C , at time t = 15 minutes the temperature value is 76.81 C, at time t = 20 minutes, the resulting temperature is 87.12 C, at t = 25 minutes the resulting temperature is 97.19 C, and the last when t = 30 minutes the temperature finally reaches 100 C. This proves that, the change in temperature is directly proportional to the change in time.

4. Conclusion

Based on the experiments that have been carried out, it can be concluded that this digital calorimeter can be used in physics practicum because it is in accordance with the analog calorimeter working system. The components are arranged in such a way and placed in a circuit box. The results of the design specifications for the digital calorimeter experimental set are as follows; The accuracy of temperature readings was obtained through repeated experiment 6 times. So, the temperature can reach 100 °C.

References

[1] Sumintono B I, Ibrahim M A and Phang F A 2010. Pengajaran Sains dengan Praktikum Laboratori UM: Perspektifdari Guru-Guru Sains SMPN di Kota Cimahi Jurnal *Pengajaran MIPA* **15** 120-127

[2] Rufiati E 2011 *Penentuan Kalor Reaksi* (Surabaya: Unair)

[3] Sukardjo 2004 *Kimia Fisika* (Jakarta: PT Rineka Cipta)

[4] Yulkifli 2013 *Sistem Sensor dan Aplikasinya* (Padang: Universitas Negeri Padang)

[5] Nurazizah E, Ramdhani M and Rizal A 2017 Rancang Bangun Termometer Digital Berbasis Sensor Ds18b20 Untuk Penyandang Tunanetra *e-Proceeding of Engineering* **4** 3294

[6] Widyastuti and Ishafit 2019 Penentuan Kalor Jenis Bahan menggunakan Metode Pendinginan Newton dan Sensor Suhu DS18B20 Berbasis Arduino *Uno Jurnal Riset dan Kajian Pendidikan Fisika* **6** 1-6

[7] Ramdhani A R and Saripudin A 2018 Radio Altimeter (Radalt) Simulator on Aircraft using Arduino UNO IOP *Conf. Ser.: Mater. Sci. Eng.* **384** 012074

Digital thermometer innovation as a body temperature measuring tool for FKIP Mataram university employees during the covid-19 pandemic

H Fuadi[1], Susilawati[1,2], A Doyan[1,2], M Taufik[2], Azhar[3] and L S Hudha[4]

[1]Master of Science Education Study Program, Mataram University, Indonesia
[2]Physics Education Study Program, FKIP, Mataram University, Indonesia
[3]Physics Education Study Program, FKIP, Riau University, Indonesia
[4]Department of Physics, Faculty of Mathematics and Natural Science, University of Mataram

E-mail: husnul.fuadi@unram.ac.id, susilawatihambali@unram.ac.id, aris_doyan@unram.ac.id, azhar@lecturer.unri.ac.id

Abstract. The impact of the Covid-19 pandemic requires that every employee in a university maintains his health. One of the most important body parts to maintain during this pandemic is body temperature. The purpose of this study was to develop a non-contact digital thermometer to measure the body temperature of FKIP employees, Mataram University. The digital thermometer development method starts from literature studies, tool specifications, hardware design, software design and thermometer testing to measure employee body temperature in the morning and afternoon. The results of the digital thermometer test showed that the employee's body temperature in the morning with an average of 33.36 ^{0}C and in the afternoon with an average of 33.96 ^{0}C. So it can be concluded that the Mataram University FKIP employees who came to work did not experience Covid-19 symptoms.

1. Introduction

The COVID-19 pandemic has had a very broad impact on the order of life in the 21st century, especially education in Indonesia [1, 2]. Since March 2020, the Indonesian government has made a policy to dismiss all educational institutions, from elementary schools to universities, not to take activities in offline classes [3, 4]. Temporary closure of educational institutions as an effort to contain the spread of the COVID-19 pandemic to millions of students in Indonesia [5, 6]. It was recorded that until December 2020 the number of positive COVID-19 patients in Indonesia reached 743,198 people, 611,097 people recovered and 22,138 people died (merdeka.com). Various policies have been issued by the Indonesian government to reduce the level of spread of the corona virus, including by imposing social distancing, physical distancing to the implementation of PSBB (large-scale social restrictions) in several areas [7, 8]. The health crisis caused by the COVID-19 outbreak has pioneered simultaneous online learning [9, 10]. The government through the ministry of education and culture continues to strive for continuous education by providing learning quota assistance to students from elementary school to university level [11]. Students are expected to be able to study, discuss, do school assignments and exams online from home [12].

The impact of the COVID-19 pandemic continues to be echoed by the government through print and electronic media [13]. Along with this, the community is expected to comply with health protocols such as maintaining distance, wearing masks and washing hands in daily activities [14]. This community habit has become a new era in the world of education called the new normal [15, 16]. Educational institutions such as universities encourage their employees not to interact directly with other residents and immediately consult the nearest health agency if they experience symptoms of high fever [17]. The university has also formed a Covid-19 task force team to control the development of the development of COVID-19 in each of their task forces [18]. To identify symptoms of high fever in an employee, the COVID-19 task force team usually uses a body temperature measuring device in the form of a gun (*thermogun*), which is fired at the forehead or palm [19, 20]. One of the weaknesses of this protocol is the need for special officers to measure the body temperature of employees and the distance between officers and employees whose body temperature is measured is only about 1 meter, this close distance actually poses a risk for Covid-19 transmission [21, 22].

Since March 2021, universities have required their employees to go to work as usual while still complying with health protocols [23, 24]. Every employee is required to check body temperature before entering a room or campus area, as stated applied to every employee at the University of Mataram. Temperature checking as an early detection of Covid-19 symptoms using a *thermogun*. The importance of checking body temperature requires university leaders to prepare medical equipment for all their employees. Efficient and effective equipment is certainly needed in dealing with these problems. Based on this, the researchers tried to make a digital thermometer that could detect the employee's body temperature quickly and precisely without touching it. A digital thermometer can be a solution to assist the leadership of the University of Mataram in controlling the body temperature of its employees without the need for a guard. This study aims to develop a digital body temperature measuring device based on Arduino Uno to be tested on FKIP employees at the University of Mataram during the New Normal period.

2. Methods

2.1. Time and place
This research started from March to June 2021, starting from problem analysis to the design of digital body temperature measuring devices and data collection of employee temperatures. This research was conducted at FKIP Mataram University with 35 employees as test subjects.

2.2. Tools and materials
The tools and materials used in this development research include; Arduino uno, LCD I2C, Gy-906 Temperature Sensor, USB cable, jumper cable, Battery plug, Breadboard, 9V battery and laptop. The components of each tool and material are presented in the image below.

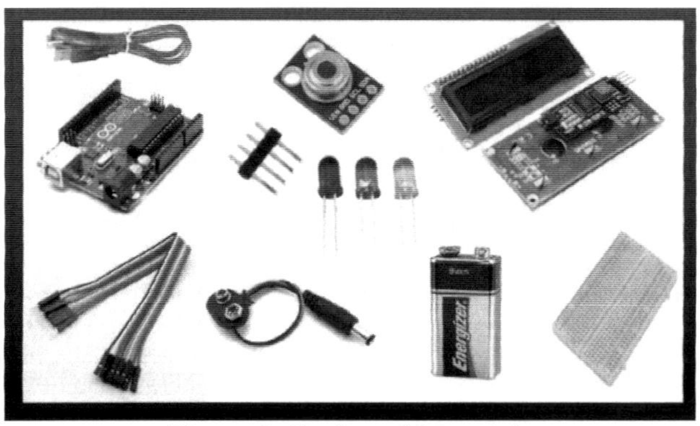

Figure 1. Tools and Materials

2.3. Digital thermometer development stages

2.3.1. Literature study

At this stage, researchers conducted an analysis of the impact of COVID-19, especially on employees of the Mataram University FKIP. All employees are required to check body temperature when entering a room or campus area. Employees who have a body temperature above the normal temperature are required to self-isolate and not come to work as usual. Based on this, it is important for educational institutions to prepare digital and non-contact body temperature detectors so that employees can know their body condition at work.

2.3.2. Digital Thermometer Specifications

At this stage the researchers began to design a digital body temperature measuring device that could be used in the FKIP campus, Mataram University. The digital thermometer blog diagram model is as shown in the image below.

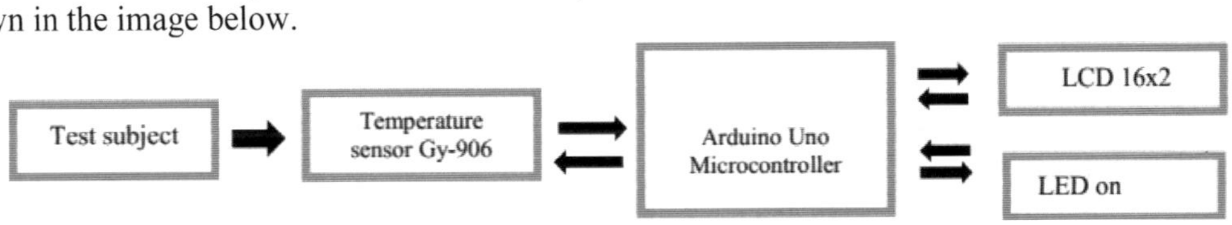

Figure 2. Digital thermometer blog diagram

2.3.3. Software design

The software design is divided into 3, namely: the algorithm on the temperature sensor, the algorithm on the LCD and the algorithm on the LED. The following is a flow diagram of the algorithm in the software.

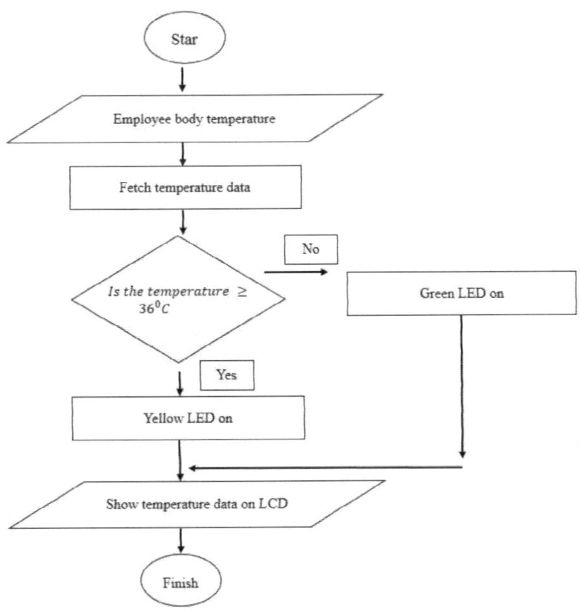

Figure 3. Flowchart of digital thermometer software

2.3.4. Hardware Design
At this stage the researcher begins to design the tool by combining or coding the tool components and materials as shown in the image below. The following is a picture of the design of the tool that has been made.

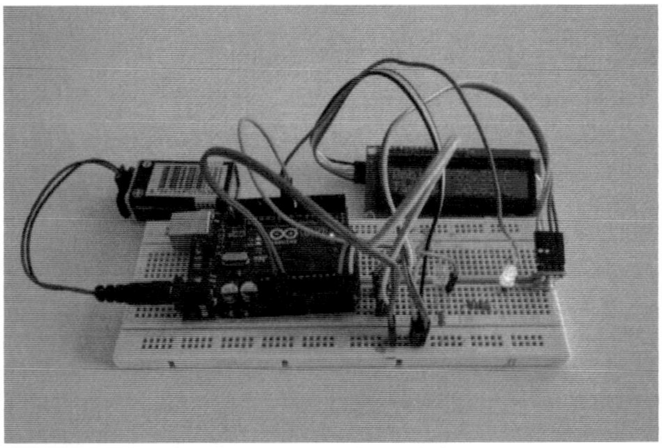

Figure 4. Non-contact digital thermometer

2.4. Digital thermometer temperature sensor trial sensor
At this stage the digital thermometer has been assembled on a bread board. The temperature sensor, LCD and LED are neatly installed with the Arduino Uno. This stage is the thermometer test stage using two objects that have different temperatures. Here is a picture of a digital thermometer test.

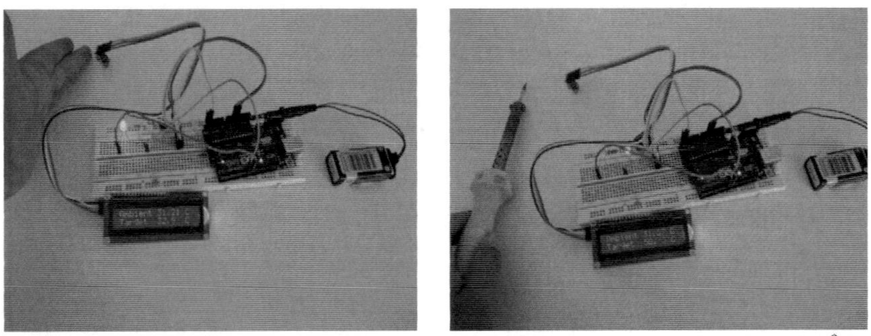

Figure 5. Measurement of objects with temperatures above and below 36 ^{0}C

2.5. Manufacture of a non-contact digital thermometer case
At this stage, the digital thermometer has been made portable so that it can be carried or used to test the temperature measurement of the FKIP staff at the University of Mataram.

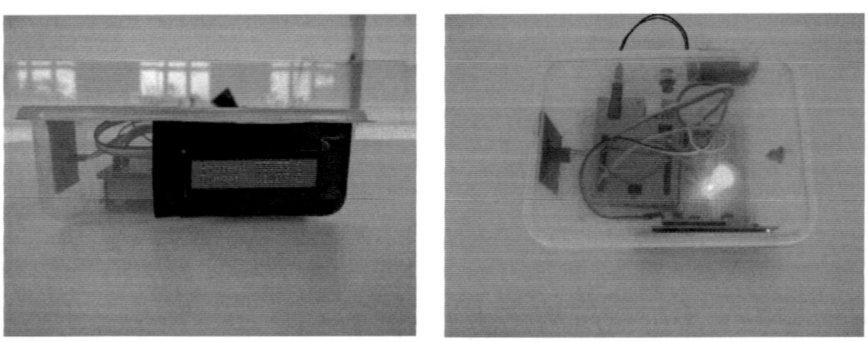

Figure 6. Non-contact digital thermometer

2.6. Data collection FKIP Mataram University Employees
Tool testing and data collection were carried out in the morning and afternoon to see the difference in temperature at that time. The number of employees who were used as test subjects in this study were 35 people consisting of 24 men and 11 women. The age range of the test subjects was from 21 to 60 years. Data collection was carried out for two consecutive days. The stages of testing and data collection on FKIP Mataram University employees are as follows;

2.6.1. Employee temperature measurement on the first day
1)Digital thermometer calibration
2). The digital thermometer is brought to the employee's room starting at 08.00 - 10.00 AM.
3). Record the results of temperature measurements and document them
4). At this stage the number of employees measured is 20 people
5). For temperature measurement at noon starting at 13.30 -15.00 PM.
6). Temperature measurements were carried out on 20 employees whose temperatures had been measured in the morning.

2.6.2. Employee temperature measurement on the second day
1). Digital thermometer calibration
2). The digital thermometer is brought to the employee's room starting at 08.00 - 10.00 AM.
3). Record the results of temperature measurements and document them
4). At this stage the number of employees measured is 15 people
5). For temperature measurement at noon starting at 13.30 -15.00 PM.
6). Temperature measurements were carried out on 20 employees whose temperatures had been measured in the morning.

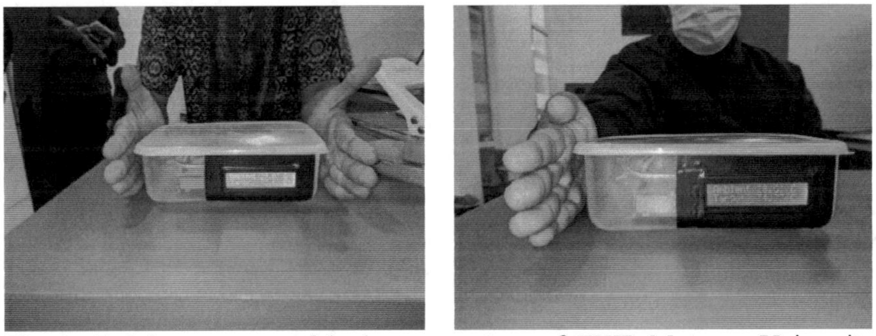

Figure 7. Measurement of body temperature of FKIP Mataram University employees.

3. Result and Discussion

3.1. Average body temperature of FKIP Unram employees
The temperature measurement of FKIP Mataram University employees was carried out in the morning and afternoon. This data collection was carried out for two consecutive days. On the first day of data collection was carried out on 20 employees in the morning starting at 08.00 - 10.00 WITA. Furthermore, repeated measurements were carried out at 13.00 – 15.00 WITA with the same object. On the second day the data was collected on 15 employees with the same treatment as the first day. Employee temperature measurements are carried out in their respective workspaces. The employee's workspace uses air conditioning so this can affect the results of temperature measurements. The results of the employee temperature measurement are divided into three categories, namely the measurement of the average temperature of the employee, the measurement based on age and the measurement

based on gender. The following is the result of measuring the temperature of the employees of FKIP, University of Mataram.

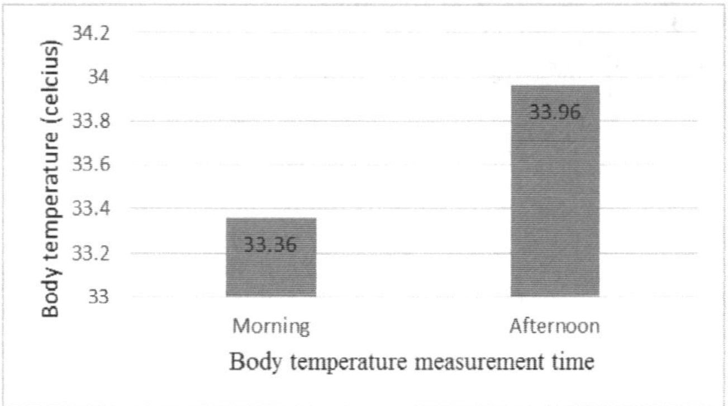

Figure 8. Average body temperature of FKIP Mataram University

3.2. The results of measuring body temperature based on the age of the employees of FKIP University of Mataram

Results of Body Temperature Measurement for FKIP Mataram University employees aged 21 – 40 years and 41 – 60 years in the morning.

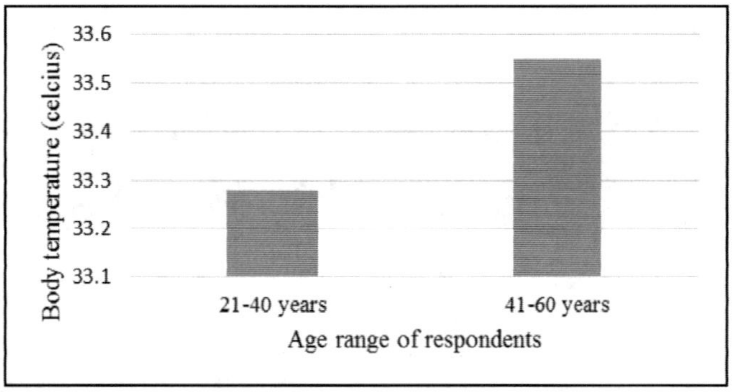

Figure 9. Average body temperature of FKIP Mataram University Employees in the Morning.

Results of Body Temperature Measurement for FKIP Mataram University employees aged 21-40 years and 41 – 60 years in the afternoon

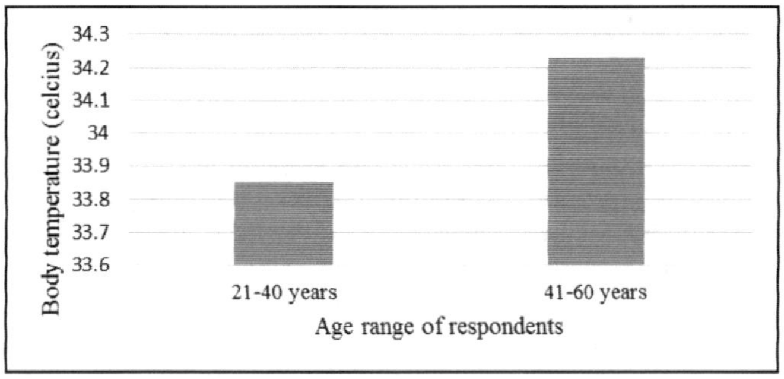

Figure 10. Average body temperature of FKIP Mataram University Employees during the Day.

Journal of Physics: Conference Series **2165** (2022) 012035 doi:10.1088/1742-6596/2165/1/012035

3.3. *The results of measuring body temperature based on the gender of the employees of FKIP University of Mataram*

Results of measuring the body temperature of gender FKIP employees at the University of Mataram in the morning.

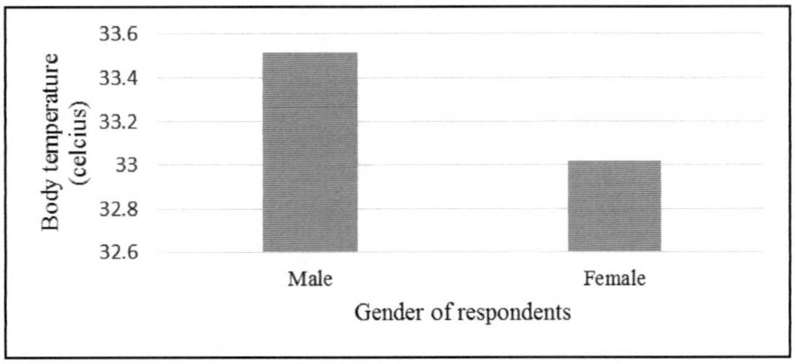

Figure 11. Average body temperature of FKIP Mataram University employees by gender in the morning.

Results of Measurement of Body Temperature of gender Employees of FKIP University of Mataram in the afternoon

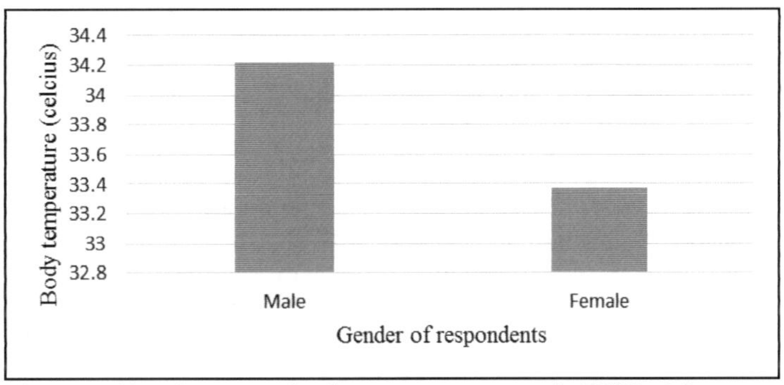

Figure 12. Average body temperature of FKIP Mataram University employees by gender during the day.

Table 1. Results of Measurement of Body Temperature of FKIP Employees, University of Mataram

	Average temperature (^{0}C)	Temperature by age (^{0}C)		Temperature by Gender (^{0}C)		Adverb of time
		(20 – 40) years	(41 – 60) years	Male	Female	
Number of	33.36	33.28	33.55	33.52	33.02	Morning
Test	33.96	33.85	34.23	34.22	33.38	Afternoon
Subjects	35	25	10	24	11	

4. Conclusion

Based on the results of measuring body temperature using a non-contact digital thermometer based on Arduino Uno, the temperature of FKIP Mataram University employees in June 2021 was below the average temperature of patients with COVID-19 symptoms. All employees who enter work are in good health and not infected with COVID-19. It is hoped that all employees will continue to comply with the health protocol and check body temperature when entering work so that it is easier to know the condition of their respective bodies when doing activities.

Acknowledgement
Thank you to the Director of Postgraduate and Head of the Master of Science Education study program at the University of Mataram. Thanks are also conveyed to the supervisors of the mechanics and thermostatistics course who helped in this research process.

References
[1] Asmuni A 2020 Problems of Online Learning during the Covid-19 Pandemic and Solutions to Solve it *Jurnal Paedagogy* **7**(4) 281-288
[2] Watrianthos R, Siregar M N H, Ardiana D P Y, Gandasari D, Purba R A, Fadhillah Y and Koryati T 2020 *Learning from Covid-19: Technological and agricultural perspectives* (Universitas Al Washliyah Labuhanbatu: Yayasan Kita Menulis)
[3] Simbolon I 2021 Increasing the Spirit of Learning Amid the Covid-19 Pandemic at the Elementary School Level in Sigumbang Village *Jpkm: Jurnal Pengabdian Kepada Masyarakat* **1**(1)
[4] Iftiyah M 2021 Building Teacher Creativity with Learning Innovations during the Covid-19 Pandemic *Prosiding Seminar Nasional Program Pascasarjana Universitas Pgri Palembang*.
[5] Aji R H S 2020 The impact of Covid-19 on education in Indonesia: Schools, skills and learning processes *Salam: Jurnal Sosial dan Budaya Syar-I* **7** (5) 395-402
[6] Heryanto T and Rochaeni A 2020 The Impact of Covid-19 on Universities in Indonesia *ResearchGate, August*.
[7] Nahdi K, Ramdhani S, Yuliatin R R and Hadi Y A 2021 *Obsession Journal: Journal of Early Childhood Education Implementation of Learning during Lockdown for PAUD Institutions in East Lombok Regency* **5**(1) 177–186
[8] Khasanah D R A U, Pramudibyanto H and Widuroyekti B 2020 Education during the COVID-19 pandemic *Jurnal Sinestesia* **10**(1) 41-48
[9] Ashari M 2020 The online learning process in the midst of anticipating the spread of the corona virus is considered not to be maximal. Taken May 26, 2020, from Pikiran Rakyat.com
[10] Darmalaksana W, Hambali R, Masrur A and Muhlas M 2020 Analysis of Online Learning during the Covid-19 WFH Pandemic as a Challenge for 21st Century Digital Leaders *Karya Tulis Ilmiah (KTI) Masa Work from Home (WFH) Covid-19 UIN Sunan Gunung Djati Bandung* 1-12.
[11] Ihsanuddin 2020 Fakta Lengkap Kasus Pertama Virus Corona di Indonesia Kompas.Com.
[12] Purwanto A, Pramono R, Asbari M, Santoso P B, Wijayanti L M, Hyun C C and Putri R S 2020 Studi Eksploratif Dampak Pandemi COVID-19 terhadap Proses Pembelajaran Online di Sekolah Dasar *Journal of Education, Psychology, and Counseling* **2** 1–12.
[13] Pujilestari Y 2020 Dampak Positif Pembelajaran Online Dalam Sistem Pendidikan Indonesia Pasca Pandemi Covid-19 **4**(1)
[14] Puspitasari R 2020 Hikmah Pandemi Covid-19 Bagi Pendidikan Di Indonesia. Diambil 26 Mei 2020, dari Institut Agama Islam Negeri Surakarta website: https://iain-surakarta.ac.id/hikmah-pandemi-covid-19-bagi-pendidikan
[15] Windhiyana E 2020 Dampak Covid-19 terhadap kegiatan pembelajaran online di sebuah perguruan tinggi kristen di Indonesia *Perspektif Ilmu Pendidikan* **34** (1) 1-8
[16] Wax R S and Christian M D 2020 Practical recommendations for critical care and anesthesiology teams caring for novel coronavirus (2019-nCoV) patients *Canadian Journal of Anesthesia*. https://doi.org/10.1007/s12630-020-01591-x
[17] Nur R Y and Annissa R 2020 Kebijakan Pemberlakuan Lockdown Sebagai Antisipasi Penyebaran Corona Virus Covid-19 *Jurnal Sosial & Budaya Syar-i FSH UIN Syarif Hidayatullah Jakarta* **7** (3) pp.227-238.

[18] Kusuma J W and Hamidah 2020 Platform Whatsapp Group dan Webinar Zoom Dalam Pembelajaran Jarak Jauh Pada Masa Pandemik Covid 19 *Jurnal Ilmiah Pendidikan Matematika* **5**(1)

[19] Kukus Y, Supit W and Lintong F 2009 Suhu tubuh: homeostasis dan efek terhadap kinerja tubuh manusia *Jurnal Biomedik: JBM* **1**(2)

[20] Cahyati, N., & Kusumah, R. (2020). Peran orang tua dalam menerapkan pembelajaran di rumah saat pandemi Covid 19. *Jurnal Golden Age*, *4*(01), 152-159.

[21] Herliandry L D, Nurhasanah N, Suban M E and Kuswanto H 2020 Pembelajaran pada masa pandemi covid-19 *JTP-Jurnal Teknologi Pendidikan* **22**(1) 65-70.

[22] Sumartiningtyas H K N 2020 Dampak Pandemi Virus Corona pada Lingkungan, Polusi Udara Global Turun. *Kompas.com* - 17/03/2020.

[23] Syaharuddin S 2020 Menimbang Peran Teknologi dan Guru dalam Pembelajaran di Era Covid-19 *Menimbang Peran Teknologi dan Guru dalam Pembelajaran di Era Covid-19*

[24] Pakpahan A K 2020 Covid-19 dan Implikasi Bagi Usaha Mikro, Kecil, Dan Menengah *Jurnal Ilmiah Hubungan Internasional* 59-64

Teaching for conceptual change on Newton's First Law

S Syuhendri[1]

[1]Physics Education Department, Universitas Sriwijaya, Palembang, South Sumatera, Indonesia

E-mail: hendrisyukur@yahoo.com

Abstract. Teaching for conceptual change used to overcome misconceptions is a teaching that explicitly intended to facilitate students to have experience conceptual change learning. The Prerequisites for this learning are dissatisfaction, intelligibility, plausibility, and fruitfulness. It was implemented a teaching approach for this purpose on Newton's First Law by a quasi experimental research method with pre-test and post-test control group design. The sample was selected based on convenience sampling. The instrument used was the Force Concept Inventory translated to Indonesian (Ind-FCI) for the domain of Newton's First Law. The independent sample U-test by $\alpha = 0.05$ was used to test the hypothesis. The effectiveness of shifting of students' conceptual understandings toward the true concepts was determined by the N-gain value. The result was the mean N-gain of the experimental and control group were 49.15% and -3.85% respectively. There was a significant difference of increase in students' conceptual understanding between those who were taught using the conceptual change approach and conventional one. Therefore, the teaching for conceptual change can be used to help students to reduce their misconceptions.

1. Introduction

Misconception is one of the problems that get attention in the education sector in Indonesia [1]. This attention is inseparable from the many misconceptions found held by students at various levels of education, including misconceptions related to physics material, and awareness of the negative influence of the misconceptions in learning. Physics as part of the natural sciences is one of the places where many misconceptions are found. This is because individuals began forming the concepts of physics in his/her mind far before coming to school. Students often describe the concepts qualitatively and relatively, and accordingly deviate from the operational definitions described by physicists [2]. Discrepancy between students' ideas and the scientific explanations of a concept is called as misconceptions [1]. For example, they understand concepts such as speed, velocity, acceleration, distance, displacement, and force differ from scientific view. This conceptualization is not correct and may be a hindrance in learning.

Various studies have shown that many misconceptions held by students. As examples, in mechanics, students think that force is needed to keep an object moving, when force stops working on the object, it will stop moving. This Aristotelian physics is still strongly held by students. Likewise, students also think that the heavier objects will fall faster. They have a view of point that the mass or weight of an object affects the speed of the objects falling in a gravitational field. [1] found that most of respondents (79.45%) on his research, consisting of prospective physics teachers, had misconception that heavier objects would fall faster. Similarly, [3] revealed that among the massive misconceptions experienced by university physics students in Newtonian mechanics topics are the heavier objects always fall faster, mass of an object make it stop moving, active force cause motion, impetus come from hit, impetus also influences the circular motion, position is undiscriminated with velocity, and motion implies force works on an object.

Moreover, [4] also reported some kinds of misconceptions in mechanics such as no force is needed to slow an object, to slow an object at a constant rate needed a decreasing force opposite its motion, and to slow an object at a constant rate needed an increasing force opposite its motion. It was also reported that physics student-teachers in Jordanian universities grasped high misconceptions related to the impetus and active force [5]. The interesting thing to know is that the same level of the misconception is experienced by respondents from different categories though. For example, there was no significant difference the level of the misconceptions between male and female students and also the students' attitude toward studying physics did not affect the level of their misconceptions and conceptual understanding in Newtonian mechanics topics [5]. All these findings revealed that misconceptions are not only common but also occur in various categories equally, both for male and female students and both have good and poor attitudes towards studying physics. Therefore, this problem needs to be a concern by physics educators.

Various studies and educational theories exhibit that misconceptions have an unfavorable impact on learning [1,6]. Misconceptions will affect the acquisition of the next material negatively. Therefore, misconceptions have to be overcome. However, the problem is not so simple. Overcoming misconceptions is not enough rely on teachers' intention to do that. Experiences show, when the teacher has conveyed the correct concept in learning and students know about it, but in the next period students will return to their old conceptions in answering the same problem. They returned to the misconceptions. In other words, misconceptions are difficult to change with traditional learning [3,7] [8,9] and need certain approach [10]. Based on this fact, it is necessary to carry out learning with an approach that aims to overcome misconceptions. Instructors need to design the effective instruction.

Based on cognitive theory, to develop the scientific understandings, instructors need first to be aware with students' preexisting conceptions. [11] put forward the theory that underlies how learning to change this conception should be carried out called as models of conceptual change [11,12]. They called it as a conceptual change model. There are four conditions that must be met in order to a conceptual change can occur in students' mind, firstly the existence of students' dissatisfaction with the conceptions they hold, and after that there is a strong belief in the truth of the substitute conceptions. This level of truth is represented by the intelligible, plausible, and fruitful nature of the substitute concept. Various strategies are derived from the model of [11]. In this study, the learning of conceptual change tested to overcome the misconception described by improving students' conceptual understanding of Newton's First Laws.

2. Methods

The method used in this research was a quasi experimental with pre-test and post-test control group design. The research was to test the effectiveness of the strategies used to induce the conceptual change in students' mind. The sample of the research chosen based on convenience sampling, comprised of two groups of Physics Education students in the Faculty of Teacher Training and Education in a state university in Indonesia. This convenience sample chosen because the participants are available and convenient to the researcher [13]. The two groups was randomly assigned as experimental and a control groups.

The data about students' conceptual understanding before and after instruction were collected by using the Indonesian version of the Force Concept Inventory (Ind-FCI) [6]. The Ind-FCI was translated from the FCI developed by Hestenes et al [14]. The FCI was designed to examine the students' conceptual understanding [15] and to probe their misconceptions. The Ind-FCI has been successfully validated [1,6]. The FCI Items related to the conceptual understanding and misconceptions in the domain of Newton's First Law are numbers 6, 7, 8, 10, 11, 17, 23, 24, and 25.

The treatment given to the experimental group was in the form of learning with a conceptual change model approach, while in the control group traditional learning was applied. The teaching materials used in the experimental group were teaching materials based on the conceptual change model. Before and after treatment, both groups were given pre-test and post-test using Ind-FCI.

The descriptive statistics of the resume of the Ind-FCI pretest and the Ind-FCI post-test were used to describe the result of the research. To see the effectiveness of the instruction used, is there an increase in learning outcomes or conceptual understanding after an instruction, it was determined the normalized gain scores (N-gain) value for the experimental and control groups. The N-gain is a valid and more effective way to measure the success of a learning. It is the ratio of actual average gain to the maximum possible average gain [16]. The N-gain <g> is calculated by the equation:

$$g >= \frac{\%<postest> - \%<pretest>}{100 - \%<pretest>} \tag{1}$$

The category to interpret the meaning of N-gain <g> is:

$(<g>) \geq 0.7$	High
$0.7 > (<g>) \geq 0.3$	Medium
$(<g>) < 0.3$	Low

Furthermore, to get the significance of the difference in increasing of students' conceptual understanding for the two groups, the hypothesis testing was carried out using the independent sample U-test by $\alpha = 0.05$.

3. Result and Discussion

The research problem is there a significant difference of the improvement of students' conceptual understanding who were taught by using conceptual change approach toward the true concepts and students' those who are taught by using the conventional approach on Newton's First Law topics. A conceptual change approach which is supported by material developed based on the conceptual change approach is also given to the experimental group during instruction. Nine of 30 items of Ind-FCI was tested to students before and after the instructions. The result of data analysis are shown in the Table 1.

Table 1. Resume of descriptive statistic of students' conceptual understanding on Newtons' First Law for experimental and control groups.

	Pre-Test	Post-Test	Gain	N-gain	Pre-Test	Post-Test	Gain	N-Gain
	Experimental Group				Control Group			
Max.	5.00	9.00	8.00	1.00	4.00	4.00	3.00	0.33
Min.	0.00	3.00	-1.00	-0.25	0.00	0.00	-3.00	-0.60
Mean	1.86	5.45	7.02	0.49	1.86	1.79	-0.07	-0.04
Mean (%)	20.71	60.61		49.15	20.69	19.92		-3.85
Std. Dev.	1.25	1.56	1.87	0.24	1.22	1.21	1.62	0.25

Tabel 1 show that the maximum score for the experimental group on the pre-test was 5 and then jumped to 9 on the post-test. In contrast, there was not any increase in the maximum score for the control group, where the score remained at 4 both for pretest to post-test. The minimum score for the experimental group also increased from 0 to 3, while for the control group the minimum score remained at 0. The N-gain for the experimental group was 49.15%, a middle category. However, the N-gain for the control group had negative value.

The comparison of N-gain for the experimental and the control groups is displayed in Figure 1. The Figure clearly shows that by applying the conceptual change learning having an impact on increase in students' conceptual understanding during the instruction. The negative gain for the control group also indicates that learning with ignoring the correct concept acquisition can not only end with no increasing in concept mastery on the students' mind but can also end in a decrease in the mastery of the concept itself. This also confirms that the instructions can cause some students' misconceptions.

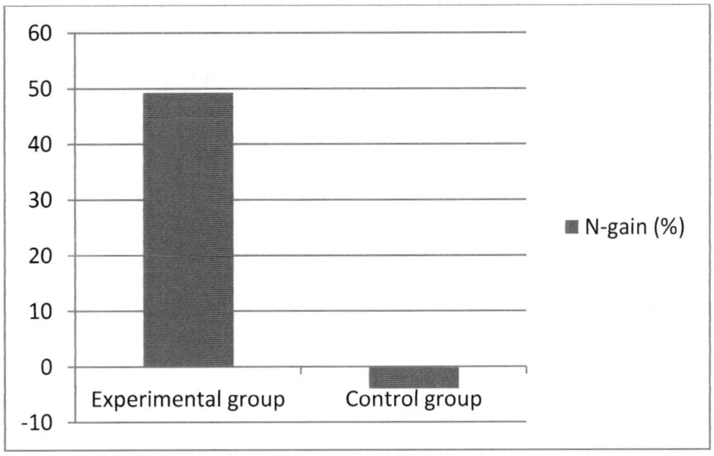

Figure 1. The average N-gain of increasing of conceptual understanding of experimental and control groups on Newton's First Law topics

Furthermore, to see the significance of the difference of the mean of N-gain between the experimental and the control groups, a statistical tests was applied. The results of the statistical test are shown in Table 2. The results of the normality test gave normal distribution for the experimental group. Then, the results of the homogeneity test showed that data were homogeneous. Based on this basic assumption, it was administered the mean difference test using the Mann-Whitney test. The Mann-Whitney test was conducted for the level of significance $\alpha/2 = 0.025$.

Table 2. The resume of the statistical test results of mean N-gain of improving of conceptual understanding for experimental and control roups on Newtons' First Law topics.

Groups	Normality test, $\alpha/2 = 0.025$		Homogeneity test		Mann-Whitney test $\alpha/2 = 0.025$		
	Sig.	Explanation	Sig.	Explanation	u-value	Sig.	Explanation
Experimental	0.169	Normal	0.853	Homogen	51.000	0.000	Sig.
Control	0.001	Not normal					

Table 2 shows that the significance values was 0.000 or sig. < α, which means that Ho is outside the reception area, and Ha is accepted. It could be concluded that there were significant differences between the average N-gain of the experimental group to control group. In other words, there were differences improvement of conceptual understanding to the true concept on the Newton's First Law concepts between the experimental group and the control group. It means that the conceptual change learning can be more effective in improving students' conceptual understanding toward the true concepts if comparing to the learning that is not based on the theory of conceptual change. More specifically, based on the pattern of acquisition of N-gain of experimental and control groups, it can be concluded that only with the conceptual change learning that instructor can promote the changing of students' conceptual understanding toward the true concepts.

The results of this research support the previous studies, such as by [3] [6] [17] [18], as well as [19] that find significant benefits of conceptual change learning for changing prior misconceptions. The increase of students' conceptual understandings in the experimental group is because in this group, the conceptual change oriented instruction, which explicitly considered students' misconceptions, was implemented. Teaching approach is very important to concern in addressing misconceptions [20]. In contrast, in the control group only implemented conventional designed instruction that were not oriented to overcome the misconceptions. This confirms that traditional learning does not provide satisfactory results in reducing misconceptions [9]. In the experimental group, learning process followed the steps proposed by [11], where students' conceptual understanding could be changed, i.e. dissatisfaction, intelligible, plausible, and fruitful.

4. Conclusion

The research confirms that the learning based on conceptual change model has been successful in promoting students' conceptual understanding toward the correct concepts. The difference learning process has provided the different results of increasing students' conceptual understanding. In conceptual change learning, the instructor has made aware about the conditions of the students' pre-conception and then prepared a specific learning to facilitate conceptual change and remove misconceptions. The conceptual change approach learning and all the learning tools developed based on conceptual change theory has been proven successful for conceptual change learning. Therefore, the instructors should pay attention to the students' initial conceptual understanding before teaching, and then carry out lessons aimed at overcoming misconceptions in their classes.

Acknowledgement

Thank you to all, students and lecturers, who have participated in this research. Thanks also to the Dean of Faculty of Teacher Training and Education, Universitas Sriwijaya who have admitted this research.

References

[1] Syuhendri S 2019 Student teachers' misconceptions about gravity In *Journal of Physics: Conference Series* **1185**012047.
[2] Motlhabane A 2016 Learner's alternative and misconceptions in physics: a phenomenographic study *Journal of Baltic Science Education* **15**(4) 424.
[3] Syuhendri S 2017 A learning process based on conceptual change approach to foster conceptual change in Newtonian mechanics *Journal of Baltic Science Education* **16**(2) 228.
[4] Wells J, Henderson R, Traxler A, Miller P and Stewart J 2020 Exploring the structure of misconceptions in the Force and Motion Conceptual Evaluation with modified module analysis *Physical Review Physics Education Research* **16**(1) 010121.
[5] Mohammad S, Khoshman J M, and Tayeh K A 2020 Jordanian Pre-Service Physics Teacher's Misconceptions about Force and Motion *Journal of Turkish Science Education* **17**(4) 528-543.
[6] Syuhendri S 2021 Effect of conceptual change texts on physics education students' conceptual understanding in kinematics *Journal of Physics: Conference Series* **1876** 012090.
[7] Fadaei A S and Mora C 2015 An investigation about misconceptions in force and motion in high school *US-china education review* **5**(1) 38-45.
[8] Bani-Salameh H N 2016 How persistent are the misconceptions about force and motion held by college students? *Physics Education* **52**(1) 014003.
[9] Rafika R and Syuhendri S 2021 Students' misconceptions on rotational and rolling motions *Journal of Physics: Conference Series* **1816**(1) 012016.
[10] Syuhendri S, Andriani N and Taufiq T 2019 Preliminary development of conceptual change texts regarding misconceptions on basic laws of dynamics *Journal of Physics: Conference Series* **1166**(1) 012013.
[11] Posner G J, Strike, K A, Hewson P W and Gertzog W A 1982 Accommodation of a scientific conception: Toward a theory of conceptual change *Science Education* **66** 211–227.

[12] Strike K A and Posner G J 1992 *A revisionist theory of conceptual change. In R A Duschl & R J Hamilton (Eds.), Philosophy of science, cognitive psychology, and educational theory and practice* (Albany, NY: SUNY Press).
[13] Creswell J W 2012 *Educational Research: Planning, Conducting and Evaluating Quantitative and Qualitative Research* (Boston: Edwards Brothers).
[14] Hestenes D, Wells M and Swackhamer G 1992 Force concept inventory *The physics teacher*, **30**(3) 141-158.
[15] Hake R R 1998 Interactive-engagement versus traditional methods: A six-thousand-student survey of mechanics test data for introductory physics courses *American journal of Physics*, **66**(1) 64-74.
[16] Hake R R 1999 Analyzing Change/Gain Score. *American Educational Research Methodology*.
[17] Kowalski P and Taylor A K 2017 Reducing students' misconceptions with refutational teaching: For long-term retention, comprehension matters *Scholarship of Teaching and Learning in Psychology* **3**(2) 90.
[18] Çil E and Çepni S 2016 The effectiveness of conceptual change texts and concept clipboards in learning the nature of science *Research in Science & Technological Education* **34**(1) 43-68.
[19] Ozkan G and Selcuk G S 2015 The effectiveness of conceptual change texts and context-based learning on students' conceptual achievement *Journal of Baltic Science Education* **14**(6) 753.
[20] Tooher H and Johnson P 2020 The role of analogies and anchors in addressing students' misconceptions with algebraic equations *Issues in Educational Research* **30**(2) 756-781.

Analysis of creeping avalanche velocity using physics of landslide approach

M Souisa[1,2,3,4], S M Sapulete[1,3,4] and A Labok[1]

[1] Physics Department, Faculty of Mathematics and Science, University of Pattimura
[2] Landslides Laboratory, Faculty of Mathematics and Science, University of Pattimura
[3] Disaster Mitigation Laboratory, Faculty of Mathematics and Science, University of Pattimura
[4] Geosciences Center, Faculty of Mathematics and Science, University of Pattimura

E-mail: thos.phys@gmail.com

Abstract. Creeping landslides are natural disasters and social disasters that often occur in the world. One of the main factors causing slope failure is rainfall and earthquakes. In this study, the constitutive model is described and analyzed for termite drift using the physics of landslide approach to deal with geomorphic changes as triggers for ground motion. The constitutive behavior of the soil-rock cover of the slope is characterized along a planar or curved slip plane and consists of different soil properties with different slope strength parameters. The morphology along the slip plane of the creeping soil is studied and analyzed to estimate the mass movement of the creeping landslide before being inhabited by rainwater and inhabited by rainwater. Calculation of the maximum velocity and runout time using the parameters of the slope geometry. The analysis shows that the geometry of the soil creeping plane on the slope and the different slope angles of the slip plane affect the velocity and runout time of the creeping landslide.

1. Introduction

The termite-type soil movement disaster that occurred in Ambon City, especially at the IAIN Campus, was caused by the entry of rainwater into the soil-rock surface of the slopes so that the soil-rock became saturated. This location is also located around the fault line so that it is very potential for termites to occur, and the soil-rock structure that makes up the sloped surface is a material loss. The type of soil rock is a loss of material so that it is saturated with water when it rains, and when it is hot, the soil rock on the surface becomes cracked or wrinkled. In the end, the soil-covered rocks that are water-type due to rain with a duration of several hours will be released and slide along the slope at high speed [1,2].

The dynamic conditions and the occurrence of slope failure in the IAIN campus area cause the type of creeping soil movement to move slowly. Although these ground motions usually move slowly, the danger and damage they can cause can be greater than that of slide-type avalanches. The losses caused by the creeping soil movement in the study area are in the form of damage to buildings and open land and cracks at certain points. Initial observations showed that the friction coefficient of the slip surface was influenced by the dynamic type, topography, transportation, lithology that formed the rock mass, and the strength of the rock mass [3].

This research uses two analytical methods. The first model, the equated mass, assumes that the landslide is a single point. The second is a two-dimensional (2D) model, in which landslides are simplified based on a special profile. Several other models for avalanche velocity usually have a more complex shape because they take into account high-velocity air fluidization, lubrication of air layers, slipping off the integrated rock during the landslide movement process, momentum transfer, buoyancy, and pore pressure [4].

When the soil-rock mass (debris) is dragged by water moving along the slope, some of it transits for a moment, while others move to the foot of the slope and settle [5,6]. This is influenced by hydraulic resistance as a very important factor affecting the landslide speed. Thus, during the movement of the landslide material, it is assumed that there is no drag and friction resistance with air, and here it is considered to have only friction and turbulence resistance. Therefore, the estimation of the mass velocity of the transiting partial creeping avalanche is still very complex and has not been studied in this study. On this basis, this study is limited to measuring the geometry of the avalanche section to make a calculation model for the estimated velocity and time of shifting of soil creeping material in dry conditions or that have not been submerged in rainwater, and wet conditions or submerged by rainwater.

2. Methods

Survey of ground motion events to determine creep velocity and total reach time. Determination of creeping velocity and time using landslide physics approach.

2.1. Study area

Field data retrieval was carried out in the ex-scratch area around the IAIN Campus in Batu Merah Village, Sirimau District, Ambon City. Geographically, the position of the research area is located at coordinates 3°40′54.24" South Latitude and 128°13′37.25" East Longitude with an altitude of (80.0 - 120.0) meters above sea level (masl), as shown in Figure 1. Geologically, the study area is composed of the Ambon Volcano Formation, and this rock unit consists of volcanic breccia lava, tuff breccia, and tuff [5,7].

Figure 1. The location of the research

2.2. The geometry of the creeping plane

Landslide of creep type is a complex natural phenomenon. From a physical point of view, this complex phenomenon needs to be modeled so that it is easy to derive the formula by assuming the creeping mass of the soil is in the form of a slab (slab/soil), following the model shown in Figure 2. From this figure, it shows the geometry of the slope, where at point A is the initial position of a block of soil mass contained potential energy (E_p) starting to move along the slope, then the potential energy is transformed into kinetic energy (E_k), and during the movement, there is energy lost due to friction (E_r). Energy losses along the soil crawl depend on different mechanisms, but this energy loss is assumed to be from friction not energy losses along the slip plane [5,8,9]. Therefore, by following the law of conservation of energy [1], then in the area of height $H_p(x)$ and distance x, resulting in:

$$mgH_A = mgH_p(x) + \int_{x_A}^{x} \mu mg\cos\beta(x)d\ell + \tfrac{1}{2}mv^2 \tag{1}$$

Where $E_{P_A} = mgH_A$ = potential energy of slip block at elevation A, m = mass of slip block (kg), μ = coefficient of friction, β = slope angle, g = acceleration due to gravity, $E_{P_P} = mgH_P(x)$ = potential energy of slip block at elevation P as a function of distance x, $E_K = \frac{1}{2}mv^2$ = kinetic energy of slip block along A to P, and E_r = energy lost during the movement due to friction. E_r can be calculated from the following Eq.(1):

$$E_r = gm\tan\phi_P(x - x_A) \tag{2}$$

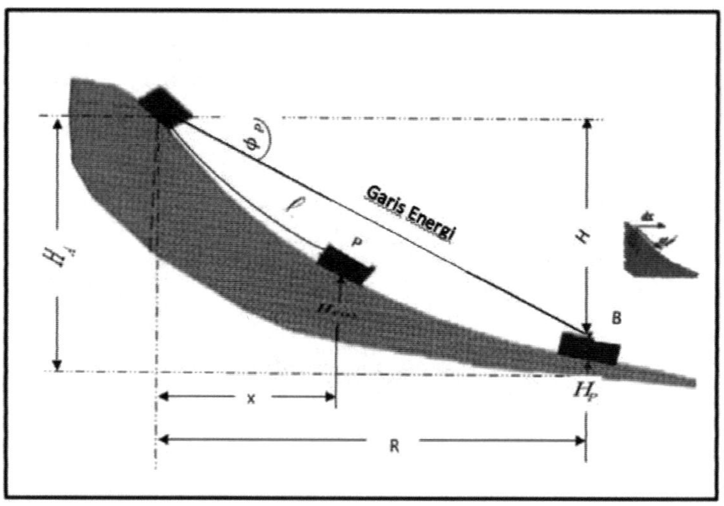

Figure 2. Slope geometry of the block mass model

Where ϕ_p = apparent friction angle. Eq. (1) can be derived a new equation obtained from the energy line equation [1,5], resulting in:

$$H_A - H_p(x) = \frac{v^2}{2g} + \tan\phi_p(x - x_A) \tag{3}$$

2.3. Estimation of the coefficient of friction of the soil creep
Based on Figure 2, if the crawling mass of the soil moves from the initial position A to the final position B under the influence of gravity following the slope, then $E_k = 0$ with $v = 0$. Thus, the change in potential energy from point A to point B must be equal to the work done by the frictional force along the path. So based on the law of conservation of energy on the d element [1], the coefficient of friction is:

$$\mu = \tan\phi_p \equiv \frac{H}{R} \equiv \frac{H}{x - x_0} \tag{4}$$

2.4. Estimation of velocity and time of runout creep
To explain the formulation of the mass movement of the soil crawler, using Figure 2 and Eq.(3). The determine the creeping velocity and time of the creeping runout, it is first assumed that the creeping material is in the form of slip-blocks. This calculation uses two approaches as follows:

- Slopes are dry, or there is no rain.
- Slopes are wet or inhabited by rainwater.

3. Result and Discussion
This section describes the estimation of the dynamics of soil creep, the estimation of the velocity and time of the soil creeping, and the three-dimensional (3D) profile of the estimation of the velocity and time of the creep runout.

3.1. Estimation of the dynamics of soil creeping
The calculation of the estimated parameters of the crawling dynamics using the geometry model of the soil crawling is shown in Figure 3. Figure 3 describes the redesign of the slope geometry model of the soil creeping by involving physical parameters. The meeting with the topographic surface will give the value of elevation or height (H) and the distance of the crawling range (R). The results of the observations found that the geometry of the slip plane was in the form of a translation with the type of soil movement being a creeping avalanche, according to the classification of the type of ground motion proposed by Varnes (1989) in [5].

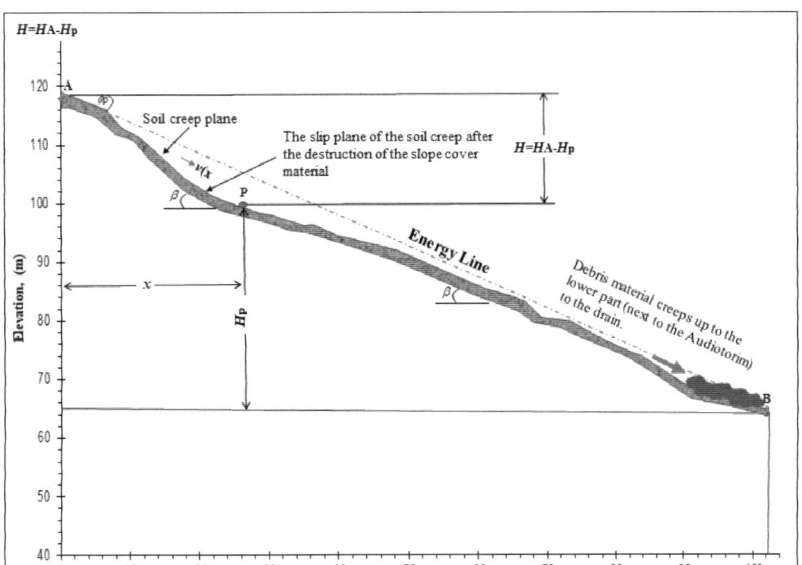

Figure 3. The geometry of the creeping plane

3.1.1. The formula for the velocity and time of the creeping reach of the soil in the condition of overhauling the slope cover soil that has not been inhabited by rainwater

Figure 3, point A is the initial position of a rock mass block that has been crushed and contains potential energy. When a block of rock mass begins to move along the slope, the potential energy is transformed into kinetic energy, and during the movement, there is energy lost due to friction. Therefore, the law of conservation of energy approach (Eq. (3)) is used to derive the equation model for the estimation of the creep velocity of the soil [1], $v(x)$ based on the slope angle on the first path ($v_1(x)$) , and the topographic elevation $H_P(x)$ for the second pass ($v_2(x)$), yielding, respectively:

$$v_1(x) = \sqrt{2g\,x(\tan\beta - \mu)} \tag{5}$$

$$v_2(x) = \sqrt{2g}\,\sqrt{(H_A - H_P(x)) - x\tan\phi_P} \tag{6}$$

where v(x) is the mass velocity of soil crawling or slip-block (m/s), $H(x) = H_A - H_P(x)$ is the change in elevation (m), $x.\tan\phi_P$ = energy line height (m), h = height difference between topography and energy line, (m), x = mass distance of soil creeping at point P (m), x_A = mass distance of soil creeping at point A (m), β = slope angle (°) and ϕ_P = coefficient of apparent friction (°).

During the slide-beam slide in the slope destruction mechanism due to soil creep in dry conditions or not yet inhabited by rainwater, the formula for total reach time is

$$\Delta t = \sqrt{\frac{R}{g\cos\beta\,(\sin\beta - \cos\beta\tan\phi)}} \tag{7}$$

3.1.2. Formulation of the speed and time of the reach of soil creeping in conditions of the destruction of the slope cover soil that has been inhabited by rainwater

In this case, the rainwater inhabits the slope, and the soil-rock mass (slide-block) is in the rainwater. Then the landslide mass moves and slides and settles at the foot of the slope. During mass movement, the avalanche experiences turbulence so that the velocity is low. To form the velocity formula, we modify Eq.(6) with the assumption that the thickness of the rigid avalanche mass is constant (λ, m), while drag and air resistance are neglected. So, using Voellmy's rheology [10], can be determined the beam-slide velocity as a function of horizontal distance [5]:

$$v_2(x) = \sqrt{2g\frac{H_A - H_p(x) - \mu x}{1 + 2x/\lambda\psi}} \tag{8}$$

Where λ = thickness of the avalanche mass, (m), and $\psi = 1/nd^{\frac{1}{3}}$ = is the turbulence parameter similar to the Chézy coefficient, n = Manning roughness coefficient (taken n = 0.035 for the type of debris) and d = depth of drag of soil crawling [5]. When these sliding blocks are moving at a constant speed, then dv/dt = 0, giving the total soil crawl time, t(s), is

$$\Delta t = \frac{R}{\cos\beta\sqrt{g\lambda\psi\,(\sin\beta - \cos\beta\tan\phi)}} \tag{9}$$

3.2. The estimation of velocity and time runout creeping

This section describes the estimation of the velocity and time of soil crawling in dry conditions (dry season) and wet conditions (rainy season).

3.2.1. The estimated velocity and time of soil creeping in dry conditions

From the estimation of the velocity and time of the crawling range based on two possibilities, this is done based on Figure 3:

- The first possibility is to calculate the soil creep along the path by taking the center of mass of the soil crawl as a representative, namely at the slope angle, $\beta = 41.7°$.
- The second possibility is taken on the creeping part with the angle of inclination, $\beta = 37.7°$.

The calculation of the crawling velocity in the study area, starting with the first trajectory as the initial position at the point $x_A = 0$ then applying Eq. (5), gives the estimation results of the velocity calculation at two different slope angles:

$$v_1(x) = \begin{cases} 1.58\sqrt{x} & \text{for} \quad \beta = 41.7^0 \\ 0.43\sqrt{x} & \text{for} \quad \beta = 37.7^0 \end{cases}$$

(10)

Where $\mu = 0.76$ is the coefficient of friction. As for the second trajectory, it is calculated using Eq. (6) so that the estimated speed of soil crawling as a function of the distance of the crawling distance is obtained

$$v_2(x) = \begin{cases} \sqrt{734.41 - 7.20x} & \text{for} \quad \beta = 41.7^0 \\ \sqrt{12.18 - 0.12x} & \text{for} \quad \beta = 37.7^0 \end{cases} \tag{11}$$

The results of the analysis of the soil creep rate v(x) and the time of the range (Figure 4), Δt is shown in Figure 5. The infographic shows that at the slope break at x = 76.0 meters and a height of 14.0 meters above sea level (masl) at the slope angle β = 41.7°, the maximum velocity of soil creeping at the research site is 13.7 m/s, and the maximum time of creeping is 14.9 seconds. While at β = 37.7° at x = 41.5 meters and a height of 55.0 masl, then at the slope break of the maximum velocity of soil crawling is 2.7 m/s, and the maximum reach time of soil crawling is 35.9 seconds.

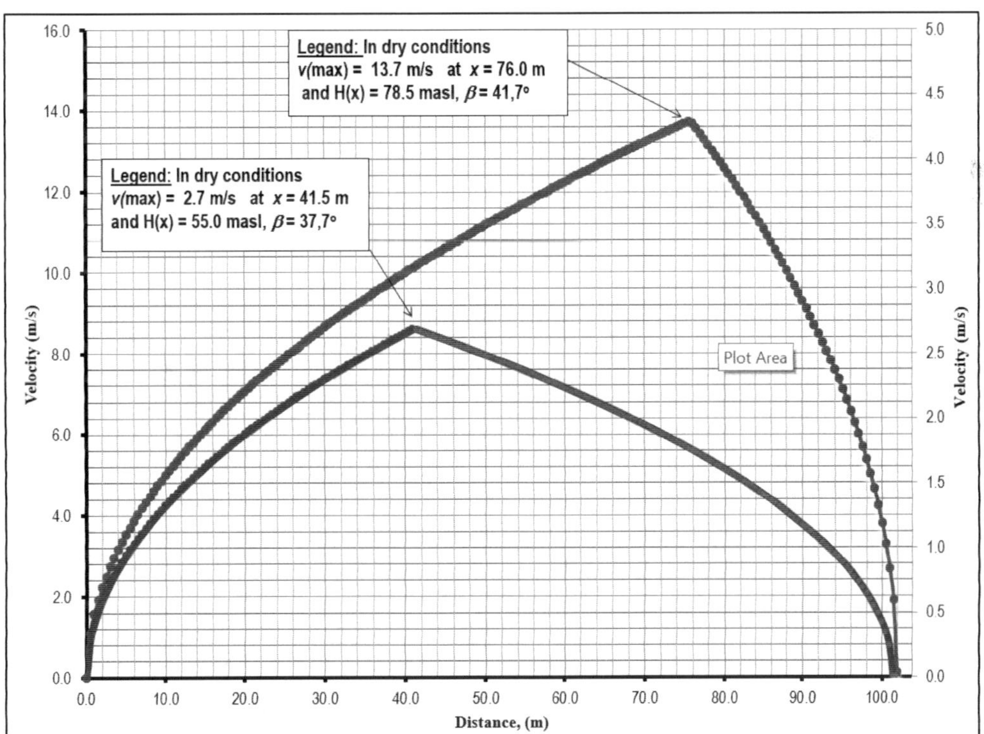

Figure 4. The creep velocity profile of the distance function in dry conditions at the angle

of the slop, β = 41,7°, and β = 37,7°.

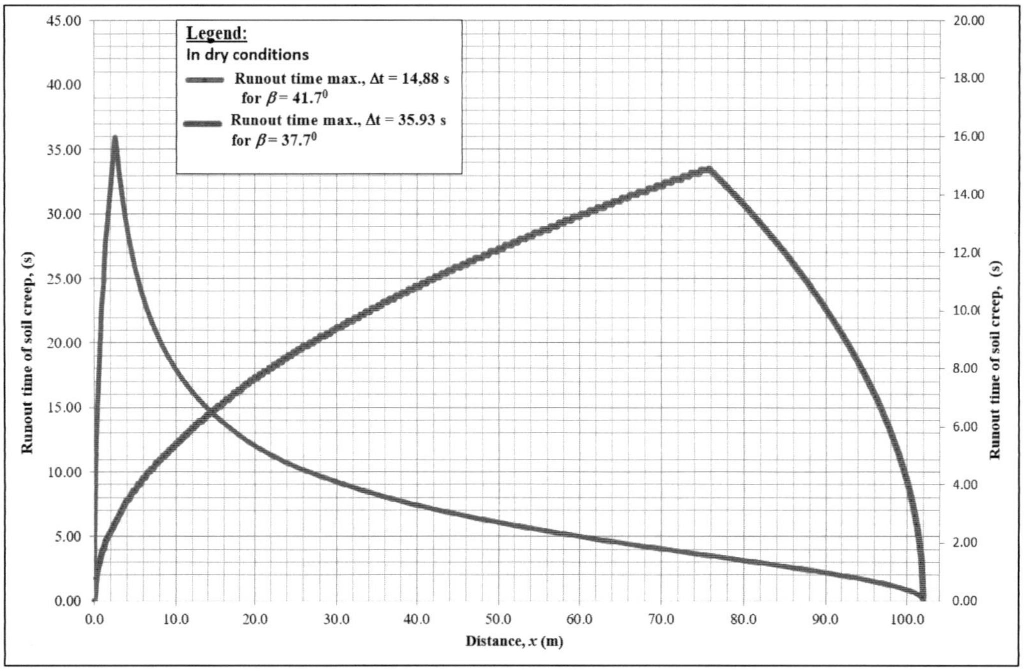

Figure 5. The runout time of the soil is a function of distance under dry conditions.

3.2.2. The estimation of the velocity and time of the creeping reach of the soil in wet conditions
Figure 3 is used to estimate the velocity and time of soil crawling in dry conditions. To determine the estimated velocity and time of the subterranean crawling, first the first path using Eq.(10) is used. As for the second line, the calculation uses Eq.(8), where the slope in wet conditions has resistance in the form of turbulence, while drag resistance and air resistance are ignored. In this condition, the equation for the estimation of the soil crawling velocity as a function of the distance range is

$$v_2(x) = \begin{cases} \sqrt{\dfrac{734.41 - 7.20x}{1 + 0.07x}} & for \quad \beta = 41.7^0 \\ \sqrt{\dfrac{12.18 - 0.12x}{1 + 0.07x}} & for \quad \beta = 37.7^0 \end{cases} \tag{12}$$

The results of the analysis of the soil creep rate $v(x)$ according to equations (10) and (12) as a function of the distance range are shown in Figures 6. The infographic shows the slope break at $x = 2.5$ meters and a height of 14.0 masl for $\beta = 41.7°$, then the maximum velocity of soil crawling is 2.6 m/s, and the maximum reach time of soil crawling is 2.7 seconds (Figure 7). As for $\beta = 37.7°$ at $x = 1.5$ meters and a height of 55.0 masl, then at the slope break position, the maximum velocity of soil crawling is 0.5 m/s and the maximum reach time of soil crawling is 7.6 seconds (Figure 7).

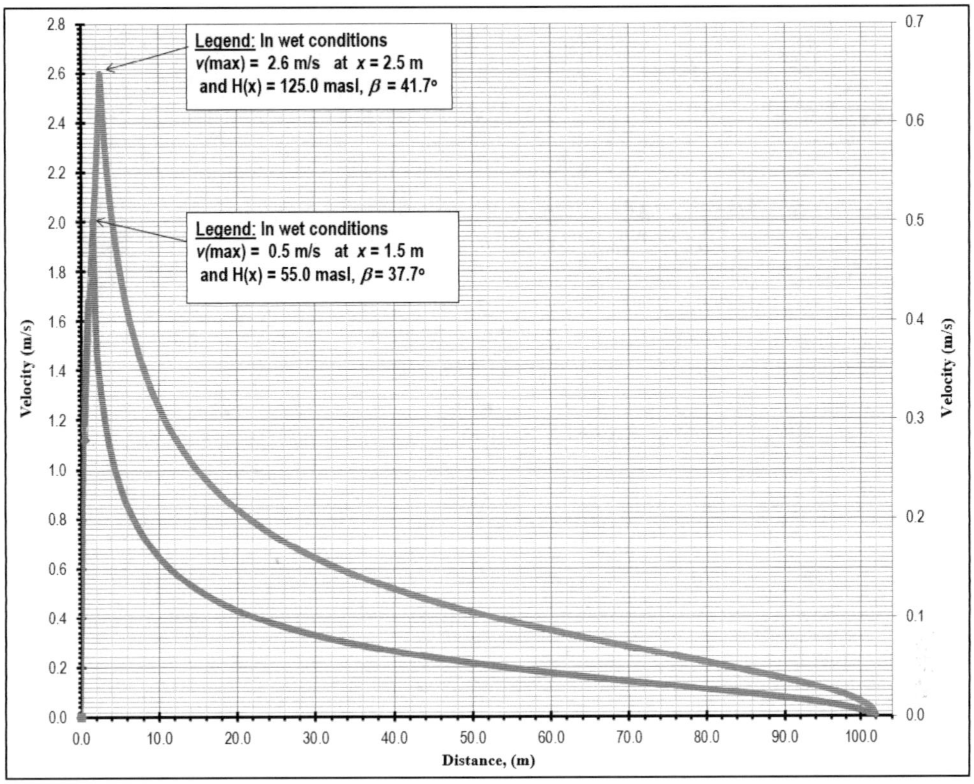

Figure 6. The creep velocity of the soil as a function of distance in wet conditions at the angle of the slope, $\beta = 41,7^{o}$, and $\beta = 37,7^{o}$.

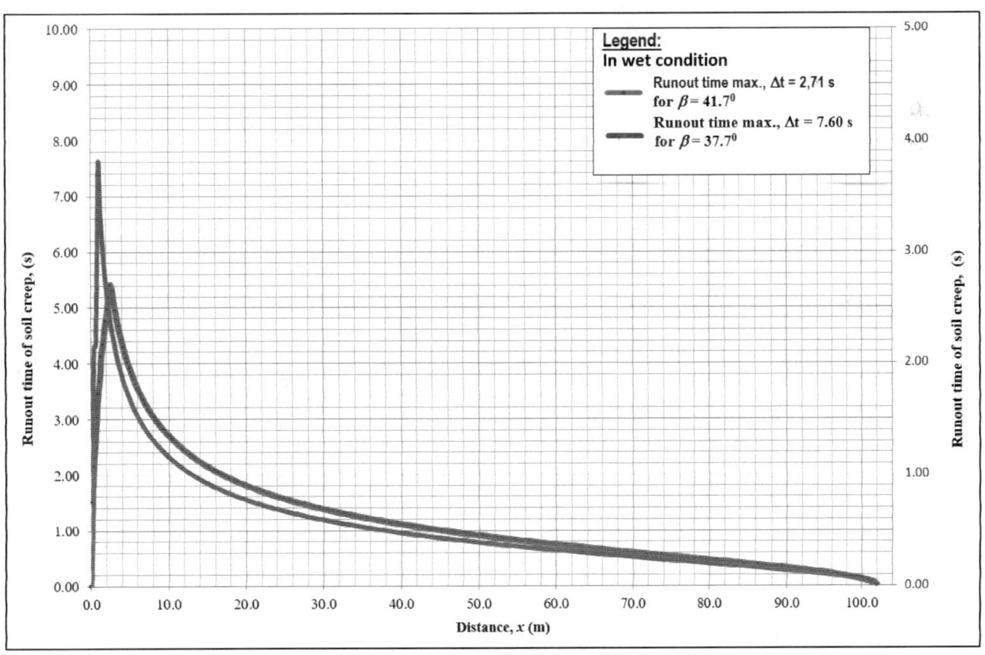

Figure 7. Time of creeping distance function of soil in wet conditions

The estimation of crawling velocity and time in the study area, only based on the geometry of the slip plane. The situation is influenced by gravity, and the Coulomb friction force acting on the soil crawling material or sliding blocks in dry conditions (not yet inhabited by rainwater). The action of these forces is used to derive the law of conservation of energy, namely potential energy and translational kinetic energy.

3.3. Profile of estimated velocity and time of subterranean creeping

For dry conditions, the 3D cross-section is shown in Figure 8 and Figure 9. The infographic on the 3D section provides connectivity between the velocity parameters as a function of the distance of the creeping range and the surface elevation of the soil creeping plane. This image shows topographical contours, showing the position of the soil creep with high to low elevation and velocity following the range according to the soil crawler plane.

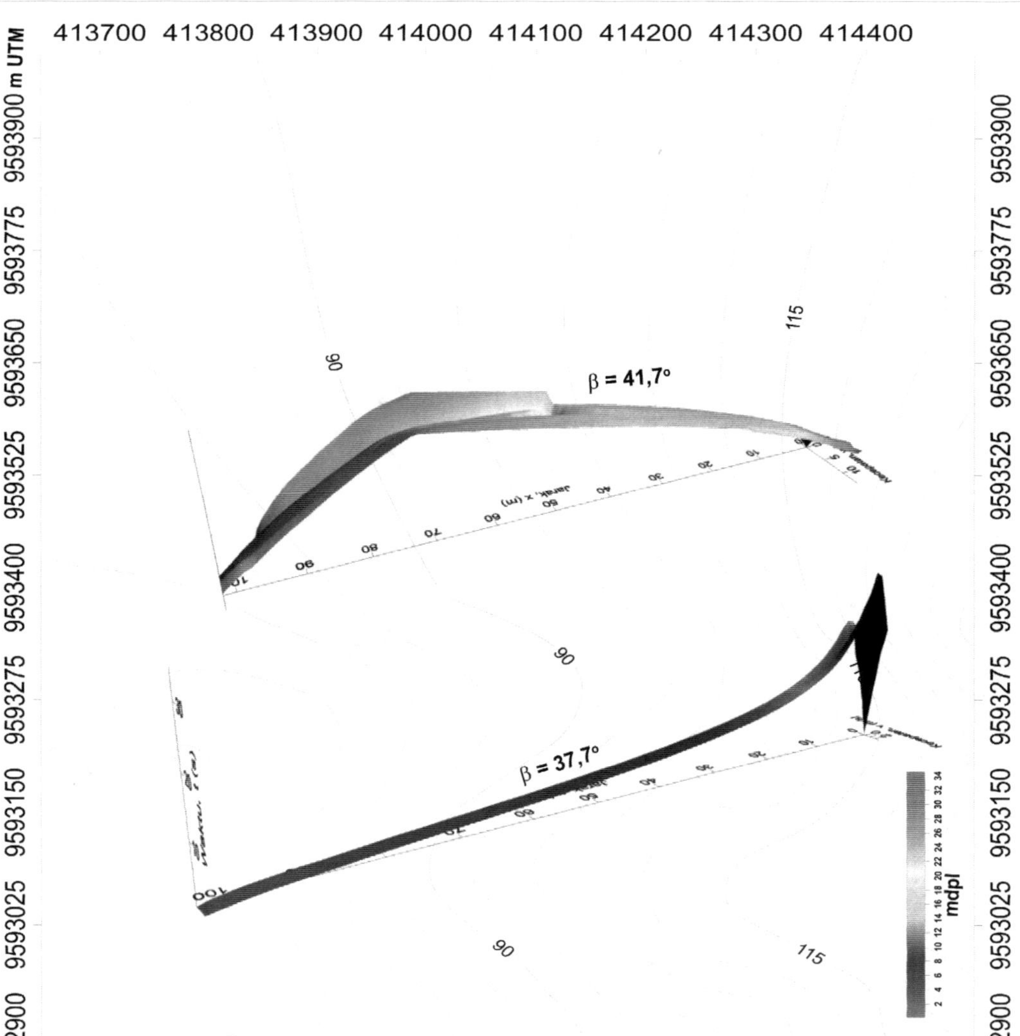

Figure 8. 3D profile of soil crawling time as a function of speed and range in dry conditions.

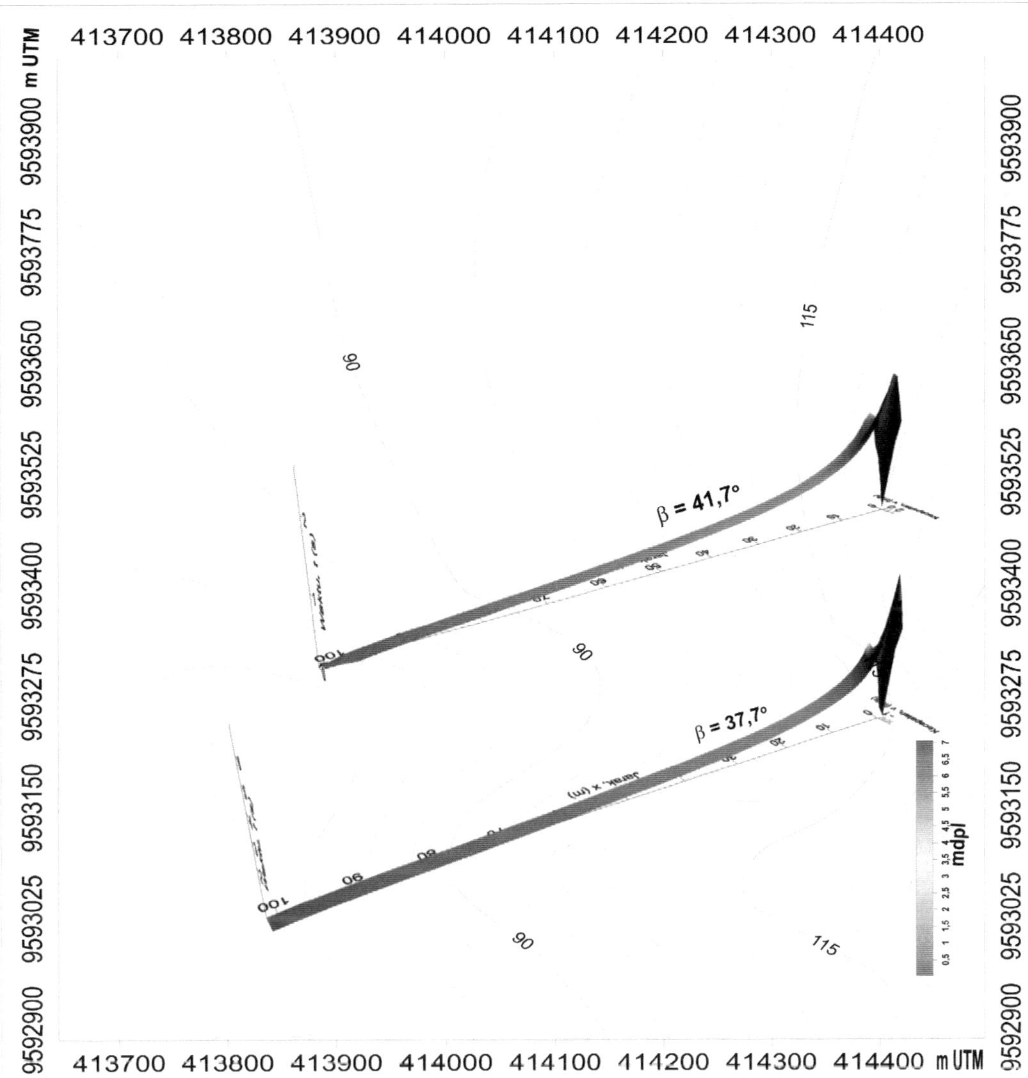

Figure 9. 3D profile of velocity as a function of distance in wet conditions

Figures 8 and 9 show the destruction of slope cover material (soil, rock, sediment, and vegetation) sliding and settling at the foot of the slope (to the right of the IAIN Campus Auditorium). This is caused by the surface cohesion and coefficient of friction decreases while the porosity increases when the soil-rock crushing occurs The factors causing the occurrence of soil creep are viewed from the avalanche physics parameters such as cohesion, porosity, and apparent friction angle or coefficient of friction. Therefore, the resisting force is less than the driving force, so the slope is unstable.

The accuracy of the estimation of the velocity and time of the crawling range depends on the measurement of the geometry of the soil crawling plane [11], the determination of the position of the former mass of the soil creep, and the determination of the apparent friction angle [1, 2]. Therefore, the results of this study can be used as a slope evaluation, and estimate the behavior of soil termites for the assessment of the prevention and mitigation of landslides in landslide-prone areas and are needed to study the mechanisms that determine the loss of energy strength after landslides soil creeping. The incidence of soil creep can certainly be handled with engineering prevention (but it cannot be permanent). This engineering effort needs to start with engineering geological investigations to delineate zones that are prone to crawling and investigate morphological conditions around the termite location so that an analysis can be carried out to take into account the morphological influence on drainage or surface and subsurface water flow that controls the termite movement. So how to design a drainage system combined with a strengthening system and vegetation planting (bio-engineering system) at the termite site. Mitigation by engineering means must be able to reduce or minimize the process of saturation by water in shale or clay. If reinforcement engineering is chosen, the reinforcement construction must be made to penetrate the strong rock layer under the clay or shale layer [12]. However, efforts to control these termites are very complex to be carried out permanently unless applying high-cost technology, this is considered uneconomical.

In general, the landslides that occur on the IAIN campus do not directly kill humans, but damaged buildings and land morphology up to several hectares. So even though it has never caused human casualties, it is very detrimental in terms of economy and comfort of living. Therefore, the IAIN campus building in the termite location should be relocated to a safer place, namely to a flat area that does not contain shale. In handling it, it is necessary to make efforts to reorganize the area by considering several geological constraints (such as the potential for termites and landslides), and supported by the enforcement of local wisdom regulations and government regulations. That is if a zone has been determined to become a sloping boundary because it is prone to crawling, landslides, and river boundaries, it is necessary to establish a building permit regulation in the boundary zone.

4. Conclusion

Based on the results of the interpretation and discussion that have been put forward, the following conclusions are:

1. An analysis of the mechanics of soil motion has been carried out to reduce the movement mechanism of the overhaul of the slope cover material in the form of creep velocity as a function of distance (Eq. 8) and the time of creeping coverage of the soil as a function of distance (Eq. 9).
2. The results of the calculation of the estimated maximum velocity of subterranean termites in dry slope conditions are 13.7 m/s with a maximum creeping time of 14.9 seconds at a slope angle = 41.7°, and for a slope angle of 37.7°, the maximum velocity of subterranean crawling obtained at 2.7 m/s with a maximum creeping time of 35.9 seconds. While the results of the calculation of the estimated maximum velocity of subterranean termites in wet slope conditions are 2.6 m/s with a maximum creeping time of 2.7 seconds at a slope angle = 41.7, and for a slope angle of 37.7, the maximum velocity of subterranean crawling obtained at 0.5 m/s with a maximum creeping time of 7.6 seconds. The movement of crushing soil crawling material in wet slope conditions slides down the slope faster than in left slope conditions.

Acknowledgement

The author would like to thank the Rector of IAIN who has permitted the research. The same thanks to the Earth Movement and Mitigation Laboratory of the Department of Physics, Faculty of Mathematics and Natural Sciences, Pattimura University who has prepared equipment for field data acquisition, and students who have assisted in the data acquisition process.

References

[1] Souisa M, Hendrajaya L and Handayani G 2018 Study on Estimates of Travel Distance, Velocity and Potential Volume of Amahusu Sliding Plane using Energy Conservation Approach in Conjunction with Geoelectric Survey *Journal of Mathematical and Fundamental Sciences* **50**(2) 166-181.

[2] Souisa M, Sapulete S M dan Souisa S O 2019 Estimasi Kecepatan dan Panjang Larian Longsor Berdasarkan Pendekatan Longsor Elm *Barekeng (Jurnal Ilmu Pengetahuan dan Terapan)*, **13**(1) 063-060.

[3] Zhang Z Q, Wu Q M, Zhang Q and Gong Z C 2013 *Estimation of Rock Mass Rating System with an Artificial Neural Network. In H. H. Wen Yu (Ed.)* (Advances in Neural Networks, Wuhan, China: Springer Berlin Heidelberg).

[4] Pudasaini S P and Miller S A 2013 The hypermobility of huge landslides and avalanches, *Engineering Geology* **157** 124-132.
[5] Souisa M 2018 Physics model movement of cover steep slopes in Ambon Island Moluccas, *Dissertation*, Institut Teknologi Bandung (Bandung), pp 1-2, 87.
[6] Souisa M, Sapulete S M and Tuarissa J C 2020 Back Calculation of Slope Stability in Batu Gadjah, Ambon City Using the Solution of Analytical and Finite Element Method Journal of *Physics: Conf. Series* **1463** 1-10.
[7] Tjokrosapoetro S, Rusmana E dan Suharsono 1994 *Geologi* Lembar Ambon, Maluku (Geological Map of the Ambon Sheet, Maluku), Lembar (Sheet) 2612-2613, Skala (Scale) 1:250.000. *Direktorat Jenderal Geologi dan Sumberdaya Mineral*, Pusat Penelitian dan Pengembangan Geologi, Republik Indonesia, pp 3-11.
[8] Jaboyedoff M and Labiouse V 2011 Technical Note : *Preliminary Estimation of Rockfall Run-out Zones*, pp 820-828.
[9] Blasio, D F V 2011 *Introduction to the Physics of Landslides* (New York: Springer Dordrecht Heidelberg London).
[10] Hungr O, Corominas J and Eberhardt E 2005 Estimating landslide motion mechanism, travel distance and velocity *Proceedings of the International Conference on Landslide Risk Management*, Vancouver, Canada, May-June, pp 99-128.
[11] Iverson, R M 2014 Debris flows: behaviour and hazard assessment *Geology Today* **30** 15-20.
[12] Sassa K 2013 International Programme On Landslides *Landslides: Global Risk Preparedness*, **1**(2) 3-24.

The effectiveness of exposure to magnetic fields of extremely low frequency 300T and 500T in inhibiting the proliferation of pathogenic bacteria to increase physical resistance of vannamei shrimp

Sudarti[1], U Qumairoh[1] and T Prihandono[1]

[1]Phiysics Education, Jember University, Indonesia

E-mail: sudarti.fkip@unej.ac.id

Abstract. Vannamei shrimp is a type of fish that rots easily, only able to survive about 6 hours at room temperature. This study aims to examine the effectiveness of exposure to an Extremely Low Frequency (ELF) magnetic field with an intensity of 300 T and 500 T in inhibiting the proliferation of pathogenic bacteria to increase the physical resistance of white vaname shrimp. The research design used a completely randomized design and was carried out at the Physics Laboratory of the Faculty of Teacher Training and Education, University of Jember. The Extremely Low Frequency (ELF) magnetic field in this study is a component of the ELF electromagnetic wave generated by the Current Transformer by minimizing the exposure to the ELF electric field to about 30 V/m. The study sample was 210 fresh vaname shrimp and divided into 7 treatment groups, namely the control group (K), the group exposed to the ELF magnetic field intensity of 300 T for 60 minutes (E-300.60'), 90 minutes (E-300.90'), 120 minutes (E-300.120'), and the group exposed to an ELF magnetic field of 500 T intensity for 60 minutes (E-500.60'), 90 minutes (E-500.90'), 120 minutes (E- 500,120'). As an indicator of vaname shrimp resistance at room temperature is the population of pathogenic bacteria, color, aroma, texture, mucus, and pH. The research data were analyzed using One Way Anova analysis. The results showed that the population of Salmonella bacteria was significantly lower than the control in the vannamei shrimp sample exposed to an ELF magnetic field of 500 T, but the population of E.Coli bacteria increased significantly in the vannamei shrimp sample exposed to an ELF magnetic field of 300 T and 500 T compared to control. Naturally, vaname shrimp stored at room temperature showed the appearance of the shrimp getting paler, and the texture getting softer, smelling rotten and slimy after the 6th hour. However, the color, smell, and texture of shrimp exposed to an ELF magnetic field of 500 T intensity for 120 minutes were able to survive at a score of 8 at the 5th hour observation and 7th at the 10th hour observation. Conclusion: Exposure to an ELF magnetic field with an intensity of 500 T was effective in suppressing the proliferation of salmonella Sp bacteria but was less effective in suppressing the proliferation of E.Coli bacteria. This condition affects the physical condition of white shrimp until it lasts at the 10th hour. The results of this study serve as the basis for developing food security technology by the ELF magnetic field.

1. Introduction

Vannamei shrimp is a type of fish that is easy to root, it only able to survive about 6 hours at room temperature. Until now, there are still many who add chemicals to maintain shrimp at room temperature. This condition will endanger the health of the people who consume it. Therefore, alternative solutions are needed to increase the resistance of vanamei shrimp that are safe for health.

Several research results report that exposure to magnetic fields Extremely Low Frequency (ELF) has the potential to increase food endurance. The ELF magnetic field is a component of ELF electromagnetic waves that have the ability to penetrate almost all materials. The interaction of the ELF magnetic field with cells is able to change the cell membrane potential, so that it will have an impact on biophysical mechanisms in cells [1]. This mechanism underlies the appearence of biological effects on living things including bacteria which are exposed to the ELF magnetic field. Several research results on the influence of the ELF magnetic field are reported. [2], the effective dose on the speed of proliferation of S. Thermophilus, L. Lactis, and L. Acidhopilusis bacteria in the process of making cream cheese is at an intensity of 5 minutes. Furthermore, the intensity of 18 mT and 20 mT had an effect on a significant decrease in the survival rate of E. coli bacteria in the urine of patients with urinary tract infections [3]. [4] , showed that the pH value in the 100 T ELF magnetic field group for 5 minutes and 20 minutes was lower than the control group in the process of making cream cheese. Then exposure to a 250 T ELF magnetic field significantly affected the growth rate of bacterial cultures [5]. Then [6], proved that the production of biomass and mannatide increased significantly in samples of -hemolytic Streptococcus in fermentation culture media exposed to magnetic fields of 400 T, 600 T, and 900 T and decreased in samples exposed to 1200 and 1500. T. Then the 100 T (50 Hz) magnetic field did not damage the DNA and therefore was not mutagenic in this test and also had no effect on the DNA damage capacity of the mutagen used [7]. Based on the results of the study, it was proven that exposure to the ELF magnetic field has the potential to inhibit the proliferation of pathogenic bacteria so that it also has the potential to increase food endurance. Therefore, this study aimed to analyze the physical resistance of white vaname shrimp due to exposure to magnetic field Extremely Low Frequency (ELF) with an intensity of 300 T and 500 T.

The next study using a different object, it used the object of shrimp meat with the type of vaname shrimp. Vannamei shrimp has advantages compared to other fishery products, namely shrimp contains a high source of protein, causing high nutritional value in shrimp. Despite the advantages, shrimp resistance is limited to only 6 hours at room temperature, causing shrimp to quickly deteriorate. Quality deterioration in shrimp is closely related to microbes decomposition and melanosis or blackspot. Pathogenic bacteria are the main factors causing the shrimp decomposition process. Pathogenic bacteria in vaname shrimp include Pseudomonas, Alcaligenes, and Escherichia coli, Salmonella. The growth of microorganisms can be affected by pH conditions. This is because each microorganism has a different pH range. The higher the pH value, the faster the food will decompose. The optimum degree of acidity for bacterial growth is between 4.6 – 7.0. While the pH value that is close to neutral is a pH that is suitable for good microbial growth. In conditions of $pH < 4$ or $pH > 11$ will experience death. Due to the activity of bacteria that can cause damage to food, the food must be treated with preservation.

Based on the results of previous researchers, it can provide a way to inhibit the decomposition of shrimp meat due to harmful bacteria. The results of previous studies showed that magnetic fields can change the reproduction and growth of acid-forming microorganisms. Thus, it can inhibit the activity of acid-forming bacteria in a solution so that it will cause the acidity to decrease slowly. If the pH decreases slowly, it will have an impact on the meat, which will make the shrimp meat not rot quickly.

The physical endurance of the shrimp can be seen through the density value. Density is the quantity of the concentration of a substance expressed in terms of mass per unit volume. In shrimp meat, volume is closely related to the amount of water content contained in the shrimp. The process of food decomposition will be easier to occur if the water content in the food is increasing. The water content in the shrimp is 70-80% and the water activity is high. This will allow condensation and will experience bacterial growth. Therefore, based on the description above, the researchers tried to conduct further research on the use of magnetic fields in the food sector, especially in shrimp preservation. The intensity used is and the duration of exposure used is 60 minutes, 90 minutes, and 120 minutes.

2. Methods

This research uses experimental research. Experimental research is comparing the effect between the experimental group, namely the group that was treated in the form of exposure to the ELF magnetic field and for 60 minutes, 90 minutes, and 120 minutes with the control group, namely the untreated group that was left exposed at room temperature. And using a completely randomized design (CRD). This research was conducted at the Advanced Physics Laboratory in the Physics Education at the Educational Sciences study program building, University of Jember in December 2020. In this study, the sample used was shrimp with good quality obtained directly from the harvest in the Muncing pond, Muncar, Banyuwangi. Determination of the sample in this study was done randomly, both the experimental group and the control group. There were 5 sample packs for the control group and 24 packs for the experimental group sample.

The tools used in this research are: magnetic field source in the Extremely Low Frequency form of CT (Current Transformer) as a generator of magnetic field Extremely Low Frequency, EMF tester type Lutron EMF-827 to measure the magnitude of the magnetic field, pH meter is used as a measuring device for pH values, pestle and mortar were used to smooth the sample and the cup was used as a medium for bacterial agar. While the materials used are: aquadest liquid used to dissolve the plastic clip sample used as a sample wrapper, a digital balance to weigh the mass of the shrimp sample, and a measuring cup as a container to see changes in the volume of shrimp.

The procedure carried out in this study is the first to prepare the tools and materials used. The experimental group was divided into two groups: group I was exposed to an ELF magnetic field of 300 T, and group II was exposed to an ELF magnetic field of 500 T. Each group consists of 12 packs of shrimp. Each shrimp consists of 5 shrimp of the same size. Then give the treatment by exposing the ELF magnetic field for 60 minutes, 90 minutes, and 120 minutes. While the control group will be left outside the room without sunlight. For measurements of density, physical conditions and pH were observed at 0 hours (before exposure), 5 hours, 10 hours, 15 hours, and 20 hours after exposure and each study was repeated 3 times. Then for the measurement of density, namely calculating the mass of shrimp as much as 2 grams, then measuring the final volume of shrimp by adding shrimp (2 grams) into 10 ml of water in a measuring cup. Then note the change in volume. Then measure the pH of the shrimp using a pH meter. The pH meter was calibrated using a buffer solution before ready to use. Then to observe the physical endurance of shrimp by using the human senses, namely the sense of smell, the sense of sight, and the sense of touch to observe the appearance, texture, and smell indicators of the shrimp. and observing the growth of Eschercia Coli bacteria by planting with Eosin Methylene Blue Agar (EMBA) media. While observing the growth of Eschercia Coli bacteria by planting withmedia Salmonella & Shigella Agar (SSA). Observations of Eschercia Coli and Salmonella sp. vanamei shrimp were carried out at 0 hours (before exposure) and after exposure at 24 hours, 26 hours, and 28 hours at 10.00 hours of planting. The next stage is the data analysis technique using descriptive data analysis techniques. The physical endurance test of vanname shrimp in this study used human senses using a scale value based on SNI-01-2728-2006 of fresh shrimp that was written on the assessment sheet. In terms of the quality of fresh shrimp in the organoleptic test to get the minimum criteria for fresh shrimp to get a value of 7.

3. Result and Discussion

3.1 Result

3.1.1 Physical endurance (appearance, smell, texture)
Observation of physical endurance (appearance, smell, texture) of white vannamei shrimp at 0 hours (before exposure) and after exposure at 5 hours, 10 hours, 15 hours, and 20th hour. Figure 1 below is a bar chart of shrimp appearance indicators that were observed at the 5th hour, 10th hour, 15th hour, and 20th hour after exposure:

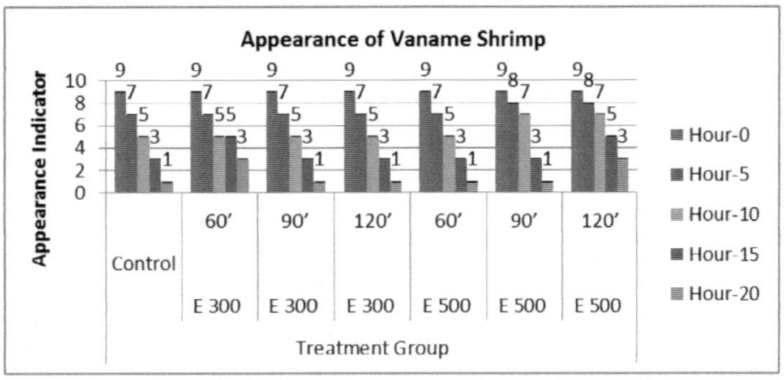

Figure 1. Appearance indicator bar chart

Based on Figure 1, it can be seen that there is a difference in the average value of appearance indicators between the control group and the experimental group. Fresh shrimp has; a clear appearance, original luminous according to species, and solid internodes which have a score of 9. Based on the results of the study, at 0 hour the shrimp were in the f resh category. Then at the 5th hour, the appearance of the shrimp began to change with the specifications intact, less clear, the light began to fade, the original color, and between the segments was solid, but the shrimp was still in the fresh category and safe for consumption. At 10 hours the control group and the experimental group (E300 T 60', E300 T 90', E300 T 120', and E500 T 60') obtained a score of 5 in the rather fresh category. However, the E500 T 90' and E500 T 120' groups have scored 7 in the fresh shrimp category. Then at the 15th hour in the control group and in all experimental groups, the appearance of the shrimp was categorized as not fresh. However, the experimental group (E300 T 60' and E500 T 120') got a higher score. Entering the 20th hour, the shrimp began to turn red and black spots appeared. Shrimp was categorized as not fresh and not proper for consumption. However, only the experimental group (E500 T 120') scored higher than the other groups. In the other group, the red shrimp are very dull.

Figure 2 below is a bar diagram of the shrimp odor indicator observed at the 5th hour, 10th hour, 15th hour, and 20th hour after exposure:

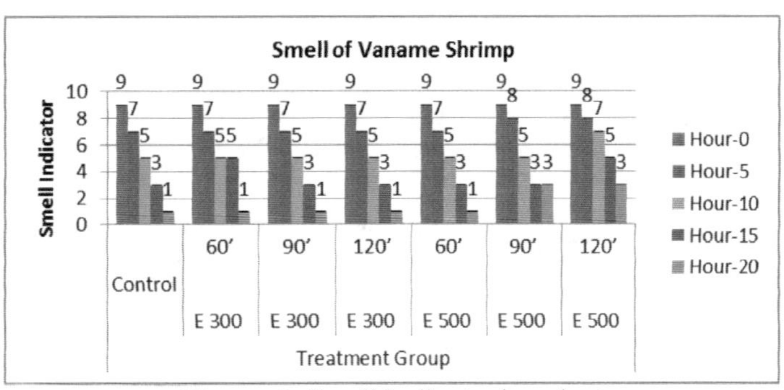

Figure 2. Smell indicator bar chart

Based on Figure 2, it can be seen that there is a difference in the average value of appearance indicators between the control group and the experimental group. The smell indicator value of the control group is smaller than the experimental group. Every hour of observation, the smell indicator value of the control group tends to decrease drastically when compared to the experimental group. Based on the results of the study, the longer the storage of vaname shrimp, the sharper of the smell produced, especially in the control group. Shrimp in the fresh category has a very fresh smell. Based on the results of research that has been done, at 0 hour the shrimp is in the fresh category. Then at the 5th hour, the smell of shrimp began to change to a fresh smell, but the shrimp was still in the fresh category and safe for consumption. At 10 hours the control group and the experimental group (E300 T 60', E300 T 90', E300 T 120', E500 T 60' and E500 T 90') began to grow indole smell with a score of 5 in the slightly fresh category. However, the E500 T 120' group scored 7 in the category of fresh shrimp with a neutral smell specification. Then at 15 hours in the control group and in the experimental group (E300 T 60', E300 T 90', E300 T 120', E500 T 60' and E500 T 90'), the smell of shrimp is in the category of not fresh with a score of 3 and already smells of indole. However, the experimental group (E500 T 120') started to grow an indole odor, obtaining a higher score. Then entering the 20th hour, the smell of indole and the stench of the shrimp got stronger. Shrimp in the category of not fresh and not fit for consumption.

Figure 3 below is a bar chart of shrimp odor indicators that were observed at the 5th hour, 10th hour, 15th hour, and 20th hour after exposure:

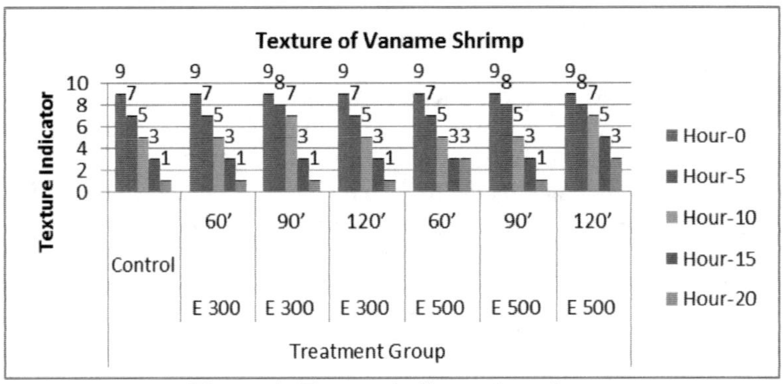

Figure 3. Texture indicator bar chart

Based on Figure 3, it can be seen that there is a difference in the average value of appearance indicators between the control group and the experimental group. The smell indicator value of the control group is smaller than the experimental group. Every hour of observation, the smell indicator value of the control group tends to decrease drastically when compared to the experimental group. Based on the results of the study, the longer the storage of vaname shrimp, the softer the texture of the shrimp, especially in the control group. Shrimp in the fresh category have a very elastic, compact, and dense texture. Based on the results of the study, at hour 0 the shrimp was in the fresh category because it got a score of 9. Then at the 5th hour, the texture of the shrimp began to change from very elastic to elastic. However, shrimp is still in the fresh category and safe for consumption. At the 10th hour the control group and the experimental group (E300 T 60', E300 T 120', E500 T 60' and E500 T 90') the texture of the shrimp became inelastic, not compact, and not dense with a score of 5 in the moderate category. However, the E500 T 120' and E300 T 90' groups scored 7 in the category of fresh shrimp with less elastic, compact, and dense specifications. Then at the 15th hour in the control group and in the experimental group (E300 T 60', E300 T 90', E300 T 120', E500 T 60' and E500 T 90'), the shrimp was categorized as not fresh because the texture of the shrimp became rather soft with a score of 3. The experimental group (E500 T 120') got a higher score with a score of 5 because the shrimp texture was better than that group. Its texture is inelastic, not compact, and not dense. Then entering the 20th hour, the texture of the shrimp is getting softer. Shrimp in the category of not fresh and not proper for consumption. So that the intensity has the highest potential in inhibiting the decrease in physical resistance of vaname shrimp on the appearance, smell, and texture parameters, namely the intensity of 500 for 120 minutes. The following is the physical condition of white vaname shrimp between the control group and the experimental group:

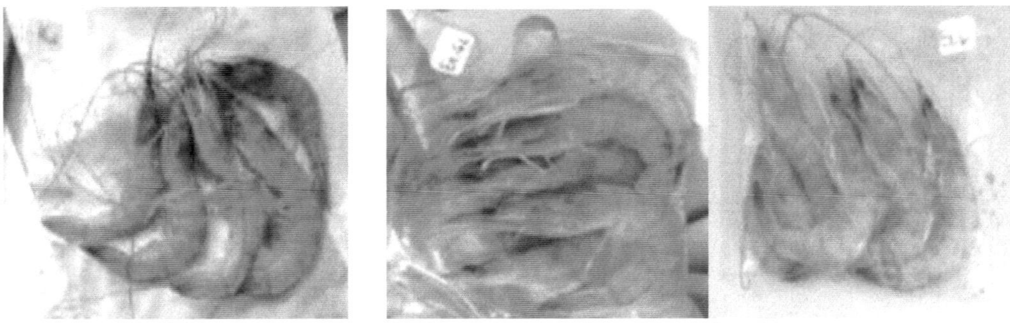

(a) (b)

Figure 4. (a) Physical condition of white vaname shrimp before exposure to the ELF magnet magnetic field, (b) and after exposure to the ELF magnet magnetic field.

3.1.2 The degree of acidity (pH) of Vannamei Shrimp

This study also aims to examine the effect of exposure to ELF fields of 300 T and 500 T in the variants of 60 minutes, 90 minutes, and 120 minutes of exposure time on the acidity degree of Vannamei shrimp. The degree of acidity (pH) is one of the indicators used to chemically observe the freshness of fish. Measurement of the degree of acidity of vaname shrimp was carried out at 0 hours (before exposure) and after exposure at 5 hours, 10 hours, 15 hours, and 20 hours. Measurement of shrimp pH using a pH meter. The pH meter was calibrated using a buffer liquid before to use. The following figure 5 is a bar chart of shrimp pH:

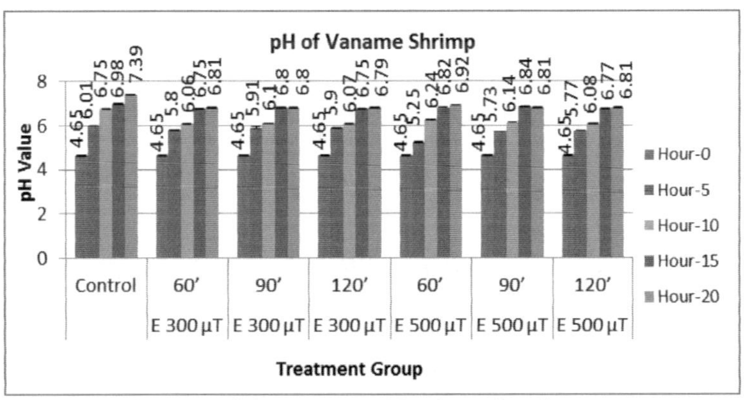

Figure 5. Diagram of the average pH value of each measurement.

Based on Figure 5, it can be seen that there is a difference in pH value between the control group and the experimental group. The pH value in the control group was higher than the experimental group. The longer the storage, the pH value of vaname shrimp will increase and the experimental group experienced a drastic increase in the pH value and the experimental group experienced an increase in the pH value by a fairly small difference. So that the intensity that has the highest potential in maintaining the increase in pH value is the intensity of 500 for 60 minutes at the 5th hour of observation.

3.1.3 Vannamei Shrimp Density

Measurement of vaname shrimp density was carried out at 0 hours (before exposure) and after exposure at 5 hours, 10 hours, 15 hours, and 20 hours. The following figure 6 is a bar chart of the average density of shrimp:

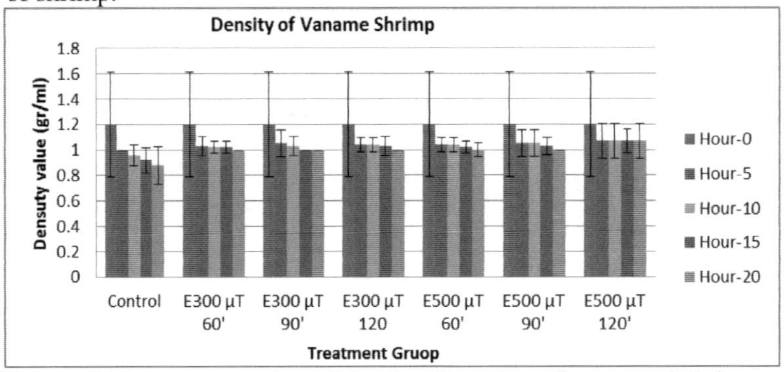

Figure 6. The average value of the density value of vannamei shrimp.

Based on Figure 6, it can be seen that there is a difference in the average value of the density of shrimp in the control group and the experimental group. The value of the density of shrimp in the control group and the experimental group had different density values for each hour of observation. And tends to decrease the density value for every hour of observation. The density value of the control group is smaller than the density value of the experimental group. Every hour of observation, the density of the control group tends to decrease significantly when compared to the average density of the experimental group. So that the intensity with the highest potential in maintaining the decrease in density value is 500 for 120 minutes.

3.1.4 Eschercia coli bacteria

Observation of the white shrimp Eschercia Coli was carried out at 0 hours (before exposure) and after exposure at 24 hours, 26 hours, and 28 hours at 10:00 hours of planting. The following figure 7 is a bar chart of the value of Eschercia Coli bacteria in white shrimp at 10.00 am planting:

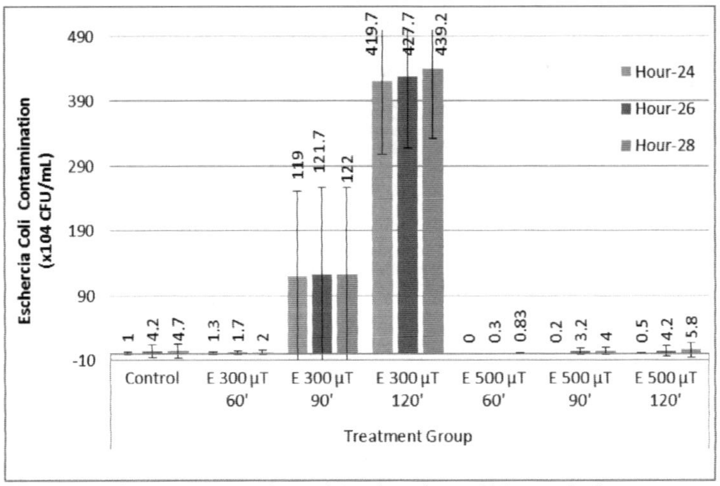

Figure 7. The average value of Eschercia Coli bacteria in vannamei shrimp planting at 10.00.

Based on Figure 7, it can be seen that there is a difference in the average value of Eschercia Coli bacteria in the control group and the experimental group. It can be seen that the control group tends to

Journal of Physics: Conference Series **2165** (2022) 012038 doi:10.1088/1742-6596/2165/1/012038

have a larger population of Eschercia Coli bacteria when placed in the control group. compare with the experimental group E 500 60'. The bacterial population increased for every hour of observation. The bacterial population of the control group was larger than the bacterial population of the experimental group. Every hour of observation, the population of the control group tends to increase significantly when compared to the population of bacteria in the experimental group. So the intensity has the highest potential to maintain the growth of Eschercia Coli bacteria, namely the intensity of 500 T for 60 minutes. Figure 8 below is the population of Eschercia Coli bacteria before and after exposure to the ELF magnetic field:

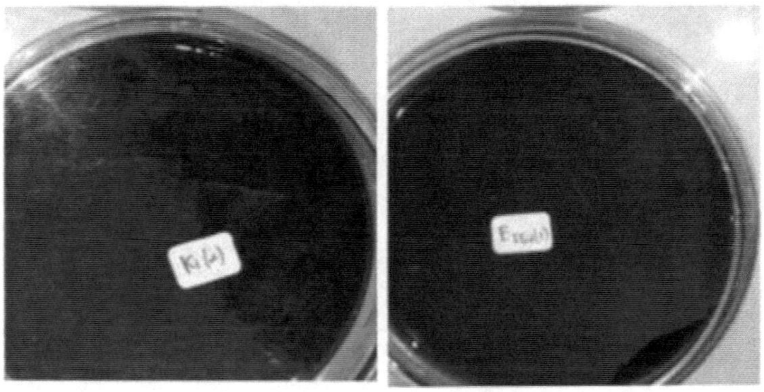

Figure 8. Population of Eschercia Coli bacteria in Vanamei shrimp before (left) and after (right) exposure to the ELF . magnetic field.

3.1.5 Salmonella Sp

Bacteria Observations of Salmonella sp. vannamei shrimp were carried out at 0 hours (before exposure) and after exposure at 24 hours, 26 hours, and 28 hours at 10.00 hours of planting. The following figure 9 is a bar chart of the value of Salmonella sp. of vaname shrimp at 10.00 am planting:

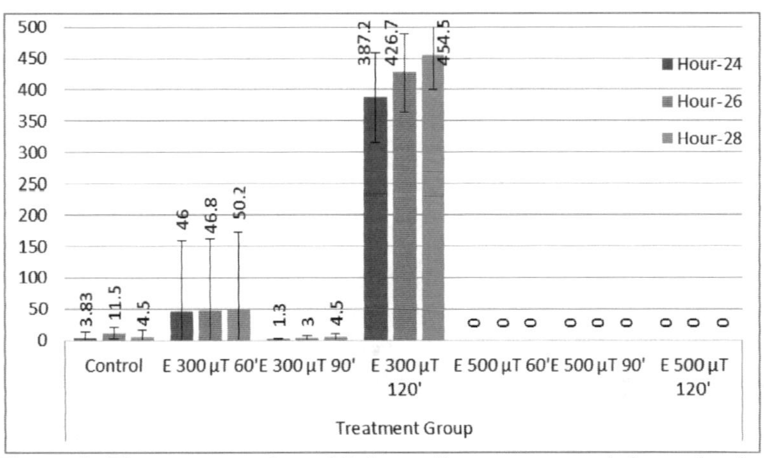

Figure 9. The average value of Salmonella sp. vaname shrimp, planting at 10.00

Based on Figure 9, it can be seen that there is a difference in the average value of Salmonella sp. vannamei bacteria in the control group and the experimental group. It can be seen that the control group tends to have a larger population of Salmonella sp. compare with the experimental group. Especially in the experimental group E 500 60', 90', and 120', the bacterial population did not grow. In the control group and the experimental group E 300 60'. 90', and 120' bacterial populations increased for every hour of observation and the control group population tended to increase significantly when compared to the bacterial population in the experimental group. So that the intensity that has the highest potential in maintaining the growth of Salmonella sp bacteria is the intensity of 500 for 60, 90, and 120 minutes. Figure 10 below is a population of Salmonella sp bacteria before and after exposure to the ELF magnetic field:

(a)

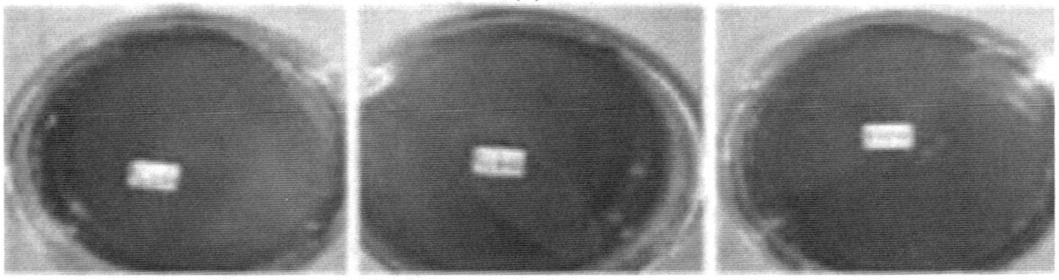

(b)

Figure 10. (a) Population of Salmonella sp bacteria in vanname shrimp before exposure to the ELF magnetic field, (b) Population of Salmonella sp bacteria in vanname shrimp after exposure to the ELF magnetic field.

3.2 Discussion

In this study, it can be seen that the longer the shrimp storage, the physical endurance (appearance, smell, and texture) of vaname shrimp continues to decrease in score. This coincided with an increase in the pH value of the higher shrimp and a decrease in the density of the shrimp. The increase in pH value and the decrease in density were caused by the faster growth of bacteria. Meanwhile, ELF-EMF affects important physicochemical processes in Gram-positive and Gram-negative bacteria [8].

Based on this research, it can be seen that the value of physical endurance indicators of appearance, smell, and texture for the control group and the experimental group has decreased in score each time. However, the decrease in score in the experimental group decreased by a fairly small difference every hour of measurement when compared to the control group which decreased drastically. Based on the appearance indicator, the experimental group (E500 T 120') scored higher than the other groups. In the other group, the red shrimp are very dull. The color change in shrimp meat from clear to red is caused by bacteria *Pseudomonas*, due to the collaboration between the pigments contained in shrimp meat. At the 20th hour of observation, the shrimp also began to grow a lot of black spots (melanosis). Melanosis formation correlates with post mortem tissue degradation caused by protease enzymes associated with the activation of PPO (polyphenoloxidase) enzymes that oxidize phenols to quinones. Quinone polymerization causes the appearance of a black pigment that gives shrimp a blackish color [9]. The spread of melanosis in shrimp was observed on the body part of the cephalothorax, which contains digestive organs that cause spoilage in shrimp [10].

Then on the smell indicator, the experimental group (E500 T 120') got a higher score. Then entering the 20th hour, the smell of indole and the stench of the shrimp got stronger. Shrimp in the category of not fresh and not proper for consumption. [11], states that chemical compounds such as proteins, fats, and carbohydrates which are broken down into ammonia, hydrogen, sulfide, ethyl mercaptan, aldehydes, aldehydes, alcohols, ketones, and carboxylic acid gases can cause a root smell in shrimp. However, at the 20th hour, only the experimental group (E500 T 120') scored higher than the other groups. So it can be seen that the experimental group E500 T 120' can maintain the freshness of shrimp. Likewise, the texture indicator of shrimp entering the 20th hour the texture is getting softer. Shrimp was categorized as not fresh and not proper for consumption. This is because of the work of enzymes, the shrimp meat becomes soft. The quality of the shrimp texture is influenced by the water content of the shrimp in it. Water content has an effect on decomposition, because meat has a high water content (68-75%) which can promote the growth of microorganisms in the meat.

Based on Figure 5, it can be seen that every hour of observation there is a change in the pH value of vaname shrimp. The change in pH value was caused by the presence of pathogenic bacteria in shrimp. After the shrimp die, the bacteria use the shrimp flesh to meet the nutritional needs so that the bacterial population increases rapidly, so that the autolysis process can take place quickly. The faster growth of bacteria can cause decomposition. Then [12], argues that the oscillation of the magnetic field on bacteria can have an impact on changes in pH values. This happens because of the inhibition of acid-forming bacteria. The occurrence of bacterial inhibition is supported by a research which done by [8], he states that the 1 mT ELF magnetic field for 2 hours can cause changes in the physicochemical properties of Gram-positive and Gram-negative bacteria so that bacterial growth slightly decreases.

Based on Figure 6, it can be seen that the density value in the control group is smaller than the experimental group. This is because exposure with the help of an ELF magnetic field can inhibit the growth of microorganisms. That way the water content of shrimp meat does not increase. If the water content does not increase, the volume will not increase either. The rapid proliferation of bacteria will cause the appearance of mucus in the shrimp. And at every hour of measurement, the density value changes. Changes in the value of the density that is getting smaller is caused by the growth of bacteria in the shrimp. The growth of these bacteria will increase the water content in the shrimp so that it affects the density of the shrimp due to changes in the volume of the shrimp. So that the presence of an ELF magnetic field will affect the growth of bacteria constantly so that it will not produce small molecular compounds containing water, then water excretion is reduced, thereby inhibiting spoilage in shrimp. This is based on [13], showing that the intensity of 730.56 T for 60 minutes can reduce the bacterial growth of milkfish up to 62% in the fifth hour after exposure after exposure compared to controls. And the intensity and duration of exposure to the ELF magnetic field for 45 minutes can drastically optimize the proliferation of Lactobacillus in the artificial luwak coffee fermentation process [14]. So this shows that there is an effect of exposure to the ELF magnetic field on physical resistance because it can inhibit bacterial activity so that bacteria will decrease. Because the number of bacteria contained in shrimp affects the physical resistance of shrimp. Meanwhile, the increase in pH value and the decrease in the density of shrimp were caused by the faster growth of bacteria. So that exposure to a magnetic field can suppress the growth of spoilage microorganisms in food. This is based on research by [12], which states that an Extremely Low Frequency (ELF) magnetic field with an intensity of 646.7μT for 30 minutes has been shown to reduce the population of Salmonella typhimurium in Gado-Gado. Then exposure to the ELF-EMF magnetic field can also show a statistically significant decrease in Staphylococcus aureus and Escherichia coli compared to controls [15]. So that shrimp exposed to a magnetic field can extend the shelf life of vaname shrimp. This is due to the inhibition of bacterial growth resulting from exposure to the ELF magnetic field. Based on this research, it can be concluded that the intensity that has a higher potential in maintaining the physical endurance of shrimp is the intensity of 500 T for 120 minutes.

4. Conclusion

The results showed that the population of Salmonella bacteria was significantly lower than the control in the vannamei shrimp sample exposed to an ELF magnetic field of 500 T, but the population of E.Coli bacteria increased significantly in the vannamei shrimp sample exposed to an ELF magnetic field of 300 T and 500 T compared to control. Naturally, vaname shrimp stored at room temperature showed the appearance of the shrimp getting paler, and the texture getting softer, smelling rotten and slimy after the 6th hour. However, the color, smell, and texture of shrimp exposed to an ELF magnetic field of 500 T intensity for 120 minutes were able to survive at a score of 8 at the 5th hour observation and 7th at the 10th hour observation. Conclusion: Exposure to an ELF magnetic field with an intensity of 500 T was effective in suppressing the proliferation of salmonella Sp bacteria but was less effective in suppressing the proliferation of E.Coli bacteria. This condition affects the physical condition of white shrimp until it lasts at the 10th hour. The results of this study serve as the basis for developing food security technology by the ELF magnetic field.

Acknowledgement
Thank you to the University of Jember Research and Community Service Institute for research funding assistance through the Reworking Grant, also thanks to the commitee of the 1st international Conference Science Physics and Education (ICSPE).

References

[1] Fei Y , Su L, Lou H, Zhao C, Wang Y and Chen G 2019) The effects of 50 Hz magnetic field-exposed cell culture medium on cellular functions in FL cells *Journal of Radiation Research* **60**(4) 424–431
[2] Sudarti, Prihandono T , Yushardi Y , Ridlo Z R and Kristinawati A 2018 Effective dose analysis of extremely low frequency (ELF) magnetic field exposure to growth of S. termophilus, L. lactis, L. acidophilus bacteria *Materials Science and Engineering* **432** (1).
[3] Mousavian-Roshanzamir S and Makhdoumi-Kakhki A 2017 The inhibitory effects of static magnetic field on Escherichia coli from two different sources at Short exposure time *Reports of Biochemistry and Molecular Biology* **5**(2) 112–116
[4] Kristinawati A and Sudarti 2016 The Influence of Extremely Low Frequency (ELF) Magnetic Field Exposure on The Process of Making Cream Cheese *Proceeding The 1st IBSC: Towards The Extended Use Of Basic Science For Enhancing Health, Environment, Energy And Biotechnology* 181–183
[5] Tessaro L W E, Murugan N J and Persinger M A 2015 Bacterial growth rates are influenced by cellular characteristics of individual species when immersed in electromagnetic fields *Microbiological Research* **172** 26–33
[6] Zhang J, Xu C, Wan Y and Gao M 2016 Effects of extremely low frequency magnetic field on production of mannatide by α-hemolytic Streptococcus *Bioelectromagnetics* **37**(5) 331–337
[7] Verschaeve L, Wambacq S, Anthonissen R and Maes A 2016 Co-exposure of ELF-magnetic fields and chemical mutagens: An investigation of genotoxicity with the SOS-based VITOTOX test in Salmonella typhimurium *Mutation Research - Genetic Toxicology and Environmental Mutagenesis* **795** 31–35
[8] Oncul S, Cuce E M, Aksu B and Inhan G A 2016 Effect of extremely low frequency electromagnetic fields on bacterial membrane *International Journal of Radiation Biology* **92** (1) 42–49
[9] Wang P A 2011 Post mortem proteolytic degradation of myosin heavy chain in skeletal muscle of Atlantic cod.
[10] Çaklı Ş, Yünlü A C, Dinçer M T, Doğan S and Koçak C 2013 Comparison of polyphenol oxidase activity in different parts of deep water pink shrimp (Parapenaeus longirostris) by using L-DOPA substrate in Marmara Sea during the fishing season E*ge Journal of Fisheries and Aquatic Sciences* **30** (1) 1–6
[11] Nirmal N and Benjakul S 2011 Retardation of quality changes of Pacific white shrimp by green tea extract treatment and modified atmosphere packaging during refrigerated storage *Journal Food Microbiologi* **149** 247–253
[12] Sudarti 2016 Utilization of Extremely Low Frequency (ELF) Magnetic Field is as Alternative Sterilization of Salmonella Typhimurium In Gado-Gado *Agriculture and Agricultural Science Procedia* **9** 317–322
[13] Sudarti S, Bektiarso S, Prastowo S H B, Prihandono T, Maryani and Handayani R D 2020 Optimizing lactobacillus growth in the fermentation process of artificial civet coffee using extremely-low frequency (ELF) magnetic field *Journal of Physics: Conference Series* **1465**(1)

[14] Sudarti, Supriadi B, Subiki, Harijanto A, Nurhasanah and Ridlo Z R 2020 A potency of ELF magnetic field utilization to the process of milkfish preservation (chanos chanos) *Journal of Physics: Conference Series* **1465**(1)

[15] Bayir E, Bilgi E, Sendemir-Ürkmez A and Hameş-Kocabaş E E 2015 The effects of different intensities, frequencies and exposure times of extremely low-frequency electromagnetic fields on the growth of Staphylococcus aureus and Escherichia coli O157:H7 *Electromagnetic Biology and Medicine* **34**(1) 14–18

CIPP vs kirkpatrick in the evaluation of physics teacher competency development programs: literature study

Muqorobin[1], Komarudin[1], A Badrujaman[1], R Arthur[1] and S Prayogi[2]

[1]State University of Jakarta, Jakarta 13220, Indonesia
[2]Mandalika University of Education, Mataram 83125, Indonesia

E-mail: muqorobin_9913920007@mhs.unj.ac.id

Abstract. This study aims to systematically review and analyze the results of research on CIPP and kirkpatrick models in the evaluation of phisycs teacher competency development programs. This literature review is specifically intended to determine the advantages and disadvantages of both evaluation models in the implementation of program evaluation. The research method uses systematic studies through four stages, namely; identification, selection, feasibility and inclusion. Search articles using the Google Scholar, ERIC, Sciencedirect and SpringerLink databases. Based on the results of the study found 1115 research articles that discuss the use of CIPP and Kirkpatrick models in the evaluation of competency development programs. The articles are mostly published in the form of proceedings and international journals and research comes from several countries. The final results showed there were 18 studies discussing the use of CIPP and Kirkpatrick models to evaluate competency development and training programs. Recommendations of the results of the study, in order to be able to combine two models namely CIPP and Kirkpatrick as an alternative evaluation model, the goal is to get an overview of the results of the evaluation of phisycs teacher competency development programs more comprehensively especially in the program that contains training components.

1. Introduction

Looking at the data of the National Examination report in High School in 2019 released by the Center for Education Assessment (Puspendik) shows that the average score in all subjects is still low. Average achievement of all subjects in Language Study Program 55.96, Natural Science Study Program 51.71 and Social Science Study Program 48.95 [1].

Furthermore, based on the report of the programme for International Student Assessment (PISA) 2018 studies show that Indonesia's ranking has decreased when compared to pisa results in 2015. The 2018 study assessed 600,000 15-year-olds from 79 countries every three years. This study compares each child's math, reading, and science performance skills. In the category of reading ability, Indonesia is ranked 74th. Indonesia's average score is 371, which is below Panama's average score of 377. Meanwhile, in the mathematics category, Indonesia is ranked 73rd with an average score of 379. Indonesia is above Saudi Arabia which has an average score of 373 [2].

Finally, a World Bank report released in August 2020 showed the reading ability scores of Indonesian students in the Programme for International Student Assessment (PISA) are projected to decline as a result of the Covid-19 pandemic. The World Bank divides the scenario into three, namely the current scenario (4 months), the optimistic scenario (6 months), and the pessimistic scenario (6 months). In the current scenario, Indonesian students lose 11 points of reading ability. While the optimistic scenario loses 16 points and pessimists lose 21 points [3].

Realizing the condition, it takes a settlement effort to improve the quality of education. One of them is through improving the quality of teacher competence, because teachers as the main key in the success of education [1]. Teacher competence is a set of abilities consisting of knowledge, skills and behaviors possessed by teachers in carrying out professional tasks. Referring to the National Standard of Education (SNP) applicable in Indonesia teacher competencies include professional competencies, pedagogic competencies, social complement and personality complement. In complementary education teachers have an important role because it directly affects the process of learning and teaching activities.

Various efforts are made to improve teacher competence including empowerment of competency development programs through integrated coaching, coaching, training and independent learning. Through these efforts, it is expected to produce improvements and improvements in the achievement of student learning outcomes. To support that, effective management of development programs must be optimized, both related to the development of the teacher's own competence and the achievement of learning objectives [1–3].

Related to that, some research results found that the development of teacher competence through training has a significant impact on the effectiveness of learning [4]. Furthermore, in the results of the study on teacher training and competency development shows that teacher training that is implemented constructively can improve the development of the profession and competence of teachers [5]. Teacher competency development programs and training can be known for their effectiveness and have a better impact in improving the quality of education when the evaluation process is carried out.

The CIPP model is an evaluation framework that covers aspects of context, input, processes and products. CIPP can be used to evaluate formative and sumative. This model provides a holistic overview of each element evaluated and is expected to provide decisions that can be used for the continuity of a program [6–8]. Through research studies with cipp model can provide a complete picture of the implementer of teacher professional development program on web-based content development [9].

While the Kirkpatrick model conceptually provides a framework for evaluation of training programs based on four levels, namely reaction, learning, behavior and results. Despite frequent criticism, this model has reliability and credibility as an evaluation model that is still relevant [10]. Evaluation research with Kirkpatrick model on teacher development and training was able to clarify the perception of principal and supervisor trainees about the results of the training process. Kirkpatrick's evaluation model is also able to determine the strengths and weaknesses of the training process for the principal [11].

There are various studies conducted by practitioners from the academic and community worlds on the use of CIPP and Kirkpatrick models in evaluating competency development programs. The results of the study can be found in various portals journals, conferences, seminars and websites, both on a national and international scale. Contemporary studies of the use of CIPP and Kirkpatric models in evaluation research can ideally be systematically collected for further analysis for the benefit of developing a more comprehensive model. The purpose of this study is to conduct a systematic literature review on the use of CIPP and Kirkpatrick models in evaluating teacher competency development programs.

2. Methods

The method used in this study is systematic literature review with data source from indexed journal articles that have been indexed by ISSN (International Standart Serial Number) and published online wesbsite. Data retrieval is done by surfing the internet and through four stages, namely through the process of identification, screening, feasibility and inclusion. First, at the identification stage, search, collection, recording and assessment of article data relevant to the research topic of five databases are often used in the search for references about program evaluation. The five databases are Google Scholar, ERIC, Sciencedirect and SpringerLink. The textual terms used as keywords in the article search are "teacher complementary development program", "training program", "CIPP evaluation model" and "Kirkpatrick evaluation model". Second, the screening stage, the use of inclusion criteria to limit time, keywords and language types. Articles used were published in 2010-2021. The articles are mostly used in proceedings and scientific journals, in English and Indonesian form and contain research topics relevant to the purpose of research, namely the use of CIPP and Kirkpatrick models in the evaluation of teacher complementary development programs. Search results are documented in a special folder on Mendeley software. Next, after creating a comparison process on file duplication, the relevant articles are organized in separate files.

Third, the feasibility stage, in the article that has been done duplication test then manually selected to hammer out a detailed assessment on the title and abstrack to ensure conformity to the problem / population, intervention, comparative, outcome and study (PICOS) to determine the selected article. The four stages of inclusion, in the inclusion of selected articles are extracted and analyzed in order to extract information about the author, the year of publication, the origin of the country of the first author, the focus of the study, research objectives, research subjects, evaluation models and research results. The exclusion criteria are in articles that focus on general studies, such as theoretical studies and other fields unrelated to competency development.

Technical analysis is done with thematic approach and prepared based on theme analysis and poured in narrative writing. Article search results are presented in the form of flowcharts systematically in accordance with the research stage (PRISMA). The validity test of the research results was conducted by asking for opinions from a team of experts through limited discussions to review the results of the study and make defects for improvement. The following is the selection process of the article.

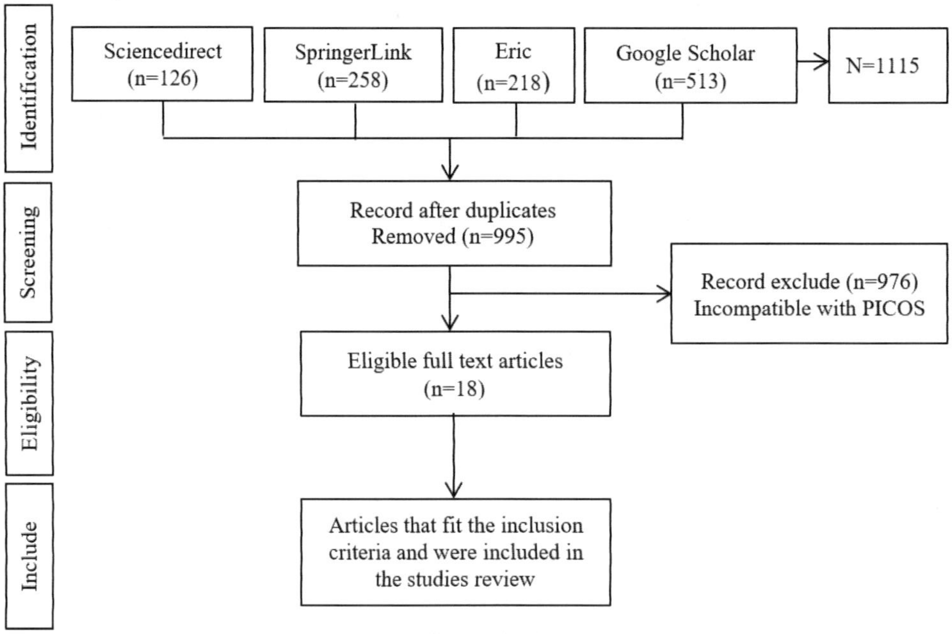

Figure 1. Article selection process

3. Result and Discussion

This literature study research was conducted through four stages, namely: identification, screening, feasibility and inclusion. Stages in the identification process are done by creating a framework according to PICOS in order (teacher competency development program, CIPP and Kirkpatrick evaluation model, alternative evaluation model as a solution and all studies). Keyword settings include the program of teacher competence development, CIPP and Kirkpatrick evaluation model, alternative evaluation model and inclusion criteria setting, as in the following table.

Tabel. 1 Inclusion criteria

	Inclusion	Exclusion
Source	Google Scholar, ERIC, Sciencedirect, SpringerLink and Elsevier.	Databases and information other
Year	July 2010 - July 2021	Studies published before or after this period
Study Type	All types of studies	-
Language	English and Indonesian	Other languages
Intervention	CIPP and Kirkpatrck evaluation models	Interventions on other models
Size of Results	Alternative evaluation model	Other
Population	Teacher complementary development program	Other
Publication Type	Scientific journals and availability of documents in conferences (free of charge) and published through the study of education and social fields	Other publications and paid

Search the online database through the following procedure; search for articles on online databases using matching keywords; using Boolean logic through relevant conjunctions (and, or); and the use of logic in systematic studies is a program of physics teachers competence development AND CIPP, Kirkpatrick model AND alternative evaluation model. The number of journals and articles in this literature study is as follows.

Table 2. Journal selection results based on PICOS and text completeness

Source Article/Journal	Initial Amount	Duplicate	Incompatible with PICOS	Journal remaining
Sciencedirect	126	113	106	4
SpringLink	258	201	198	3
Eric	218	198	195	3
Google Scholar	513	483	477	8
	1115	995	976	18

Based on the table above, initially the number of articles that can be collected as many as 126 from Sciencedirect, 258 from SpringLink, 281 from Eric and 513 from Google Scholar in a systematic literature review of the 1115 articles selected, based on the PICOS framework there are 18 articles that fit the criteria and are subsequently included. The results of this study also showed that during the period 2010-2021 there were 18 publications of research articles on the use of CIPP and Kirkpatrick models in evaluating teacher competency development programs. The focus of comparative development in this research article varies from education and training programs, workshops and induction programs. This study seeks information on the use of CIPP and Kirkpatrick evaluation models and which can be used for the development of a more holistic evaluation model in assessing teacher competency development programs. The distribution of research results covers several countries; Indonesia, Pakistan, United Kingdom, USA, Greece, Romania, Republic of Ireland, Turkey, Malaysia, Iran and China. Most of the research results come from Indonesia. There are several variations of critical notes from researchers related to the use of CIPP and Kirkpatrick models in the evaluation of teacher competency development programs.

Some critical notes are affirming the reliability of the evaluation model, identifying some of the weaknesses of the evaluation model operation and the idea of finding the interrelationship of the two evaluation models [10,12]. The following table describes the characteristics of the articles contained in the journal.

Table 3. Characteristics of the article

1st Author	Country 1st Author	Research Objectives	Competency Development Program	Evaluation Model	Research Results
[12]	Romania	Review the new paradigm of Kirkpatrick's evaluation model in VET professional training for competency improvement.	Training in the form of short courses	Kirkpatrick	Kirkpatrick's evaluation model has many suitability in various training areas and provides an overview of training objectives.
[10]	Pakistan	Design and propose the design of cipp and Kirkpatrick's combined evaluation model in the teacher development and development program.	Competency development and training	CIPP & Kirkpatrick	There is a new strategy in the practice of evaluation of development and training programs, which emphasizes on detailed exploration of the context, objectives, financing, effective programs and outputs of the program.
[13]	Indonesia	Knowing the achievement of objectives and effectiveness of training program to improve teacher competence.	Teacher competency improvement training	CIPP	Evaluation research with CIPP model concludes that the implementation of training is declared successful from the aspects of context, input, process and product.
[9]	Turkey	Evaluating professional development programs in developing web-based content.	Professional development programs	CIPP	Development programs designed not through needs analysis and achievement of results are less than optimal because some aspects of inputs and processes do not meet the minimum criteria.
[14]	Indonesia	Evaluate teacher training programs (PPG).	Teacher training program	CIPP	In terms of context, inputs, processes and products show conformity in the assessment of PPG program implementation.
[15]	Indonesia	Discover the effects of online training in improving teacher PCK STEM competency.	Online competency development training	CIPP	The implementation of the training program shows the suitability of various aspects through the CIPP evaluation model. The implementation of the training has not been able to develop and interpret well in stages 3 and 4.
[15]	USA	Evaluating training programs conducted by professionals.	Professional training	Kirkpatrick	Kirkpatrick's adapted model is very effective for evaluating training and education programs.
[16]	UK	Evaluating the effectiveness and assessing the achievement of training program objectives. Investigating the effects of	Teacher training	Kirkpatrick	Through evaluation with four stages can verify the usefulness of education and training programs in improving knowledge and competence.

1st Author	Country 1st Author	Research Objectives	Competency Development Program	Evaluation Model	Research Results
[11]	Iran	the implementation of education and training programs.	Teacher education and training	Kirkpatrick	Kirkpatrick's model can describe the effectiveness of the workshop implementation from the process and behavior aspects of the participants.
[17]	Iran	Evaluating the effect of conducting workshops on teaching and learning methods.	Teaching and learning method workshops	Kirkpatrick	The demands of competency development needs encourage paradigm changes and evaluation implementation processes.
[18]	Ireland	Analyzing conceptual and contextual studies of education and training program evaluations. Reviewing the effectiveness of the training model with Kirkpatrick.	Education and training programs Training	CIPP	There are obstacles in the implementation of training that can affect the effectiveness of training. The implementation of the Kirkpatrick model was unable to answer some of the questions related to components in the induction program.
[19]	Malaysia	Testing the effectiveness of using kirkpatrick models to evaluate teacher induction programs.	Teacher induction program	Kirkpatrick	The evaluation results describe changes in year-over-year improvement and skills as well as changes in self-teaching.
[7]	Greece	Evaluate teacher professional training programs in developing evidence-based learning instructionals.	Programming program professional teacher	Kirkpatrick	All stages of Kirkpatrick's training program showed good condition and in accordance with standards.
[20]	USA	Analyzing stages in Kirkpatrick on character education trainees .	Character education training program	Kirkpatrick	Training shows normal conditions on all context components, inputs and processes, while product components need improvement.
[21]	Indonesia	Evaluating tiered basic training in PAUD teachers.	Tiered training	Kirkpatrick	The evaluation results through the CIPP model provide a generally good and sufficient picture of the program results on the product.
[22]	Indonesia	Know the evaluation stage of teacher competency improvement training program.	Competency improvement training.	CIPP	The training program is not in accordance with evaluation standards because it only shoots two stages, namely the beginning and middle of the training.
[23]	Indonesia	Assess the development of the training program gradually.	Training programs	CIPP	
[24]	Indonesia			CIPP	

Based on table 3 shows that cipp model tendency is used to evaluate complementary development programs whose form varies, namely: 1) Professional development programs. 2) Competency development training. 3) Workshop. While on the Kirkpatrick model the evaluation focus is more on: 1) Complementary improvement training. 2) Education and training programs for increased competence. 3) Induction program for teachers. In general, both evaluation models are able to uncover and describe the implementation of the development program in accordance with the stages and aspects of the evaluation model.

Referring to the article data in the table is also obtained information that both models have advantages and disadvantages according to the focus characteristics of the object evaluated. Cipp model has the advantage that it is suitable to evaluate the competency development program comprehensively, because evaluators can freely evaluate the program even though the program has not been started and at the same time can evaluate during the program [10]. Unlike the Kirkpatric model, it does not provide evaluator flexibility to evaluate training programs and the evaluation process can be done only when the program is implemented. In the implementation of the CIPP model is also very suitable to be used to evaluate programs that have been designed and planned well, because it pays attention to aspects of context, inputs, processes and products of the program [9,13,19,21,26].

The shortcomings in the CIPP model of information coverage are focused on the information needed only by decision makers and are unable to dig deeper into the important issues of competency development programs, any information should refer to standards and there is no opportunity for wider interachievement alternatives. Another drawback of the CIPP model is that it is not able to dig deeper into training-based development programs such as cost aspects, training processes and assessment programs and implementation takes a relatively long time.

Kirkpatrick's model, meanwhile, has its advantages; coverage in the evaluation of competency development programs is more comprehensive because it covers aspects of knowledge, skills and attitudes of participants who are the subjects of the program [12,21]. Kirkpatrick's model is also very relevant to be used to vaccinate teacher competency development programs such as educational and training activities [11,12,17,18,22,27].

The shortcomings in the Kirkpatrick model depend heavily on all stages of the evaluation. When evaluated only in stages 3 and 4, the interpretation process cannot be developed properly, due to limitations of reference on the rationale of the program [16]. In addition, on this model is also very dependent on the setting of the implementation of the training program, if there are internal obstacles then it very easily affects the quality of the results of the program [7]. Lastly, the weakness of the Kirkpatrick model is that it pays less attention to the context and input aspects in the implementation of the program as well as the slowness of measuring results over a long period of time.

4. Conclusion

The results of the study with systematic literature review showed that the last research article in 2021 and earlier in 2010. Geographically, the results of research from various countries in the Continent of Asia, Europe and Africa. Articles resulting from the publication of proceedings and international articles. The process in article search uses four databases namely; Google Scholar, ERIC, Sciencedirect and SpringerLink are majority English speaking, which can be opened in full text and free of charge. The results of the study obtained as many as eighteen research articles that discuss the cipp and Kirkpatrick models in the evaluation of teacher competency development programs. The scope of development forms includes short training activities, professional development, induction programs, workshop programs, education and training. Recommendations of research results is necessary to develop and develop a mixed model in order to have a comprehensive focus of study.

Acknowledgement

The writing of this scientific work is supported by the Education Research and Evaluation Study Program (PEP) of The State University of Jakarta. The author thanked all lecturers and compatriots, who actively reviewed and provided constructive criticism for the improvement of the quality of this article.

References

[1] Masrifah M, Setiawan A, Sinaga P and Setiawan W 2019 The content quality of teacher's pedagogical and professional competence standards of senior high school physics teacher guide books *Journal of Physics: Conference Series* **1157**

[2] Desimone L M, Porter A C, Garet M S, Yoon K S and Birman B F 2002 Effects of Professional Development on Teachers' Instruction: Results from a Three-year Longitudinal Study *Educational Evaluation and Policy Analysis* **24** 81–112

[3] Kuijpers J M, Houtveen A A M and Wubbels T 2010 An integrated professional development model for effective teaching *Teaching and Teacher Education* **26** 1687–94

[4] Rahman F, Nabi Bux J, Akhter Y, Saeed Ul Hasan C and Ajmal M 2011 Relationship Between Training Of Teachers And Effectiveness Teaching *International Journal of Business and Social Science* 150–60

[5] Morais A M, Neves I P and Afonso M 2005 Teacher training processes and teachers' competence - A sociological study in the primary school *Teaching and Teacher Education* **21** 415–37

[6] Aziz S, Mahmood M and Rehman Z 2018 Implementation of CIPP Model for Quality Evaluation at School Level: A Case Study *Journal of Education and Educational Development* **5** 189

[7] Saad M and Mat Dr N B 2013 Evaluation of Effectiveness of Training And Development: The Kirkpatrick Model *Asian Journal of Business and Management Sciences* **2** 14–24

[8] Lestari S, Rosana D and Supahar 2017 The Development of CIPP Evaluation Model Instruments on the Application of Science Project Learning Assessment *4th ICRIEMS Proceedings* 57–64

[9] Yurdakul B, Uslu Ö, Çakar E and Yildiz D G 2014 Evaluation of the Professional Development Program on Web Based Content Development *Educational Sciences: Theory & Practice* **14** 1427–37

[10] Muhammad Maqsood Khalid 2012 Exploring the link between Kirkpatrick (KP) and context, input, process and product (CIPP) training evaluation models, and its effect on training evaluation in public organizations of Pakistan *African Journal of Business Management* **6** 274–9

[11] Alsalamah A and Callinan C 2020 Key barriers to training effectiveness for female head teachers in saudi arabia: A qualitative survey *Athens Journal of Education* **7** 397–416

[12] Moldovan L 2016 Training Outcome Evaluation Model *Procedia Technology* **22** 1184–90

[13] Umam K A and Saripah I 2018 Using the Context, Input, Process and Product (CIPP) Model in the Evaluation of Training Programs *International Journal of Pedagogy and Teacher Education* **2** 19

[14] Bukhori Dalimunthe M, Wahyu Nugrahadi E, Fitrawaty F and Zaimah Dalimunthe R 2020 Evaluation of Teacher Professional Training Program with Contextual, Input, Process, and Product (CIPP) Models 168–74

[15] Ipa J, Hasanah S S and Permanasari A 2021 Online Professional Development for Improving Teacher ' s STEM PCK Competence , Can It Be An Alternative ? (An Evaluation Using the CIPP Model) **5** 162–71

[16] Kennedy P E, Chyung S Y, Winiecki D J and Brinkerhoff R O 2014 Training professionals' usage and understanding of Kirkpatrick's Level 3 and Level 4 evaluations *International Journal of Training and Development* **18** 1–21

[17] Mahmoodi M, Rashtchi M and Abbasian G R 2019 Evaluation of in-service teacher training program in iran: Focus on the Kirkpatrick model *Education and Self Development* **14** 20–38

[18] Heydari M R, Taghva F, Amini M and Delavari S 2019 Using Kirkpatrick's model to measure the effect of a new teaching and learning methods workshop for health care staff *BMC Research Notes* **12** 1–5

[19] McNamara G, Joyce P and O'Hara J 2010 Evaluation of adult education and training programs *International Encyclopedia of Education* 548–54

[20] Grammatikopoulos V, Tsigilis N, Gregoriadis A and Bikos K 2013 Evaluating an induction training program for Greek teachers using an adjusted level model approach *Studies in Educational Evaluation* **39** 225–31

[21] Mazur R and Woodland R H 2017 Evaluation of a cross-cultural training program for Pakistani educators: Lessons learned and implications for program planning *Evaluation and Program Planning* **62** 25–34

[22] Fatimah M 2017 Evaluation on Training Program of Integrating Character Education in Civics Education for the Teachers Learning Community **100** 180–5

[23] Christy R Evaluation Of Tiered In Order To Increase PAUD Teacher Competence In Medan City 226–30

[24] Ngadi F and Anu Z 2020 Evaluasi Program Pelatihan Peningkatkan Kompetensi Pendidik Paud PENDAHULUAN Salah satu strategi peningkatan mutu pendidik PAUD yang telah diberlakukan selama ini melalui pengembangan gugus . Upaya peningkatan mutu pendidik seperti yang dipersyaratkan dal **1** 30–43

[25] Mote K 2017 The Developing of CIPP Evaluation Model on the Evaluation of Early Childhood Educator Training in Early Children Education Development Center for Early Children and Community Education North Sumatera **104** 138–43

[26] Powell B and Conrad E 2015 Utilizing the CIPP Model as a Means to Develop an Integrated Service-Learning Component in a University Health Course *Journal of Health Education Teaching* **6** 21–33

[27] Alsalamah A and Callinan C 2021 Adaptation of Kirkpatrick's four-level model of training criteria to evaluate training programmes for head teachers *Education Sciences* **11**

Flipped classroom using e-module to improve understanding of light concepts: needs analysis of e-module development to empower scientific explanation

Cari[1], M Nasir[1], W Sunarno[1] and F Rahmawati[1]

[1]Sebelas Maret University, Indonesia

E-mail: muhammad_nasir@student.uns.ac.id

Abstract. This study aims to determine the effectiveness of the flipped classroom using e-module to improve understanding of the light concept, analyze the trend of increasing student concept understanding, and analyze the weaknesses of e-module instructional activities. This research method is the concurrent embedded mix method. The research design is a one-group pretest-posttest design. The subjects involved 22 students of IAIN Palangka Raya and Hamzanwadi Selong University. The sample was selected by cluster random sampling. Quantitative data were collected using a two-tier test. Qualitative data were collected through interviews and observations. Quantitative data analysis technique using t-test. Qualitative data were analyzed descriptively qualitatively. The results showed that: 1) There was a significant increase in student's conceptual understanding after the implementation of the flipped classroom using e-module, 2) The trend of increasing concept understanding in the low category was 18%, the medium category was 68%, and 14% in the high category, and 3) The weakness of this e-module is that there are no hand-on and scaffolding-oriented instructional activities to empower explanations. scientific in elaborating concepts.

1. Introduction

The learning process is a process of delivering messages from the source of the message to the recipient of the message. The process of delivering this message can be done synchronously and asynchronously [1]. So that the use of technology becomes a very important consideration to improve the effectiveness and quality of the learning process [2]. Besides that, the curriculum based on the Indonesian National Qualifications Framework emphasizes problem-solving using Science and Technology.

Utilization of technology in learning can use the flipped classroom learning strategy, namely learning that minimizes direct teaching from the teacher, but maximizes indirect teaching with material support that can be accessed anytime by students [3]. The flipped classroom gives students to prepare for lessons by watching videos, listening to broadcasts, reading articles, discussions, or group work [4]. The flipped classroom is an instruction that used accessed at home before entering class, the classroom becomes a place to solve problems and develop concepts [5].

Technology-based learning strategies that can be used by educators in the learning process in the current digitalization era to increase the activeness and independence of students are flipped classrooms using e-modules. Through e-module, educators can share material synchronously or asynchronously. The advantage of the e-module is that it is equipped with audio, video, images, and animations that can help students elaborate concepts in learning. The flipped classroom strategy using e-modules is indispensable in online learning in the era of the COVID-19 pandemic, especially in physics teaching materials [6], [7].

Diagnostics of students' initial concepts need to be done before learning begins as a reference for designing teaching materials that suit the needs of students. Low initial concept understanding causes students to have difficulty connecting other concepts related to the concept being studied [8]. Like the case of fiber optic material, because fiber optic material is closely related to the concept of light, an inventory of trends in understanding the concept of light is very important before teaching fiber optic material. The results of research by Yoanita & Akhlis (2015) show that the lowest level of understanding of the concept of light occurs in indicators of knowing the definition, nature of light, and understanding the conditions for the process of reflection and refraction of light [9].

The results of a preliminary study conducted by researchers on the trend of understanding the concept of light among students of IAIN Palangka Raya and Hamzanwadi Selong University showed that 50% of the items about the concept of light were not understood by students. Many indicators of understanding the concept of light that are still not understood by students include the concept of changes in the magnitude and direction of refractive rays when passing through a medium that has a different refractive index, conditions for the occurrence of critical angles, conditions for total internal reflection, and the meaning of the mathematical equation for attenuation of light waves.

The results of a preliminary study on the trend of understanding the concept of light are used as a basis for researchers to develop e-modules. After learning through flipped classrooms using e-module, it is hoped that there will be an increase in understanding of the concept of light. This study, in addition to measuring the increase in student's conceptual understanding after learning through flipped classrooms using e-modules, but also taking inventory of the causes of student difficulties in answering questions of understanding concepts, and evaluating the weaknesses of instructional activities in e-modules as consideration for compiling instructional activities on e-modules. The next step is to empower students' scientific explanations. To train students to be able to make the right decisions based on information and data, students need to be trained to argue using scientific explanations.

2. Methods

This research was conducted by using the exploratory sequential design [10]. In this study, qualitative research is a large part of this research and quantitative data is used as data to support research results. The research design is a one-group pretest-posttest design. The subjects of this study involved 22 students of IAIN Palangka Raya and Hamzanwadi Selong University. The sampling technique is cluster random sampling, which is a sampling technique where a researcher divides the population into several separate groups, then from these clusters, several samples are chosen randomly. Quantitative data were collected through a test instrument for understanding the concept of light in the form of multiple-choice and answer reasons. Qualitative data about the reasons why students had difficulty answering certain light concepts were collected through interviews. Qualitative data on instructional weaknesses in the e-module were collected through document observations and student activities in the learning process. Obtained from the results of interviews. The quantitative data analysis technique uses a t-test to measure the effect of flipped classroom treatment using e-modules to improve students' understanding of the concept of light. Before the t-test, the analysis prerequisite test was conducted, namely the normality test and data homogeneity. Qualitative data were analyzed descriptively qualitatively using normalized gain [11].

Table 1. Criteria for Normalized Gain Score

Criteria	Percentage Level of Mastery
$gain \geq 0.7$	High
$0.3 \leq gain < 0.7$	Medium
$gain < 0.3$	Low

Qualitative data analysis to describe the trend of increasing student understanding of concepts, mapping the indicators of questions that are still difficult to answer by respondents. The concept map of this research as shown in Figure 1.

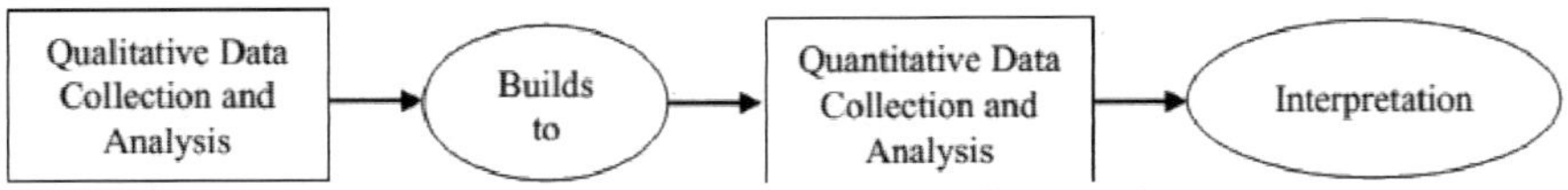

Figure 1. The concept map of research

3. Result and Discussion

3.1. Effectiveness of flipped classroom using e-module to improve understanding of the concept of light

The results of the analysis prerequisite test consist of the results of the normality test in Table 2 and the homogeneity of the data in Table 3.

Table 2. Tests of Normality

	Kolmogorov-Smirnov		
	Statistic	df	Sig.
Post-test	.146	22	.200
Pre-test	.200	22	.022

The significance value (p) of the post-test data on the Kolmogorov-Smirnov test is 0.2 ($p > 0.05$), so the post-test data is normally distributed. The significance value (p) of the pre-test data on the Kolmogorov-Smirnov test was 0.02 ($p < 0.05$), so the pre-test data were not normally distributed

.

Table 3. Test of Homogeneity of Variances

Levene Statistic	df1	df2	Sig.
.793	1	42	.378

The significance value of the test of homogeneity of variances is 0.37 ($p > 0.05$), so the pre-test and post-test data variances are the same or homogeneous.

The results of descriptive data analysis of data containing the mean, number of samples, standard deviation, and standard error of the mean are shown in Table 4.

Table 4. The Results of Descriptive Data Analysis

		Mean	N	Std. Deviation	Std. Error Mean
Pair 1	Post-test	66.5909	22	17.20924	3.66902
	Pre-test	36.1364	22	17.85736	3.80720

T-test analysis was conducted to see the significance of the difference between the pre-test and post-test data. The significance of these differences as a way to test the effect of the flipped classroom assisted by e-module on increasing students' understanding of concepts. The results of the t-test are shown in Table 5.

The significance value (2-tailed) was 0.000 ($p < 0.05$). So that the results of the pre-test and post-test experienced a significant change. Based on the descriptive statistics in Table 3, shows that the average pre-test is 36.13 and the average post-test is 66.59, so it is proven that the post-test is higher. The results of this study are in line with the results of research by Nafaida, Halim & Rizal (2015) showing a significant difference between students' understanding of the light refraction concept before and after the application of the phet-based module [12]. E-modules are effective in conveying an overview of the required material [13]. The e-module is effectively flipped classroom-oriented in increasing student motivation and learning outcomes [14]. Students show higher confidence in their ability to achieve the specified goals after completing the e-module [15]. The survey results have shown that students find the learning experience more comfortable, interesting and effective, and very comfortable given the much greater activity expectations in flipped classroom-oriented e-module learning[16].

Table 5. Paired Samples Test

Paired Differences		t	df	Sig. (2-tailed)
95% Confidence Interval of the Difference				
Lower	Upper			
22.71710	38.19199	8.185	21	.000

The advantage of the flipped classroom is that students have the opportunity to learn independently and working together to solve problems in class [3]. The advantage of the flipped classroom are 1) have a strong factual knowledge base, 2) understand facts and ideas in a coherent and organized manner, and 3) organizes new learning in a way that aids retrieval and application [17]. Flipped classroom learning can allow students to relate new content to their schemas [18]. Flipped classrooms can promote independent learning and improve social connectedness and student performance [19],[20], [21]. The results of Mok's research (2014) show that student feedback about flipped classroom learning is generally very positive with many respondents considering it effective and useful for learning [22]. The flipped classroom must be presented cognitively and practically, teacher skills in designing material using various equipment and to transform the material with a learning management system [23], [24].

3.2. The trend of increasing students' understanding of light concepts

The trend of increasing students' light concepts after being taught through a flipped classroom assisted by e-modules can be seen in Figure 2.

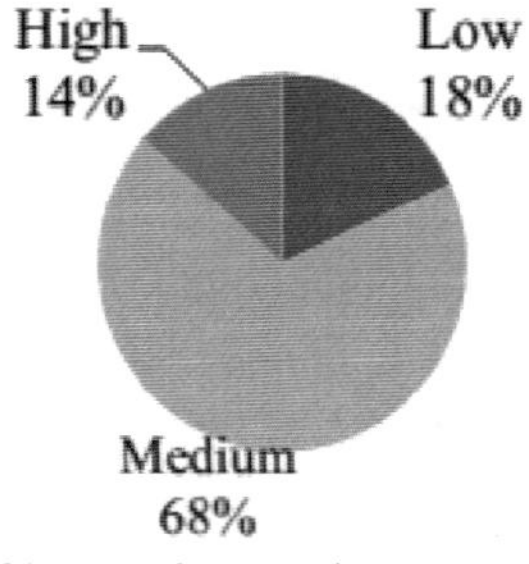

Figure 2. The trend of increasing student concept understanding.

The increase in students' understanding of concepts after the implementation of the flipped classroom using e-modules was mostly in the moderate category as much as 68%. while the increase in understanding of concepts in the high category is very small, namely 14%, so it is necessary to refine the flipped e-module so that most of the increase in understanding the concept of light can increase in the high category. But on the other hand, the Flipped Classroom strategy-oriented e-module can reduce the increase in understanding the concept of light with a low category of only 18%.

The data from the analysis of the achievement of each indicator of understanding the concept of light after being taught through flipped classroom using e-module is shown in Figure 3.

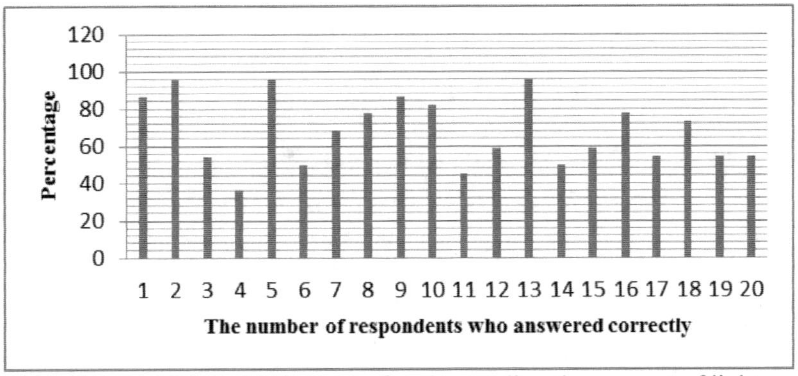

Figure 3. Achievement of indicators of understanding the concept of light.

The concepts that respondents still find difficult after being treated using e-modules are question number 4 (only 36% could answer correctly), question number 11 (only 45% could answer correctly), question number 6, and 14 (only 50% can answer correctly).

Question number 4 deals with interpreting the meaning of the representation of the direction of the rays, the correct concept that the direction of the rays is a representation of the direction of wave propagation because the direction of the rays is in the same direction as the direction of wave propagation. Questions and samples of student answers in question number 4.

Question: The direction of the rays is a representation of....
An example of a student's scientific explanation of the answer claim to question number 4 is:
- The direction of wave propagation is a representation of the direction of the rays.
- Because the direction of the ray is always perpendicular to the direction of ray propagation.

Question number 4 only 1 student is correct out of 22 respondents on the pre-test because only 22% of respondents believed that the wavefront was a plane perpendicular to the direction of wave propagation. After learning using e-modules there was an increase in the number of students who were able to answer question number 4 as many as 8 students. The scientific explanation for the claim for the answer to question number 4 does not yet refer to the structure of the scientific explanation in the form of claim, evidence, reasoning, and rebuttal, but only consists of claim and reasoning.

Question number 11 deals with giving an example of a critical angle representation, the correct concept that a critical angle is an angle of incidence that produces an angle of refraction of 90° to the normal line or parallel to the interface between the two mediums. Questions and samples of student answers in question number 11.

Most of the students incorrectly answered question number 11 on the pre-test because 56% of respondents thought that the angle of incidence was greater than the critical angle when the refracted ray was parallel to the boundary plane of the medium. After learning using e-modules, there was no significant increase in critical angle concept understanding because there was only 1 student who answered correctly. The scientific explanation claims the second and third answers are less precise. The first trend of scientific explanation meets the three structures of scientific explanation, namely claim, evidence, and reasoning, but the rebuttal component has not yet emerged.

Question: The critical angle in Figure 4 is shown by….

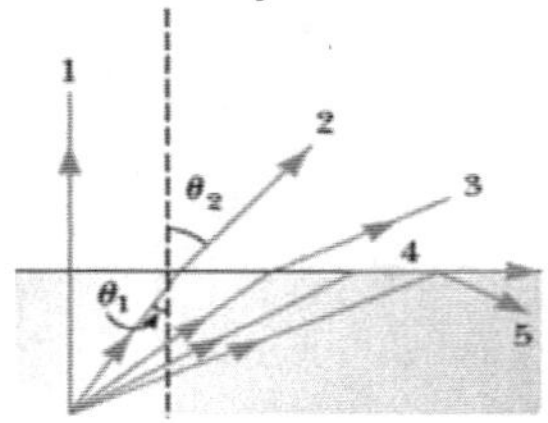

Figure 4. The propagation of light from a dense medium to a less dense medium in problem number 11

An example of a student's scientific explanation of the answer claim to question number 4 is:

- The critical angle is the angle of incidence that produces an angle of refraction of 90° or the angle of the incident ray from a denser medium to a less dense medium whose angle of refraction is parallel to the boundary between the two media. So from the figure, the critical angle is shown in Number 4

Question number 6 deals with making inferences from diffuse reflection events, the correct concept that angles of incidence and angle of reflection are equal because Snell's law of reflection still applies. Because the plane is not flat so that the normal line position of each plane segment is not the same so that the direction of the angle of incidence and reflection of each segment of the plane is not the same, but each segment of the plane angle value of the angle of incidence is equal to the angle of reflection. Questions and samples of student answers in question number 6.

Question: The statement that is incorrect about reflection in Figure 5 is….

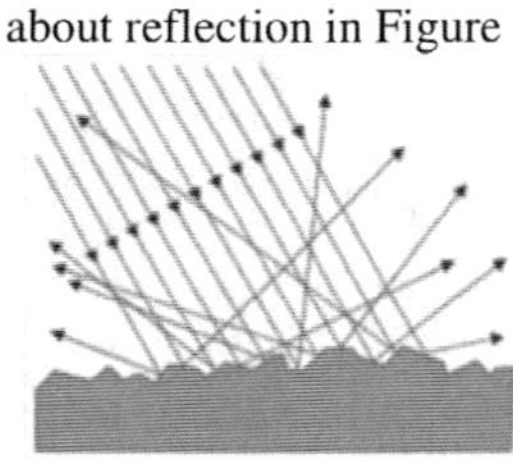

Figure 5. The diffuse reflection in question number 6

An example of a student's scientific explanation of the answer claim to question number 6 is:

- Because each segment of the roughness of the plane has its normal line direction. So, even though the direction of the rays is irregular, Snell's law still applies to every segment of the roughness of the plane.
- Diffuse reflection is a reflection that occurs when a light source falls on an object with a rough surface so that light is reflected in all uncertain directions.

Some students were wrong in answering question number 6 on the pre-test because 89% of respondents agreed that the angle of incidence and angle of reflection was not the same on an uneven plane. Respondents assume that Snell's law does not apply to an uneven plane because it is imagined with a random light direction due to the uneven surface of the plane. They do not realize that the normal lines on the uneven plane are not in the same direction. After learning using e-module there was an increase in the number of students who were correct as many as 4 people. The first trend of scientific explanation meets the three structures of scientific explanation, namely claim, evidence, and reasoning, but the rebuttal component has not yet emerged. The second and third scientific explanations do not meet the structure of scientific explanations.

Question 14 deals with identifying the conditions for total internal reflection to occur, the correct concept that the conditions for total internal reflection to occur if (1) light comes from a denser medium to a less dense medium ($n_1 > n_2$), and (2) the angle of incidence must be greater than the critical angle ($i > \theta$). Questions and samples of student answers in question number 14.

Question: If n represents the index of refraction, i represents the angle of incidence, and θ represents the critical angle, then total internal reflection occurs if....
An example of a student's scientific explanation of the answer claim to question number 6 is:

- Total internal reflection only occurs if the light is directed from a medium with a certain refractive index to a medium with a lower refractive index, if $n_1 > n_2$ it will result in $\text{Sin}\ \theta > 1$
- The refractive index is not equal and the angle of incidence is equal to the critical angle

Some students were wrong in answering question number 14 on the pre-test because the sub-chapter had not been taught by the teacher while studying at high school. Question number 14 can only be answered correctly by 4 students in the pre-test. After learning using e-module there was an increase in the number of students who were able to answer correctly as many as 7 students. The scientific explanation given to support the claimed answer is not correct. The first reason is wrong in the equation Sin > 1, the second reason is wrong in the statement that the angle of incidence is equal to the critical angle.

The indicators of understanding the concept of light that is still difficult for students after being treated with an e-module oriented flipped classroom are (1) Interpreting the meaning of the representation of the direction of light, (2) Providing examples of critical angle representations, (3) Making Snell's law inferences from diffuse reflection events, and (4) Identifying the conditions for total internal reflection to occur.

3.3. Weakness Analysis of Material-Oriented E-modules

Indicators interpreting, giving examples, inference, and identification need to be improved again through appropriate learning activities. Indicators identify, infer, interpret, and provide examples related to decision-making skills through scientific explanations supported by information and data. The weakness of this e-module is that there are no hands-on activities and concept elaboration to train students' scientific explanation skills. Therefore, the e-module needs activities that bridge the hands-on activities through the inquiry learning model and concept elaboration to empower scientific explanations through argumentation structure scaffolding.

Research by Hsu, Chiu, Lin, & Wang (2015) on the Argumentative Scientific Inquiry System with structured argumentation scaffolding shows that there is a significant increase in scientific explanation skills [25]. Guided inquiry can improve students' conceptual understanding of light concepts [26]. Thus, to achieve a balance in achieving conceptual understanding and scientific explanation skills, in the next flipped classroom-oriented e-module development, it is necessary to consider guided inquiry accompanied by argumentation scaffolding.

4. Conclusion

There is a significant difference in student's understanding of light concepts after the implementation of the flipped classroom using e-module. The trend of increasing understanding of the concept of light in the low category is 18%, the medium category is 68% and the high category is 14%. Scientific explanation to support the claim of answers to questions do not meet the structure of scientific explanations in the form of claims, evidence, reasoning, and rebuttal, but only includes claim answers and reasoning. The weakness of this e-module is that there are no hands-on and scaffolding-oriented instructional activities to train students' scientific explanations in elaborating concepts. The implication for improving this e-module is to consider guided inquiry learning models and argumentation scaffolding techniques.

Acknowledgment
I would like to express my gratitude to the Science Education Doctoral Study Program of Sebelas Maret University.

References

[1] Harpaz Y 2005 Teaching and learning in a community of thinking *Journal of Curriculum and Supervision* **20** 136-157.
[2] Budiman H 2017 Peran teknologi informasi dan komunikasi dalam pendidikan *Al-Tadzkiyyah: Jurnal Pendidikan Islam* **8** 31-43.
[3] Mehring J 2018 The flipped classroom *Innovations in flipping the language classroom* 1-9
[4] Milman NB 2012 The flipped classroom strategy: What is it and how can it best be used? *Distance Learning* **9** 77-85.
[5] Tucker B 2012 Online instruction at home frees class time for learning *Education Next* **12** 1-7.
[6] Anindhyta C, Widha S and Sri B 2020 Is the android digital web module based on flipped classroom needed by teachers and high school students in pati district during the covid-19 pandemic? *Proceedings of the 4th International Conference on Learning Innovation and Quality Education* 1-6.
[7] Fahmy AFR, Sukestiyarno S and Scolastika M 2019 Mathematical literacy based on student's self-regulated learning by flipped classroom with Whatsapp module *Unnes Journal of Mathematics Education Research* **8** 125-132.
[8] Putri FM 2015 Pengaruh penerapan kombinasi metode inkuiri dan reciprocal teaching terhadap capaian pemahaman konsep siswa *Edusains* **7** 18-26.
[9] Yoanita P and Isa A 2015 Pengembangan e-diagnostic test untuk identifikasi tingkat pemahaman konsep siswa smp pada tema optik dan penglihatan *Unnes Science Education Journal* **4** 1-8.
[10] Creswell JW 1994 Research design *Academia.edu* 155-179
[11] Hake RR 1998 Interactive-engagement versus traditional methods: A six-thousand-student survey of mechanics tests data for introductory physics courses *American Journal of Physics* **66** 64-74.
[12] Nafaida R, Abdul H and Syamsul R 2015 Pengembangan modul berbasis phet untuk meningkatkan pemahaman konsep dan motivasi belajar siswa pada materi pembiasan cahaya *Jurnal Pendidikan Sains Indonesia* **3** 181-185.
[13] Shimamoto D 2012 Implementing a flipped classroom: An instructional module TCC Conference 1-9.
[14] Hidayati A, Andra S and Raimon E 2020 Development of e-module oriented flipped classroom strategies in computer network learning *Jurnal RESTI (Rekayasa Sistem Dan Teknologi Informasi* **4** 429-437.
[15] Hoffmann JA and Rachel WT 2017 Flipped classroom module on the shock for medical students *MedEdPORTAL* **13** 1-5.
[16] McCabe BA 2000 Flipped learning in a civil engineering module: student and instructor experiences *Irish Journal of Technology Enhanced Learning* **4** 1-15.
[17] National Research Council 2000 *How people learn: Brain, mind, experience, and school (Expanded Editio* (Washington, DC, USA: National Academy Press).
[18] Shuell T 1986 Cognitive conceptions of learning *Review of Educational Research* **56** 411-436.
[19] Jdaitawi M 2019 The effect of flipped classroom strategy on students learning outcomes *International Journal of Instruction* **12** 665-680.
[20] Akcayır G and Murat A 2018 The flipped classroom: A review of its advantages and challenges *Computers & Education* **126** 334-345.
[21] Triatmojo W and Priyadi M 2021 Implementation of Flipped Classroom on Experiences in Online Learning During Pandemic Covid-19 for a Project-Base Vocational Learning Guide *Journal of Physics: Conference Series* **1842** 012019.

[22] Mok HN 2014 Teaching tip: The flipped classroom *Journal of Information Systems Education* **25** 1-7.
[23] Ozdamli F and Gulsum A 2016 Flipped classroom approach *World Journal on Educational Technology: Current Issues* **8** 98-105.
[24] Prasetyo BD, Suprapto N and Pudyastomo R N 2018 The effectiveness of flipped classroom learning model in secondary physics classroom setting *Journal of Physics: Conference Series* **997** 012037.
[25] Hsu CC, Chiu C H, Lin C H and Wang T I 2015 Enhancing skill in constructing scientific explanations using a structured argumentation scaffold in scientific inquiry *Computers & Education* **91** 46-59.
[26] Maulidati R, Yusrizal and Mursal 2015 Penerapan metode pembelajaran inkuiri terbimbing untuk meningkatkan pemahaman konsep dan keterampilan berfikir kritis siswa SMP pada materi cahaya *Jurnal Pendidikan Sains Indonesia (Indonesian Journal of Science Education* **3** 71-84.

Synthesis of low viscosity of polymethylhydrosiloxane using monomer of dichloromethylsilane

V F Arini[1], U Fauziah[1], D G Auliya[1], S Setiadji[2,3], Fitrilawati[1] and Risdiana[1]

[1] Department of Physics, Faculty of Mathematics and Natural Sciences, Universitas Padjadjaran, Jl. Raya Bandung-Sumedang km 21 Jatinangor, Sumedang, West Java, Indonesia 45363
[2] Doctoral Course Program, Department of Chemistry, Faculty of Mathematics and Natural Sciences, Universitas Padjadjaran, Jl. Raya Bandung-Sumedang km 21 Jatinangor, Sumedang, West java, Indonesia 45363
[3] Department of Chemistry, Faculty of Science and Technology, UIN Sunan Gunung Djati Bandung, Jl. A. H. Nasution No. 105 Cibiru, Bandung, West Java, Indonesia

E-mail: fitrilawati@phys.unpad.ac.id

Abstract. Polydimethylsiloxane (PDMS) is a vitreous humour substitutes in vitreoretinal surgery to treat retinal detachment. PDMS was synthesized from the monomer octamethylcyclotetrasiloxane (D4). However, D4 is difficult to obtain in Indonesia. So, we need an alternative to replace PDMS. One type of siloxane polymer that can be a potential substitute as PDMS is polymethylhydrosiloxane (PMHS). In this study, PMHS was obtained from the hydrolysis-condensation process of dichloromethylsilane (DCHS). The hydrolysis process produces monomer that is ready for the condensation process to produce PMHS. Monomer and polymer samples were characterized to determine viscosity, surface tension, refractive index, and functional groups. The monomer sample has a viscosity in the range of 28.7-36.4 mPa.s. The surface tension and refractive index are in the range of 17-18 mN/m and 1.3954-1.3958. The polymer sample has a viscosity of 560-1130 mPa.s with a surface tension of 19 mN/m and a refractive index in the range of 1.3966-1.3970. From the FTIR measurements, all major functional groups in both monomers and polymers of PDMS were observed with the addition of Si-H functional groups which characterize PMHS. These results indicate that PMHS with low viscosity has been successfully obtained by hydrolysis-condensation method of DCHS.

1. Introduction

The eye is a very important optical system in the human body. One of the important parts of the eye is vitreous humour. Vitreous humour is eye fluid that fills about 2/3 of the eye volume, which is located between the lens and retina [1, 2]. Vitreous humour serves to maintain the shape of the eye and keep the retina in place during eye movement [1, 3]. Retinal detachment is one of the abnormalities in the vitreous humour, where the neurosensory retina is separated from the retinal epithelium, causing holes in the eye [1, 4]. Retinal detachment can be caused by several factors, including changes in the physical properties of the vitreous humour, people with high myopia which makes the retina thinner and become easily torn (above 6 diopters), have a family history, have had cataract surgery, and have serious injuries [5]. Retinal detachment can be treated with vitreoretinal surgery with a vitrectomy technique [1]. Vitrectomy is a surgical procedure, where the vitreous humour is cut and removed, then replaced with an artificial replacement fluid [1]. Since 1962 until now, polydimethylsiloxane (PDMS) is a material that is widely used as a substitute for vitreous humour in retinal surgery in human eyes [2, 4-6].

PDMS was obtained by synthesizing octamethylcyclotetrasiloxane (D4) monomer by ring-opening polymerization (ROP) method [5]. However, D4 is expensive and difficult to obtained in Indonesia. Therefore, it is necessary to find other materials to substitute the damaged vitreous humour. One type of siloxane polymer that can be a potential substitute as PDMS is polymethylhydrosiloxane (PMHS). PMHS is a hybrid polymer consisting of repeating units of silicone and oxygen bonds in the polymer chain and has bonds between silicone with methyl and hydrogen [7]. PMHS has the advantage of being optically transparent, stable, and non-toxic [8, 9]. PMHS is also a colorless free flowing liquid that is soluble in most organic solvents [9]. However, PMHS is commercially available in very low viscosity [9]. For the purpose of substituting vitreous humour, PMHS is required to have a minimum viscosity of 1000 mPa.s. Therefore, PMHS need to be synthesized in order to qualify the minimum viscosity.

PMHS polymers can be obtained by synthesizing of dichloromethylsilane (DCHS). The chemical structure of DCHS is CH_3HSiCl_2. DCHS has silicone atom bonds with two chlorine atoms, one methyl and one hydrogen atom. PMHS synthesis from DCHS material can be carried out by hydrolysis-condensation method [7, 10]. The hydrolysis method is the process of breaking and decomposing compounds using water [11]. Water will break down DCHS compounds that produce monomer and unnecessary side product of hydrochloric acid (HCl). The chemical reaction in the hydrolysis process can be seen in Equation 1. The silicone bond with 2 chlorine atoms owned by DCHS must be removed and replaced with a silicone bond with hydroxyl functional group (OH) so that it can be polymerized into PMHS by the condensation method. The condensation method is a polymerization process formed from functional groups such as OH, NH_2, and COOH [12]. Therefore, the condensation polymerization process can occur because the monomer already has a bond with the OH functional group from the previous hydrolysis process. A new silicone oil of PMHS will be formed after the polymerization process.

$$CH_3SiHCl_2 + 2H_2O \rightarrow CH_3SiH(OH)_2 + 2HCl \tag{1}$$

In this paper, we synthesize PMHS from DCHS material by hydrolysis-condensation method. The monomer yield will be compared with commercial D4 monomer and properties of PMHS polymer will be compared with low viscosity commercial of PDMS.

2. Methods

Figure 1 shows an illustration of the experimental method. The monomer synthesis is carried out by the hydrolysis method. The reaction was carried out by mixing 25 ml of DCHS with 100 ml of dichloromethane (DCM) as solvent. Then added 12.5 ml of milli-Q water. The synthesis process was carried out at a temperature of 35 °C for 3 hours using a magnetic stirrer with a stirring speed of 200 rpm. The monomer solution and the side product of HCl were separated using a separatory funnel and a clear solution was taken. Then the DCM solvent was separated by a rotary evaporator for 15-30 minutes at 40 °C. After this process, a monomer sample will be formed. In the polymerization process by natural condensation method, the monomer sample is aged at a temperature of 15-20 °C. The viscosity of the monomer will change gradually and will be monitored by measuring the viscosity periodically.

The synthesized samples, both monomers and PMHS polymers, were characterized using Perkin Elmer Spectrum 100 FTIR spectrometer to detect functional groups. Viscosity was measured using SEKONIK VISCOMATE viscometer model VM-10A-MH. Refractive index was measured using AS ONE I-500 refractometer (Brix 0 ~ 90%), and surface tension was measured using a surfgauge. For comparison, commercial D4 and commercial PDMS products were also characterized.

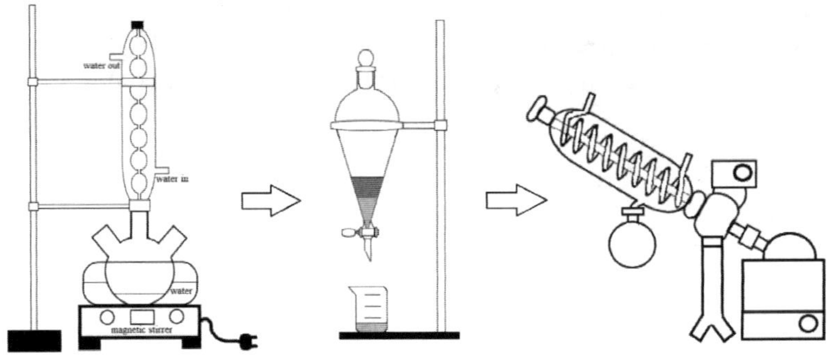

Figure 1. Experimental method illustration

3. Result and Discussion

Table 1 shows the value of viscosity, refractive index, and surface tension of monomer (M) and polymer (P). For monomer samples, the values of viscosity are higher than commercial D4. The values of monomer are 36.4, 28.8 and 28.7 mPa.s, respectively while 2.25 mPa.s for commercial D4 [13]. In the monomer synthesis process, it is estimated that linear chain is formed, not cyclic chain as characteristic of commercial D4 [13]. The values of surface tension of the monomer samples are in the range of 17-18 mN/m which is similar to commercial D4. Likewise, the value of the refractive index of the monomer sample is also similar to commercial D4, namely 1.3954, 1.3958 and 1.3954, where the refractive index of commercial D4 is 1.3950 [13]. The properties of monomers similar to the properties of commercial D4 is one of the considerations that would make PMHS can be polymerized.

The viscosity values of all samples changed after 90, 76 and 72 days, respectively. The viscosity values changed from 36.4, 28.8 and 28.7 mPa.s to 560, 620 and 1130 mPa.s. The viscosity changed as the sample polymerizes into long chains. There is one sample which has a viscosity value above 1000 mPa.s which is appropriate with low viscosity of PDMS [2, 4]. For S-3, the viscosity of 1130 mPa.s was obtained after the sample reacted for 72 days. For S-1 and S-2, the value of viscosity is still below the commercial value of PDMS after polymerization for 90 and 76 days. In this study, the polymerization time was not well controlled. It is because there is still acid content of HCl which has not been separated completely in the hydrolysis process.

The surface tension of all polymer samples obtained the similar value with commercial PDMS. The value of surface tension of polymer is 19 mN/m, while for the commercial PDMS is 20 mN/m [5]. This surface tension will affect the tensile strength of a surface with another surface, so that the higher the surface tension, the emulsification can be avoided [6, 14].

The refractive index of all polymer samples obtained the similar value with that of commercial PDMS. The value of refractive index of polymer are 1.3970, 1.3966 and 1.3970, while for the commercial PDMS is 1.4044 [5]. This refractive index value is close to refractive index of vitreous humour [1]. The refractive index is very influential on the visual process of the human eye. If the refractive index value of the replacement fluid is different with the normal refractive index value, it will have an impact on changing the path of light to the retina and causing hypermetropia and myopia [14, 15].

Table 1. Characteristic results of monomer and polymer samples

Sample	Viscosity (mPa.s)		Surface tension (mN/m)		Refractive index	
	M	P	M	P	M	P
S-1	36.4	560	18	19	1.3954	1.3970
S-2	28.8	620	17	19	1.3958	1.3966
S-3	28.7	1130	17	19	1.3954	1.3970

Note: (M) = Monomer, (P) = Polymer

The FTIR spectra and functional groups of monomer sample can be seen in Figure 2 and Table 2. For S-3, it is found that the spectra of main functional group were similar with spectra of commercial D4. The vibrations of Si–O–Si, Si–CH_3, Si–C and C–H are at almost the same wave number as commercial D4. The difference between the monomer sample and commercial D4 was only the additional functional group Si-H at a wave number of 2169.7 cm^{-1}. The presence of additional Si-H functional group is caused by the characteristics of the DCHS material which has silicone bonds with hydrogen.

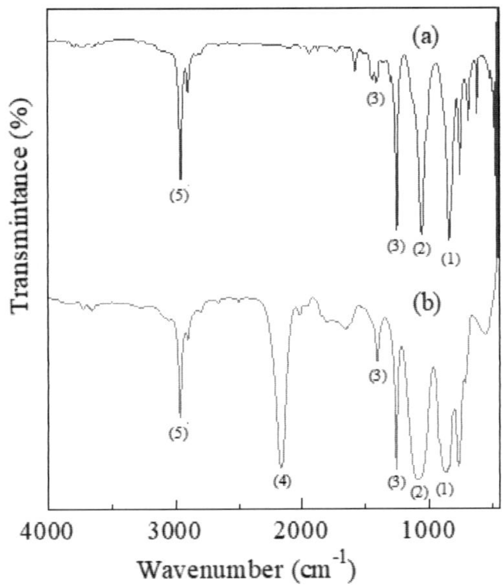

Figure 2. IR spectra of (a) commercial d4, (b) monomer sample

Table 2. Functional group of commercial D4 and monomer sample [13]

No	Functional group	Wavenumber (cm^{-1})	
		Commercial D4	Monomer Sample
(1)	Si-C	843.1	858.6
(2)	Si-O-Si	1058.5	1092.5
(3)	Si-CH_3	1254.3, 1412.3	1262.7, 1408.1
(4)	Si-H	-	2169.7
(5)	C-H	2900.7, 2958.9	2905.5, 2966.6

The FTIR spectra and functional groups of polymer sample can be seen in Figure 3 and Table 3. For S-3, it is found that the spectra of main functional group were similar with spectra of commercial PDMS. The vibrations that similar with commercial PDMS are Si–O–Si, Si–CH_3, Si–C and C–H. The difference between the polymer sample and commercial PDMS was only the additional functional group Si-H at a wave number of 2136.6 and 2202 cm^{-1}. This characteristic is caused by the material properties of the DCHS.

Because the characteristics of the PMHS are similar with the characteristics of commercial PDMS and vitreous humour, it is expected that PMHS can be a substitute for vitreous humour without emulsification or other eye irritation after being injected into the eye.

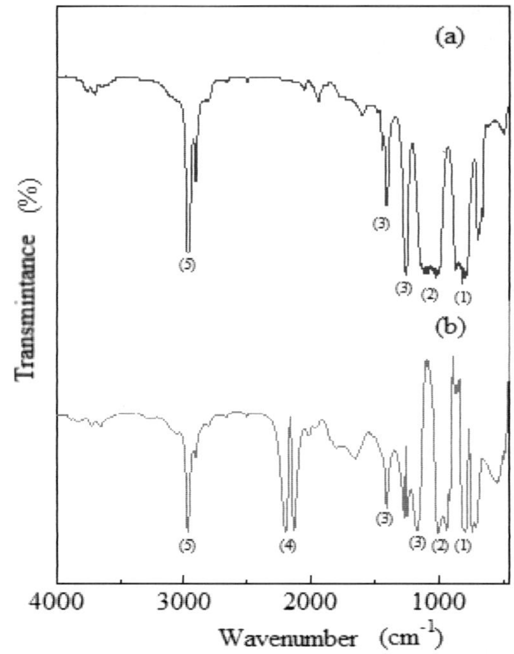

Figure 3. IR Spectra of (a) Commercial PDMS, (b) Polymer sample

Table 3. Functional group of commercial PDMS and polymer sample [5]

No	Functional group	Wavenumber (cm^{-1})	
		Commercial PDMS	Polymer Sample
(1)	Si-C	792 – 823.8	792.2
(2)	Si-O-Si	1022.8, 1111.9	1007
(3)	Si-CH_3	1263, 1412.3	1171.4; 1408.2
(4)	Si-H	-	2136.6; 2202
(5)	C-H	2905.5, 2971.7	2969.9

4. Conclusion

We have successfully synthesized a low viscosity of PMHS using the DCHS. In the hydrolysis process, the results of the three monomer samples had high viscosity values compared to commercial D4. The surface tension and refractive index values of monomer samples are similar to commercial D4. The monomer sample has the same main functional group as commercial D4, which are vibration of Si–O–Si, Si–CH_3, Si–C and C–H. The difference in the FTIR characterization between the monomer sample and commercial D4 is the existence of a Si-H functional group at a wave number of 2169.7 cm^{-1}, which is due to the characteristics of the DCHS. In polymer synthesis, PMHS has been successfully obtained with a viscosity value of 1130 mPa.s which corresponds to the low viscosity of PDMS which can be a substitute for vitreous humour in vitreoretinal surgery. The surface tension and refractive index values are similar to commercial PDMS. PMHS has the same main functional group as commercial PDMS. The difference in the FTIR characterization between PMHS and commercial PDMS is the existence of a Si-H functional group at a wave number of 2136.6 and 2202 cm^{-1}, which is caused by the characteristics of the DCHS material.

Acknowledgement

Authors would like to thank Kemenristek-DIKTI of Indonesia for financial support in the scheme of Fundamental Research (Penelitian Dasar) 2021, contract no. 1207/UN6.3.1/PT.00/2021. This work was also partially supported by Academic Leadership Grant of Universitas Padjadjaran 2021, contract no. 1959/UN6.3.1/PT.00/2021.

References

[1] Swindle K E and Ravi N 2007 *Expert Rev. Ophthalmol.* **2** 255
[2] Donati S, Caprani S, Airaghi G, Vinciguerra R, Bartalena L, Testa R, Mariotti C, Porta G, Simonelli F and Azzolini C 2014 *BioMed Reasearch Int.* **2014** 351804
[3] Alovisi C, Pacino C, Sanctis U and Eandi C 2017 *Journal of Ophthalmology* **2017** 3172138
[4] Foster W J 2008 *Expert Rev. Ophthalmol.* **3** 211
[5] Setiadji S, Fitrilawati, Fauza A N, Ardi A, Novianti R M, Syakir N, Waslaluddin, Rahayu I, Kastasasmita A S and Risdiana 2019 *Material Science Forum* **966** 189
[6] Giordano G G and Refojo M F 1998 *Prog. Polym. Sci.* **23** 509
[7] Chruściel J J, Fejdyś M and Fortuniak W 2018 *Polymers* **10** 484
[8] Lee S J, Goedert M, Matyska M T, Ghandehari E M, Vijay M and Pesek J J 2008 *J. Micromech. Microeng.* **18** 025026
[9] Yadav M S, Singh A S, Agrahari A K, Mishra N and Tiwari V K 2019 *ACS Omega* **4** 6681
[10] Seyferth D, Prud'Homme C and Wiseman G H 1982 *Inorg. Chem.* **22** 2163
[11] Mardina P, Prathama H A and Hayati D M 2014 *Konversi* **3** 2
[12] Flory P J 1946 *Chem. Rev.* **39** 137
[13] Setiadji S, Sumiyanto E, Fitrilawati, Syakir N, Noviyanti A R, Rahayu I and Risdiana 2020 *Key Engineering Materials* **860** 234
[14] Setiadji S, Sumiyanto E, Seprinandar P D J, Farida H, Fitrilawati, Syakir N and Risdiana 2019 *Jurnal Material dan Energi Indonesia* **09** 8
[15] Fitrilawati, Fauza A N, Ardi A, Novianti R M, Syakir N, Kartasasmita A S and Risdiana 2018 *Journal of Physics: Conf. Series* **1080** 012016

Characterization of manufacturing indicator electrodes coated PVA-enzyme coated 1X GA and 3X GA 2.9 % coated PVC-KTpClPB coated using potentiometry method

A Hakim S[1], S Mihardi[1], A Rais[1], Willy[1] and Y T H Panjaitan[1]

[1]Physics Department, Faculty of Mathematics and Natural Sciences, Universitas Negeri Medan, Medan, Indonesia

E-mail: abdhakims@unimed.ac.id, mihardiunimed@gmail.com, rais.rayyan71@yahoo.com, willysibarani45@gmail.com, yakobthp99@gmail.com

Abstract. Research has been carried out on the manufacture of tungsten indicator electrodes with a diameter of 1 mm coated with PVA-Enzyme coated with PVC-KTpClPB using the biosensor potentiometer method. It has a sensitivity of 19,096 mV/decade, a detection range of 1.10^{-5} - 5.10^{-4} M, a detection limit of 1.10^{-5} M and a correlation coefficient of $R^2 = 0.9431$. This detection range is very small, only a difference of one order to the negative power. On the basis of this small range, the PVA-Enzyme and PVA-Enzyme-GA solutions have been characterized by UV-vis showing a very clear difference in the width of the absorbance peak at wavelength. The indicator electrodes B1-4 and B3-4 have been analyzed using UV-vis, FTIR, SEM-eds and XRD. The method used is the biosensor poteniometric method, urease enzyme immobilization technique, which analytes urea in determining sensitivity, detection range and detection limit, as well as variable signal analysis of 60 signals/second. The results obtained from the indicator electrode B1-4 have a sensitivity of 46.67 mV/decade, a detection range of 10^{-4} - 10^{-2} M and a detection limit of 10^{-4} M, $R^2 = 99.62$ %. The indicator electrode B3-4 has a sensitivity of 38.16 mV/decade, a detection range of 10^{-6} - 10^{-4} M and a detection limit of 10^{-6} M, $R^2 = 95.42$ %. The best results on electrodes B1-4.

1. Introduction

The manufacture of indicator electrodes with the biosensor potentiometric method of urease enzyme immobilization technique. The substrate material is a PVA polymer for immobilizing the urease enzyme and PVC water-resistant material or a solution as a water-soluble PVA-Enzyme protector. Research has been carried out on indicator electrodes coated with PVA-enzyme coated with PVC-KTpClPB with a sensitivity of 19,069 mV/decade in a detection range of 1.10^{-5} - 5.10^{-4} M with a detection limit of 1.10^{-5} M and a correlation coefficient of $R^2 = 0.9431$ [1]. This electrode analysis has been characterized using SEM-EDX [2]. Synthesis of buffer solution KH_2PO_4 +KCl [3]. The sensitivity, detection range and detection limit are problems in this study. First, I analyzed this small detection range by adding Glutaraldehyde (GA) variation (2.6 – 3.0) % into a solution of 0.0500 Gr PVA-Enzyme. The best result of UV-Vis analysis was obtained at GA 2.9%. After analyzing the solution, the indicator electrodes were made in two ways. The first method was that the indicator electrodes were coated with PVA-E-GA (1x, 2x, and 3x) and then coated with PVC-KTpClPB 1x were analyzed using XRD types A1-4, A2-4 A3-4. The way the two indicator electrodes were coated with PVA-E (1x, 2x, and 3x) were coated with GA and then coated with PVC-KTpClPB 1x were analyzed using XRD types B1-4, B2-4, B3-4. The best analysis results in the second method are indicator electrodes B1-4 and B3-4. The best result of XRD analysis used SEM-EDS advanced analysis. The results of SEM morphology analysis showed that the electrodes denoted B1-4 had more pores and the electrodes denoted B3-4 had large pores but there were cracks or fractures. The best results of the next B1-4 and B3-4 analysis were analyzed using FTIR [4]. Based on the four-step sequence of analysis [5; 6] this is where the researchers continued to analyze the indicator electrode using a potentiometric cell with the biosensor potentiometric method of the urease enzyme immobilization technique. This analysis will characterize the sensitivity, detection range and detection limit of samples B1-4 and B3-4. The potentiometer cell uses an Ag/AgCl reference electrode and indicator electrodes B1-3 and B3-4 with a buffer solution of KH_2PO_4 + KCl with a concentration of 0.001 M pH 7.5. This solution has been analyzed [7]; [8].

2. Methods

2.1. Material

The material consists of standard urea 56 180, enzyme EC 3.5.1.5 (Urease) U4002, tungsten diameter 1.0 mm 267 562 99.99%, Phosphate Buffer KH_2PO_4, PVA [-CH_2CHOH-]n, PVC (CH_2CHCl)n, potassium tetrakis 4-chlorophenyl borate $(ClC_6H_4)_4BK$, tetrahydrofuran C4H8O, KCl, Glutaraldehyde derived from Sigma-Aldrich and method. As well as the tools used with the cell potentiometer, the cell potentiometric method, the biosensor, the urease enzyme immobilization technique. Potentiometer (Keithley 199 DMM, USA), tungsten indicator electrode (W), Ag/AgCl MF-2052 RE-5B reference electrode in a microcomputer assembled electrochemical cell (ADI Powerlab instruments, Australia), magnetic stirrer and flow injection (FIA).

2.2. Manufacture

The indicator electrodes are made in two notations B1-4 and B3-4. The materials used are the same, only different in the first layer of immobilization of the urease enzyme on the PVA-Enzyme-GA substrate, while the second layer is the same, namely PVC-KTpClPB.

The composition of the ISE (ion selective electrode) polymer membrane is [9] 1% by weight of ionophore (I), polymer matrix (PVC): plasticizer (1:2) was used for the first layer of PVA-Enzyme. Polymer synthesis in the development of ISE sensors according to [10] with a ratio of I:PVC:KTpClPB:Plasticizer 10:165:5:330, for the second layer of PVC-KTpClPB. Glutaraldehyde concentration of 2.5-3% [11] results of the analysis of the optimal concentration of 2.9% glutaraldeyde for urease immobilization.

PVA 0.0350 g was dissolved in 10 mL of hot water to cool in a glass tube. PVC 0.0350 g + KTpClPB 0.0500 g dissolved in 10 mL THF in a glass tube and covered with a plastic/aluminum foil cover. 2.9% GA solution is 0.29 g GA in 10 mL of distilled water. To coat the indicator electrode, dip into each of the instructions solutions B1-4 and B3-4.

3. Result and Discussion

Likewise in Figure 2, analysis of the signal display from the B3-4 indicator electrode with a potentiometer cell, namely the signal display at a 2k:1 scale in the image on the left and the signal display at a frequency of 10000 Hz. If you pay close attention to Figure 2, there is a time response to the 0.001 M KCl phosphate buffer solution pH 7.5 with urea molarity variations ranging from 10^{-7} - 10^{-2} M, see Figure 3b.

Figures 3a for B1-4 and 3b for B3-4 have a time response to the potentiometer cell voltage ranging from 10^{-7} - 10^{-1} M urea variation in KH_2PO_4 buffer solution with 0.001 M KCl pH 7.5. Analysis of images 3a and 3b can be used as a guide to determine sensitivity, detection range and detection limit.

On the basis of the relationship of -log (urea concentration) M to the potentiometer voltage of the potentiometer cell. Figure 3a in the researcher's analysis there is only a range of 10^{-4} -10^{-2} M.

According to signal analysis, the variables in Figures 4a and 4b have symmetry in the molarity of urea 10^{-6} - 10^{-2} M in phosphate buffer KH_2PO_4 and KCl 0.001 M pH 7.5. This symmetry range is used as the following conclusions. Analysis of this variable signal was taken as many as 60 signals/second each from the cell signal of the indicator electrode potentiometer B1-4 and B3-4.

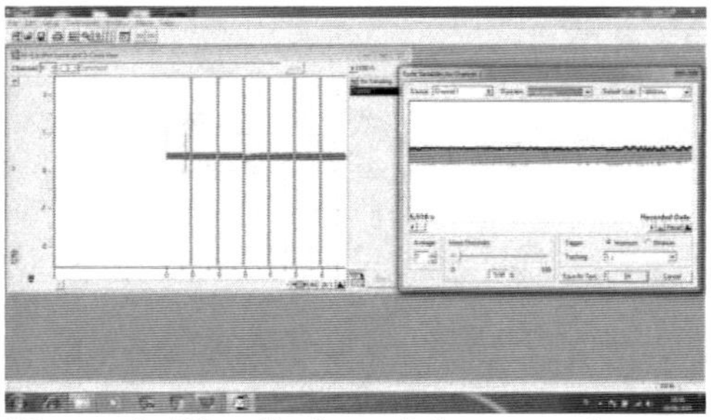

Figure 1. The display of the potentiometer cell signal and the display of the variable analyzer at a frequency of 10000Hz from the indicator electrode B1-4.

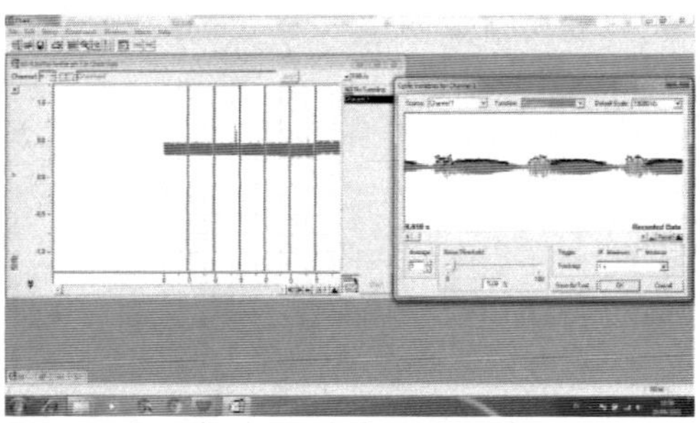

Figure 2. Display of the potentiometer cell signal and display of the variable analyzer at a frequency of 10000Hz from the indicator electrode B3-4.

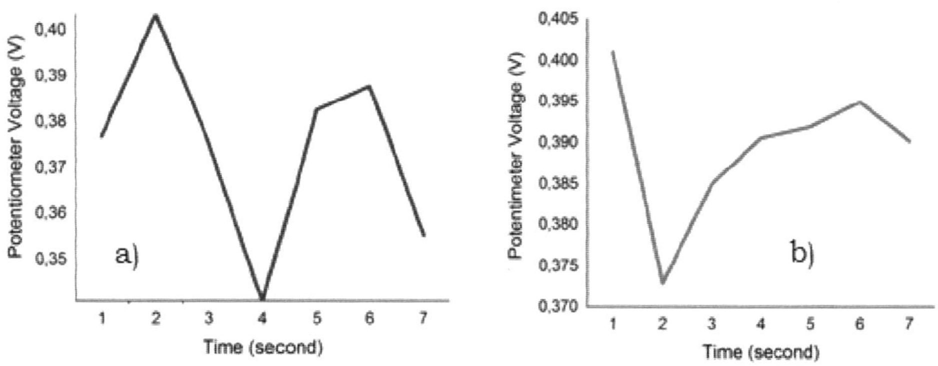

Figure 3. The response time to the potentiometer cell voltage ranged from 10^{-7} - 10^{-1} M urea variation in KH_2PO_4 buffer solution with 0.001 M KCl pH 7.5. B1-4 a), B3-4 b).

First for the B1-4 indicator electrode, the analysis pays attention to Figures 3a and 4a, it can be seen in Figures 5a and 5b that the sensitivity is 46.67 mV/decade, the detection range is 10^{-4} - 10^{-2} M and the detection limit is 10-4 M, R^2 = 99.62%. Probability of linear response time only lies in the molarity of urea 10^{-4} - 10^{-2} M, others do not. Second, for the B3-4 indicator electrode, the analysis looks at Figures 3b and 4b, it can be seen in Figures 6a and 6b, the sensitivity is 36.67 mV/decade, the detection range is 10^{-6} - 10^{-2} M and the detection limit is 10^{-6} M. There are several opportunities for linearity to be obtained due to the relationship between the -log concentration of urea (M) to the potentiometer voltage is not perfectly linear.

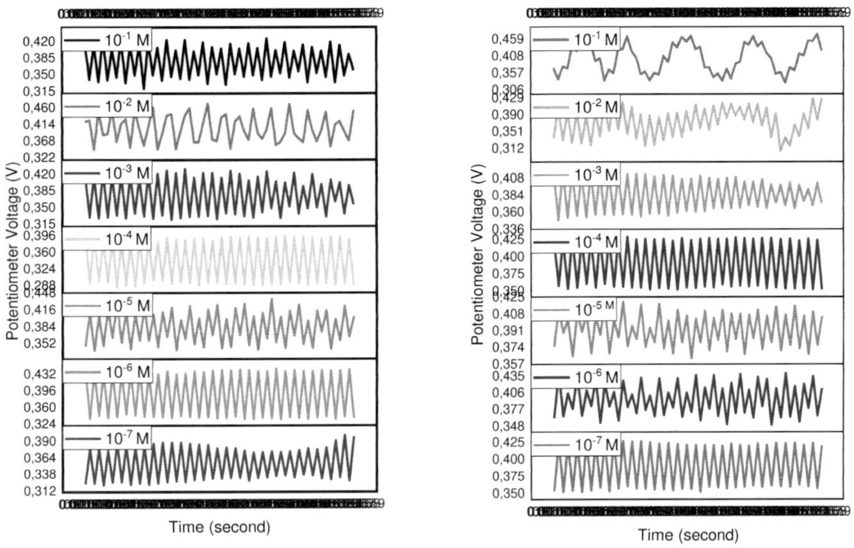

(a) (b)

Figure 4. Variable signal analysis of indicator electrodes B1-4 and B3-4.

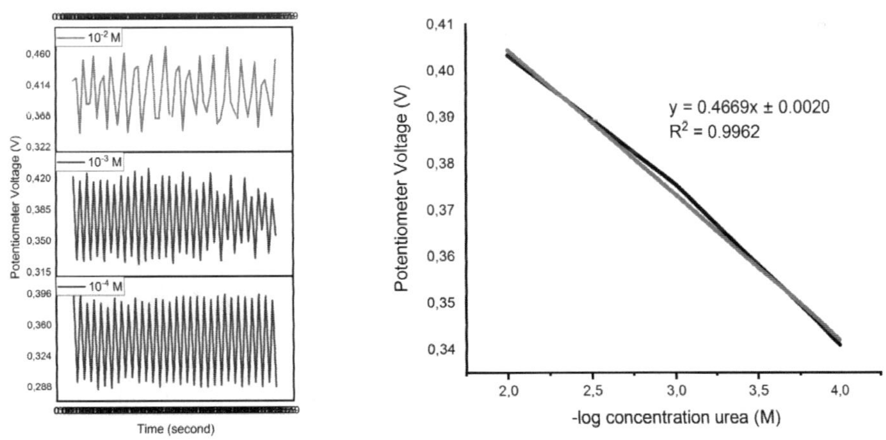

(a) (b)

Figure 5. Analysis of the variable signal from the potentiometer cell and the sensitivity is 46.67 mV/decade, the detection range is 10^{-4} - 10^{-2} M and the detection limit is 10^{-4} M, R^2 = 99.62 %. B1-4.

Analysis of images 6a and 6b, images 7a and 7b as well as images 8a and 8b there are differences in the sensitivity of the detection range and detection limit. The greatest sensitivity was obtained at

Journal of Physics: Conference Series **2165** (2022) 012042 doi:10.1088/1742-6596/2165/1/012042

38.16 mV/decade at 10^{-6} - 10^{-4} M urea molarity for indicator electrode B3-4. This is also confirmed by the analysis of the variable signal from the potentiometer cell. The symmetry of the signal is also found in the urea molarity of 10^{-6} - 10^{-4} M, and the detection limit is 10^{-6} M, $R^2 = 95.42$ %.

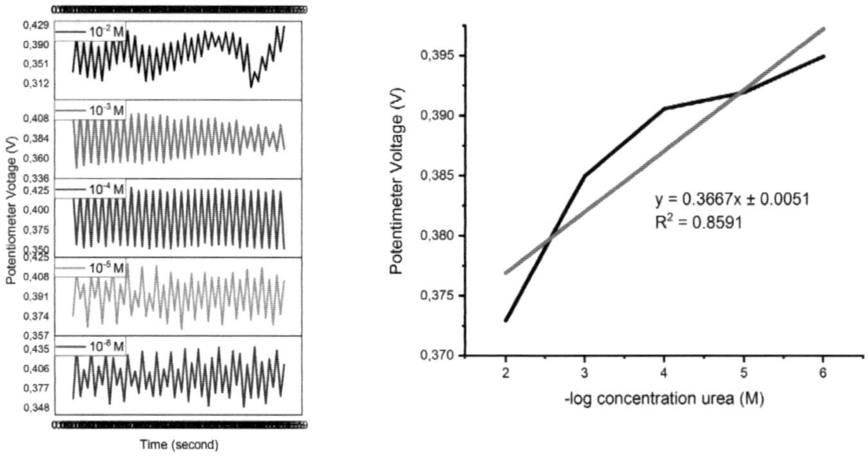

(a) (b)

Figure 6. Analysis of variable signal and sensitivity is 36.67 mV/decade, detection range is 10^{-6} - 10^{-2} M and detection limit is 10^{-6} M, $R^2 = 85.91$ %. B3-4.

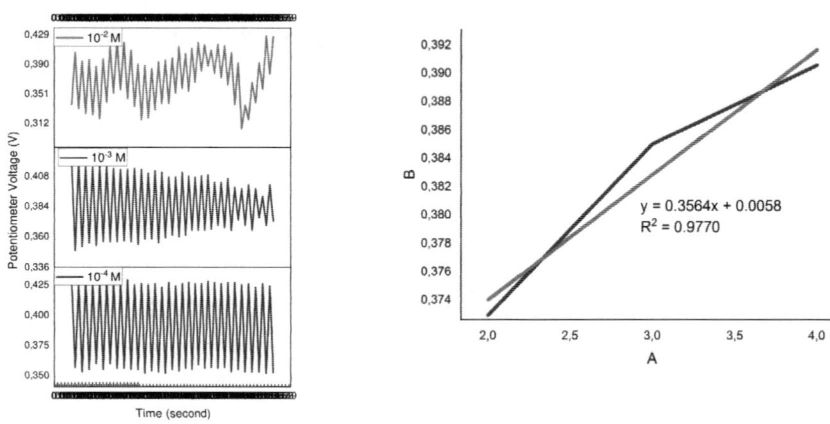

(a) (b)

Figure 7. Variable signal analysis and sensitivity 35.64 mV/decade, detection range 10^{-4} - 10^{-2} M and detection limit 10-4 M, $R^2 = 97.70$ %. B3-4.

The symmetry of the signal [7] is very dependent on the incoming signal and outgoing signal from the redox process, that the incoming signal is not the same as the outgoing signal with a total signal data of 400 signals/second [12]. Signal symmetry [13] can be seen in Figure 1 and Figure 2, Figure 4a and Figure 4b, Figure 5a, 6a, 7a and 8a. The voltage between the oxidation and reduction peaks [14] is theoretically 59 mV for a reversible reaction, as can be seen in Figures 5a and 8a. Voltage greater than 100 mV is the peak of nonsymmetric oxidation and reduction indicating a nonreversible (irreversible) reaction, as can be seen in Figures 6a and 7a.

It has been found that the PVA electrode membrane has a sensitivity of 19.7 mv/decade with a range of 5.10^{-7} - 10^{-2} M by [15]. The sensitivity value varies depending on the range and limit of detection [16] and [17].

The development of a biosensor system follows [18], namely (1) Selectivity, (2) Sensitivity, (3) Response linearity, namely the concentration range of the target analyte to be measured, (4) Reproducibility of signal response, samples having the same concentration analyzed several times should give the same response, (5) Fast response time and recovery time for reusability of the biosensor system, (6) Stability and operating life.

Conductive polymer is a chemical sensor in the application of metal electrodes coated with a conductive polymer film used for potentiometric measurements [19]. Conductive polymers are organic materials that have electrical and optical properties similar to those of metallic conductors and semiconductors.

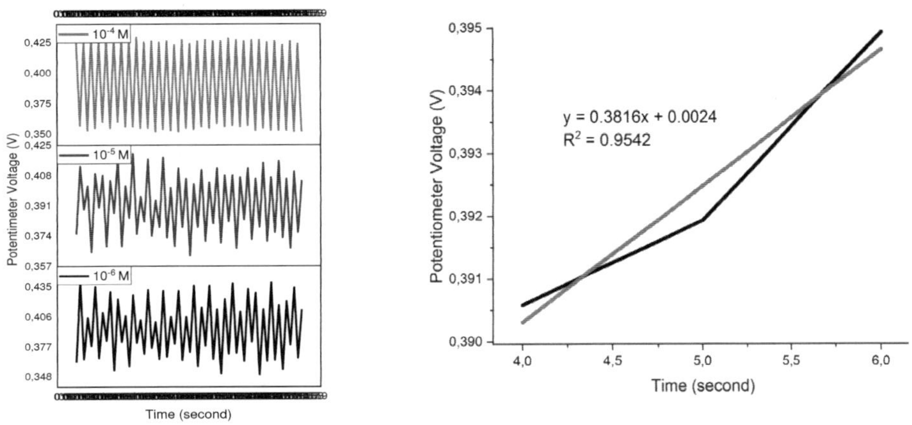

(a) (b)

Figure 8. Variable signal analysis and sensitivity 38.16 mV/decade, detection range 10-6 - 10-4 M and detection limit 10-6 M, R2 = 95.42 %. B3-4.

The immobilization technique was developed based on three important mechanisms, namely (a) physical adsorption, (b) covalent immobilization, (c) streptavidin-biotin immobilization. Achieving high sensitivity and selectivity requires minimization of nonspecific adsorption and stability [20].

On this basis, the researchers developed a modified PVA-enzyme coated with GA with an immobilization technique using a potentiometric biosensor method, coated again with PVC-KTpClPB to protect the electrode from the KH2PO4 +KCl buffer solution. In order for a long lasting indicator electrode to be used.

Research has been carried out on PVA-enzyme electrodes coated with PVC-plasticiser 19,069 mV/decade in the range from 1.10^{-5} - 5.10^{-4} M with a detection limit of 1.10^{-5} with R^2 = 94.31% using 0.001 M KCl phosphate buffer 0.001 M pH 7.5 [1].

The indicator electrode coated with PVA-enzyme coated GA coated with PVA-KTpClPB increased its detection range to two orders of B1-4, detection range was 10-4 - 10-2 M and B3-4 detection range was 10-6 - 10-4 M. So it is true that GA can increase the detection range.

4. Conclusion

Based on the analysis of variable signal, sensitivity, detection range and detection limit, there is a relationship between the symmetry of the variable analysis signal with sensitivity, detection range and detection limit to determine the best indicator electrode. The results of the analysis and research showed that the indicator electrode B1-4 had a sensitivity of 46.67 mV/decade, a detection range of 10^{-4} - 10^{-2} M and a detection limit of 10^{-4} M, R^2 = 99.62 %. B3-4 sensitivity is 38.16 mV/decade, detection range is 10^{-6} - 10^{-4} M and detection limit is 10^{-6} M, R^2 = 95.42 %. This step is also corroborated by analysis and characterization of UV-vis, FTIR, SEM-EDS and XRD from previous studies. Thus the best indicator electrode sample is B1-4.

Acknowledgement
The author would like to thank the Chancellor of the State University of Medan who has provided research grants APPLIED RESEARCH, the leadership of the LPPM State University of Medan, the Chemical Laboratory of the State University of Medan so that the research data can be analyzed and the author can write this article first.

References

[1] Hakim S A , Sembiring T, Tarigan K, Sebayang K, Situmorang M, and Noer N M 2019 Characterization of Membrane PVA-Enzyme Coated PVC-KTpClPB as Urea Sensor with Potentiometric Method *Rasayan J. Chem.* **12**(2) 780-786

[2] A Hakim S, Krista T, Krista S, Timbangen S and Manihar S 2018 The Characteristics Membrane Pva-Enzyme And Coating Pvc-Plasticizer With Sem-Edx *International Journal of Scientific & Engineering Research* **9**

[3] A Hakim S, Krista T, Manihar S and Timbangen S 2018 Synthesis of Urea Sensors using Potentiometric Methods with Modification of Electrode Membranes Indicators of ISE from PVA- Enzymes Coating PVC- KTpClPB *Journal of Physics: Conf. Series* **1120** 012024

[4] A Hakim S, Sembiring T, Sebayang K, Situmorang M, Tarigan K, Sembiring K, Mihardi S and Waruwu E N 2020 Characterization of indicator electrodes using the potentiometric biosensor method as urea sensor with XRD, SEM-EDS, UV-Vis and FTIR *ICTAP 2020*

[5] A Hakim S, Krista T, Situmorang M and Sembiring T 2019 Characterization Membrane Composition Of PVA-Enzyme Coating PVC-KTpClPB As Urea Sensor With UV-VIS, SEM-EDX and XRD IOP *Journal of Physics: Conf. Series* **1317** 012042

[6] A Hakim S, Tarigan K, Situmorang M, Sembiring T and Ritonga F S 2018 Characterization of Electrodes Membrane PVA-Enzyme Coating PVC-KTPCLPBas Urea Sensors With SEM-EDX, XRD and UV-Vis *AISTSSE 2018*

[7] A Hakim S, Situmorang M, Sebayang K, Sembiring T, Tarigan K, Mihardi S, Noor N M and Elnovreny J 2021 Characterization Of Urea Analyte Solution and Electrolyte Solution Using Non-Immobilized Electrode Indicators and Pva-Enzyme Coated PVC-KTpClPB Immobilization *Rasayan J. Chem.* **14**(2) 1356-1368.

[8] Victor H S, A Hakim S 2021 The Use of Tungsten in Potentiometry to Detect Pospat Baffer and Urease Enzyme *Jurnal Penelitian Pendidikan IPA* **7**(3) 325-330

[9] Górski L, Klimaszewska D, Pietrzak M, Malinowska E 2007 Enzymatic detection of glucose using fluoride-selective electrodes with polymeric membranes *Original Paper Anal Bioanal Chem* **389** 533-539

[10] Vlascici, D, Eugenia Fagadar Cosma E F, Pica E M, Cosma V, Otilia Bizerea O, Gheorghe Mihailescu G and Olenic L 2008 Free Base Porphyrins as Ionophores for Heavy Metal Sensors *Sensors* **8** 4995-5004

[11] Kayastha A M and Srivastava P K 2001 Pigeonpea (Cajanus cajan L) Urease Immobilized on Glutaraldehyde-Activated Chitosan Beads and Its Analytical Applications *Applied Biochemistry and Biotechnology* **96** 43-52

[12] Basile A, Bhatt A I and O'Mullane A P 2015 Stabilizing lithium metal using ionic liquidsfor long-lived batteries *Nature communications* 1-11

[13] E Kim, Y Liu and G F Payne 2017 Catechol-Based Hydrogel for Chemical Information Processing *Biomimetics* **2**(11)

[14] Rahman M M, Li X B, Lopa N S, Ahn S J and Lee J J 2015 Electrochemical DNA Hybridization Sensors Based on Conducting Polymers *Sensors* **15** 3801-3829

[15] Ibupoto Z H, Shah S M U A, Kimleang K and Willander M 2012 Electrochemical L-Lactic Acid Sensor Based on Immobilized ZnO Nanorods with Lactate Oxidase *Sensors* **12** 2456-2466

[16] Elhag S, Ibupoto Z H, Nur O and Willander M 2014 Incorporating ß-Cyclodextrin with ZnO Nanorods: A Potentiometric Strategy for Selectivity and Detection of Dopamine Sensors **14** 1654-1664

[17] Thakur M S and Ragavan K V 2013 Biosensors in food processing *J Food Sci Technol* **50**(4) 625-641

[18] Braham Y, Barhoumi H, Maaref A, Bakhrouf A and Renault N J 2014 Proteus Mirabilis Bacteria Biosensor Development Based on Modified Gold Electrode with 4-Carboxyphenyl Diazonium Salts for Heavy Metals Toxicity Detection *Sensors & Transducers* **27** 29-38

[19] Nimse S B, Song K, Sonawane M D, Sayyed D R and Kim T 2014 Immobilization Techniques for Microarray: Challenges and Applications *Sensors* **14**(12) 22208-22229

Usage of e-portfolio as an assessment tool in physics learning

Sarwandi[1,2], Basuki Wibawa[1] and Redisa Wibawa [1]

[1]Departement of Educational Technology, Postgraduate Program, State University of Jakarta
[2]Human Resourse Development, Ministry of Transportation

E-mail: sarwandi.diklathub@yahoo.co.id

Abstract. An e-portfolio is a digital data collection of trainees that can advance learning by providing a way to organize, archive, and display individual or group work. The purpose of writing this article is to explain the use of e-portfolios as an assessment tool in training activities. E-portfolio as an assessment tool is more effective than standardized assessment tests which are many and complex because they facilitate the learning process and assessment. The method used is a literature study. As a result, the e-portfolio assessment using the LMS device was used as a learning medium, especially to store all of the participants' learning activities during the training. E-portfolio assessment can support the learning process well, because it can assess the data collection of the work of trainees during the training process, both cognitive, affective and psychomotor assessments.

1. Introduction

The world of education today forces us to optimize the use of information and communication technology in learning activities. The main information and communication technology on which it is based is digital media technology through a web-based internet network. Assessment is a central to educational practice, an important component of education, an essential component in the learning and teaching environment, and the heart of formal higher education [1]. In this process, assessment become significant for educational practices, it should be implemented considerably in Training Institution. There has been a shift in concepts and methods in an assessment system; from traditional assessment to digital assessment. Nowadays, technology plays significant roles in the process of assessment, especially, to evaluate students learning experiences and also to measure their progress academically. That is mean we can not avoid anymore about development of technology because the global demands of industry 4.0 and the fast growth of information technology (IT) infrastructure, very promising in the online education [2] included in the e-portofolio.

During the Covid-19 pandemic, almost all Education Institutions carried out online learning processes both directly (synchronously) and indirectly (asynchronously), so that all learning materials and learning assignments were stored in digital form or electronic documents called electronic portfolios.

"...Electronic portfolios can be used as a tool for assessing scientific literacy in evolutionary theory relevant to the industrial revolution 4.0" [3]. E-portfolios can be used effectively as an assessment in learning, can improve the different learning styles of students and support independent learning [4]. The e-portfolio can be a versatile pedagogical tool and career benefit that allows students to reflect on academic performance, get feedback, and complete assignments to get an assessment and showcase it to potential employers [5]. Electronic Portfolio (or e-Portfolio) is one of the innovative e-learning tools that has proven to be very useful in monitoring student progress and increasing Educational output [6].

Based on some of the opinions above, the author intends to explain the electronic portfolio or e-portfolio as an assessment tool in learning activities. The e-portfolio is expected to make it easier for lecturer and cadets to access learning materials and assignments anytime anywhere. Another convenience is that the use of e-portfolios can help lecturer monitor learning activity in accessing and collecting assignments or reports, as well as to encourage participants to be more active and more creative in the learning process.

1.1. Definition of e-Portfolio and Assessment

An e-Portfolio is a collection of student work built electronically on the Web or on recorded media such as CDs, DVDs, Zip disks, etc. [6]. In a broad sense, Kunnari and Laurikainen define e-Portfolio as "student's digital work and learning space to collect, create, share, collaborate, reflect on learning and competence, and store assessments and evaluations [7].

Bruckner states that in education an electronic portfolio is a collection of student work that can advance learning by providing a way to organize, archive, and display work [8]. Furthermore, according to Spector an e-portfolio (spelled ePortfolio) is a digital collection of artifacts collected and organized to show the achievements of individuals, groups, or organizations [9]. From some of the opinions above, it can be concluded that e-portfolio is a collection of student work that is built electronically to organize, archive, reflect on learning, store assessments and evaluations and display work in advancing learning both individually, in groups and in organizations.

The definition of assessment according to Arifin is a systematic and continuous process or activity to collect information about the process and learning outcomes of students in order to make decisions based on certain criteria and considerations [10]. A broader definition of assessment put forward by Cronbach and Stufflebeam, that the assessment process is not just measuring the extent to which goals are achieved, but is used to make decisions.

According to Haryati that assessment is a term that includes all methods commonly used to determine student learning success by assessing the performance of individual students or groups [11]. Based on the opinion above, it can be concluded that assessment is a comprehensive process of collecting information which is carried out continuously to determine the ability or success of students in learning by assessing student performance both individually and in group activities.

1.2. Definition of e-Portfolio and Assessment

Based on the fact that e-portfolios as an assessment tool provide an advantage in encouraging lifelong learning by making it possible to collect, monitor, and study student experiences [12]. The use of e-portfolios effectively can be carried out for various types of assessments, namely as a formative assessment tool which is believed that e-portfolios are more effective than standard assessment tests considering the many and complex aspects of learning [13].

The portfolio-based valuation model refers to a number of basic principles of valuation. The basic principles of the assessment are process and outcome assessment, periodic and continuous assessment, fair assessment, and assessment of the social implications of learning [14]. In the process of teaching and learning activities or *kegiatan belajar mengajar* (KBM), the process of testing or assessment is a component that is no less important than other processes. Assessment or testing or often called assessment has many models, such as class-based assessment, benchmarking, and portfolio.

However, evaluating using an e-portfolio is advantageous and convenient because as a repository for large amounts of data, it provides the necessary space for digital media artifacts such as high-resolution graphic images and digital video. E-portfolios continue to become more and more common and are very attractive to millennial students who are very comfortable with computer-based media.

2. Methods

The writing of this article uses a qualitative approach with a literature study method that focuses on the results of writing related to the topics discussed. A literature study is a comprehensive overview of the research that has been done on a specific topic to show the reader what is already known about the topic and what is not known, to seek rationale from research that has been done or for further research ideas. Literature studies can be obtained from various sources, both journals, books, documentation, internet and libraries. This article describes the use of e-portfolios as an assessment in Physics learning at the Aircraft Engineering Study Program in Indonesian Aviation Polytechnic Curug Ministry of Transportation.

3. Result and Discussion

Based on a study of several articles related to e-portfolios, it can be explained that e-portfolio as an assessment tool in the Physics learning process for cadets of the Aircraft Engineering study program in Indonesian Aviation Polytechnic Curug, Ministry of Transportation have a positive impact on the development and learning outcomes of cadets. The results of this study are explained according to the following description:

3.1 Device used

E-portfolios are widely used as an assessment tool in the learning process, especially in the current pandemic conditions, almost all educational institutions, both public and private schools apply online learning. In the online learning process, each educational institution or institution uses different platforms or tools internet-based learning.

The devices commonly used to support the learning process include the Learning Management System (LMS), Edmodo, Facebook, Google Classroom and Whatsapp. This device is often the choice for education providers and teaching staff. However, in the implementation of learning at the Aircraft Engineering study program in Indonesian Aviation Polytechnic Curug, the organizers use the LMS as a medium in the learning process and assessment.

3.2 Procedure for using e-portfolio

Before starting the online learning process, educators and students must have an account to enter the system that has been prepared by the organizer or training committee. After having an account, teachers and students can enter and carry out online learning activities. The steps taken to enter and start the learning process are:

1) Login to the device you are using
2) Enter and choose a predetermined menu
3) Online learning process
4) Finish learning log out or exit the device.

3.3 Implementation of E-portfolio

E-portfolio is software that helps students to present their mastery and skills in a richer way, in this case using the help of technology. The application of e-portfolios with platforms or online learning tools, there are menus that are used for the implementation of the learning process. The menus contain e-portfolio materials for lecturers and cadets that can be accessed anytime and anywhere.

The e-protofolio that is used as material in the learning process includes learning materials, assignments, attendance and question and answer columns. During the online learning process, e-portfolio materials will be stored and documented in the digital devices used and can be viewed by lecturers and cadets at any time.

3.4 E-Portfolio Assessment

E-portfolio as an assessment tool in training can provide direct feedback and reflection for both lecturers and cadets. The e-portfolio that is a source of assessment of cadets in the physics learning process includes the tasks submitted, attendance, activity and creativity, discipline, communication techniques and test results. All of these sources are used to assess the cognitive, affective and psychomotor abilities of cadets.

From the explanation of the results above, that the e-portfolio in the physics learning process for Cadets of the Aircraft Engineering study program in Indonesian Aviation Polytechnic Curug, Ministry of Transportation is very appropriate. The internet-based digital learning tools used are Learning Management System (LMS) devices. This device is used as a learning medium, especially to store all participant learning activities. During the learning process, the activities carried out, including the submitted assignments, are in the form of an electronic portfolio that is stored as an assessment material for lecturers and educational institution.

This is in accordance with the opinion of Bruckner that Learning Management System (LMS) is software used to deliver, track, and manage education and learning [8]. For example, the LMS tracks attendance, time for assignments, and student progress. Lecturers can also post announcements, grade assignments, check learning activities, and participate in class discussions. Cadets can submit their work, read and respond to discussion questions, and take quizzes.

The use of e-portfolios with LMS devices is expected to support the learning process well, although at the begining of the learning implementation, participants still experience many problems due to lack of knowledge and skills of participants. According to Kicken in Lukitasari, it is often found that students are still confused and have an impact on not understanding the use of these facilities properly [15]. From the statement of Kicken and his friends, it also happened to the trainees who were new to and using the device. Therefore, cadets need to know and understand the technical use of LMS devices according to the procedures presented. So that in its implementation, assistance is needed to find out the obstacles and difficulties experienced in order to be able to use the device properly as a learning medium.
An E-Portfolio for assessment is a collection of work that demonstrates achievements in a field, program, or project. E-portfolios for assessment can measure soft skills, such as critical thinking and problem solving skills. Like the E-portfolio for learning, the components of the E-portfolio for assessment mostly contain a process of reflection and self-assessment [16]. Therefore, e-portfolio as a tool for assessment is expected to be able to assess the data collection of the work of cadets during the learning process, both cognitive, affective and psychomotor assessments.

4. Conclusion

An e-portfolio is a digital data collection of cadets that can advance learning by providing a way to organize, archive, and display the work of individuals, groups and organizations. Assessment by using an e portfolio is profitable and convenient because it is a place to store large amounts of data, providing the necessary space for works produced with digital media such as high-resolution graphic images and digital video. The e-portfolio assessment uses an LMS device that is used as a learning medium, especially to store all cadets learning activities during education and learning. E-portfolios with LMS devices can support the learning process well, because they can assess the data collection of the work of cadets during the learning process, both cognitive, affective and psychomotor assessments.

Acknowledgment
The author would like to thank for the Aircraft Engineering study program in Indonesian Aviation Polytechnic Curug, Ministry of Transportation, Departement of Educational Technology State University of Jakarta and all those who helped the realization of this article.

References

[1] Wibawa B and Paidi 2019 The Development of Blended Learning Based on Handphonr for Computer System Subject on XI Grade of SMKN 1 Bengkulu City *Humanities & Social Sciences Reviews* **7**(3), 497–502

[2] Wuisan, Imelda P, and Wibawa B 2019 *International Journal of Recent Technology and Engineering (IJRTE)* **8** 2S9

[3] Prastiwi M S, Kartowagiran B and Susantini E 2020 Assessing using technology: Is electronic portfolio effective to assess the scientific literacy on evolution theory *International Journal of Emerging Technologies in Learning* ***15***(12), 230–243

[4] Yastibas A E, Yastibas G C 2015 The use of e-portfolio-based assessment to develop students' self-regulated learning in English language teaching *Procedia-social and behavioral sciences* **176**: 3-13

[5] Syzdykova Z, Koblandin K, Mikhaylova N and Akinina O 2021 Assessment of E-Portfolio in Higher Education *International Journal of Emerging Technologies in Learning* ***16***(2), 120–134

[6] Tatnall A 2020 Encyclopedia of Education and Information Technologies *Encyclopedia of Education and Information Technologies*. https://doi.org/10.1007/978-3-030-10576-1

[7] Kunnari I and Laurikainen M 2017 Collection of Engaging practices on ePortfolio Process. Häme University of Applied Sciences, Hämeenlinna. Retrieved from https://drive.google.com/file/d/0BxEnFq7yUumMUGV2V2VxVmNaNFU/view

[8] Bruckner, M 2015 *Educational Technology* https://www.researchgate.net/publication/272494060.

[9] Michael S J 2015 *The SAGE Encyclopedia of Educational Technology* (SAGE Publication, Inc. United States of America)

[10] Arifin Z 2012 *Evaluasi Pembelajaran (*Direktorat Jenderal Pendidikan Islam Kementerian Agama)

[11] Haryati M 2009 *Model & Teknik Penilaian pada Tingkat Satuan Pendidikan* (Gaung Persada. Jakarta)

[12] Hui Y K 2017 *The Role of E-Portfolio for Smart Life Long Learning. In International Conference on Smart Education and Smart E-Learning* (Springer, Cham) pp. 327-356.

[13] Sharifi M, Soleimani H and Jafarigohar M 2016 E-portfolio evaluation and vocabulary learning: Moving from pedagogy to andragogy *British Journal of Educational Technology* **48**(6): 1441-1450 . https://doi.org/10.1111/bjet.12479

[14] Budimansyah D 2002 *Model Pembelajaran dan Penilaian Portofolio* (PT. Genesindo. Bandung)

[15] Lukitasari M, Handhika J and Murtafiah W 2016 Potensi E-Portofolio sebagai asesmen Alternatif di Perguruan Tinggi *Prosiding Seminar Nasional Hasil Penelitian*, 56–59.

[16] Ibrahim N and Wargahadibrata R. A H 2016 Pemetaan Fungsi Platforn E-Portofolio Untuk Perkuliahan di Jurusan Kurikulum dan Teknologi Pendikan Fakultas llmu Pendidikan Universitas Negeri Jakarta *JTP - Jurnal Teknologi Pendidikan*, ***18***(3), 202–214.

Impact of The STEM approach with formative assessment in PjBL on students' critical thinking skills

Parno[1*], D A Nur'aini[1], S Kusairi[1] and M Ali[2]

[1]Physics Education, Universitas Negeri Malang, Indonesia
[2]School Education, Universiti Teknologi Malaysia, Malaysia

E-mail: parno.fmipa@um.ac.id

Abstract. Critical thinking skills (CTS) are skills that need to be developed in the 21st century. However, building students' CTS through a STEM approach with formative assessment in project-based learning (PjBL-STEM-AF) is rarely done. This study aims to investigate the impact of STEM-AF in PjBL in constructing students' CTS in the topic of fluid statics. This study used a quasy experiment with a pretest-posttest control group design with students of class XI in Malang, Indonesia, as subjects. PjBL-STEM-AF model and the PjBL model was applied to the Experimental and Comparison classes, respectively. Students made two projects, namely a pressure element HARTL and a simple boat project. The research instrument was 9 essay items of critical thinking skills test with a reliability of 0.625. The data were analyzed by t-test, N-gain, and d-effect size. The results showed that the CTS in the Experimental class with a mean of 65.19 and an increase of 0.52 (medium) was significantly higher than the comparison class with 52.36 and 0.35 (medium). Experimental class' students have different levels of improvement on the likelihood and uncertainty analysis indicator. However, the two classes had almost the same increase in the indicator of problem-solving and decision-making, and an increase in the low category on the hypothesis testing indicator. The value of effect size at 1.13 (Very Large category) suggests that PjBL-STEM-AF learning should be done more frequently in the field. The Experimental class students gave a more positive learning response than the Comparison class. It is recommended in future studies to add an "Art" aspect to the STEM approach to develop students' CTS further.

1. Introduction

The 21st century requires communication skills, creativity, critical thinking, and the ability to manage digital information [1]. Science education should develop 21st-century skills such as critical thinking, problem-solving, and information literacy [2]. The 2013 Indonesian curriculum also aims to hone 21st-century thinking skills such as critical, creative thinking skills, and solving problems [3]. These skills are the process of thinking and behavior of students when studying content in learning. It appears that students need critical thinking skills in learning science.

Physics subjects require critical thinking skills (CTS). Critical thinking requires high-level thinking to analyze the right reasons for making decisions during solving a problem [2]. This skill refers to three cognitive complexities, namely analyzing, evaluating, and creating [4,5]. The application of CTS requires students to ask and answer their questions [6]. Therefore, CTS is not only about the process of gathering information and knowledge, but is considered as the ability to think clearly and rationally [7].

Several studies show that students still have all or some of the CTS indicators that are below expectation. The indicators of student analysis, evaluation, explanation, and conclusion drawing are still low [8]. Evaluation indicators [9], and analytical skills in physics [10] are also still low. This is due to the difficulty of students in connecting learning outcomes with actual phenomena [9]. In static fluid material, students only achieve 20% critical thinking skills [11], or less than 30% [12], or less than 50% [13]. Meanwhile, static fluid material presents daily events, solving mathematical equations, and applying technology [14]. Several studies including 5E-peer instruction learning [15], and discovery learning [16] have been carried out to build critical thinking, but the learning has not covered all the characteristics of static fluid material, and has not provided opportunities for teacher-student feedback during the learning process. Students can find problems by studying the concept of static fluid by observing the surrounding environment and solving them based on relevant theories and concepts [17]. CTS can be trained by implementing an integrated active learning project [12] and providing problem situations [18]. Learning feedback through formative assessment can help teachers and students to find the best way to learn [19,20] so that the learning process becomes perfect [21]. It appears that in static fluid material CTS can be stimulated to develop through solving everyday contextual problem situations in the form of making projects as technology miniatures in project-based learning integrated STEM approach with formative assessment (PjBL-STEM-AF).

STEM (science, technology, engineering, and mathematics) is an alternative learning approach that requires students to make miniature technology products and use them as learning media. The preparation of implementing a particular learning model that is integrated with STEM is very important to achieve learning objectives [22]. For example, the application of the STEM integrated project-based learning model (PjBL-STEM) can provide experiences related to the integration of knowledge and its application [23]. PjBL-STEM challenges, motivates, and demands students to be analytical in the process of improving higher order thinking skills [24]. It is clear that through PjBL-STEM students have experience making products as their own learning media.

In the STEM process, an assessment of PjBL-STEM learning is usually given including a summative assessment to ensure students meet the objectives [24]. This assessment is carried out at the end of the lesson to evaluate the learning that has been carried out and the time for the assessment has been set [25]. However, the assessment given only at the end of the lesson did not reveal students' thinking skills [26]. In other words, summative assessment can only measure the achievement of learning objectives, but without improving students' critical thinking skills. Therefore, a formative assessment is needed during the learning process so that it can be a means to help students apply knowledge.

Formative assessment is an appropriate assessment to help students apply knowledge during the learning process [25]. Through formative assessment, teachers can provide appropriate responses to students' learning progress. The form of the teacher's response gives is adjusted to the level of students' learning progress. Responses from teachers that will help improve students' CTS. Provision of formative assessment can be done through reports and student journals [26]. However, the PjBL-STEM model with formative assessment to build critical thinking skills on static fluid materials is still rarely done. Therefore, this study aims to investigate the effect of the STEM approach with formative assessment in the PjBL model to build students' critical thinking skills.

2. Methods

This type of research is a quasi-experimental with pretest-posttest non-equivalent group design [27]. The experimental class (n=30) applied 7 PjBL-STEM model syntax [25] with 5 domains in formative assessment [28], while the comparison class (n=22) applied 5 PjBL model syntax [29]. The subjects of this study were students of class XI SMAN 1 Malang, Indonesia, which was determined by cluster random sampling technique. The act of learning static fluid material is carried out in 2 rounds. The first round presents the sub-materials of pressure, hydrostatic pressure, meniscus, capillarity, and viscosity with products in the form of pressure element HARTL, while the second round presents the sub-materials of Pascal's and Archimides' laws with products in the form of simple boat.

The treatments given to the two classes differ, but some treatments are the same. The same treatments include (1) making the same engineering product, namely a pressure element HARTL project in the first round, and a simple boat project in the second round, (2) contextual problems in the first and second rounds of learning, each of which is a diver who died due to decompression, and road access to school was blocked by a stretch of river, and (3) syntax activities 1, 2, 3&4, 5, 6, and 7 in PjBL-STEM-AF correspond to syntax activities 1, 2, 3, 4 , 5, and 6 in PjBL, and (4) using the same length of time. The different activities between the two classes include the last syntax (1) the experimental class carries out self-assessment and peer-assessment, while the comparison class conveys difficulties in doing projects, (2) the experimental class carries out more quantitative activities in the laboratory than the comparison class, (3) the experimental class does project up to the step of improving the project that has been tested and collects reports on the results of project trials, while the comparison class only goes to project trials and collects project creation video files, and (4) the experimental class performs calculations and analyzes the result data more observations than the comparison class. The forms of formative assessment activities that are only given to the experimental class include (1) submission of project assessment criteria, (2) discussions to find the advantages and disadvantages of problem solutions, (3) the division of the second round of groups is made based on the scores of the first round of quizzes, (4) discussing design a group project with other groups, (5) write Lesson Tweets containing material that has been understood during the lesson, (6) discuss the completion of quiz questions, and (7) draw conclusions on the overall learning from the sub-materials of the first meeting to the last.

This study uses The Fluid Statics' Critical Thinking Skills Test instrument. This instrument is in the form of 9 essay questions containing indicators of critical thinking skills: reasoning, hypothesis testing, argument analysis, likelihood and uncertainty analysis, and problem-solving and decision-making [30] with reliability of 0.625. Pretest and posttest data were analyzed by independent t-test to determine the presence or absence of significant differences [31], N-gain to determine the level of improvement [32], and effect size to determine the strength of practical implementation in schools [31].

3. Result and Discussion

The data on students' CTS experimental and comparison classes are presented in Table 1 below.

Table 1. Data summary of students' critical thinking skills before and after learning

Classes	N	Before learning		After learning	
		Average	SD	Everage	SD
Experiment	30	28.27	8.22	65.19	11.20
Comparison	22	26.77	7.83	52.36	11.67

It appears that students of both classes have almost the same CTS before learning. However, after learning the experimental class students have a higher CTS than the comparison class. This means that the STEM approach with formative assessment has a positive impact on increasing students' CTS.

The results of the analysis of normality, homogeneity, and difference tests on the pretest and posttest data are presented in Table 2 below.

Table 2. The normality and homogeneity data, and different est result

	Classes	Kolmogorov-Smirnov test (Sig.)	Levene Test (Sig.)	t Test (Sig.)
Pre-test	Experiment	0.200 (normal)	0.844 (homogen)	0.509 (no differences)
	Comparison	0.200 (normal)		
Post-test	Experiment	0.200 (normal)	0.968 (homogen)	0.000 (differences)
	Comparison	0.200 (normal)		

It appears that the test requirements for normality and homogeneity of the data are met so that the different test can use the independent t-test. The test results of differences in pretest data showed that the students of the two classes had no different initial critical thinking skills. This means that if there is a difference in critical thinking skills at the end of the lesson, then the difference is solely caused by the existence of different learning actions between the experimental and comparison classes. The results of the posttest data difference test showed that the experimental class students had significantly better critical thinking skills than the comparison class. This means that the STEM approach with formative assessment in PjBL has a significant positive impact in building students' critical thinking skills on static fluid material. The results of this study are in accordance with previous research that there was an increase in students' critical thinking skills after applying the PjBL-STEM model [33].

The critical thinking skills of experimental class students can outperform the comparison class because of the role of the STEM approach with formative assessment. Besides playing a role in developing thinking skills, PjBL-STEM also builds cooperative learning skills [34]. During the project creation process, the teacher monitors project progress and student learning progress in groups by ensuring each group works collaboratively to complete the project [35]. Through discussions during project trials, students are trained to develop reasoning indicators, test hypotheses in the form of interpreting relationships between variables, and analyze possibilities and uncertainties to asses the achievement of STEM aspects. Moreover, the experimental class students followed up the trial by taking steps to improve the projects that had been tested and submitting reports on the results of project trials. In addition, as many as 6 formative assessment activities, which have been mentioned in the method, were only carried out by experimental class students. One of these activities is discussing the design of the group's project with other groups. As a regulator of the discussion process, the teacher's role is to help connect the one student's response to another [36].

The results of increasing critical thinking skills of experimental and comparison class students are presented in Table 3 below.

Table 3. The results of the N-gain calculation of experimental and comparison class students

Classes	N-gain (category)	N–gain of learning (category)	
		Cyclus 1	Cyclus 2
Experiment	0.52 (medium)	0.47 (medium)	0.80 (high)
Comparison	0.35 (medium)	0.36 (medium)	0.32 (medium)

It appears that the experimental and comparison class students have increased critical thinking skills at the same level. But in line with the results of the independent t-test, PjBL-STEM-AF learning has a higher increase than PjBL. PjBL-STEM-AF learning can produce N-gains above the active learning class average of 0.48 [37], while PjBL learning can only produce N-gains that tend to be close to the low category threshold. The improvement of students' critical thinking cannot be separated from the provision of formative assessments based on the development of knowledge and skills, discussion results, and the results of student observations during the learning process [35]. Assessment activities in the form of providing feedback in this study are based on the analysis of quiz scores and student self-assessment. Students' difficulties are discussed together in class and become a form of feedback based on knowledge and observations. In addition, students make projects up to steps to improve projects that have been tested and submit reports on the results of project trials.

Table 3 also shows that the increase in N-gain in the experimental class is higher than the comparison class, both in learning rounds 1 and 2. In this case, the provision of formative assessment in PjBL-STEM gives better results than only in the form of a summative assessment [38]. Providing formative assessment in the form of feedback can increase the quality of PjBL-STEM implementation in the classroom [39]. In fact, in the second round of learning material, STEM learning approach with formative assessment was able to increase one level above PjBl learning. Some forms of formative assessment that are only carried out in the experimental class include students writing down any material that has been well understood, and conveying the overall learning conclusion, starting from the sub-materials of the first meeting to the last.

The improvement of each student's critical thinking indicator in the experimental and comparison classes is presented in Table 4 below.

Table 4. The results of calculating the N-gain for each critical thinking indicator

Classes	Critical thinking skills indicators				
	Reasoning	Hypothesis testing	Argument analysis	Likelihood and uncertainty analysis	Problem-solving and decision-making
Experiment	0.67 (medium)	0.19 (low)	0.62 (medium)	0.44 (medium)	0.47 (medium)
Comparison	0.45 (medium)	0.14 (low)	0.48 (medium)	-0.09 (low)	0.46 (medium)

It appears that the experimental class students have an increase in N-gain greater than the comparison class on all critical thinking indicators, especially on reasoning and argument analysis indicators. In fact, different levels of improvement occurred in the likelihood and uncertainty analysis indicators. However, both classes have almost the same improvement in the problem-solving and decision-making indicators, and the low category increase in the hypothesis testing indicator.

The difference in improvement in the reasoning indicators between the two classes occurred in the completion of item number 2. This item has the aim of "Presenting the Archimedes Law lab activity, students can explain the Archimedes Law concept based on the results of the lab activity". In laboratory cases, eggs can experience 3 phenomena, namely sinking, neutral buoyancy, and floating in water that is added with different levels of salt solution. Before learning, students are only able to interpret one phenomenon. However, after learning the comparison class students were only able to interpret two phenomena, while the experimental class students were able to interpret the three phenomena. This happened because in making the project the experimental class students took steps to improve the project that had been tested, while the comparison class students only went to the project trial step. In addition, the experimental class students discussed the design of their group projects with other groups. The activities of designing projects, evaluating, and communicating project designs involve lots of ideas, creativity, and critical thinking skills [25] so as to be able to train students' reasoning indicators.

The difference in improvement between the two classes also occurs in the argument analysis indicator. In this case, the experimental class students could complete questions number 4 and 5 better than the comparison class students. Item number 4 has the objective of "Presenting an argument regarding the condition of objects in the water, students can correct errors in the argument". Experimental class students were able to analyze that the argument "Grant states that wood floats because it is 'lighter than water' and stones sink because it is 'heavier than water" is an inaccurate conclusion statement because actually this is not related to mass, but involves density of the object. Meanwhile, the comparison class students only stated that Grant's argument was inaccurate or that the mass did not affect floating or sinking. Item number 5 has the objective of "Two pictures of dams with different shapes and an argument are presented, students can analyze the truth of the arguments put forward". In this case, the experimental class students were able to assess the credibility of the source of information that the argument "According to Risa a dam in the form of an upright beam is more effective than a beam with a larger base in retaining water because the pressure at all points on the water is the same" is not true because this is actually related to The principle of hydrostatic pressure is proportional to the depth of the water. While the comparison class students were only able to judge that the statement was not true because of the influence of hydrostatic pressure.

Students of both classes had almost the same improvement in items number 1 and 8 for the problem-solving and decision-making indicators. In item number 1 which aims "A paper clip experiment is presented, students can find the source of the problem and plan possible solutions based on the instructions given". Students of both classes were able to do the problem-solving and decision-making about “Ani conducted an experiment by putting a paper clip into a vessel filled with water using a toothpick. It turns out that paper clips can float on the surface of water. What should Ani do to make the paper clip sink to the bottom of the water without touching it?” by heating water in a vessel or dissolving a surfactant so that the surface tension of the water is reduced so that the paper clip sinks. It appears that the students of both classes are able to identify the parts of the problem and adjust the solution plan. Item number 8 has the aim of "Presenting a fragment of the hydrostatic pressure experiment step, students can determine the appropriate practicum objectives". In this case, based on several practicum steps provided, students can recognize the parts of the problem and adjust the solution plan by accurately formulating the practicum objectives, namely to determine the effect of fluid depth on hydrostatic pressure. Students can also develop their hypothesis, namely that the greater the difference in the height of the liquid surface in the two U pipes, the greater the hydrostatic pressure measured.

In item numbers 1 and 8 above, students have the least difficulty among the other questions. Based on the results of the interviews, this happened because more than 50% of students had studied this material before. In addition, both classes have the same steps in project creation, namely up to the project pilot step. Critical thinking skills in problem-solving and decision-making are developed through identifying emerging ideas and developing one reasonable and creative solution [40]. Asking important questions related to the idea or concept being studied is a way of learning through the design being developed [41]. It is clear that the critical thinking skills that are trained in the activities of designing, evaluating, and communicating projects are problem-solving and decision-making.

In item number 7, experimental class students have a much greater increase in critical thinking than comparison class students, which occurs in the indicators of likelihood and uncertainty analysis. This item has the objective of “Presenting an oil drop experiment, students can calculate the magnitude of the velocity of oil in the air”. In this case, students are required to calculate the possible values expected from the experiment, namely analyzing known quantities, writing equations to find the terminal velocity, and finding the exact value of the terminal velocity. Before learning most of the students did not know anything about this problem. The results of the interviews showed that around 74.9% of students had never studied this material at all. After learning the comparison class students were only able to recognize known quantities, while the experimental class students were able to continue even though only to determine the terminal velocity formula. The results of the interview showed that about 24.1% of students had studied the sub-material, but could not solve the problem because they did not understand or did not remember. It appears that this item is quite difficult for students. In fact, both classes conducted problem-based learning. This learning is seen as an appropriate activity in developing critical thinking skills [42]. The process of problem identification and investigation has given students an idea of the purpose of making the project [25]. Activities undertaken to train critical thinking skills are asking fundamental questions based on problem identification, analyzing information, and proposing solutions. Asking questions to understand an event is a form of reasoning ability, choosing the information needed and making conjectures is a form of ability to analyze possibilities and uncertainties [40]. However, the STEM approach with formative assessment causes experimental class students to be better able to develop indicators to analyze possibilities and uncertainties.

Students of both classes increase critical thinking in the low category, which occurs in the hypothesis testing indicator represented by item number 3. This item has the goal of "Presenting 3 pictures of water in a vessel with different upper cross-sectional area, but the same lower cross-sectional area, students can prove that the magnitude of the downward hydrostatic force by the water at the bottom of the three vessels is the same”. In this case the students of both classes have not been able to test the hypothesis so that they cannot draw valid conclusions from the available information. Students should be able to conclude that the downward hydrostatic force is the same and prove it through the equation $F=pA=\rho ghA$. It appears that students still think that hydrostatic pressure is influenced by the value of cross-sectional area [43] or the fluids' volume [44,45].

Calculation of effect size produces d-effect size = 1.13 (Very Large category). This shows that the implementation of PjBL-STEM-AF actions in the experimental class, and PjBL in the comparison class is highly recommended to be carried out in schools. The actions of these two actions have a very large impact on students in building critical thinking skills.

The results of the questionnaire on students' responses to the learning process are presented in Table 5 below.

Table 5. Positive students' response for learning process in experiment and comparison classes

Classes	Students' response		Total (%)
	Agree (%)	Very Agree (%)	
Experiment	43.64	51.82	95.46
Comparison	65.68	26.36	92.05

It appears that the experimental class students gave a slightly higher positive response than the comparison class. The comparison class is not a control class because it is taught using the PjBL model. It means that in general students feel comfortable in both classes. However, in the Very Agree option, the experimental class students gave a much higher response than the comparison class. This means that the PjBL-STEM approach learning with formative assessment in the experimental class is able to provide a higher sense of comfort than PjBL in the comparison class. This result follows the response to research that states that students positively respond to the application of STEM in PjBL [46]. Also, learning physics with the STEM approach ensures that students are more comfortable than without the STEM approach [47].

4. Conclusion

Before learning the experimental and comparison class students, each had an average critical thinking ability score of 28.27 and 26.77 so that both had the same initial state. After learning PjBL-STEM-AF students had an average critical thinking ability score of 65.19 which was significantly higher than PjBL students in the score of 52.36. The improvement of critical thinking ability of experimental class students N-gain = 0.52 (medium) is also better than the comparison class N-gain = 0.35 (medium). In the review of critical thinking ability indicators, the experimental class' students have different levels of improvement on the likelihood and uncertainty analysis indicator. However, the two classes had almost the same increase in the indicator of problem-solving and decision-making, and an increase in the low category on the hypothesis testing indicator. The value of effect size at 1.13 (Very Large) suggests that PjBL-STEM-AF learning should be done more frequently in the field. The Experimental class students gave a more positive learning response (Very Agree) than the Comparison class. It is recommended in future studies to add an "Art" aspect to the STEM approach to further develop students' CTS.

References

[1] Higgins S 2014 Critical thinking for 21st-century education: A cyber-tooth curriculum? *Prospects* **44** 4 559–574

[2] Beers S Z 2011 *What are the skills students will need in the 21 st century?* (London: ASCD Author). Retrieved from https://cosee.umaine.edu/files/coseeos/21st_century_skills.pdf

[3] Khaira N 2018 Pengaruh Pembelajaran STEM Terhadap Peserta Didik pada Pembelajaran IPA, *Seminar Nasional MIPA IV* (pp. 233–237).

[4] Howard L W, Tang T L P and Jill Austin M 2015 Teaching Critical Thinking Skills: Ability, Motivation, Intervention, and the Pygmalion Effect *Journal of Business Ethics* **128** 1 133–147

[5] Ku K Y L, Ho I T, Hau K T and Lai E C M 2014 Integrating direct and inquiry-based instruction in the teaching of critical thinking: an intervention study *Instructional Science* **42** 2 251–269

[6] Gupta T, Burke K A, Mehta A and Greenbowe T J 2015 Impact of guided-inquiry-based

instruction with a writing and reflection emphasis on chemistry students' critical thinking abilities *Journal of Chemical Education* **92** 1 32–38

[7] Kahlke R and Eva K 2018 Constructing critical thinking in health professional education *Perspectives on Medical Education* **7** 3 156–165

[8] Susilowati, Sajidan and Ramli M 2017 Analisis keterampilan berpikir kritis siswa madrasah aliyah negeri di Kabupaten Magetan. In *Seminar Nasional Pendidikan Sains 2017* **21** 223–231

[9] Priyadi R, Mustajab A, Tatsar M Z and Kusairi S 2018 Analisis Kemampuan Berpikir Kritis Siswa SMA Kelas X MIPA dalam Pembelajaran Fisika *Jurnal Pendidikan Fisika Tadulako* **6** 1 53–55

[10] Sari, A. L. R., Parno, & Taufiq, A. (2016). Kemampuan Berpikir Kritis Dan Pemahaman Konsep Fisika Siswa SMA pada Materi Hukum Newton *Seminar Pendidikan IPA Pascasarjana UM* 88-99

[11] Pamungkas Z S, Aminah N S and Nurosyid F 2019 Analysis of student critical thinking skill in solving fluid static concept based on metacognition level *International Conference on Mathematics and Science Education* **1153** 1–6

[12] Puspita I, Kaniawati I and Suwarma I R 2017 Analysis of Critical Thinking Skills on the Topic of Static Fluid *International Conference on Mathematics and Science Education* **895** 1–4

[13] Adi A S, Sugiyanto and Rusilowati A 2018 Identifikasi Profil Kesulitan Belajar Fisika Topik Fluida Statis pada Siswa SMA di Kabupaten Demak *Unnes Physics Education* **7** 1 1–6

[14] Young H D and Freedman R A 2016 *Sears and Zemansky's University Physics with Modern Physics* San Francisco: Pearson Addison Wesley

[15] Sundari P D, Parno and Kusairi S 2018 Students' Critical Thinking Ability in Integrated Learning Model *Jurnal Kependidikan: Penelitian Inovasi Pembelajaran* **2** 2 348-360

[16] Wartono, Hudha M N and Batlolona J R 2018 How are the physics critical thinking skills of the students taught by using inquiry-discovery through empirical and theorethical overview? *Eurasia Journal of Mathematics, Science and TechnologyEducation* **14** 2 691–697

[17] Sulasih, Suparmi A and Sarwanto 2017 Profile of student critical thinking ability on static fluid concept. In *International Conference on Science and Applied Science 2017* **909** 1–8

[18] Belecina R R and Ocampo J M 2018 Effecting Change on Students' Critical Thinking in Problem Solving *EDUCARE: International Journal for Educational Studies* **10** 102 109–118

[19] Sadler D R 1989 Formative assessment and the design of instructional systems *Instr. Sci.* **18** 119–144

[20] Isaacs T, Zara C, Herbert G, Coombs S J and Smith C 2013 *Key Concepts. In Educational Assessment* London: SAGE Publication Ltd.

[21] Arifin M 2017 Analysis of The Indonesian High School Curriculum Focusing Upon The English Program *IDEAS: Journal on English Language Teaching and Learning, Linguistics and Literature* **5** 2 134-145

[22] Stohlmann M, Moore T and Roehrig G 2012 Considerations for Teaching Integrated STEM Education *Journal of Pre-College Engineering Education Research* **2** 1 28–34

[23] Lou S J, Shih R C, Diez C R and Tseng K H 2011 The impact of problem-based learning strategies on STEM knowledge integration and attitudes: An exploratory study among female Taiwanese senior high school students *International Journal of Technology and Design Education* **21** 2 195–215

[24] Capraro R M, Capraro M M and Morgan J R 2013 *STEM Project-Based Learning: An Integrated Science, Technology, Engineering, and Mathematics (STEM) Approach Second Edition* (Boston: Sense Publishers)

[25] Morgan J R, Moon A M and Barroso L R 2013 *Engineering Better Projects: STEM Project-Based Learning An Integrated Science, Technology, Engineering, And mathematics (STEM) Approach* Netherland: Sense Publisher

[26] Kane S N, Mishra A and Dutta A K 2016 Preface: International Conference on Recent Trends in Physics (ICRTP 2016) *Journal of Physics: Conference Series* **755** 1 1–6

[27] Cohen L, Manion L and Morrison K 2018 *Research Methods in Education 8th Edition* New York: Routledge-Taylor & Francis Group

[28] Ramsey B S and Duffy A 2016 *Formative Assessment in the Classroom* USA: Education First

[29] Hugerat M 2016 How teaching science using project-based learning strategies affects the classroom learning environment *Learning Environments Research* **19** 3 383–395

[30] Tiruneh D T, De Cock M, Weldeslassie A G, Elen J and Janssen R 2017 Measuring Critical Thinking in Physics: Development and Validation of a Critical Thinking Test in Electricity and Magnetism *International Journal of Science and Mathematics Education* **15** 4 663–682

[31] Morgan G A, Leech N L, Gloeckner G W and Barrett K C 2011 *SPSS for introductory statistics: Use and interpretation Fourth Edition* London: Lawrence Erlbaum Associates

[32] Hake R R 1998 Interactive-engagement versus traditional methods: A six-thousand-student survey of mechanics test data for introductory physics courses *Am J Phys* **66** 1 64–74

[33] Sumardiana, Hidayat A and Parno 2019 Kemampuan Berpikir Kritis pada Model Project Based Learning disertai STEM Siswa SMA pada Suhu dan Kalor *Jurnal Pendidikan: Teori, Penelitian, dan Pengembangan* **4** 7 874–879

[34] Eslami Z and Graver R 2015 *English Language Learners and Project-Based Learning.* In *STEM Project-Based Learning An Integrated Science, Technology, Engineering, And mathematics (STEM) Approach.* Netherland: Sense Publisher 119-128

[35] Sahin A 2013 *STEM Project-Based Learning: Specialized Form of Inquiry-Based Earning.* In *STEM Project-Based Learning An Integrated Science, Technology, Engineering, And mathematics (STEM) Approach* Netherland: Sense Publisher

[36] Kim Hee-jeong 2017 Teacher Learning Opportunities Provided by Implementing Formative Assessment Lessons: Becoming Responsive to Student Mathematical Thinking *International Journal of Science and Mathematics Education* **17** 2 341–363

[37] Jackson J, Dukerich L and Hestenes D 2008 Modeling Instruction: An Effective Model for Science Education *Science Educator* **17** 1 10-17

[38] Pryor C R and Kang R 2013 *Project-Based Learning: An Interdisciplinary Approach For Integrating Social Studies With Stem Introduction.* In *STEM Project-Based Learning An Integrated Science, Technology, Engineering, And mathematics (STEM) Approach.* Netherland: Sense Publisher 129-138

[39] Capraro R M and Corlu M S 2013 *Changing Views On Assessment For Stem Project-Based Learning Introduction.* In *STEM Project-Based Learning An Integrated Science, Technology, Engineering, And mathematics (STEM) Approach.* Netherland: Sense Publisher 109-98

[40] Tiruneh D T, Weldeslassie A G, Kassa A, Tefera Z, Cock M and Elen J 2016 Systematic design of a learning environment for domain- specific and domain-general critical thinking skills *Educational Technology Research and Development* **64** 3 481–505

[41] Reeve E M 2013 *Implementing Science , Technology , Mathematics , and Engineering (STEM) Education in Thailand and in ASEAN. The Institute for the Promotion of Teaching Science and Technology (IPST).*

[42] Kek M Y C A and Huijser H 2011 The power of problem-based learning in developing critical thinking skills: Preparing students for tomorrow's digital futures in today's classrooms *Higher Education Research and Development* **30** 3 329–341

[43] Wijaya P C, Koes S and Muhardjito 2016 The Diagnosis of Senior High School Class X Mia B Students Misconceptions about Hydrostatic Pressure Concept Using Three-Tier *Jurnal Pendidikan IPA Indonesia* **5** 1 14—21

[44] Matthew G, Adam M, Bazan Z and Wagner D J 2013 Exploring student difficulties with pressure in a fluid *AIP Conference Proceedings* **1513** 154-157

[45] Loverude M E, Heron P R L and Kautz C H 2010 Identifying and Addressing Student Difficulties with Hydrostatic Pressure *American Journal of Physics* **78** 75-85

[46] Tseng K H, Chang C C, Lou S J and Chen W P 2013 Attitudes towards science, technology, engineering and mathematics (STEM) in a project-based learning (PjBL) environment *International Journal of Technology and Design Education* **23** 1 87–102
[47] Teevasuthonsakul C, Yuvanatheeme V, Sriput V and Suwandecha S 2017 Design Steps for Physic STEM Education Learning in Secondary School *J. Phys.: Conf. Series* **901** 012118

Concept understanding of students through core physics learning tools based on guided inquiry assisted by PhET virtual media

Susilawati [1,2], A Doyan[1,2], Wahyudi[1], S Ayub[1], and J Ardhuha[1]

[1]Physics Education, Faculty of Teacher Training and Education, University of Mataram, Lombok, West Nusa Tenggara, Indonesia.
[2] Master of Science Education Program, University of Mataram, Lombok, West Nusa Tenggara, Indonesia.

E-mail: susilawatihambali@unram.ac.id, aris_doyan@unram.ac.id

Abstract. This research is a development research that aims to test the effectiveness of the inquiry model-based core physics learning device with the help of PhET virtual media in improving students' conceptual understanding. The device was developed using a 4D model which includes define, design, develop and disseminate. The sample of this study was all undergraduate students who took core physics courses which consisted of two classes, namely class A (experimental class) and class B (control class) which amounted to 50 people. The results of the pre-test and post-test were analyzed using the t-test at a significance level of 5%, while the increase in understanding of the concept used the N-gaint test. The results of the t-test show that the t-count value is greater than the t-table. This means that core physics learning tools using an inquiry model with the help of PhET virtual media affect students' understanding of concepts. Analysis of increasing understanding of the concept of d using the N-gain test. The N-gain analysis shows that students' understanding of concepts has increased the use of these devices in the high category. This means that students' understanding of concepts is effectively improved using an inquiry model with the help of PhET virtual media.

1. Introduction

Physics is a science that studies matter and energy and the relationship between them [1]. In physics learning, abstract concepts cannot be taught directly; for example, in core physics courses, quantum physics, and solid matter physics [2]. Learning abstract concepts, of course, requires media that can present abstract concepts more concretely. The media can be in the form of images, audio, video, or animation so that the concepts presented are clearer and easier to understand [3]. Thus, physics learning is adjusted to the characteristics of the material presented [4].

The implementation of core physics learning requires systematic scenarios so that the concepts presented can be maximally absorbed by students to achieve learning objectives [5]. Core physics concepts are obtained through expert observations and experiments. Core physics learning can be done by applying a learning model that actively involves students and guides students to make discoveries. The learning model that can activate students is the guided inquiry model [6].

The guided inquiry model can activate students in learning activities to be used in improving students' conceptual understanding [7] [8]. This follow the opinion [9] which states that learning the basic concepts and principles of physics material is suitable for using a guided inquiry model.

In addition to using the guided inquiry model, efforts to increase students' conceptual understanding can be done by using moving images to explore. The learning media can measure, collect, and analyze data from the measurement results [10]. One of the media that can be used in learning physics is PhET Simulation. The media can help students understand concepts that seem abstract. In addition, the media is interactive to make students interested and explore a material [11]. Guided inquiry-based learning tools and PhET virtual media are expected to increase students' understanding of physics concepts. Therefore, it is necessary to research by applying the guided inquiry learning model assisted by PhET virtual media.

2. Methods

Core physics learning tools using an inquiry model with the help of PhET virtual media were developed using a 4D model design, namely define, design, develop and disseminate [12]. The sample of this study was all undergraduate students who took core physics courses which consisted of two classes, namely the experimental class (class A) and the control class (class B) which consisted of 50 people. The experimental class consisted of 25 people, while the control class consisted of 25 people. Before the core physics learning tools were given, all students were tested for knowledge initially by giving a pre-test. Furthermore, testing students' conceptual understanding through post-test. The test given is in the form of a conceptual understanding test consisting of 10 questions and includes all indicators of concept understanding (interpreting, giving examples, classifying, summarizing, drawing references, comparing, and explaining). The results of the pre-test and post-test were analyzed using the t-test with a significance level of 0.05. The analysis of increasing concept understanding was carried out using the N-gain (g) test (equation 1) [13] [14]. The results of the N-gain calculation obtained are matched with the N-Gain table. The N-Gain values consist of low ($g < 0.30$), medium ($0.70 > g > 0.30$), and high categories ($g > 0.70$) [15].

$$\text{N} - \text{gain} = \frac{S_{post} - S_{pre}}{S_{max} - S_{min}} \quad (1)$$

3. Result and Discussion

Concept understanding data is obtained by giving tests to students in the form of concept understanding questions. The pre-test was carried out before the core physics material was given, while the final test was given after the core physics material was given. The instrument for understanding the concept consists of 10 questions consisting of 1 question for the atomic nucleus sub-chapter, 1 question for the radioactivity sub-chapter, 2 questions for the Radioactivity decay sub-chapter, 2 questions for the nuclear reaction sub-chapter, 2 questions for the Radiation interaction sub-chapter, 2 questions Radiation detector subchapter. The collected concept understanding data is then analyzed to get the value of the effectiveness of the learning device.

Table 1. T-test results for each sub-material.

Sub-Material	t-count	t-table
Atomic Nucleus	11.59	1.67
Radioactivity	8.92	
Radioactivity Decay	7.93	
Core Reaction	8.56	
Radiation Interaction	7.98	
Radiation Detector	8.62	

The analysis of concept understanding in this article consists of an analysis of the understanding of the concept of each sub-material and each indicator of the understanding of the concept. The first analysis is an analysis of the understanding of the concept of each sub-material. Table 1 shows that the value of t-table is smaller than t-count. These results indicate that there is an influence of core physics learning tools with guided inquiry models on students' understanding of concepts [16] [17][18].

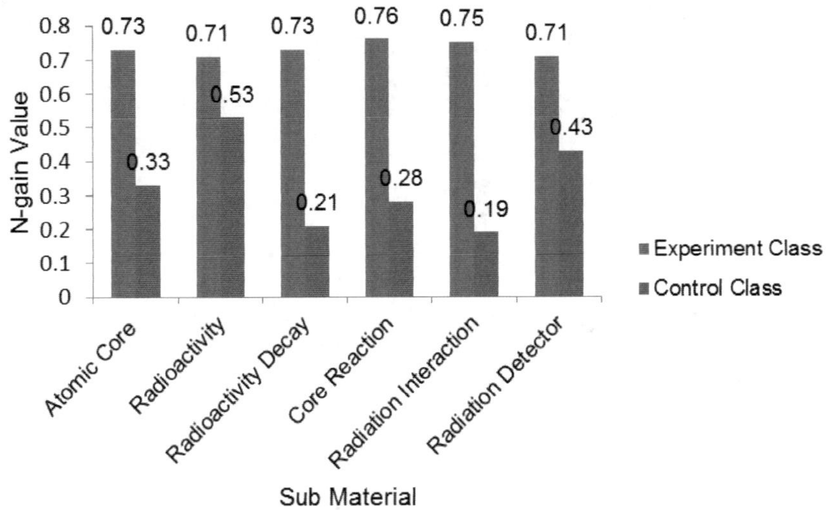

Figure 1. Comparison of the N-gain value of each sub material.

Figure 1 clearly shows that for the experimental class on the sub-atomic nucleus, radioactivity, decay of radioactivity, nuclear reactions, radiation interactions, radiation detectors have N-gain values above 0.70. This means that the value of increasing concept understanding is in the high category for the experimental class. Meanwhile, for the control class in the atomic nucleus sub-material, radioactivity and radiation detectors are in the medium category ($0.70 > g > 0.30$). In the sub-materials of Radioactivity Decay, Nuclear Reaction, Radiation interaction, the N-gain value is categorized as low ($g < 0.30$). These results indicate that increasing the understanding of the control class (B) concept is lower than that of the experimental class (A). This is due to the use of inquiry-based core physics learning tools with PhET virtual media in the experimental class. These results are consistent with research [19][20][21] which shows that the use of PhET virtual media can improve students' conceptual understanding.

Table 2. T-test results for each indicator of concept understanding

Indicators	t-count value	t-table value
Interpret	12.01	
Give an Example	6.78	
Classify	5.64	
Summarize	3.29	1.67
Interesting Reference	5.13	
Compare	7.78	
Explain	3.42	

The second analysis is the analysis of each indicator of concept understanding. Indicators of understanding this concept include interpreting, giving examples, classifying, summarizing, drawing references, comparing, explaining. The results of the t-test for each indicator of concept understanding are shown in Table 2. Based on Table 2, it is clear that the value of t-count is greater than t-table. This means that using core physics learning tools with guided inquiry models affects students' conceptual understanding [21].

Data on student learning outcomes (N-gain) for the experimental class and the control class for each indicator are shown in Figure 2. Based on this figure, the experimental class obtained an N-gain score with a high category on all understanding indicators. For the control class, the N-gain score in the low category was obtained on the indicators of interpreting, classifying, drawing references, and comparing, while giving examples, summarizing, and explaining were in the medium category.

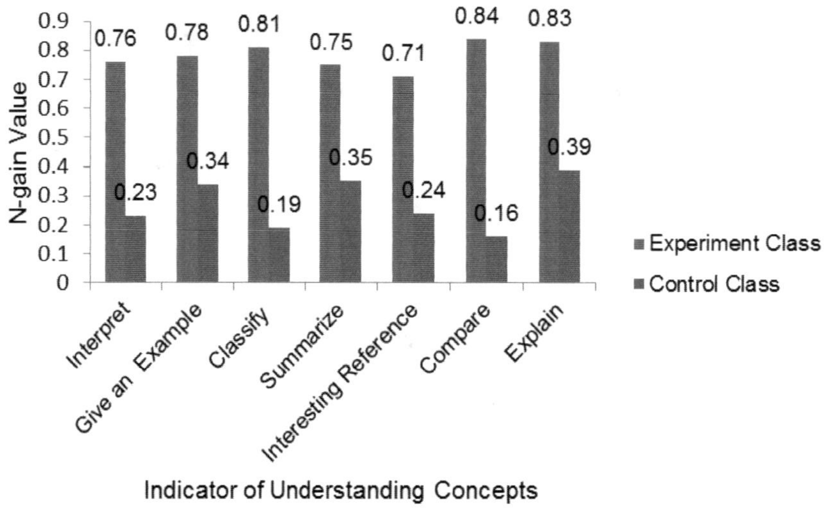

Figure 2. Comparison of N-gain Values for each Indicator of Concept Understanding.

Based on the results of the analysis that has been carried out, it shows that students who are taught using core physics learning tools using an inquiry model with the help of PhET virtual media have a better understanding of concepts. This happens because students are given the freedom to tinker with the content contained in PhET, such as adding or removing bonds and lone pairs [22]. Thus a lot of experience can be obtained by students from these activities, thus enabling students to gain meaningful experiences, and arouse students' curiosity about the concepts they are learning [23].

4. Conclusion

Based on the study result, it can be concluded that the core physics learning device using an inquiry model with the help of PhET virtual media can improve students' understanding. This is indicated by the value of t-table is smaller than t-count. In addition, the N-gain value for the experimental class is in the high category, while the experimental class is in the medium and low category.

Acknowledgement

Thanks to the Dean of FKIP, Head of the PMIPA Department, and Head of the Physics Education Study Program and all parties so that this research can be carried out properly.

References

[1] Kanginan M 2016 *Fisika Untuk SMA/MA Kelas X* (Bandung: Erlangga).

[2] Kusdiastuti M, Harjono A, Sahidu H and Gunawan G 2017 Pengaruh Model Pembelajaran Inkuiri Berbantuan Laboratorium Virtual Terhadap Penguasaan Konsep Fisika Peserta Didik *Jurnal Pendidikan Fisika Dan Teknologi* **2** 85–96.

[3] Khasanah B U, Doyan A, Gunawan, Susilawati, Kartini, Hakim S and Muliyadi L 2019 Analysis Validation of Learning Media Quantum Phenomenon *Jurnal Penelitian Pendidikan IPA* **5** 189-193.

[4] Rahayu S 2017 Model Simulasi dalam Mata Kuliah Strategi Pembelajaran Fisika *Jurnal Pendidikan Fisika dan Teknologi* **1** 102-118.

[5] Kartini, Doyan, A, Kosim, Susilawati, Khasanah B U, Hakim S and Muliyadi L 2019 Analysis of Validation Development Learning Model Attainment Concept to Improve Critical Thinking Skills and Student Learning Outcomes *Jurnal Penelitian Pendidikan IPA* **5** 185-188.

[6] Istiqamah N, Doyan A and Taufik M 2016 Pengaruh Model Pembelajaran Discovery dan Inkuiri Terbimbing Berbasis Eksperimen Terhadap Hasil Belajar Fisika dan Sikap Ilmiah Siswa *Jurnal Penelitian Pendidikan IPA* **2** 27-38.

[7] Susilawati, Doyan A, Harjono A and Kosim 2019 Penerapan Model Pembelajaran Inkuiri Berbasis Media Virtual Program Java Pada Guru Fisika Dan Siswa SMA *Jurnal Pengabdian Masyarakat Sains Indonesia* **1** 4-10.

[8] Lewa I W L, Susanto H and Marwoto P 2017 Implementasi Model Pembelajaran Inkuiri Terbimbing pada Siswa SMA *Jurnal Pendidikan Fisika dan Teknologi* **7** 400–404.

[9] Susilawati, Susilawati and Sridana N 2015 Pengaruh Model Pembelajaran Inkuiri Terbimbing Terhadap Keterampilan Proses Sains Siswa *Jurnal Tadris IPA Biologi FITK IAIN Mataram* **8** 27-36.

[10] Ekawati Y, Haris A and Amin B 2015 Penerapan Media Simulasi Menggunakan PhET (Physics Education And Technology) Terhadap Hasil Belajar Fisika Peserta Didik Kelas X SMA Muhammadiyah Limbung *Jurnal Pendidikan Fisika Unismuh* **1** 35-48.

[11] Astutik S and Prahani B K 2018 The Practicality and Effectiveness of Collaborative Creativity Learning (CCL) Model by Using PhET Simulation to Increase Students' Scientific Creativity *International Journal of Instruction* **11** 409-424.

[12] Fraenkel J R, Wallen N E and Hyun H H 2012 *How to Design and Evaluate Research in Education (8th ed.)* (New York: McGraw-Hill).

[13] Doyan A, Gunawan, Susilawati, Khasanah B U, and Muliyadi L 2020 The Effectiveness of Quantum Phenomenon Learning Media with Think Pair Share Model Implementation on Understanding Concept of Students *Journal of Physics: Conference* Series **1521** 022037 1-6.

[14] Doyan A, Susilawati, Kosim, Wardiawan Z, Hakim S, Muliyadi L and Hamidi 2020 The Development of Physics Module Oriented Generative Learning to Increase the Cognitive Learning Outcomes and Science Process Skills of the Students *Journal of Physics: Conference Series* **1521** 022059 1-6.

[15] Susilawati, Doyan A and Ayub S 2019 Perbedaan Keterampilan Generik Sains antara Model Pembelajaran Berbasis Masalah dengan Inkuiri Terbimbing Ditinjau dari Kemampuan Berpikir Kritis Siswa SMA *Jurnal Pendidikan Fisika dan Teknologi* **5** 16-24.

[16] Susilawati, Doyan A, Artayasa P, Soeprianto H, Harjono A and Kartini 2019 Effectiveness of Scientific Learning Guided Inquiry Devices Based on Real Media to Improve Understand Concept and Skills Process of Science Students *The 2nd International Conference on Elementary Education* **2** 517-524.

[17] Ramlee N, Rosli M S, Saleh N S 2019 Mathematical HOTS Cultivation via Online Learning Environment and 5E Inquiry Model: Cognitive Impact and the Learning Activities," *iJET* **14** 140-151.

[18] Verawati N N S P, Hikmawat, Prayogi S 2020 The Effectiveness of Inquiry Learning Models Intervened by Reflective Processes to Promote Critical Thinking Ability in Terms of Cognitive Style *iJET* **15** 212-220.

[19] Habibi, Jumadi, Mundilarto 2020 PhET Simulation as Means to Trigger the Creative Thinking Skills of Physics Concepts *iJET* **15** 166-172.

[20] Rais A A, Hakim L and Sulistiawati 2020 Pemahaman Konsep Siswa melalui Model Inkuiri Terbimbing Berbantuan Simulasi PhET *Physics Education Research Journal* **2** 1-8.

[21] Haryadi R and Pujiastuti H 2020 PhET simulation software-based learning to improve science process skills *J. Phys.: Conf. Ser.* **1521** 022017.

[22] Rizkiana F and Apriani H 2020 Simulasi Phet: Pengaruhnya Terhadap Pemahaman Konsep Bentuk dan Kepolaran Molekul *Jurnal Inovasi Pendidikan Sains* **11** 1 -7.

[23] Moore E B, Chamberlain J M, Parson R and Perkins K K 2014 PhET Interactive Simulation: Transformative tools for Teaching Chemistry *Journal of Chemical Education* **91** 1191 -1197.

Geophysical Investigation and Geochemical Study of Sediment along the Coastal Area in Kota Belud Sabah, Malaysia

B Musta[1,2*], **M A Asat**[1], **S Y Ling**[1] **and H Saleh**[1]

[1]Geology Program, Faculty of Science and Natural Resources, Universiti Malaysia Sabah, 88400 Kota Kinabalu, Sabah, Malaysia

[2]Small Islands Research Centre, Universiti Malaysia Sabah, 88400 Kota Kinabalu, Sabah, Malaysia

*E-mail: babamus@ums.edu.my

Abstract. This article discusses the geophysical investigation of sediment and the status of heavy metals pollution in the coastal sediment from Usukan Beach, Kota Belud Sabah Malaysia. Study area is located at the northwest coast of Sabah which is bounded by the South China Sea and made up of Crocker Formation which age from Late Eocene to Early Miocene, Wariu Formation with Middle Miocene age and Quaternary Alluvium. This study using geoelectrical method namely electrical resistivity (IR) and induced polarization (IP) to measure the thickness of sediment and the parent rock lithology respectively. Whereas the heavy metal contents were measured using the Inductively Coupled Plasma-Optical Emission Spectrometry (ICP- OES). Three sets of subsurface cross-sections were conducted for the geophysical survey and covered the depth around 37 meters for all the survey line. The result of geophysical survey shows that the alluvium quaternary deposits in the low-lying flat area have a thickness up to 13 metres and overlie the mudstone dominated rock of the Crocker Formation. The geochemical analysis shows a decreasing ranking order of elemental concentrations Fe>Mn>Zn>Cu>Pb>Cr>As. The analyses show all elements are within the background values and acceptable standard limits of the Sediment Quality Guidelines (SQG) by USEPA for coastal sediments. The assessment of heavy metals revealed only minimal degree of pollution in the coastal sediments.

1. Introduction

Kota Belud is situated about 70 km to the northwest of Sabah, Malaysia. The area mainly consists of Crocker Formation and Wariu Formation [1], [2]. The age of Crocker Formation is Late Eocene to Early Miocene, whereas Wariu Formation is Middle Miocene [3]. The Crocker Formation comprises of interbedded sandstone, siltstone, and shale units. The Wariu Formation consists of a mixture of chaotic rock deposits from different origins and ages, including chert-spilite, limestone, breccia, and ophiolitic rocks from [4], [5], [6]. The Quaternary alluvium is originated from the weathering of the underlying Crocker Formation and Wariu Formation. The alluvium is transported from higher terrains to the coastal area and is mixed with corals along the beach shoreline.

Most of the small Islands around the Kota Belud area are built up by the Crocker Formation and overlay by the coral reef. The distribution of the coral reef in the offshore islands such as Pulau Layang-Layang also shows the same geomorphological pattern in the Kota Belud coast which is controlled by the lithology and structural geology [7]. Hence the chemical and physical weathering processes of parent rock will influence the distribution of heavy metals on the coastal sediment and the offshore islands. The redistribution of heavy metals in coastal sediments and small islands are also controlled by the current's movement, intertidal waves, and benthic organism activities. This signifies that the sediments serve as a reservoir and a secondary source of the metal pollution ecosystem [8], [9]. The fine-grain sediment can act as a reservoir to entrapped and to adsorb the heavy metals in thecoastal area [10].

Coastal sediments are sensitive indicators for assessing heavy metals contamination from terrestrials such as parent rock materials, including from anthropogenic inputs such as agricultural, and domestic wastes [11], [12], [13]. Kota Belud is considered as one of the active agricultural activities, especially in paddy field cultivation. Once the heavy metal concentrations from the parent rock materials exceeded the range of acceptable background limits or are enriched by anthropogenic sources, they may become highly toxic even at very low concentrations[14]. The standard limits established in Sediment Quality Guidelines (SQG) [15] are used to determine whether the total metal concentrations are within the allowable concentration limits and monitor the eco-toxicity level or quality of the coastal sediments [16], [17]. Therefore, the objectives of this research are to estimate the coast sediment thickness as well as the parent rock materials and to determine the status of sediment quality in coastal of Kota Belud, Sabah.

2. Materials and Methods

2.1. Geophysical survey

The geophysical survey was performed along the coastal region from Usukan beach, which lies within a latitude of 6^0 17' 40" N to 6^0 36' 50" N and longitude of 116^0 16' 50" E to 116^0 33' 0" E (Figure 1). The thickness of the coast sediment and the lithology of the parent rock was estimated using electrical resistivity (ER) and induced polarisation (IP), respectively. Imaging referred to the production of subsurface cross-section based on theprocessed field data and thus represented the geophysical characteristic of the subsurface materials. Three (3) survey stations were chosen on the alluvium quaternary and, each survey line is 200 meters in length with an electrode spacing of 5 meters. Using this arrangement, the maximum depth of penetration is up to 40 meters deep. The ABEM Terrameter LS was used for the survey, enabling the two methods to be surveyed concurrently.

2.2. Geochemical study

A total of 53 sediment samples were collected for the geochemical study from four (4) sampling stations, using a core sampler and PVC cores with a diameter of 10cm and a length of 100cm. The core samples were closed tightly using the Styrofoam cork to avoid contamination before being extruded using the horizontal extrusion method to remove the pieces from the PVC cores. The samples were then sliced and divided into 10cm thick subsamples, which were kept in airtight polyethylene bags before being taken to the lab for further analysis.

The heavy metal contents were measured using the Inductively Coupled Plasma-Optical Emission Spectrometry (ICP-OES) Perkin Elmer Optima 5300 DV model at the <63μm fine grain fraction. The minerals in the sediments were dissolved using aqua regia digestion, and then filtered using a 0.45m Whatman filter membrane [18]. The laboratory analyses for physico-chemical properties of sediment such as pH, moisture content (MC), organic matter (OM) and particle size distribution (PSD) were also measured using the British Standard Method [19].

2.3. Sediment quality analysis

The contamination levels of As, Cr, Cu, Mn, Pb, and Zn elements in the sediment were evaluated using the Igeo, EF, CF, mCd and PLI parameters. To assess the sediment quality, the values were compared with the SQG as suggested by US EPA legislation.

Geoaccumulation Index (Igeo). Geoaccumulation index (Igeo) were used to compare the contamination level of sediment with that of the background concentration by [20], whereby the

Journal of Physics: Conference Series **2165** (2022) 012046 doi:10.1088/1742-6596/2165/1/012046

world average concentration of respectivemetal in shale is selected as the background value. Igeo classification was used in this study [21].

Enrichment Factor (EF). EF is an indicator to distinguish the heavy metals originating from either anthropogenic or natural inputs. The iron (Fe) is chosen as the reference element in this study to create the regional geochemical baseline of heavy metal. EF classification was used in this study [22].

Contamination Factor (CF) and Modified Contamination Degree (mCd). Contamination Factor (CF) were used to evaluate the extent of sediment contamination in coastal environment due to toxic substances. The sum of contamination factor (ΣCF) was also calculated using the modified contamination degree (mCd) equation introduced by [23].

Pollution Load Index (PLI). PLI was a method derived from the mean of contamination factors (CF) for every metal present in the study area to natural background value as proposed by [24].

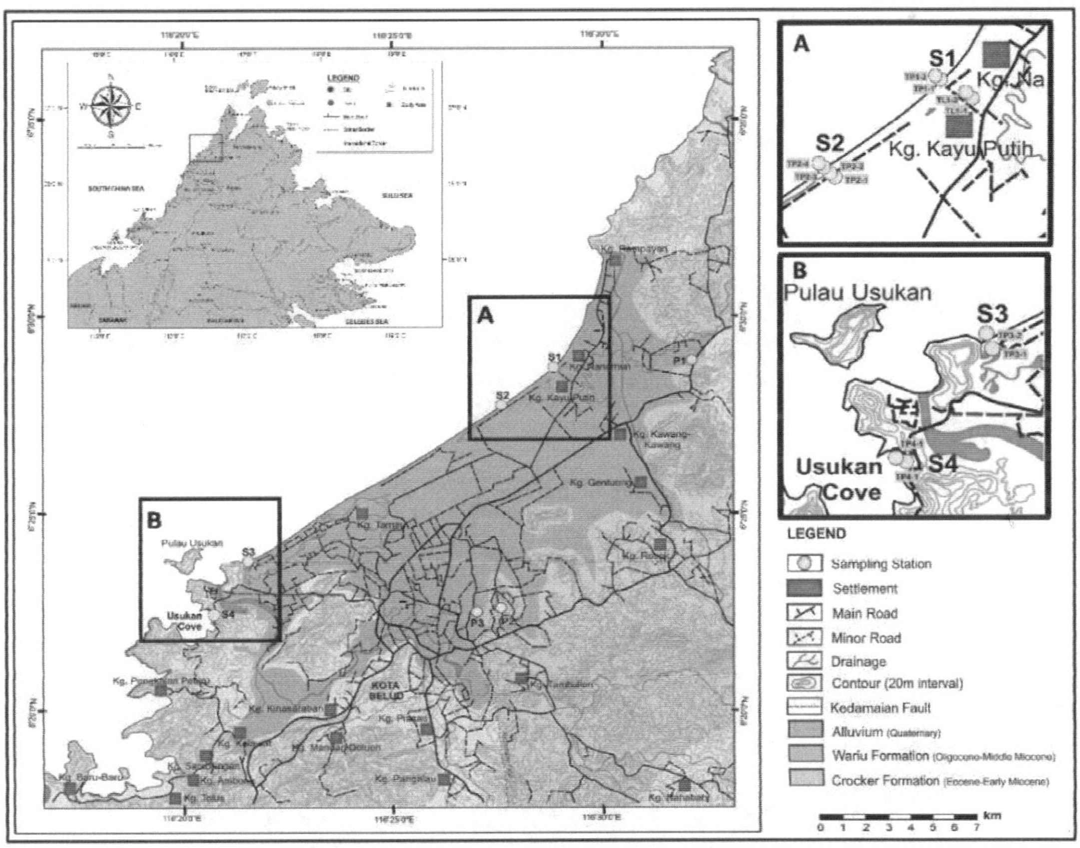

Figure 1 Geological map of the study area and sampling stations in Kota Belud, Sabah.(Source: Minerals and Geoscience Department of Malaysia, 2015)

Journal of Physics: Conference Series **2165** (2022) 012046 doi:10.1088/1742-6596/2165/1/012046

3. Result and Discussion

3.1 Geophysical Survey

Three (3) sets of subsurface cross-sections were obtained from the geophysical survey. Depth of investigation acquired for the surveys is around 37 meters. The result shown here represents a survey line perpendicular to the coastline and is spread towards landward from left to right. Three zones of resistivity are identified for each station,as shown in Figures 2, Figures 3, and Figures 4.

High resistivity zones are represented by resistivity values ranging from 80 to 100 Ωm. This zone has a maximum thickness of 6 meters in all stations. Resistivity values of 80 to 20 Ωm are classified as medium resistivity zone. This zone has a thickness around 7 meters for stations 1 and 2, whereas for station 3 it is thinner at around 4 meters. Finally for the low resistivity zone, the values are 20 Ωm and below. This zone is the thickest of all stations with a general thickness of 27 meters.

The cross-section of induced polarization (IP) showed chargeability in range of 0 to 16 msec. Clear distinction of high chargeability zone and the low can be observed from the cross-section at all stations. Zone of high chargeability of clay or fine parent rock material are in the range of 10 to 16 msec while the zone of medium and low chargeability of coastal alluvium is in the range of 10 to 2 msec and below 2 msec respectively.

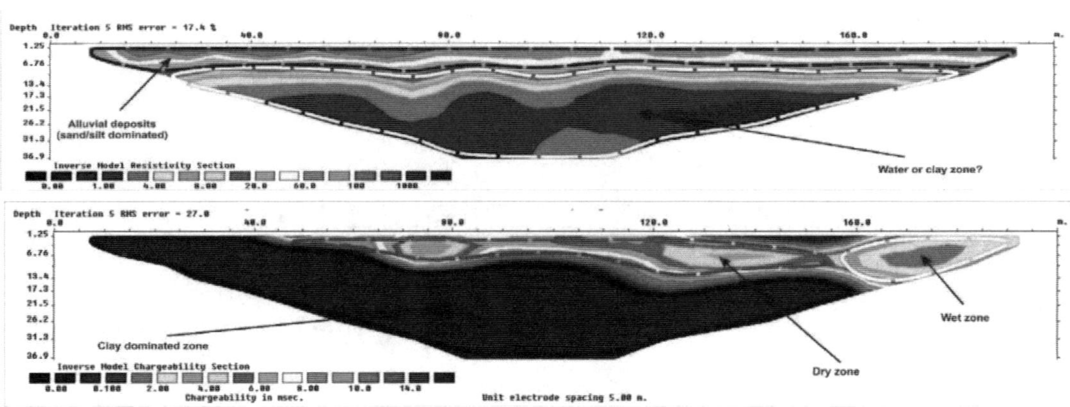

Figure 2. Cross-section of the processed ER and IP data for station 1.

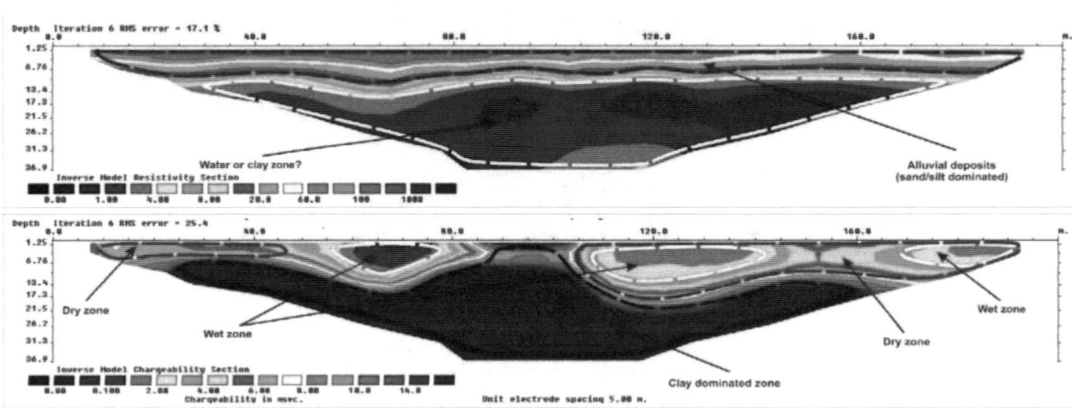

Figure 3. Cross-section of the processed ER and IP data for station 2.

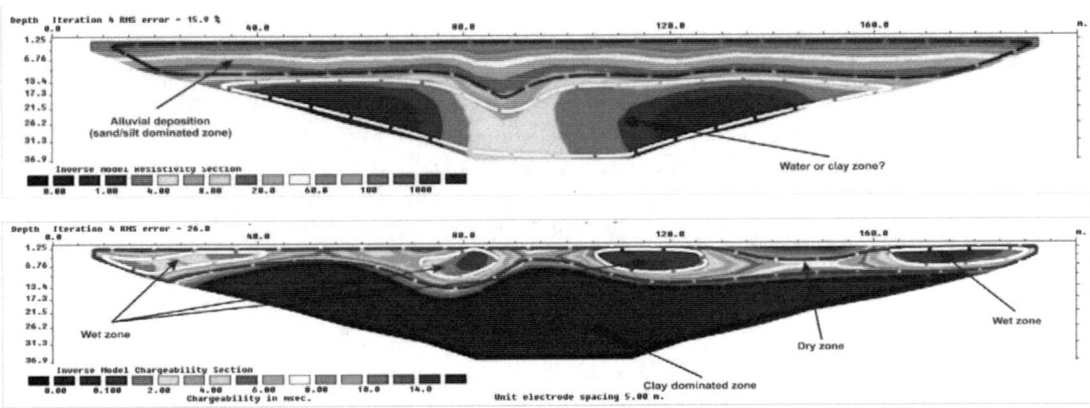

Figure 4. Cross-section of the processed ER and IP data for station 3.

3.2 Geochemical Study

Physico-chemical properties. The result of the geochemical analysis of sediment is shown in Table 1. The result indicates that the geochemical parameters determined are the arithmetic mean and standard deviation for pH (7.29±0.98), organic matter (1.27±1.89%) and texture of the clay (3.2±7.6%), silt (3.4±7.6%), and sand (93.3±15.1%) in the sediments. The results of pH analysis show an increase of pH from 5.7 (acidic) to 8.5 (alkaline) as the samplings are heading from inland into the sea. The ophiolitic parent materials from the Wariu Formation and bicarbonate materials coral deposits significantly influence the physico-chemical properties of coast sediments. The pH of water can be increased by dissolving bicarbonate materials from shelled organisms and corals[25], [26]. The lowest organic matter is 4.99%, whereas the highest is 6.73%. Sediments with higher clay percentages have greater organic content due to organic the binding ability of organic particles on the delicate clay surfaces by adsorption-desorption processes [27] [28]. The sediment texture from the study site were varies from sandy, sandy loam, and sandy clay loam texture. The percentage of clay and silt range from 0.05% to 25.3%, whereas the sand ranges from 49.4% to 99.7%. Higher clay and silt percentage is found in inland region (samples TL) which form a low resistivity zone, whereas higher sand percentage is found in coastal region (samples SP and TP) which form the high resistivity zone.

Table 1. Average values for the physico-chemical properties of coastal sediments.

Samples	SP2	SP3	TP1-1	TP1-2	TP2-1	TP2-2	TP2-3	TP2-4	TP3-1	TP3-2	TP4-1	TP4-2	TL1-1	TL1-2	Range	Mean ± SD
pH	5.7	6.5	7.1	7.9	7.3	7.8	7.9	7.5	7.3	7.7	8.4	8.5	6.1	6.4	5.7-8.5	7.29±0.98
MC %	5.0	5.8	25.0	25.8	24.0	23.8	15.3	25.3	26.2	39.9	25.9	27.9	35.4	48.8	4.9-48.8	25.3±12.9
OM %	0.3	0.3	0.3	0.3	0.2	0.2	0.2	0.1	0.3	0.8	1.6	1.8	4.7	6.7	0.1-6.7	1.27±1.89
Sand %	98.7	98.0	99.1	99.7	99.1	99.6	99.7	99.6	98.4	98.8	99.8	98.7	67.8	49.4	49.4-99.7	93.3±15.1
Silt %	0.55	1.07	0.46	0.20	0.05	0.30	0.25	0.25	1.02	0.91	0.10	0.91	16.1	25.3	0.05-25.3	3.4±7.6
Clay %	0.76	0.92	0.40	0.15	0.05	0.15	0.05	0.15	0.61	0.30	0.15	0.36	16.1	25.3	0.05-25.3	3.2±7.6

Heavy Metal Content. Table 2 shows the total mean concentration of Fe (17,804±6,513 ppm), As (5.2±3.1 ppm), Cr (6.23±3.26 ppm), Cu (20.7±15.5 ppm), Mn (299.8±109.2 ppm), Pb (9.3±6.5 ppm) and Zn (39.3±9.8 ppm) in the fine grain <63μm fraction. The ranking order for elemental concentrations is Fe>Mn>Zn>Cu>Pb>Cr>As. The heavy metal concentrations are compared with the standard limits for coastal sediments based on the Sediment Quality Guidelines (SQG) established by USEPA [15]. The distribution of heavy metals in coastal sediments is greatly influenced by natural and anthropogenic factors [29], [30]. Any contaminants are transported via drainage system into coastal and marine environment that serve as natural sink for the accumulation of heavy metal burden. The mean enrichment factor suggested Cr and Pb are from a mixture of both natural and anthropogenic origins, while Zn, As, Cu and Mn are contributed from natural or lithogenic origin.

Table 2. Total concentration of heavy metals in sediment and their background value.

Samples	Total Mean Concentration (ppm)						
	Fe	**As**	**Cr**	**Cu**	**Mn**	**Pb**	**Zn**
SP2	20,051	3.7	81.1	13.1	325	6.8	40.5
SP3	16,291	2.1	69.6	10.5	228	5.9	43.7
TP1-1	17,399	9.7	91.1	9.9	375	2.5	43.5
TP1-2	22,076	12.2	-	11.4	541	4.3	43.4
TP2-1	16,130	4.1	86.4	10.8	259	8.2	36.1
TP2-2	18,810	7.8	-	10.8	376	9.1	40.3
TP2-3	17,060	7.5	-	9.7	371	6.2	39.2
TP2-4	17,841	6.7	90.7	10.8	318	8.1	40.6
TP3-1	19,678	3.4	96.1	15.2	246	7.6	51.4
TP3-2	20,745	4.8	79.0	31.4	320	7.3	51.3
TP4-1	4,476	3.4	9.0	18.7	137	8.1	19.4
TP4-2	5,373	3.2	8.2	40.7	103	9.6	22.1
TL1-1	29,829	1.6	-	60.2	344.7	28.1	-
TL1-2	23,496	2.4	74.4	35.9	255.1	18.9	-
Mean ± SD	17,804 ± 6513	5.2 ± 3.1	62.3 ± 32.6	20.7 ± 15.5	299.8 ± 109.2	9.3 ± 6.5	39.3 ± 9.8
SQG by US EPA	**No Pollution**	<3	<25	<25	<300	<40	<90
	Moderate	38	25-75	25-50	300-500	40-60	90-200
	High	>8	>75	>50	>500	>60	>200

Sediment Quality Assessment. The result of sediment indices such Igeo, EF, CF, mCd and PLI for As, Cr, Cu, Mn, Pb and Zn elements are given in Table 3. The results for all elements show Igeo<0, which indicates very low contamination in the sediments. The mean EF values show either EF<1 indicating nil or very low enrichment of As (0.93), Cu (0.87), Mn (0.82), and Zn (mean 0.99) or 1≤EF<3 for Cr (1.90) and Pb (1.01) indicating minor enrichment of anthropogenic inputs. The mean EF is ranked as Mn<Cu<As<Zn<Pb<Cr which suggest that Cr is the most enriched element.

Table 4 shows the mean CF values of heavy metals were ranked Mn (0.35) <As (0.40) <Zn (0.41) <Cu (0.46) <Pb (0.47) <Cr (0.76). The mCd values are <1.5 suggesting very low degree of contamination in the study area. The PLI also revealed only mild pollution (0≤PLI<1) in the coastal sediments. The mean EF values show As, Cu, Mn and Zn are EF <1, which describe none to very low enrichment, whereas Cr and Pb are 1 ≤EF< 3 describing minor enrichment of anthropogenic inputs. This also means As, Cu, Mn and Zn are exclusively of lithogenic (crust) origin, which is from the parent soil-forming rocks. The EF between 1 to 3 for Cr and Pb shows the mixed origin from both natural and anthropogenic sources [31]. The mCd values also sum up the factor analysis to <1.5 which suggests very low degree of contamination by anthropogenic sources in the Kota Belud coastal area. The PLI revealed all sampling stations show mild pollution (0≤PLI<1), indicating only low deterioration of the sediment quality [24]. The overall values of mCd and PLI indicate that anthropogenic sources have a minor impact on the levels of heavy metal pollution in coastal sediments.

Table 3. Igeo and EF parameters based on Turekian and Wedepohl as reference [20].

Samples	Geoaccumulation Index (Igeo)						Enrichment Factor (EF)					
	As	Cr	Cu	Mn	Pb	Zn	As	Cr	Cu	Mn	Pb	Zn
SP2	-2.38	-0.74	-2.37	-1.97	-2.14	-1.81	0.68	2.12	0.68	0.90	0.80	1.00
SP3	-3.19	-0.96	-2.69	-2.48	-2.34	-1.71	0.48	2.24	0.67	0.78	0.86	1.33
TP1-1	-1.01	-0.57	-2.77	-1.77	-1.88	-1.71	2.02	2.75	0.60	1.20	1.11	1.24
TP1-2	-0.68	-	-2.56	-1.24	-1.72	-1.72	2.00	-	0.54	1.36	0.97	0.98
TP2-1	-2.24	-0.64	-2.65	-2.30	-2.28	-1.98	0.93	2.81	0.70	0.89	0.90	1.11
TP2-2	-1.32	-	-2.65	-1.76	-1.89	-1.82	1.50	-	0.60	1.11	1.01	1.07
TP2-3	-1.39	-	-2.79	-1.78	-1.97	-1.86	1.59	-	0.60	1.21	1.06	1.14
TP2-4	-1.54	-0.57	-2.64	-2.00	-2.03	-1.81	1.36	2.67	0.64	0.99	0.97	1.13
TP3-1	-2.51	-0.49	-2.15	-2.38	-1.89	-1.47	0.63	2.56	0.81	0.69	0.97	1.30
TP3-2	-2.03	-0.77	-1.10	-2.00	-1.65	-1.47	0.83	2.00	1.59	0.86	1.09	1.23
TP4-1	-2.54	-3.91	-1.85	-3.22	-3.59	-2.88	0.22	0.08	0.35	0.14	0.11	0.17
TP4-2	-2.59	-4.05	-0.73	-3.63	-2.79	-2.69	0.18	0.06	0.64	0.09	0.15	0.16
TL1-1	-3.57	-	-0.16	-1.89	-0.10	-	0.20	-	2.12	0.64	2.22	-
TL1-2	-3.05	-0.86	-0.91	-2.32	-0.67	-	0.36	1.66	1.60	0.60	1.90	-
Mean	**-2.15**	**-1.36**	**-2.00**	**-2.20**	**-1.92**	**-1.91**	**0.93**	**1.90**	**0.87**	**0.82**	**1.01**	**0.99**

Table 4. CF, mCd and PLI parameters for the coastal sediment

Elements	Contamination Factor (CF)						ΣCf	mCd	PLI
	As	Cr	Cu	Mn	Pb	Zn			
SP2	0.29	0.90	0.29	0.38	0.34	0.43	2.63	0.44	**0.40**
SP3	0.16	0.77	0.23	0.27	0.30	0.46	2.19	0.37	**0.32**
TP1-1	0.75	1.01	0.22	0.44	0.41	0.46	3.29	0.55	**0.49**
TP1-2	0.94	-	0.25	0.64	0.46	0.46	2.74	0.46	**0.50**
TP2-1	0.32	0.96	0.24	0.31	0.31	0.38	2.51	0.42	**0.37**
TP2-2	0.60	-	0.24	0.44	0.40	0.42	2.11	0.35	**0.40**
TP2-3	0.57	-	0.22	0.44	0.38	0.41	2.02	0.34	**0.39**
TP2-4	0.52	1.01	0.24	0.37	0.37	0.43	2.93	0.49	**0.44**
TP3-1	0.26	1.07	0.34	0.29	0.40	0.54	2.90	0.48	**0.43**
TP3-2	0.37	0.88	0.70	0.38	0.48	0.54	3.34	0.56	**0.53**
TP4-1	0.26	0.10	0.42	0.16	0.12	0.20	1.26	0.21	**0.19**
TP4-2	0.25	0.09	0.90	0.12	0.22	0.23	1.82	0.30	**0.22**
TL1-1	0.13	-	1.34	0.41	1.40	-	3.27	0.55	**0.56**
TL1-2	0.18	0.83	0.80	0.30	0.94	-	3.05	0.51	**0.51**
Mean CF	**0.40**	**0.76**	**0.46**	**0.35**	**0.47**	**0.41**			

4. Conclusion

The study area is made up by the Crocker Formation, Wariu Formation, and alluvium Quaternary. The result of geophysical survey shows that the alluvium quaternary deposits at the study site have a thickness of up to 13 metres and overlay the oldest rock of the Crocker Formation, which is dominated by fine material of clay and mudstone. The geochemical analysis shows a decreasing ranking order of elemental concentrations Fe>Mn>Zn>Cu>Pb>Cr>As. The analyses show all elements are within the background values and acceptable standard limits of the Sediment Quality Guidelines (SQG) by USEPA for coastal sediments. The sediment quality indices are highly correlated to the physico-chemical properties such as the acidic to alkaline conditions (range 5.6-8.5), moisture content (mean 25.3%), organic matter (mean 1.27%) and texture (mean sand 93.3%, silt 3.4%, clay 3.2%) that affect the accumulation of the metals from inland to the coastal sediments. Geophysical analysis shows that the clayey alluvial deposits from inland region made up of a low resistivity zone, whereas the sandy alluvial deposits from coastal region made up a high resistivity zone which affects the mobility of heavy metals. The assessment of heavy metals from both inland and coastal region revealed only minimal degree of pollution in the sediments. The mCd and PLI values suggest that Kota Belud marine ecosystem is mildly polluted due to very low degree of contamination by anthropogenic sources.

Acknowledgement
We are thankful to Universiti Malaysia Sabah for providing laboratory facilities at the Faculty of Science and Natural Resources, Universiti Malaysia Sabah, and for financial support from the MOSTI funds with CodeTR@M001-2019.

References

[1] Tongkul, F 1990 Structural style and tectonic of Western and Northern Sabah. *Bulletin of Geol. Soc. of Malaysia* **27** 227-239.
[2] Mineral and Geosciences Department of Malaysia (JMG) 2015 *Geological Map of Sabah (4th Ed.)* The Director General of Minerals and Geoscience Malaysia.
[3] Tahir S, Siong K V, Musta B and Asis J 2017 Facies and Sandstone Characteristics of The Kudat Formation, Sabah, Malaysia *Geological Behavior* **1(2)** 20-25
[4] Tjia H D 1988 Accretion tectonics in Sabah: Kinabalu suture and East Sabah accreted terrane *Bulletin of Geol. Soc. of Malaysia* **22** 237-251
[5] Tongkul F 2006 The structural style of Lower Miocene Sedimentary Rocks, Kudat Peninsula, Sabah *Bulletin of the Geol. Soc. of Malaysia* **49** 119-124
[6] Junaidi, Asis and Jasin B 2012 Aptian to Turonian radiolaria from the Darvel Bay Ophiolite Complex, Kunak, Sabah *Bulletin of Geol. Soc. of Malaysia* **58** 89-96
[7] Hutchison C S and Vijayan V R 2010 What are the Spratly Islands? *Journal of Asian Earth Sciences* **39** 271 – 385
[8] Kalnejais L H, Martin W R and Bothner M H 2010 The release of dissolved nutrients and metals from coastal sediments due to resuspension *Marine Chemistry* **121**(1-4) 224-235
[9] Villaescusa-Celaya J A, Gutierrez-Galindo E A and Flores-Munos G 2000 Heavy metals in the fine fraction of coastal sediments from Baja California (Mexico) and California (USA) *Environmental Pollution* **108** 453-462
[10] Vallius H, Ryabchuk D and Kotilainen A 2007 Distribution of heavy metals and arsenic in soft surface sediments of the coastal area off Kotka, northeastern Gulf of Finland, Baltic Sea *Geological Survey of Finland Special Paper* **45** 33–48
[11] Shirani M, Afzali K N, Jahan S, Strezov V and Soleimani-Sardo M 2020 Pollution and contamination assessment of heavy metals in the sediments of Jazmurian playa in southeast Iran *Scientific reports* **10(1)** 1-11
[12] Wuana R A and Okieimen F E 2011 Heavy metals in contaminated soils: a review of sources, chemistry, risks and best available strategies for remediation *International Scholarly Research Notices* **2011**
[13] Jayamurali D, Varier K M, Liu W, Raman J, Ben-David Y, Shen X and Gajendran B 2021 An Overview of Heavy Metal Toxicity *Metal, Metal Oxides and Metal Sulphides for Biomedical Applications* 323-342

[14] U.S. Environmental Protection Agency (USEPA) 1977 *Guidance for the Pollutional Classification of Great Lakes Harbor Sediments*, Region V, Chicago, Illinois.

[15] Birch G 2018 A review of chemical-based sediment quality assessment methodologies for the coastal environment *Marine pollution bulletin* **133** 218-232

[16] Abolfazl N and Ahmad I 2012 Sediment quality assessment of Klang Estuary, Malaysia. *Aquatic EcosystemHealth & Management* **15(3)** 287-293

[17] U.S. Environmental Protection Agency (USEPA), 2014. Method 6010D (Revision 4): Inductively coupled- plasma atomic emission spectrometry, Washington, DC: Environmental Protection Agency

[18] BSI. 1990. BS1377: 1990 *British Standard Methods of Tests for Soils for Civil Engineering Purposes London* British Standard Institution (BSI).

[19] Turekian K K and Wedepohl K H 1961 Distribution of the elements in some major units of the earth's crust *Geol Soc Am Bull* **72**(2) 175-92

[20] Müller G 1969 Index of geoaccumulation in the sediments of the Rhine River *Geojournal* 2 108–118.

[21] Herut B and Sandler A 2006 Normalization methods for pollutants in marine sediments: review and recommendations for the Mediterranean *IOLR Report H* **18**

[22] Taylor S R 1964 Abundance of chemical elements in the continental crust: a new table. *Geochim CosmochimActa* **28** 1273–1285

[23] Hakanson L 1980 Ecological risk index for aquatic pollution control: A sedimentological approach *Water. Res.* **14** 975–1001

[24] Tomlinson D L, Wilson J G, Harris C R and Jeffrey D W 1980 Problems in the assessment of heavy- metal levels in estuaries and the formation of a pollution index *Helgoländer meeresuntersuchungen* **33(1- 4)** 566-575

[25] Arslan H and Demir Y 2013 Impacts of seawater intrusion on soil salinity and alkalinity in Bafra Plain, Turkey *Environmental monitoring and assessment* **185(2)** 1027-1040

[26] Zhang W, Chen X, Tan H, Zhang Y and Cao J 2015 Geochemical and isotopic data for restricting seawater intrusion and groundwater circulation in a series of typical volcanic islands in the South China Sea *Marine pollution bulletin* **93(1-2)** 153-162

[27] Hassink J and Whitmore A P 1997 A model of the physical protection of organic matter in soils *Soil Science Society of America Journal* **61(1)** 131-139

[28] Gui Y, Zhang Q, Qin X and Wang J 2021 Influence of Organic Matter Content on Engineering Propertiesof Clays *Advances in Civil Engineering* **2021** 1-11

[29] Qi S, Leipe T, Rueckert P, Di Z and Harff J 2010 Geochemical sources, deposition, and enrichment of heavy metals in short sediment cores from the Pearl River Estuary, Southern China *Journal of marine systems* **82** S28-S42

[30] Yakovlev E Y, Zykova E N, Zykov S B, Malkov A V and Bazhenov A V 2020 Heavy metals and radionuclides distribution and environmental risk assessment in soils of the Severodvinsk industrial district, NWRussia *Environmental Earth Sciences* **79** 1-16

[31] Arumugam A 2016 Ecological risk assessment of the Miri coast, Sarawak, Borneo: A biogeochemical approach (Doctoral dissertation, Curtin University).

The use of magnetic data from *swarm satellite* as an efforts to estimate the potential of iron sand in the south Bawean Island area

T Widyawati[1], A Susilo[1] , Wiyono[1], Sunaryo[1], A B P Wardoyo[1], Karimah[1]
[1]University of Brawijaya , Malang

E-mail: Tututwidyawati26@gmail.com

Abstract. This study describes the use of magnetic data obtained through the Swarm satellite and launched by the Europan space agency to estimate iron sand deposits in the southern part of Bawean Island, Gresik Regency. Data processing in this study was carried out using Microsoft Excel and Oasis Montaj software. The interpretation process in this study was carried out in two stages, namely the qualitative interpretation stage using contour residual anomaly maps and quantitative interpretation using 2D and 3D modeling. The interpretation results show that the aeromagnetic data used can be used as initial data in estimating sand deposits in the southern part of Bawean Island, Gresik Regency.

1. Introduction

Bawean Island is a small island located in the Java Sea. Administratively, Bawean Island is an island that belongs to the Gresik Regency, East Java Province. The distance between Bawean Island and the city of Gresik is about 120 Kilometers or 80 nautical miles. Geographically, Bawean Island is located at 5°00'00"- 6°00'00" South Latitude and 111°00'00"- 114°00'00" East Longitude [1],[2],[3],[4].

Based on regional geology [5] the rock composition in the study area is as follows. Qa is a type of alluvial rock formation consisting of gravel, gravel, sand, silt, and clay. Qa units are river and coastal deposits, generally not compact and easy to lose. Qv is a type of rock formation from the Balibak volcano which consists of alternating lava, volcanic breccia, and tuff. Tmps is a type of cobblestone formation consisting of quartz sandstone with claystone and peat inserts. Toml is a type of limestone rock formation consisting of reef limestone, clastic limestone, and local crystal limestone.

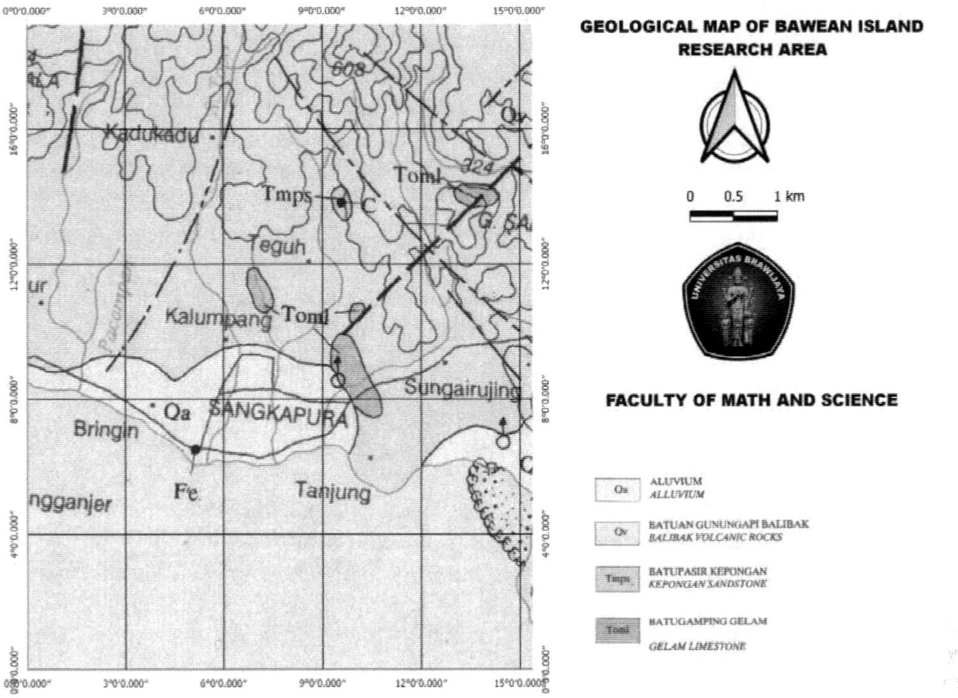

Figure 1. geological map of the study area

Bawean Island is an island that has various potential mineral and energy resources that can be utilized. Some of the potential mineral and energy resources of Bawean Island according to [5] include andesite and basalt igneous rocks, sand, gravel, limestone, calselite, kaolin, clay, iron sand, coal, and oil seepage

Iron sand is one of the mineral resources in Bawean Island which is the focus of this research. this is because iron sand is a material that is widely used in industry [6]. Some of the mineral content contained in iron sand has made it a material of high economic value [7]. In general, iron sand is found in coastal areas. Iron sand is formed from the process of destruction by weather, surface water, and waves from the original rock that already contains iron minerals (Mineral Inventory Directorate Team, 2005). Based on [5] the potential for iron sand on Bawean Island is in alluvium (Qa) rock formations.

Research related to iron sand on Bawean Island has also been carried out by [8]. The results of the research that have been carried out show that the amount of iron sand distribution in the Bawean Island area is relatively small. Therefore, the purpose of this study was to determine the distribution of iron sand in the Bawean Island area with a wider coverage using magnetic data obtained from satellites. The application of magnetic surveys has also been carried out by several previous researchers as follows [9-14].

Several benefits can be obtained in this research. among others are as follows. The results of the research that has been done can be used to determine the distribution map of an area that is suspected to contain iron sand potential in the Bawean Island area. The resulting distribution map can be used as initial knowledge before exploration is carried out. The resulting distribution map can also be used as supporting data for the application of other methods in revealing the potential for iron sand in the research area. In addition, the results obtained in this study can also be used as additional literature for further research.

Journal of Physics: Conference Series **2165** (2022) 012047 doi:10.1088/1742-6596/2165/1/012047

2. Methods

The method used in this study is a magnetic method which in its application utilizes the magnetic properties of rocks. The magnetic method measures variations in the magnetic field on the earth's surface. The occurrence of variations in the magnetic field on the earth's surface is caused by variations in the distribution of magnetic objects beneath the earth's surface. There are several corrections that must be made in the magnetic method to get the total magnetic field anomaly value [15-18], namely

a. Diurnal correction

The correction is used to eliminate deviations in the magnetic field intensity caused by differences in the time of data collection and the effect of sunlight in a day. The diurnal correction can be calculated by the following equation.

$$H_D = \frac{(t_n - t_{aw})}{(t_{ak} - t_{aw})} (H_{ak} - H_{aw}) \tag{1}$$

b. IGRFcorrection

IRGF correction is a correction that is used to eliminate the effects of magnetic forces originating from the internal earth. Generally the IRGF correction of the magnetic method is provided by the NOAA (National Oceanic Atmospheric Administration) Geophysical data center. However, in this study the IGRF correction can be known through the IGRF menu which is already available in the montaj oasis software.

After the diurnal correction and IGRF correction have been carried out, the total magnetic field anomaly value can be determined through the following equation.

$$\Delta H = H_n - (H_D + H_{IGRF}) \tag{2}$$

2.1 Data Acquisition

Research using magnetic data has been carried out in the southern part of Bawean Island, Gresik Regency, East Java Province. Magnetic data in this study has been obtained through the Swarm satellite launched by the europan space agency on July 10, 2021. The data was taken with an area of 4 × 6 Km with coordinate boundaries 112°624633 East Longitude - 112°67813 East Longitude and 112°623335 East Longitude - 112° 67753 east longitude, 5°819555 south longitude - 5°82019 south longitude and 5°854498 south longitude - 5°85632 south longitude. Magnetic data is taken with the distance between the points is 220m. This study using magnetic data aims to determine the alleged potential for iron sand located on Bawean Island, Gresik Regency, East Java Province.

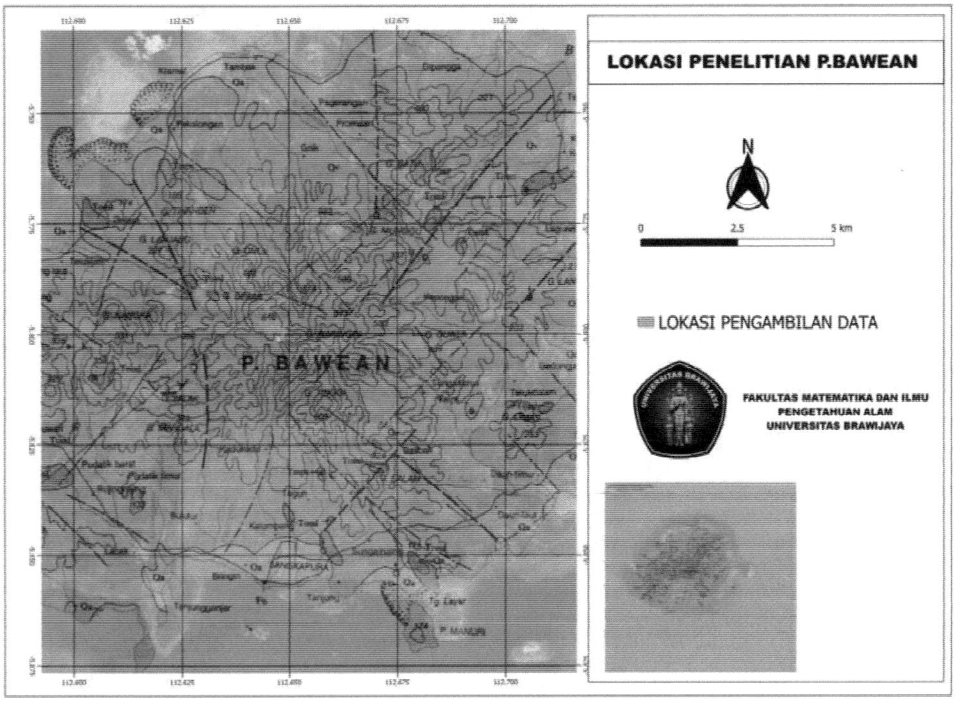

Figure 2. P. Bawean research location map

2.2 Data processing
Magnetic data processing in this study was carried out in several stages as follows. The parameters of the magnetic data obtained in this study are the intensity of the magnetic field, the position of latitude, longitude, altitude, and time. The data that has been obtained will then be carried out with daily correction and IRGF correction processes to obtain the total magnetic field anomaly value. This process is carried out using Microsoft Excel software.

After obtaining the value of the magnetic field anomaly, then the contour map of the total magnetic anomaly is made. Then the total magnetic anomaly is separated to obtain regional anomalies and residual anomalies. The process of separating the total magnetic anomaly was carried out using a Butterworth filter. All the contouring processes are carried out using the oasis montaj software.

2.3 Interpretation
The interpretation stage in this study was carried out in two ways, namely qualitative interpretation and quantitative interpretation. The qualitative interpretation in this study is an interpretation based on the contour of the residual anomaly map that has been processed using oasis montaj software. The information obtained from the qualitative interpretation that has been carried out is in the form of estimates of rock types in the research area that have been grouped based on the value of the magnetic anomaly.

The quantitative interpretation in this study is based on the results of 2D and 3D modeling. The information obtained in the quantitative interpretation in this study is in the form of a number stating the estimated depth and volume of the research target, namely rock deposits containing iron sand. To get the results of 2D modeling in this study, it is done by determining the slicing on the contour of the residual anomaly map that is adjusted to an area that has anomaly according to the target in the study. The area with high anomaly is the target area in this research. It is based on the research target which is iron sand with strong magnetic properties

2D modeling in this study can be used to determine the type and depth of a rock in the study area. The rock types in this study have been interpreted based on the susceptibility values shown in a (Table 1) as follows.

Table 1. Magnetic susceptibility price table

Jenis	Suseptibilitas × 10^{-3} (SI)	
	Kisaran	Rata-rata
	Sedimentary	
Dolomite	0 – 0,9	0,1
Limestone	0 – 3	0,3
Sandstone	0– 20	0,4
Shales	0,01 – 15	0,6
Av. 48 sedimentary	0– 18	0,9
	Metamorphic	
Amphibolite		0,7
Schist	0,3 – 3	1,4
Phylite		1,5
Gneiss	0,1 – 25	
Quartzite		4
Serpentine	3– 17	
Slate	0– 35	6
Av. 61 metamorphic	0– 70	4,2
	Igneous	
Granite	0– 50	2,5
Rhyolite	0,2 – 35	
Dolorite	1– 35	17
Augite-syenite	30 – 40	
Olivine-diabase		25
Diabase	1 – 160	55
Porphyry	0,3 – 200	60
Gabbro	1– 90	70
Basalts	0,2 – 175	70
Diorite	0,6 – 120	85
Phyroxenite		125
Peridotite	90– 200	150
Andesite		160
Av. Acidic igneous	0– 80	8
Av. Basic igneous	0,5 - 97	25

To obtain a 3D volumetric model in this study, inverse modeling was carried out on the residual anomaly contour. 3D modeling is modeling used to determine the thickness and direction of continuity of the rock type that is the target of research.

3. Result and Discussion

3.1. Anomali medan magnetic total

The total magnetic anomaly map (TMI) in the magnetic method is a contour anomaly map that shows variations in the magnetic field below the ground surface.

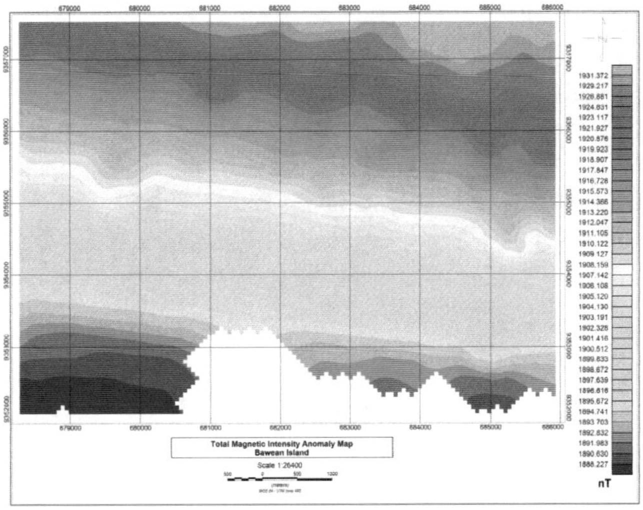

Figure 3. Total magnetic anomaly (TMI) contour map

Figure 3 shows the contour map of the magnetic anomaly of the study area. The research area, in this case, is dominated by magnetic anomalies ranging from high anomalies to low anomalies. The value of the total magnetic anomaly generated in the study area, in this case, is in the range of 1888,227 nT to 1931,372 nT. The difference in the value of the magnetic anomaly, in this case, is caused by differences in the mineral content below the surface. Analysis of the total magnetic anomaly contour map in this way has also been carried out by several researchers [19][20][21]. The magnetic anomaly that is the target in this study is a high magnetic anomaly. this is adjusted to the target of the research in the form of iron sand which has high magnetic properties. so it can be assumed that an area that has iron sand potential is an area that has a high magnetic anomaly as well.

3.2. Anomaly separation

Separation of the total magnetic field anomaly was carried out to eliminate the influence of magnetic objects on the surface. The results obtained from the separation process of total magnetic field anomalies are regional anomalies and residual anomalies. The residual anomaly in this study is used as the basis for the qualitative interpretation resulting from the reduction of the total magnetic field anomaly with its regional anomaly. The use of residual anomalies as a basis for qualitative interpretation with the aim of finding iron sand prospects in an area was also carried out by [22]. The following is the residual anomaly contour map data of the research area.

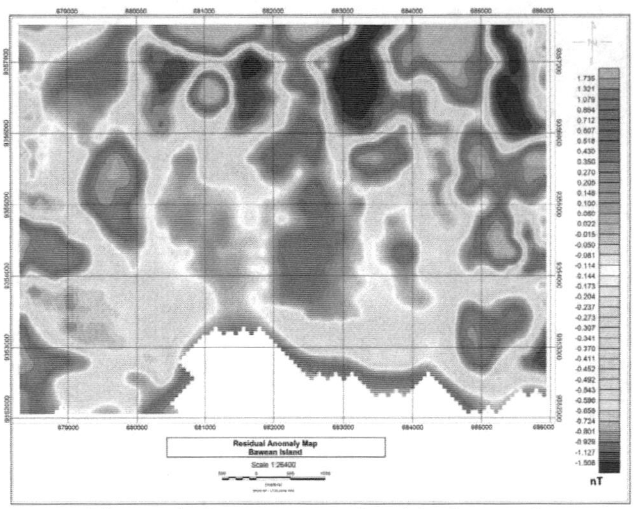

Figure 4. Contour map of residual anomaly

The magnetic field anomaly that is the target in this study is a magnetic field anomaly with a high value. This is because iron sand is naturally ferromagnetic (magnetic properties with a strong magnetic attraction). Magnetic anomalies with high value are thought to indicate the presence of magnetic minerals contained in rocks below the surface. So in this case it can be assumed that the area that has the potential for iron sand is an area that has a fairly high magnetic anomaly field as well. Based on the contour data of the residual anomaly map, it shows that the high anomaly values are grouped for the value of the magnetic field intensity which is in the range of -1.508 nT to 1.735nT. Magnetic anomaly values based on Figure 4 in this case are grouped into three groups of anomalies, namely low anomalies, moderate anomalies, and high anomalies. The low anomaly, in this case, is on a blue to green color scale with a magnetic anomaly value of -1.508 to -0.144 nT. The moderate anomaly, in this case, is on a yellow to orange color scale with a magnetic anomaly value of -0.114 to 0.430 nT. The high anomaly, in this case, is on the red to purple color scale with a magnetic anomaly value of 0.518 to 1.736 nT.

It should be noted that the residual anomaly map contour data in Figure 4 is the residual anomaly map contour data in the form of a dipole. If the data in the form of a dipole is used as data interpretation, it will be difficult. To make this easier, the data must be converted into a data in the form of a monopole through a reduction process to the poles. This happens because the map contour of the residual anomaly that has been monopolized can be seen more clearly, so that it will simplify the data interpretation process. The interpretation process using contour maps that have been monopolized to simplify the interpretation process is also carried out by [23]. The following is the residual anomaly contour map data which is already in monopole form.

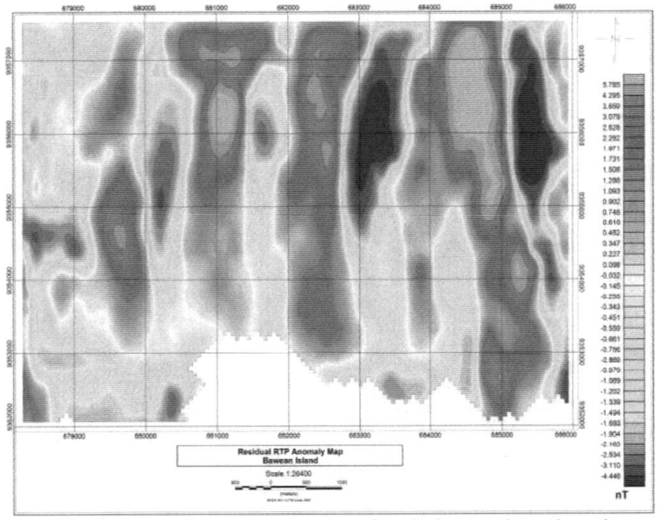

Figure 5. Residual anomaly map contour after being reduced to the poles (RTP)

To determine an area that is suspected to contain iron sand potential in this study, it is done by overlaying the geological map of the research area with the contour of the residual anomaly map. The results of the overlay between the geological map of the research area and the contour of the magnetic anomaly map, in this case, are shown in a picture as follows.

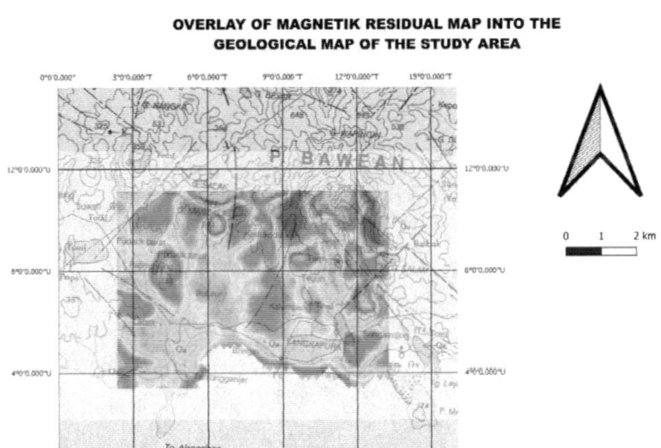

Figure 6. Geological map overlay of research area with residual anomaly map contour

Figure 6 shows that the magnetic anomaly distribution pattern is from minimum (low) to maximum (high) in the study area. Areas with high magnetic anomalies that have been targeted in this study as well as the suitability of areas where iron sand is generally located around coastal areas have been found in a Qa formation rock. Based on [5] the Qa rock formation is a type of rock consisting of gravel, gravel, sand, mud, and clay. The Qa rock formations in this case are deposits located in rivers and beaches which are generally not compact and easily separated. So, based on this it can be assumed that iron sand deposit is a material contained in the Qa rock formation.

3.3. 2D modeling

Before making a 2D modeling, first, a slice is drawn above the contour of the residual anomaly map. Slice withdrawal in this study is determined in an area that is estimated to have iron sand deposits. An estimate of an area that has iron sand deposits in this case is indicated by the area in the Qa formation and adjusted for the location of the potential presence of iron sand which is generally around the coast and areas that have high magnetic anomalies. Thus, 2 slices are drawn with line 1 leading from southeast to northwest and line 2 leading from south to north which is shown in a figure (Figure 7) as follows.

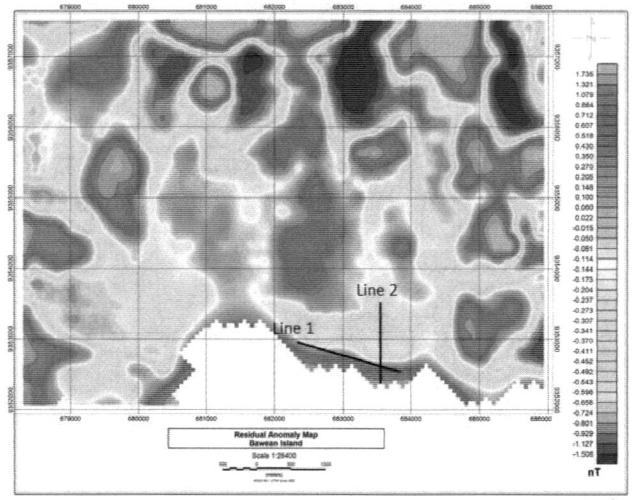

Figure 7. Slice drawing above the residual anomaly map contour

Based on the drawing of the slice on line one, a 2D cross-section is generated as follows.

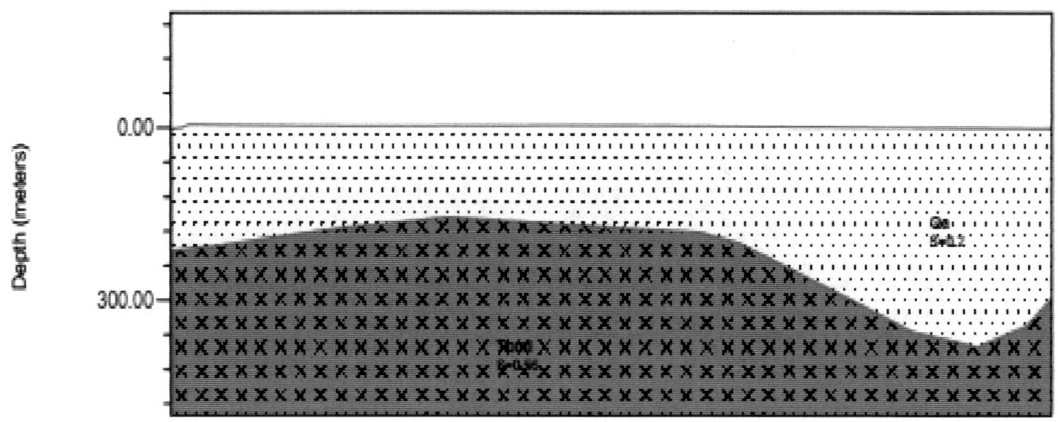

Figure 8. 2D modeled map in line 1

Figure 8 shows that, in the 2D modeling results, there are 2 types of rock formations arranged along line 1. The magnetic susceptibility values scattered along line 1 are in the range of 0.20 to 0.56. The Qa rock formation which is thought to contain iron sand is located at a depth of 0 - 350 meters from the earth's surface with a susceptibility value of 0.20.

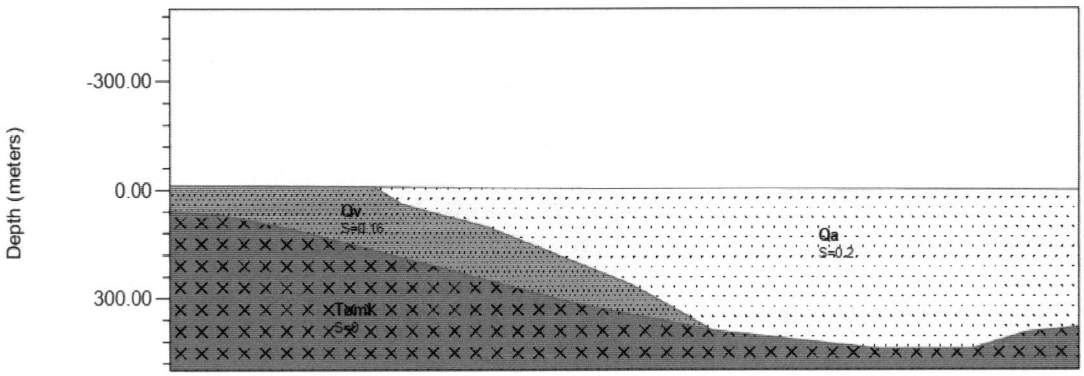

Figure 9. 2D modeled map on line 2

Figure 9 shows the types of rock formations arranged along track 2. There are 3 types of rock formations scattered along track 2 with magnetic susceptibility values in the range of 0 to 0.20. The Qa rock formation which in the study is suspected to contain iron sand is at a depth of 0 - 400 meters from the earth's surface with a susceptibility value of 0.20.

3.4. 3D modeling

Figure 10 is the result of 3D modeling which shows the subsurface model of the research area. The results of 3D modeling in this case provide information related to the volume of the suspected rock type containing iron sand. there are tools in the oasis montaj software that can be used to determine the volume that is suspected to be a type of rock containing iron sand. The same thing has been done by previous researchers [24] who used the results of 3D modeling to determine the volume of the research target. The results of 3D modeling in this study indicate the volume of the Qa rock formation which is thought to contain iron sand in it is 1.165819511 km^3. Analysis of the results of 3D modeling to identify the distribution of iron ore has also been carried out in research [25]. From the 3D map pattern in (Figure 10) it can also be seen that the alleged iron sand deposit is located on the south side of the study area that crosses from west to east.

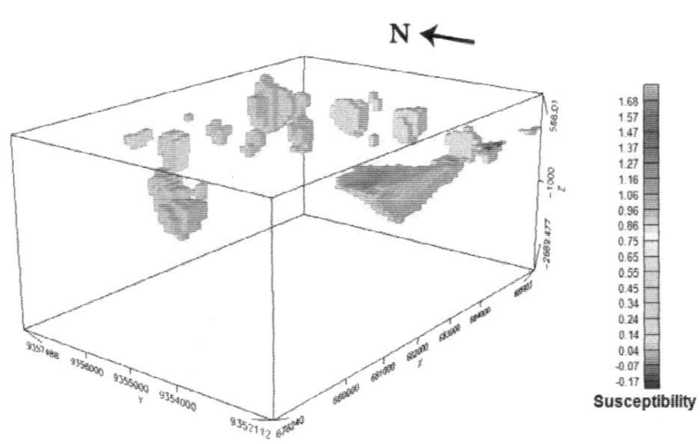

Figure 10. 3D modeled map

4. Conclusion

Based on the discussions that have been carried out, it can be seen that iron sand is thought to be in the Qa rock formation. The depth of the research target in this study has been shown by the results of 2D modeling. Based on these data, it can be seen that the Qa rock formation which is suspected to contain iron sand is at a depth of 0 to 400 meters with a susceptibility value of 0.20. The results of 3D modeling in the study have provided information regarding the estimated volume of the research target. Based on the 3D data, it can be seen that the volume of the rock formation that is thought to contain iron sand is 1.165819511 km^3 which is located on the south side of the study area across from west to east. Magnetic data in this study is initial data to determine the distribution of magnetic anomalies in the study area. Further research needs to be done so that efforts to determine the potential of iron sand in the area can provide more accurate results. Some of the methods that can be applied are resistivity methods, polarization induction methods, geochemical methods and so on.

Acknowledgement

The authors are grateful to the european space agency for the release of the magnetic data, used in this study.

References

[1]. Albab A, Aryanto N C D 2017 *Bulletin .Of the marine Geology* 32 L1
[2]. Sukandar, Dewi C S U and Handayani M 2017 *Jurnal. Ilmu – ilmu perairan. Pesisir dan perikanan.* 6 L3
[3]. Aesthetika N M 2012 *Jurnal. Ilmu. Komunikasi* 1 L1
[4]. Hasan M F, Santoso W D, Dewi K T 2012 *Proc. Ann. Con. and Exh* (Yogyakarta)
[5]. Aziz S, Hardjoprawiro, Mangga S A 1998 *Geologi Lembar Bawean dan Masalembo Jawa*
[6]. Lestari T V and Sehah A 2020 '*Bulletin. of Scientific. Contribution* **18** p. 10.
[7]. Bijaksana S 2002 *Jurnal Geofisika* (Bandung).
[8]. Karimah I 2018 *Jurnal Fisika Indonesia* **22** L 1
[9]. Holden E J. et al 2012 'Identifying structural complexity in aeromagnetic data: An image analysis approach to greenfields gold exploration', *Ore Geology Reviews* **46** pp. 47–59. doi: 10.1016/j.oregeorev.2011.11.002.
[10]. An Z, Qingyun D 2016 *Journal of Applied Geophysics* **135** pp 204–211. doi: 10.1016/j.jappgeo.2016.10.013.
[11]. Assran A S M , E Qassas, R A Y, Yousef M H M 2019 *Egyptian Journal of Petroleum* **28**(1) pp. 61–69. doi: 10.1016/j.ejpe.2018.11.002.
[12]. Olasunkanmi, N. *et al.* 2020 'Interpretation of high resolution aeromagnetic data of Kaoje and its environ, western part of the Zuru Schist belt, Nigeria: implication for Fe–Mn occurrence', *Heliyon*, **6**(1), p. e03320. doi: 10.1016/j.heliyon.2020.e03320.
[13]. Opeloye, S. A. *et al.* 2021 'Palaeoenvironmental reconstruction and oolitic ironstone mapping of the Agbaja Ironstone Formation in the Nupe Basin, North-central Nigeria: Insights from sedimentological and aeromagnetic analyses' *Results in Geophysical Sciences* **5** p. 100010. doi: 10.1016/j.ringps.2021.100010.
[14]. Adewumi T and Salako K A 2018 ' Delination of mineral potential zone using high resolution aeromagnetic data over part of Nasawara State, North Central, Nigeria' *Egyptian Journal of Petroleum* **27** 759–767
[15]. Telford W M, Geldart L P and Sheriff R E 1982 *Applied Geophysics* Second Edition (New York : Cambridge University Press)
[16]. Osinowo O O and Falufosi M O. 2018 'Magnetic and very low frequency electromagnetic (VLF-EM) investigation for gold exploration around lhale in Bunnu-Kabba area of kogi, north-central Nigeria' *Contribution to Geophysics and Geodesy* **48**(3)
[17]. Susilo A and Sunaryo 2018 'Investigation of Sidoarjo mud vulcano ("LUSI") impact on the subsurface using geomagnetic method at Sidoarjo district, Indonesia' *Disaster Advances* **11**(3).
[18]. Susilo A. et al. 2017 'Investigation of jabung temple subsurface at Probolinggo, Indonesia using resistivity and geomagnetic methods' *International Journal of Geomate* **13**(74 – 80).

[19]. Broto S and Putranto T T 2011 'Application of geomagnetic methods in geothermal exploration' *Journal of Engineering* **32** (1)
[20]. Nuha and Avisena, 2012 'Modeling the subsurface structure of the Songgoriti hot spring area in Batu City based on geomagnetic data' *Neutrino Journal* **4** (2)
[21]. Nurdin N H. et al. 2017 'Identification of the distribution pattern of subsurface rock intrusion using the geomagnetic method in the Jenelata River, Gowa Regency' *Geocelebes Journal* **1** (1)
[22]. Ramadhan A. et al. 2014 'Estimation of iron ore potential in Sepoteng Hamlet, Sungai Betung District, Bengkayang Regency using the geomagnetic method' *Prisma Physics* **2**(3)
[23]. Junursyah G M L and Rahmat W . 2019 ' The potential of iron sand deposit in the grabag area and its sorrounding based on geo-magnet data' *Journal Of Geology And Mineral Resources* **2** (2)
[24]. A Jaingot et al. 2015 ' 3D model of hematit mineral based on geomagnetic data in the Uekuli Village Tojo Una – una Regency' *Promine Journal* **3**(1)
[25]. Rachmawiana et al. 2013' Magnetic 3D modeling using MAG3D to identify the distribution of iron ore in the "RAM-UNILA" area. *Journal of Exploration Geophysics* **1**(1)

How do additional instructions change the answer? Study of pre-service physics teachers' misconception about buoyancy

Maison[1], R Asma[2], A Doyan[3] and L Saputri[1]

[1]Physics Education Study Program, Universitas Jambi, Indonesia
[2]State High School 1, Jambi City, Indonesia
[3]Physics Education Study Program, FKIP, University of Mataram, West Nusa Tenggara, Indonesia

E-mail: maison@unja.ac.id

Abstract. The misconceptions experienced by pre-service physics teachers are essential to know; However, these are not easy to identify, let alone to remediate. This study aims to find out how additional instruction changes pre-service physics teachers' responses and whether or not it affects the misconceptions of pre-service physics teachers. Respondents were tested using the online Four-Tier Buoyancy Instrument (FTBI) twice, namely at Time 1 and 2. Pre-service physics teachers were given additional instructions between the two measurements by reminding them of the principle of balance on a stationary object or an object moving in a straight line with constant velocity. Quantitative data from the two areas of the instrument are then analyzed and compared to determine the differences in respondents' answers. In-depth interviews were conducted with respondents who had different answers to the instruments. The results showed that a small proportion of respondents had changed their answers to be the correct answers on several FTBI items. The interview results showed that it seems there was a tendency to change the respondents' cognitive structure from locally coherent to global coherent. However, this change does not necessarily reduce pre-service physics teachers' misconceptions about buoyancy significantly.

1. Introduction

Continuous experiences and interactions between students and the surrounding environment form some of these students' knowledge, attitudes, and skills. When they learn a concept, their interpretation can be influenced by experience and information obtained from the environment. In certain circumstances, it can lead to a conception different from those held by scientists, which is called a misconception [1]. Many types of research related to students' misconceptions have been carried out using various qualitative and quantitative techniques and instruments. Quantitatively, the instruments used continue to develop, starting from the form of ordinary multiple-choice, two-tier, three-tier, and four-tier [2], in the form of paper and web-based such as [3].

Since the research conducted by [4] using a four-tier misconception instrument, there have been many studies conducted by other researchers to develop and apply a four-tier format instrument in diagnosing misconceptions including [5], [6], [7], and [8]. Misconception research on fluid topics has also been published using four-tier instruments, for example, [9] on floating and sinking and [10] on density. They researched pre-service science teachers and found 74 misconceptions on the concept of floating and sinking and 48 misconceptions on density. The presence of misconceptions is undoubtedly disturbing and hampers the student learning process [11]. It certainly needs to be a concern for pre-service physics teachers because they will become teachers, so they should not have misconceptions about the science they will teach later. They need to be given treatment to reduce the misconceptions they experience.

Can additional instruction change or improve students' understanding of concepts on the topic of buoyancy, and does this increase also significantly reduce their misconceptions? This study aims to determine how additional instruction given to students relates to correct answers and misconceptions.

2. Methods

Researchers selected a sample of 96 pre-service physics teachers of Universitas Jambi in Indonesia randomly. Respondents are members of the population who have studied Basic Physics in which there is material on Archimedes' Law. Data was collected using an online instrument consisting of five items in a four-tier format on buoyancy. The instrument was used sequentially twice, namely at Time 1 and Time 2, like the pre-test and post-test. Between Time 1 and Time 2, pre-service physics teachers were given a small additional instruction about the principle of balance of forces acting on objects. This principle applies to all things that are at rest or moving in a straight line. Although this additional instruction is not explicitly related to buoyancy, it is hoped to add insight to pre-service physics teachers to think more broadly in viewing a physical phenomenon. In general, the technical data collection and treatment were (1) taking the test at Time 1, (2) small additional instructions, (3) taking the test at Time 2, and (4) interviewing pre-service physics teachers who had different answers between Time 1 and Time 2.

The process of scoring quantitative data on test results in this study was carried out quickly because the online instrument that we designed (web-based) was equipped with a program to directly calculate correct scores and misconception scores for Tier-1, Tier-1&3, and All Levels. Besides being able to find out the understanding and misconceptions experienced by respondents, this application can also display False Positive (FP), False Negative (FN), and Lack of Knowledge (LK) scores for each pre-service physics teacher, class, or all students as soon as they submit an answer. It facilitates the process of identifying pre-service physics teachers' misconceptions and, at the same time, providing information about which students need further interviews (based on data that there are differences in their answers and willingness to be interviewed).

Following the instrument's characteristics in a four-tier format, the scoring for correct answers has three categories: valid scores on Tier-1, Tier-1&3, and All-Tier. If a student chooses the correct answer on Tier-1, he will be given a score of 1 for Tier-1, and if he also chooses the correct reason for the answer (it is on Tier-3), he will also get a score of 1 for Tier-1&3. Meanwhile, for All-Tier, pre-service physics teachers can get a score of 1 if he has the correct answer, is sure of the answer, has the right reason, and is also confident of that reason. The correct answer for the All-Tier category shows that a student knows a concept. If a student has the correct answer and the reason is wrong, the student is categorized into False Positive (FP). While the student chooses the right reason, but the answer is incorrect, the student is classified as False Negative (FN). Furthermore, if pre-service physics teachers are unsure of the answers or explanations, pre-service physics teachers are grouped into people who do not know the concept (LK).

This misconception instrument has one correct answer and reason for each item, and the rest are answer choices and reasons that contain misconceptions. The misconception assessment process uses a different key with the correct answer key, as shown in Table 1. A misconception can be identified or explored using one or more items as shown in types M1, M2, M3, and M5. However, the maximum score for pre-service physics teachers' misconceptions of each kind remains one. Therefore, the number of misconceptions scores must be divided by the number of measuring items. For example, for M1, there are four items – namely Item 1, Item 2, Item 4, and Item 5. If a student has two answers that match the key for M1, then the student's misconception score is $(1 + 1)/4 = 0.5$.

Table 1. Answer keys for scoring misconception.

Misconception		Item Answer Keys
Type	**Description**	
M1	Pre-service physics teachers have the concept that the closer to the surface, the buoyant force is smaller.	1.1.a; 1.2.a; 1.3.a; 1.4.a 2.1.a; 2.2.a; 2.3.b; 2.4.a 4.1.b; 4.2.a; 4.3.d; 4.4.a 5.1.b; 5.2.a; 5.3.d; 5.4.a
M2	Pre-service physics teachers have the concept that the closer to the surface, the buoyant force is greater.	1.1.b; 1.2.a; 1.3.b; 1.4.a 2.1.b; 2.2.a; 2.3.d; 2.4.a 4.1.a; 4.2.a; 4.3.b; 4.4.a 5.1.a; 5.2.a; 5.3.b; 5.4.a
M3	Pre-service physics teachers have the concept that buoyancy is determined by hydrostatic pressure.	1.1.a; 1.2.a; 1.3.d; 1.4.a 2.1.a; 2.2.a; 2.3.a; 2.4.a
M4	Pre-service physics teachers have the concept that the smaller the mass of an object, the greater the buoyant force.	3.1.a;.3.2.a; 3.3.a; 3.4.a
M5	Pre-service physics teachers have the concept that the buoyant force is not determined by the volume of the submerged object but is determined by the total volume of the object.	3.1.c; 3.2.a; 3.3.c; 3.4.a 5.1.a; 5.2.a; 5.3.a; 5.4.a

The comparison of repeated measurement scores at Time 1 and Time 2 for correct and misconception scores were determined using the Wilcoxon Signed Rank Test. In this case, we compared (1) pre-service physics teachers' correct scores before and after receiving additional instruction and (2) pre-service physics teachers' misconception scores before and after receiving additional instruction on the topic of buoyancy.

3. Result and Discussion

The study results are described as the percentage of correct scores and misconceptions based on the data obtained at Time 1 and Time 2. In addition, the rate of false positives, false negatives, and lack of knowledge is also seen.

3.1. Correct Answers

After scoring Tier-1, Tier-1&3, and All-Tier for Time 1 and Time 2, data on correct scores for both measurements were obtained.

Table 2. Percentage of correct answers with and without additional instructions.

Item	**Tier-1**		**Tier-1&3**		**All-Tier**	
	Time 1	Time 2	Time 1	Time 2	Time 1	Time 2
Item 1	16.67	23.96	15.63	20.83	15.63	19.79
Item 2	10.42	15.63	8.33	12.5	8.33	12.5
Item 3	16.67	14.58	6.25	10.42	6.25	9.38
Item 4	7.29	7.29	3.13	4.17	2.08	4.17
Item 5	10.42	12.5	6.25	7.29	6.25	7.29
Mean	12.29	14.79	7.92	11.04	**7.71**	**10.63**

The correct scores percentage of pre-service physics teachers answer – for example, in Time 1 – is highest on Tier-1 with an average of 12.29 and the lowest on All-tier with an average score of 7.71. This is due to the difference in the way of scoring as explained in the method section, wherein Tier-1 pre-service physics teachers are given a score of 1 if the answer is correct regardless of other Tiers. The correct score of the all-tier is the lowest because pre-service physics teachers get a score of 1 if the answers and reasons are correct and have confidence in the answers and reasons. Therefore, the All-tier score is more appropriate to determine how pre-service physics teachers understand the concept because it is more accurate. If we look at the percentage of correct scores, pre-service physics teachers have the highest score on Item 1 and the lowest score on Item 4. The difference of correct scores between Time 1 and Time 2 can be seen in Figure 1.

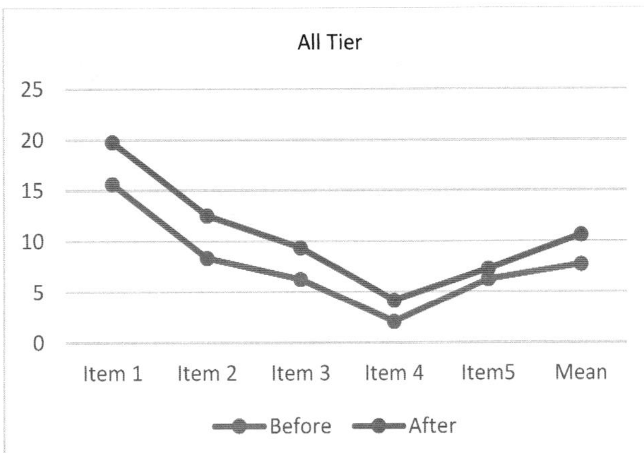

Figure 1. Correct answers percentage before and after treatment.

A Wilcoxon Signed Rank Test revealed a statistically significant increase in pre-service physics teachers' correct answers following additional instruction in the test, $z = -1.98$, $p < 0.05$, with a small effect size ($r = 0.14$). The mean on the correct score increase ($M = 7.7$) to post additional instruction ($M = 10.6$). The results of this statistical test indicate an increase in the percentage of pre-service physics teachers' correct scores on the All-tier. This finding aligns with [12] research which uses hands-on activities in learning to increase students' understanding of floatation concepts. Whether this increase in the rate of correct scores at the same time has reduced the misconceptions that occur, it is necessary to look at the results of the analysis of pre-service physics teachers' misconceptions.

3.2. Misconception

After scoring the items using Item Answer Keys in Table 1, the percentage of student misconceptions for each type in Time 1 and Time 2 is obtained, as shown in Table 3.

Table 3. Percentage of misconception with and without additional instructions.

Misconception	Tier-1		Tier-1&3		All-Tier	
	Time 1	Time 2	Time 1	Time 2	Time 1	Time 2
M1	45.05	44.79	6.51	4.69	5.73	4.69
M2	47.66	47.4	20.83	18.23	19.79	17.45
M3	50	52.6	25.52	25.56	24.48	26.56
M4	63.54	61.46	63.54	61.46	59.38	57.29
M5	40.63	38.8	14.32	16.67	14.32	15.63
Mean	49.38	49.01	26.14	25.52	**24.74**	**24.32**

Like the correct score data, the highest score percentage is in Tier-1, and the lowest is in All-Tier. Based on the assessment criteria, the All-tier needs to be a concern and can be used to decide whether a student has misconceptions or not. A review of the All-Tier for each type of misconception shows that pre-service physics teachers experience the least misconceptions in type M5 and the most in type M4 with a percentage of 59.38% at Time 1 and 57.29% at Time 2. M4 misconceptions are measured using question number 3, which is four-tier, as shown in Figure 2.

3.1 A boat full of cargo floats on the surface of the water. If some of the boat's cargo is removed, then
   a) the buoyant force on the boat increases
   b) the buoyancy force on the boat decreases
   c) the buoyant force on the boat does not change

3.2 What is your level of confidence in the answers above?
   a) sure
   b) not sure

3.3 The reasons for the answers above are:
   a) the weight of the boat decreases, so the buoyancy increases
   b) the boat becomes lighter; its position rises so that the buoyancy is reduced
   c) buoyancy is determined by the size of the boat, not by the load

3.4 What is your level of confidence in the reasons above?
   a) sure
   b) not sure

Figure 2. Question number 3 of FTBI.

The results of the All-tier analysis show that 59.38 percent or as many as 57 pre-service physics teachers who choose option a at Time 1 have the conception that if the boat's load is reduced, the buoyant force acting on the boat increases (M4). This conception contradicts or differs from the physicists' conception, where reducing the load will cause the part of the boat immersed in the water to decrease so that the buoyant force acting on the boat is also reduced. The analysis results in Time 2 showed a slight decrease in the percentage to 57.29 percent or as many as 55 people who still had misconceptions about the same concept. So, two people were no longer categorized as having misconceptions; these two people were then interviewed. The interview results turned out to support the quantitative findings that at Time 2, one person had changed the answer to the correct answer but for the wrong reasons (the misconception score was zero). Another person became unsure of the solution. In the analysis, the student was no longer categorized as experiencing misconceptions but lacked knowledge.

Furthermore, respondents who choose option c in question number 3 are categorized as having other types of misconceptions; namely, the buoyant force is not determined by the volume of the submerged object but is determined by the total volume of the object (M5). Pre-service physics teachers assume that the magnitude of the buoyant force acting on the boat is constant. This conception is very different from Archimedes' Law which states the magnitude of the buoyant force experienced by an object while in a fluid is equal to the weight of the fluid separated by the object and is determined by the volume of the submerged object. Overall, the analysis results of misconceptions between Time 1 and Time 2 can be seen in Figure 3.

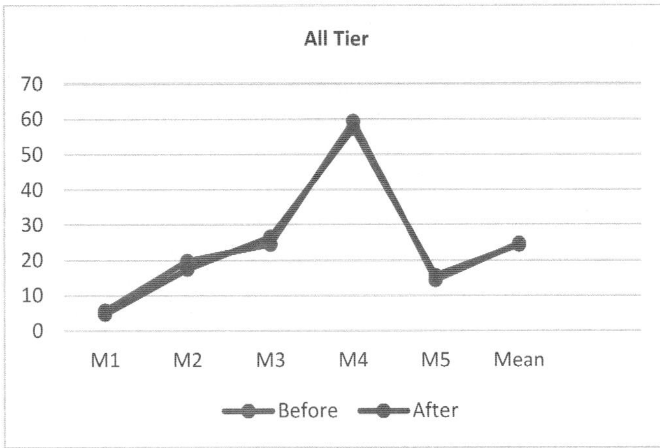

Figure 3. Misconception score before and after treatment.

A Wilcoxon Signed Rank Test revealed a non-statistically significant decrease (from $M = 24.7$ to $M = 24.3$) in pre-service physics teachers' misconceptions following additional instruction in the test, $z = -0.58$, $p > 0.05$. These results indicate that the misconceptions experienced by pre-service physics teachers about buoyancy tend not to change. Most respondents still have the wrong conception that (1) the reduction in the load on the boat causes the buoyant force acting on the boat to increase, (2) the size of the load does not determine the buoyant force acting on the boat, but by the size of the boat, because the size of the boat is fixed, the buoyant force of the boat is also constant, (3) the buoyant force is constant because the density of water and the density of the boat are constant. Then, why did the percentage of pre-service physics teachers' correct answers increase significantly? It is necessary to analyze the categories of false positives, false negatives, and lack of knowledge.

3.3. False Positive, False Negative, and Lack of Knowledge

As explained in the method section, if a student has the correct answer and the reason is false, the student is categorized into False Positive (FP). If the answer is wrong, but the reason is correct, it is classified as False Negative (FN). Furthermore, if pre-service physics teachers are unsure of the answers or explanations, they are grouped into people who do not have knowledge (LK) even though the answers and reasons are correct. The results of data analysis for these three categories are shown in Table 4.

Table 4. Comparison of FP, FN, and LK values at Time 1 and Time 2

Time	False Positive	False Negative	Lack of Knowledge
Before (Time 1)	4.16	1.30	24.22
After (Time 2)	2.08	0	4.95

The results of the data analysis shown in Table 4 show a decrease in the percentage of FP, FN, and LK. The results of interviews with respondents who in Phase 1 were categorized as false-positive pre-service physics teachers (FPS), false-negative pre-service physics teachers (FNS), and lack of knowledge pre-service physics teachers (LKS) are as follows:

> In Stage 1, I chose the correct answer but with the wrong reason. After the additional instructions in Step 2, I tried to use the principle of equilibrium of forces acting on objects so that I knew the correct explanation. (FPS1)

> I changed my answer to the correct answer, but I am still not sure about my answer. (FNS1)

> My answer and reasoning in Stage 1 were the same as in Stage 2, but I finally believed in the answers; after additional instructions, I tried to use the concepts I had learned to answer the questions. (LKS1)

Based on the interview above, it is known that FPS1 and LKS1 are contributors to the increase in the percentage of correct scores, while FNS1 is not because they are not sure of the answer. This data shows that the rise in the rate of correct scores comes from the FP, FN, and LK groups, so that the increase in the percentage of correct scores does not automatically represent a decrease in student misconceptions. The results of this study indicate that misconceptions tend to be stable and difficult to change, as found by previous researchers [11][13].

4. Conclusion

Measurement and remediation of misconceptions are not easy to do. It takes the appropriate instrument to diagnose a misconception and, of course, a more serious effort to change it. The treatment or instruction may increase pre-service physics teachers' correct answers to an item related to a concept being tested, but it is not necessarily significant in changing pre-service physics teachers' misconceptions.

Acknowledgment

This research was conducted using research funds from the University of Jambi in Indonesia through a scheme provided by the research institute.

References

[1] Clement J, Brown D E and Zietsman A 1989 Not all preconceptions are misconceptions: Finding 'anchoring conceptions' for grounding instruction on students' intuitions, *Int. J. Sci. Educ.*, **11** 5, pp. 554–565

[2] Gurel D K, Eryilmaz A and McDermott L C 2015 A review and comparison of diagnostic instruments to identify students' misconceptions in science, *Eurasia J. Math. Sci. Technol. Educ.* **11** 5, pp. 989–1008

[3] Maison, Kurniawan W, Kurniawan D A and Hanum A 2021 "Application of a web-based four-tier misconception instrument on the topic of work and energy," *Proc. 3rd Green Dev. Int. Conf. (GDIC 2020)* **205** 2020, pp. 206–208

[4] Caleon I S and Subramaniam R 2010 Do students know what they know and what they don't know? Using a four-tier diagnostic test to assess the nature of students' alternative conceptions *Res. Sci. Educ.* **40** 3 pp 313–337

[5] Kaltakci-Gurel D, Eryilmaz A and McDermott L C 2017 Development and application of a four-tier test to assess pre-service physics teachers' misconceptions about geometrical optics," *Res. Sci. Technol. Educ.*, **35** 2, pp. 238–260

[6] Maison M, Safitri I C and Wardana R W 2020 Identification of misconception of high school students on temperature and calor topic using four-tier diagnostic instrument," *EDUSAINS* **11** 2, pp. 195–202

[7] Maison, Kurniawan D A and Widowati R S 2021 The quality of four-tier diagnostic test misconception tnstrument for parabolic motion *J. Pendidik. dan Pengajaran* **54** 2 pp. 359–369

[8] Maison, Asrial, Susanti N, Effrita A and Tanti 2021 Identification of students' misconception about light using a four-tier instrument *J. Phys. Conf. Ser.* **1876** 1

[9] Kiray S A, Aktan F, Kaynar H, Kilinc S and Gorkemli T 2015 A descriptive study of pre-service science teachers' misconceptions about sinking-floating," *Asia-Pacific Forum Sci. Learn. Teach.* **16** 2

[10] Kiray S A and Simsek S 2021 Determination and Evaluation of the Science Teacher

Candidates' Misconceptions About Density by Using Four-Tier Diagnostic Test," *Int. J. Sci. Math. Educ.* **19** 5 pp. 935–955
[11] Hammer D 1996 More than misconceptions: Multiple perspectives on student knowledge and reasoning, and an appropriate role for education research *Am. J. Phys.* **64** 10 pp. 1316–1325
[12] Unal S 2008 Changing students' misconceptions of floating and sinking using hands-on activities *J. Balt. Sci. Educ.* **7** 3, pp. 134–146
[13] Hasan S, Bagayoko D and Kelley E L 1999 Misconceptions and the certainty of response index (CRI)," *Phys. Educ.* **34** 5 pp. 294–299

Development of length expansion temperature detector based on arduino uno

Azhar[1], Siswadi[2], A Doyan[2,3], Susilawati[2,3] M Taufik[3] and L S Hudha[4]

[1]Physics Education, FKIP, Riau Univesity, Riau, Indonesia
[2]Master of Science Education Program, University of Mataram, Lombok, West Nusa Tenggara, Indonesia.
[3]Physics Education, FKIP, University of Mataram, Lombok, West Nusa Tenggara, Indonesia.
[4]Department of Physics, Faculty of Mathematics and Natural Science, University of Mataram

E-mail: azhar@lecturer.unri.ac.id, siswadi.ampenani@gmail.com, aris_doyan@unram.ac.id, susilawatihambali@unram.ac.id

Abstract. Length expansion is an increase in the length of an object due to receiving heat. If a solid object is heated then the object will expand in all directions, in other words the length increases the length of an object because it receives heat. A tool for comparing the length expansion of various metals is the Musschenbrock. To measure the temperature of an object can use a sensor. The sensor used in this research is the LM 35 based on Arduino Uno. The development method for this Length expansion tool temperature detector starts from tool design, hardware design, software design, tool calibration and tool testing. From the results of research using an Arduino Uno-based temperature detector, it can be said that this research was successful because the values obtained were almost the same as the existing theory. So, this detector is suitable for use in laboratories or schools as a learning medium to determine the length expansion of metals.

1. Introduction

Along with the rapid development of science and technology today, this life cannot be separated from physics. Physics is one of the abstract science subjects [1]. Physics is one of the main pillars of science that can provide an understanding of natural phenomena, which studies objects in nature, symptoms, natural events, and interactions of objects in the universe. Physics is also a collection of facts, concepts, laws or principles, equations and theories that must be studied and understood. Physics is widely implemented in everyday life such as electronic devices, motor vehicles and so on [2]. Basically physics was born and developed by conducting experiments. The experimental set is a tool that can be used to assist in the process of determining the parameters to be measured in research. In order for the research results to be more accurate, an experimental set that has a high level of accuracy and precision is needed. To get a high level of precision and accuracy, one of them is by designing and manufacturing experimental sets digitally. One of the experimental sets that will be designed and made is the Length expansion experimental set [3].

Expansion is a physical change of a substance with the characteristics of becoming longer, wider, or increasing in volume due to heating of the substance. Expansion is the increase in the length of an object due to receiving heat. In the Length expansion, the width and thickness values are very small compared to the length of the object. So the width and thickness are considered non-existent [4]. The coefficient of expansion of the length of an object is the ratio of the increase in length to the initial length of the object per unit increase in temperature. If a solid object is heated then the object will expand in all directions , in other words the length increases the length of an object because it receives heat. The tool to compare the length expansion of various metals is Musschenbrock [5]. The length expansion of an object is influenced by several factors, namely the initial length of the object, the coefficient of Length expansion and the magnitude of the temperature change [6]. In general, the size of an object will change when the temperature changes [7]. Thus the coefficient of Length expansion of a substance can be defined as the relative change in the length of a substance divided by the degree of change in temperature.

To measure the temperature of an object can use a sensor. The sensor is a device that receives a stimulus and responds with an electrical signal [8]. A temperature sensor is a tool for detecting or measuring the temperature of an object, space or certain system which is then converted into electrical quantities [9]. To make a temperature sensor we can use a microcontroller. Microcontroller is an electronic component that can be programmed and has the ability to execute programmed steps [10]. One of the microcontrollers that can be used is Arduino. Arduino is a microcontroller that can function as the control center of a system as a whole and is displayed on the LCD screen [11] . Arduino is a microcontroller board based on ATmega 328. This device has 14 input/output pins consisting of 6 pins that can be used as PWM outputs, 6 analog inputs, a 16 MHz crystal oscillator, a USB connection, a power jack, an ICSP head, and a reset button. Arduino is capable of supporting a microcontroller so that it can be connected to a computer or laptop using a special USB cable. To operate Arduino, you can use a USB cable connected to a computer or other power source such as a battery [12, 13] . Based on this, the researchers tried to make a temperature detector based on Arduino Uno . This detector can be a solution to help practicum at school or in the laboratory to support the teaching and learning process.

2. Methods

Time and place

This research was conducted from May to June 2021 at Bidayatul Hidayah Islamic Middle School Mataram.

Temperature Detector Development Stage

1) Tool Dcsign Stagc

a) Hardware Design

There are 2 hardware designs of the system, namely the design of the system builder electronics circuit and the mechanical design of the system. The electronic circuit design of the system builder is usually described in the form of a block diagram. The block diagram of system consists of several circuits, including the power supply circuit, the LM35 sensor circuit, the Arduino Uno microcontroller circuit, and the LCD circuit described in Figure 1 and Figure 2.

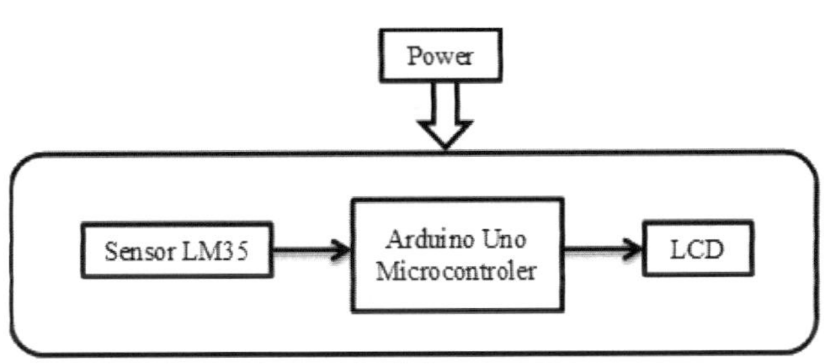

Figure 1. System Block Diagram

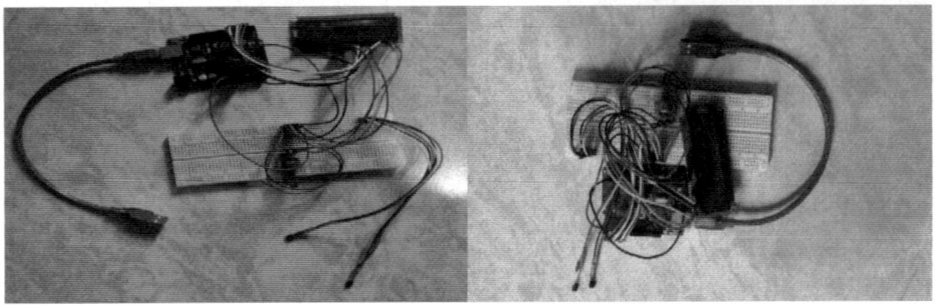

Figure 2. Hadware Design

b) Software Design

The software design of the digital expansion set experiment system can be seen in Figure 3.

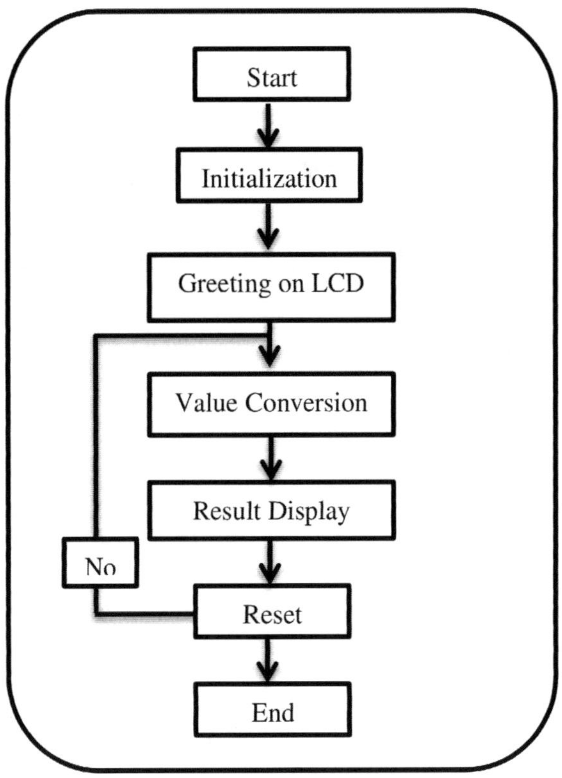

Figure 3. Sofware Design

2) Tool Calibration Stage

Before the tool is used, it is necessary to calibrate the tool first so that the tool used can run and be used properly and correctly.

Journal of Physics: Conference Series **2165** (2022) 012049 doi:10.1088/1742-6596/2165/1/012049

3) Tool Trial Stage

Testing the tool is done by placing the Temperature Sensor on the heated metal. Then see the amount of temperature obtained through the LCD. The temperature detector is placed on three different metals namely Brass, Stainless and Aluminum.

3. Result and Discussion

In the activities that have been carried out based on the design and testing of the Length expansion device temperature detector, the following data were obtained:

Table 1. Data from Observation Group 1

No	Object	L_0 (cm)	L_t (cm)	L (cm)	T_1 (°C)	T_2 (°C)	T (°C)	(cm/ ° C)
1	**Aluminum**	20	20.3	0.3	31.5	88.8	57.3	0.000262
2	**Brass**	20	20.2	0.2	31.5	85.7	54.2	0.000185
3	**Iron**	20	20.1	0.1	31.5	72.3	40.8	0.000123

Table 2. Data from Observation Group 2

No	Object	L_0 (cm)	L_t (cm)	L (cm)	T_1 (°C)	T_2 (°C)	T (°C)	(cm/ ° C)
1	**Aluminum**	20	20.3	0.3	31.5	87.8	56.3	0.000266
2	**Brass**	20	20.2	0.2	31.5	86.1	54.6	0.000183
3	**Iron**	20	20.1	0.1	31.5	73.2	41.7	0.00012

Table 3. Data from Group 3 Observations

No	Object	L_0 (cm)	L_t(cm)	L (cm)	T_1 (°C)	T_2 (°C)	T (°C)	(cm/ °C)
1	**Aluminum**	20	20.3	0.3	31.5	88.9	57.4	0.000261
2	**Brass**	20	20.2	0.2	31.5	86.5	55	0.000182
3	**Iron**	20	20.1	0.1	31.5	72.8	41.3	0.000121

Expansion Length is an increase in the length of an object due to the influence of changes in temperature or an increase in the size of an object due to receiving heat. In this Length expansion if divulging a metal rod length L is increased by (ΔL), the length will increase by $\Delta L = L a \Delta$, with a is a constant called the coefficient of expansion as Length. Coefficient a The unit has a "per degree" and depends on the material (material), where $a = \Delta L /$ L. ΔT we see that a an increase in the fraction of the length per unit change in temperature. Although a bit fickle together temperatures, for most practical purposes, a may be considered as constant for a particular material.

This study aims to determine the expansion that occurs in solid objects and how solid objects can expand. This assessment uses 3 metals, namely aluminum, brass and iron. After going through the assembly of tools and programming on the Ardiuno Uno, a tool (see figure 2) is produced that is able to measure temperature in real time using the LM35 temperature sensor. The results of sensor readings can be seen on the LCD screen which is designed to use a delay of 1 second. This means that every second, a temperature change will be displayed according to the measured conditions.

This study begins by measuring each initial length of the object, namely the three objects have the same length of 20 cm with the same initial degree of 31.5 °C. Aluminum, brass and iron / stainless is heated using spirit for 15 minutes. Based on the observations from this experiment, it turned out that each object experienced a different temperature change. This proves the absorption of heat or heat in the metal. Based on the results of the research from the three groups of researchers, the results were not too much different. Group 1 matches α for aluminum amounted to 0.000262, while the brass of 0.000185 and 0.000123 of iron. Group 2 matches α for aluminum amounted to 0.000266, while brass and iron amounted to 0.000183 at 0.00012. As for the 3 group matches α for aluminum amounted to 0.000261, while the brass of 0.000182 and 0.000121 of iron. From the results of research using an Arduino Uno-based temperature detector, it can be said that this research was successful because the values obtained were almost the same as the existing theory. So, this detector is suitable for use in laboratories or schools as a learning medium to determine the length expansion of metals. The weakness of this tool has not been able to determine the length of the metal directly. This weakness can be overcome by developing tools towards a more complex or perfect direction in order to get more precise and maximum results.

4. Conclusion

Base on the results of research using an Arduino Uno-based temperature detector, it can be said that this research was successful because the values obtained were almost the same as the existing theory. So, this detector is suitable for use in laboratories or schools as a learning medium to determine the length expansion of metals. The weakness of this tool has not been able to determine the length of the metal directly. This weakness can be overcome by developing tools towards a more complex or perfect direction in order to get more precise and maximum results.

References

[1] Subiantoro A W 2009 *Pentingnya Praktikum Dalam Pembelajaran IPA. Pelatihan Pengembangan Praktikum IPA Berbasis Lingkungan Bagi Guru-Guru MGMP IPA SMP Kota Yogyakarta* (Universitas Negeri Yogyakarta)

[2] Jumiati 2016 Pembuatan Alat Praktikum Termoskop Guna Menjelaskan Radiasi Kalor Berbasis Teknologi Murah Dan Sederhana

[3] Meiza, Nofsi, Yulkifli, Kamus, Zulhendri 2017 Pembuatan Set Eksperimen Muai Panjang Digital Berbasis Mikrokontroler Atmega 328 *Pillar of Physics* **10**

[4] Chang 2012 *Fisika Untuk Universitas* (Bandung: Bina Cipta)

[5] Waluyo 2009 *Fisika Dasar* (Semarang: Adi Prasetya)

[6] Waluyo 2021 *Fisika Untuk Sains dan Tekhnik* (Jakarta: Erlangga)

[7] Winnamo 2007 *Penuntun praktikum Fisika Dasar* (Banjar Baru: Universitas Lambung Mangkurat)

[8] Yulkifli 2011 *Sensor Fluxgate* (Padang: Universitas Negeri Padang)

[9] Sirait, M 2009 Pengaturan Temperatur Ruangan Dengan Menggunakan Sensor Suhu LM 35 Berbasis Mikrokontroler AT89S51 *Skripsi* (Medan: Universitas Sumatera Utara)

[10] Yohandri 2013 *Mikrokontroler Dan Antar Muka* (Padang: Universitas Negeri Padang)

[11] Nasri 2010 *Penuntun Praktikum Fisika Dasar. (*Jambi: Universitas Jambi)

[12] Kumar, NS, Vuayalakshmi , B., Prarthana , RJ, & Shankar, A. (2016, November). Sistem peringatan sampah pintar berbasis IOT menggunakan Arduino Uno *Konferensi Wilayah 10 IEEE 2016 (TENCON)* 1028-1034

[13] Ismail N, Wahid S and Ahmad N 2020 Pelacakan Kendaraan RFID Berbasis Arduino untuk Keamanan Rumah *Jurnal Fisika: Seri Konferensi* **1529** 2.

Development of Low-Cost IoT based data acquisition system for real-time monitoring of volcanoes seismicity

D R Santoso*, S Maryanto, T Putro, and A Priswanto

Physics Department, Brawijaya University, Jl. Veteran 65145 Malang, Indonesia.

E-mail: dieks@ub.ac.id

Abstract. This paper describes how to develop a low-cost data acquisition (DAQ) system for real-time remote monitoring of volcano seismicity. The DAQ system is developed based on IoT platform, consists of several module devices both hardware and software, and installed in two places called seismic-stations (SS) and observation-post (OP). In the implementation, the SS is placed on the body of the volcano meanwhile the OP is usually placed at a certain distance from the SS with more adequate electrical and communication facilities. In this design, the SS consists of seismic sensors and a digitizer, whereas the OP consists of a computer server and ADSL modem. Communication between SS and OP conducted by using a 5 GHz Wi-Fi telemetry system. At the OP, the data is processed to produce the form of digital seismograms, hypocentres, arrival times, and signal magnitudes in SAC or mini seed file formats by Earthworm software. Database server for cloud access is created using Winston Wave Server (WWS) and MySQL Data Base Management Systems (DBMS). The seismic data can be accessed in real time remotely by users using the SWARM application via website access.

1. Introduction

Implementation of IoT technology currently is not only in the manufacturing industry, but also in other fields [1-4]. In environmental monitoring [5–10], IoT allows users to observe the changes in environmental conditions easily in order to determine any further actions should be undertaken [11, 12]. The capabilities offered by IoT can revolutionize the technology of environmental observation including volcanoes observation to be more efficient [13].

Indonesia is one of countries that are vulnerable to volcanic eruption disasters. Intense monitoring of eruption-prone volcanoes is needed to provide early warning before eruptions occur so it can reduce losses [14]. Volcano seismicity is one of the vital signs to be observed. Increased seismic activity means the early symptom of an eruption [15]. In Indonesia, most of the seismographs used are still local monitoring, it means that observation data is accessed only at the OP, and cannot be accessed in real time from long distance. Monitoring data are sent periodically to other places using VSAT (Very Small Aperture Terminal). This is where the internet (IoT) becomes important role to be implemented in the volcanoes monitoring system, especially in Indonesia.

Data transmission through internet are widely used by ADSL (Asymmetric Digital Subscriber Line) and Fiber Optics (FO) technology. However, the ADSL is the best choice for volcano monitoring system due to the ability in data distribution. Although the ADSL has a lower speed than FO, it does not an obstacle because seismic data tend to have relatively small data packet capacity [16]. Seismic data transmission using ADSL can be done by building a server-based data storage system. Database makes it easy for us to process, change, protect the data, and provide easy access of data for people around the world [17]. The availability applications of database server such as MySQL, and also data acquisition software for seismic such as Earthworm (by ISTI) and Winston Wave Server (WWS, by USGS) can be used to build seismic data storage [18], and freely for educational purposes. Seismic data can be accessed by users using the SWARM application (Seismic Wave Analyse and Real-time Monitoring).

2. Methods

In this project, the Data Acquisition (DAQ) system is developed based on IoT technology, consists of several module devices both hardware and software, and installed in two places called seismic-stations (SS) and observation-post (OP). The SS is seismograph system. It consists of a power supply, seismic sensors, amplifier and filter, digitizer, ethernet interface, switch hub, Wi-Fi transceiver. Block diagram of SS section is given in Fig.1. The OP section (Fig.2) consist of Wi-Fi transceiver, switch hubs, ADSL modems, and database server computers. Data received from the SS will be processed by Earthworm software, then the data will be stored into a database server. The Database Management System (DBMS) used is MySQL. The database integrated with Winston Wave Server as a provider of real-time seismic data and web servers.

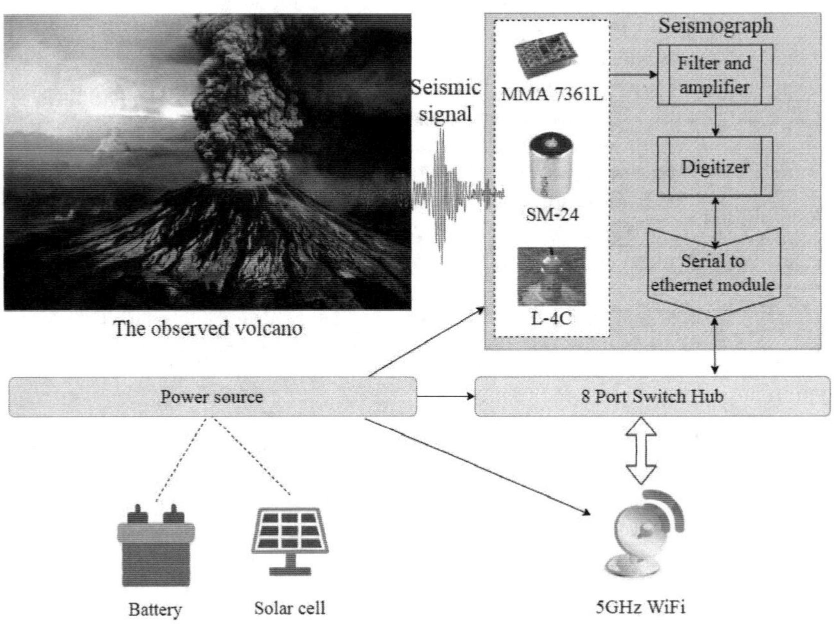

Figure 1. Block diagram of seismograph system (SS)

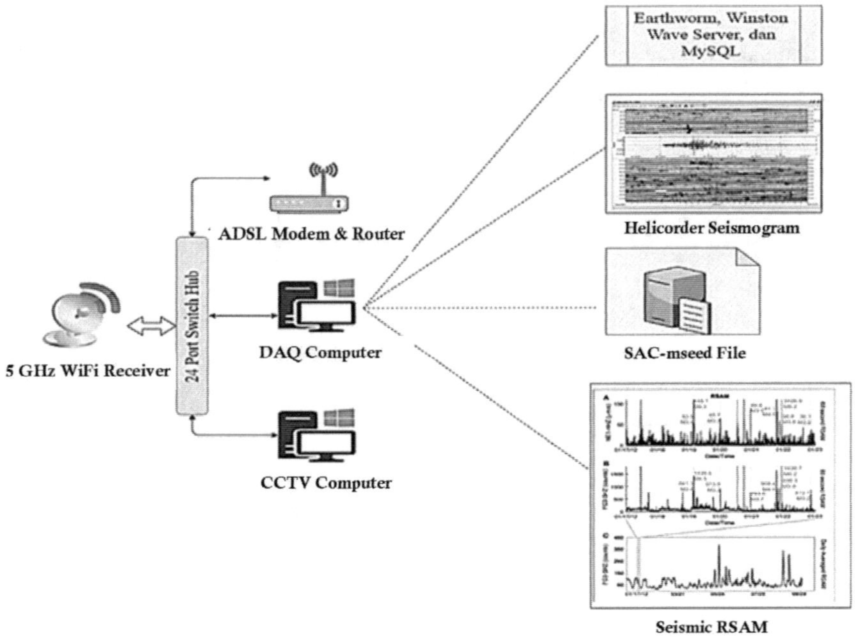

Figure 2. Block diagram of receiver and server system (OP)

2.1. Seismic Sensors

The seismic sensors are used in this project consists of three kinds of sensors, MEMS MMA7361L accelerometer, SM24 Geophone sensor, and L4C Geophone sensor. Table 1 shows the specifications of the sensor.

Table 1. Specification of seismic sensor

Sensor	Specification	Value	Unit
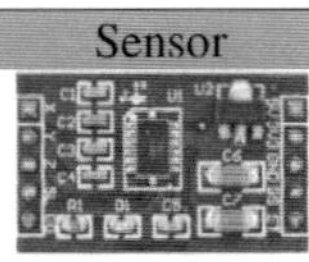	Voltage operation	2.2-3.6	V
	Sensitivity	0.8	V/m/s^2
	Bandwidth	DC to 800	Hz
MMA 7361L	Current consumption	400	uA
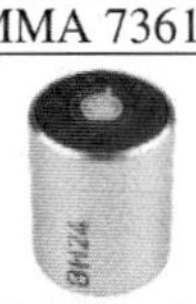	Natural frequency	10	Hz
	Typical spurious frequency	240	Hz
	Sensitivity	28.8	V/m/s
	Damping resistance	1339	Ohm
SM24 Geophone			
	Natural frequency	1	Hz
	Peak to Peak Coil Travel	6.25	mm
	Sensitivity	127	V/m/s
	Damping resistance	4434	Ohm
L4C Geophone			

The MEMS MMA7361L is acceleration sensor, and now widely used in research of seismic sensor because of their performance, compactness, and low cost. The geophone sensor has high sensitivity, but they rarely detect seismic signals with large amplitude, only a few milli-volts to dozens of milli-volts. So, it is necessary to strengthen the output in order can be read clearly. Placement of sensors in nature in addition to detecting seismic signals, sensor will also detect other signals, this signal is referred to as background noise such as a tree in the wind and the steps of a passing animal. For volcanic seismic monitoring, the signals that are usually detected are tremor, volcanic earthquake, eruption earthquake, and local tectonic earthquake. The filter is designed to have a working range from 0.5 Hz to 25 Hz that commonly used is a combination of high pass and lowpass filters that create a bandpass filter, then filter combined with a differential amplifier circuit to produce high gain above 40 dB.

2.2. Digitizer

The block diagram of digitizer design shown in Fig.3. It has Analog to Digital Converter ADS1115 that provide 16 bits accuracy. It can provide high dynamic range or signal to noise ratio (SNR) around 90dB, with sampling rate up-to 860 sps. The digitizer has bipolar input with input level around +/- 5 volt, this is commonly used in seismological instrument. High accuracy timestamping of seismic data is provided by GPS model Ublox-Neo 6m is enough to get high accuracy timing.

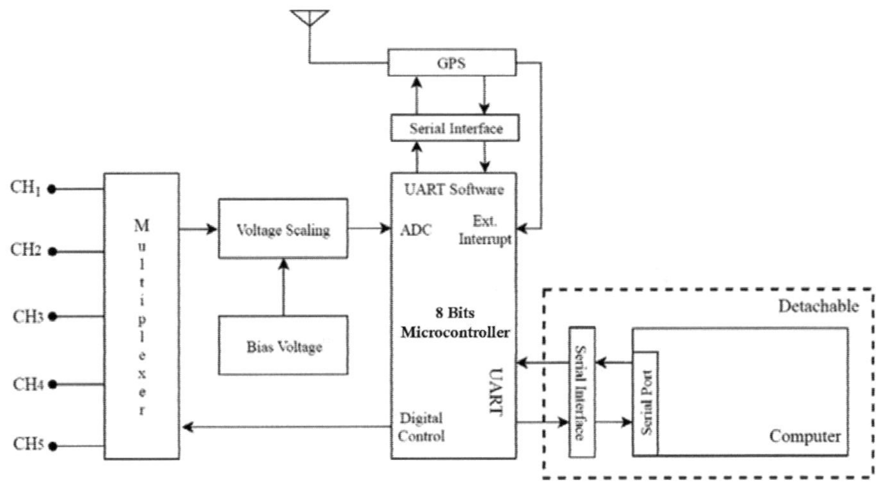

Figure 3. Block diagram of seismic data acquisition

2.3. Earthworm-Based Automatic System for Real-Time Data Processing

The Earthworm system software was created in Virginia since 1993 and now is separate project of Instrument Software Technology Inc. (ISTI). Earthworm work based on Transport Ring concept, which shared memory with specified size is implemented. Transport Ring store queue messages from system module where the messages content can be packet of seismic waveforms, seismic arrival times, result of determination of earthquake source parameters and magnitude. The message as input data can be read by other module connected to shared memory [18].

In this project an application called “isdaqadsend” that runs on Windows OS was developed to exchange data with Earthworm system. This application demultiplexing data from digitizer than re-assemble to data packet that suitable with the Earthworm which is called as TYPE_TRACEBUF2. The data packed will send to the one of transport rings, then data will process by other module to get information about the earthquake. Wave_serverV provides a network-based service of trace data. It acquires Earthworm trace data messages for specified channels, and maintains a disk-based circular buffer for each channel, then offers a network service capable of supplying specified portions of trace data from specified channels. Archman module save trace data to disk in various supported formats. Export-generic represent the Earthworm long-distance communication modules that provide a rapid, secure message exchange facility between two or more Earthworm or other processing systems. Any link capable of supporting the TCP/IP protocol suite can be used. Table 2 show the modules that are used in Earthworm for automatic system for real-time data processing.

Table 2 Earthworm modules that used for automatic data processing

Module Name	*Output Message*	*Information Content*
isdaqadsend	TYPE_TRACEBUF2	Waveform data
wave_serverV	TYPE_TRACEBUF2	Waveform data
archman	SAC_FILE_FORMAT	Waveform with SAC format
export_generic	TYPE_ACK	Waveform for exchange data
ew2rsam	TYPE_TRACEBUF2	Statistics of event
ew2ssam	TYPE_VTABULARDATA	Statistics of event
heli_ewII	TYPE_TRACEBUF2	Digital seismogram

2.4. IoT- based system using ADSL network and MySQL Database Management System

Implementation of IoT based on ADSL technology and Database server using MySQL only require two important devices, they are the ADSL modem and the computer as a server. The ADSL modem is configured to forward clients that contact the modem's public IP address and port to the data centre on the server computer. One important feature possessed by an ADSL modem to be able to do that is the Network Address Translation (NAT) feature. The basic role of NAT is translating Public IP networks into private networks and vice versa. All hosts that are behind NAT can be directly accessed via the internet that contacts Public IP and Modem ports [19]. Figure 4 show the network topology for internet communication. In this case, a private IP address is mapped with a Public IP address which's the address of WWS and website.

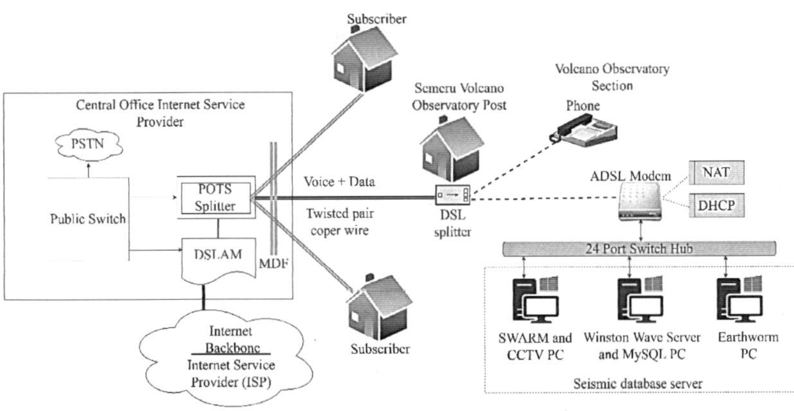

Figure 4. Network topology for internet-based long-distance communication

3. Results and Discussion

3.1. Hardware System

This signal conditioning circuit for signal's amplifying and filtering has 5 input channels, set in bandwidth of 0.5 Hz - 25 Hz and fixed gain of 40 dB. The digitizer is used as a signal processing, which converts analog seismic signals into digital data. The digitizer consists of ATMEGA328 microcontroller IC as the main controller. It is low power, very low price, and is easily available in electronic stores. The type of ADC used in this digitizer module is the ADS1115. It is a 16-bits resolution with maximum sampling rate of 860 sps. It is able to give a dynamic range of 84dB (14-bit effective number of bits).

The signal amplitude that can be sampled is dual polarity +/- 5volt. The sampling speed can be chosen from 10sps to 100sps. For timestamp and location marker purposes, Ublox NEO-6M type GPS module is used. This module has a fairly high level of timing accuracy of around 100uS. However, in this design the accuracy achieved is around <5ms, and it is good enough for Tectonic and Vulcanic seismic data. Figure 5 is a photo of the signal conditioning circuit and digitizer hardware.

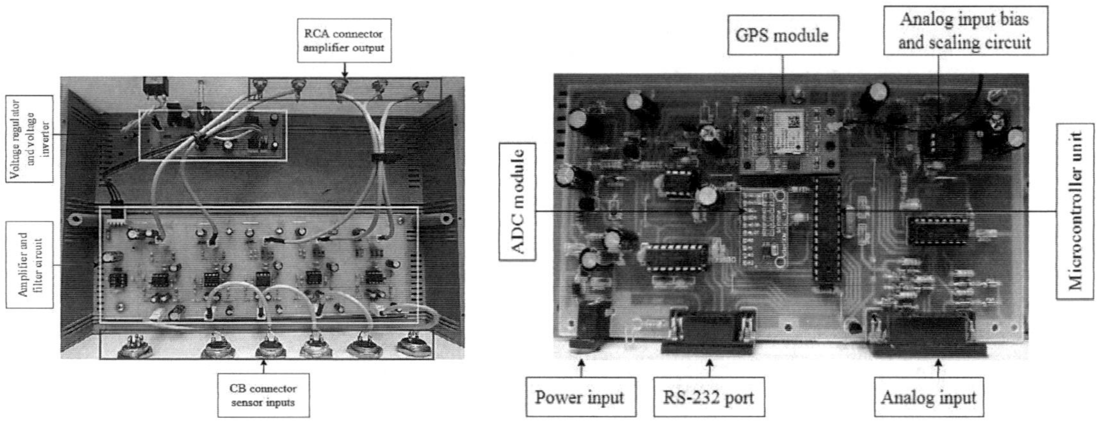

Figure 5. Photo of Signal Conditioning and digitizer hardware

3.2. Seismograph Station System Installation (Field Application)

To test the performance of the developed system, we installed the equipment in one of existing the Semeru volcano station, i.e. Leker station, East Java Indonesia. It has latitude -8.137444, longitude 112.485962, and altitude 1045.800 m. The location is about 10 km from the top of Semeru volcano, and about 4 km from the observation post [20]. Figure 6 is a photo of seismic station installation in the Leker area, and a monitoring computer as a data receiver.

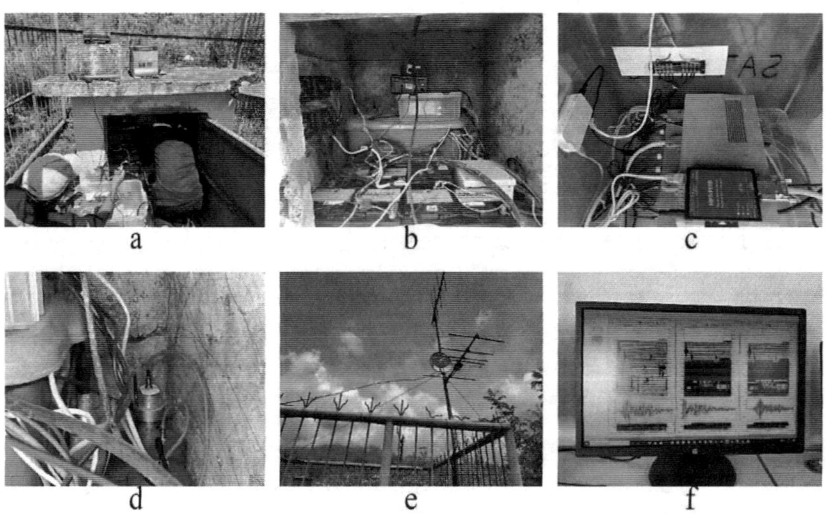

Figure 6. Installation of the system at the Leker seismograph station. (a) Leker station bunker; (b) power source; (c) system hardware (d) seismic sensors; (e) digital and analog radio transmitter; (f) Computer Monitoring

3.3. Remote Monitoring System

Data communication through ADSL lines with 512 kbps speed upstream provided by the server can supply seismic data in real time for up to 4 clients effectively. Data is accessed via SWARM by using the WWS data source by entering the Winston address which is "semerumcs.servequake.com" and port 15020 (only for testing). The address is a representation of the Public IP of the ADSL modem which is then translated into a private IP in the local server section. The server address is obtained using the assistance of a third party, using Dynamic Domain Name Server (DDNS) from noip.com. The noip.com server will monitor every change in public IP on the ADSL modem. If there is reconnecting on the modem, the noip.com server will automatically change the IP address at semerumcs.servequake.com. This makes it very easy compared to accessing an IP address from an ADSL modem.

Seismic monitoring systems that have been installed at the Leker Station can record several earthquake events, such as volcano-tectonic type A (VT-A), Long-Period (LP) event, local tectonics, eruption earthquake, and tremors. Because of this system is compatible with SWARM analysis software (USGS), earthquake events that occur can be directly analysed at that time. So that information related to earthquake magnitude, frequency, and so on can be obtained in a short time. Figure 7 is an example of a seismogram of 24-hour seismic monitoring which shows many seismic events where the data is recorded by the L-4C seismometer, the most sensitive sensor. While the other two sensors, the Geophone SM-24 and MEMS Accelerometer MMA7361L is cannot record due to the relatively low earthquake energy and earthquake frequency factors that are only dominant bellow 4Hz. There are about 3 times the type A volcano-tectonic earthquake (VT-A) ,2 times Long-Period (LP) event, 3 times tremor, and 7 times earthquake eruption. Where the highest intensity of LP earthquake occurred at 10:22:13 UTC, with a peak amplitude count of 7061 count and duration about 110 seconds. To be able to convert digital count units into physical units such as mm/s (millimeter per second) or um/s (micrometer per second), it is necessary to know the correction of the instrument, including sensor sensitivity, local effect amplification, instrument strengthening, and volts/bits of the digitizer.

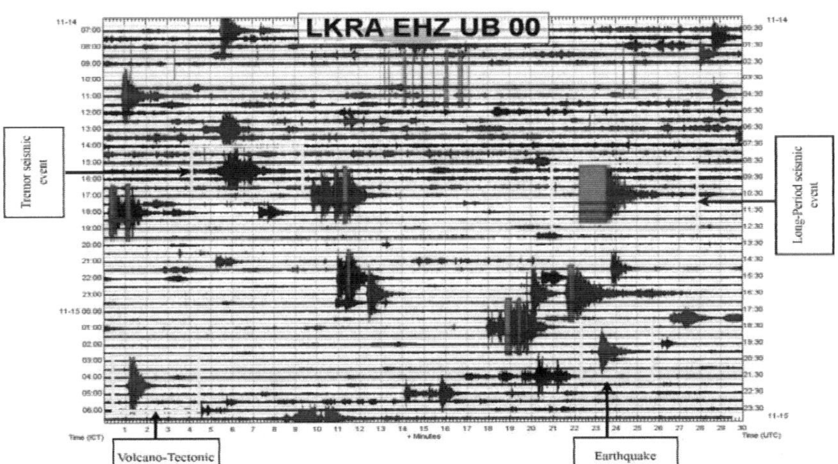

Figure 7. Seismogram of seismic events

Figure 8 shows details of the Long-Period (LP) event waveform that occurred at 10:22:13 UTC. The dominant frequency spectrum of earthquake is Long-Period (LP) is in the narrow frequency band between 2 and 5 Hz [21]. It is very difficult to determine the earthquake source because there is no clear difference in arrival time between the Primary (P) and secondary (S) waves [22]. Another seismic event recorded by a seismograph is eruption earthquake, tremor, and volcano-tectonic type A (VT-A). The type usually originates from beneath volcanoes with a depth of 1km - 20km. The waveform of this type is the same as a shallow tectonic earthquake in general, with the time of arrival of primary and secondary earthquakes that can be clearly distinguished [23]. When this earthquake occurs, as soon as the primary wave P arrives, then it arrives in the secondary wave phase S which will cause a change in the wave frequency and earthquake energy. The frequency of the S wave will be smaller than the P wave, but the S wave has a higher energy [24]. Figure 9 is a waveform of type A (VT-A) volcano-tectonic earthquake.

Furthermore, an earthquake eruption (Fig.10) is an earthquake accompanied by a Strombolian eruption or other (larger) explosive eruption. Most of these signals can be identified by the occurrence of airwaves caused by sonic impulses during an explosion, when the gas extends from inside the crater and is expanded and then elapsed when close to the crater hole. The tremor (Fig.11) is a vibration that occurs continuously in a few minutes or several hours, where the initial vibration and the end of vibration are not clearly visible. The tremors that are mostly recorded by seismograph stations are a type of volcanic tremor, whereas harmonic tremors are rarely recorded by seismograph.

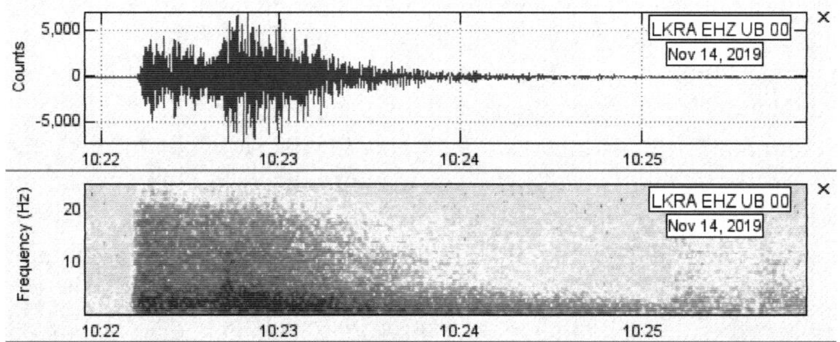

Figure 8. Waveform and frequency spectrogram of Long-Period (LP) event

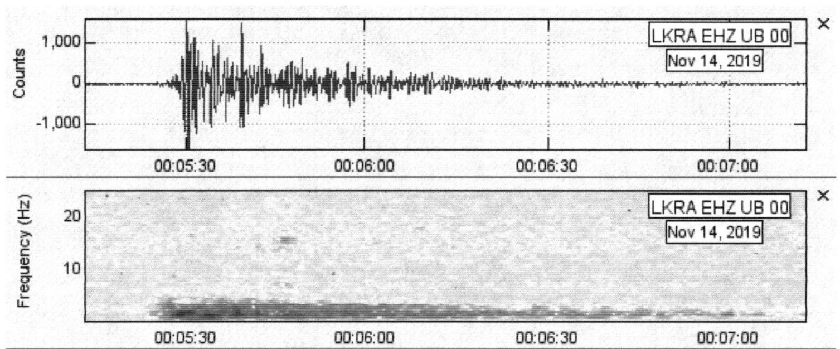

Figure 9. Waveform and frequency spectrogram of VT-A event

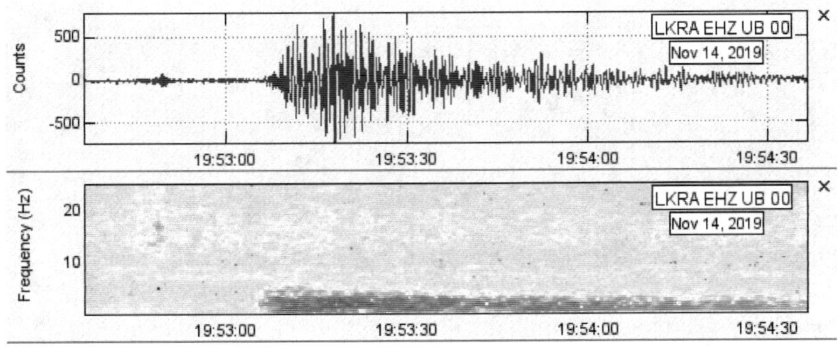

Figure 10. Waveform and frequency spectrogram of earthquake eruption event

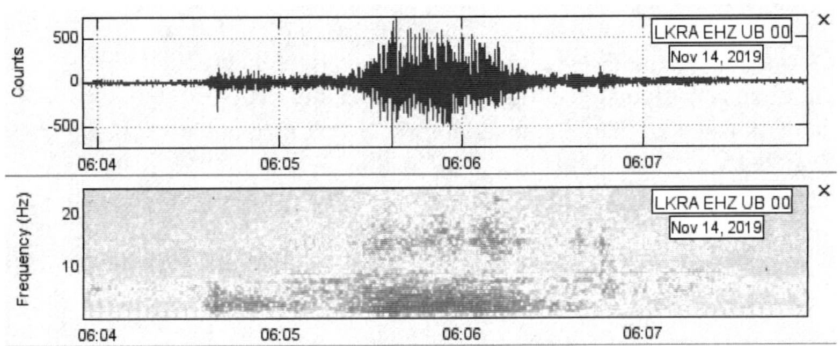

Figure 11. Waveform and frequency spectrogram of tremor event

4. Conclusion

The use of IoT platform in the development DAQ and signal processing systems will provide many benefits such as simplifying equipment, increasing performance, and reducing procurement costs. In this study, a low-cost IoT-based DAQ system dedicated to remote real-time monitoring of volcanic seismicity has been developed and implemented. The DAQ system that is built consists of several devices, both hardware and software. The combination of an ADSL-based internet platform and a seismic database server, makes volcanic seismic activity accessible from anywhere via an internet connection. This developed system has the ability to be used as a network system to monitor several volcanoes at once, which will be very useful for countries that have many volcanoes like Indonesia.

Acknowledgements

This research was funded by the HAPPU-LPPM Brawijaya University research grant, through contract number 538.3.3/UN10.C10/PN/2021, and is a continuation of research project funded by the Faculty of Natural Sciences Brawijaya University with contract number 37/UN10. F09/PN/2020.

References

[1] Tehrani Y, Atarodi S 2019 Design & implementation of a high precision & high dynamic range power consumption measurement system for smart energy IoT applications, *Meas. J. Int. Meas. Confed.* **146** (2019) 458–466. https://doi.org/10.1016/j.measurement.2019.06.037.

[2] Pawar P, TarunKumar M, Vittal K P, 2019 An IoT based Intelligent Smart Energy Management System with Accurate Forecasting and Load Strategy for Renewable Generation, *Measurement.* **107187**. https://doi.org/10.1016/j.measurement.2019.107187.

[3] Mekala M S and Viswanathan P 2019 CLAY-MIST: IoT-cloud enabled CMM index for smart agriculture monitoring system, *Measurement* **134** 236–244

[4] Elijah O, Rahman T A, Orikumhi I, Leow C Y, Hindia M N 2018 An Overview of Internet of Things (IoT) and Data Analytics in Agriculture: Benefits and Challenges *IEEE Internet Things J.* **5** 3758–3773.

[5] Garg U K C S, Hilton J, Aryal J, Forbes-Smith N 2019 Cloud Computing in natural hazard modeling systems: Current research trends and future directions *Int. J. Disaster Risk Reduct.* **38** (2019) 101188. https://doi.org/10.1016/j.ijdrr.2019.101188.

[6] El Moulat M, Debauche O, Mahmoudi S, Brahim L A, Manneback P, Lebeau F 2018 Monitoring System Using Internet of Things for Potential Landslides *Procedia Comput. Sci.* **134** (2018) 26–34. https://doi.org/10.1016/j.procs.2018.07.140.

[7] Kshirsagar R, Mudhalwadkar R P, Kalaskar S 2019 Design and development of IoT based water quality measurement system *Proc. Int. Conf. Trends Electron. Informatics, ICOEI* **2019** 1199–1202. https://doi.org/10.1109/ICOEI.2019.8862663.

[8] Mois G, Folea S, Sanislav T 2017 Analysis of Three IoT-Based Wireless Sensors for Environmental Monitoring *IEEE Trans. Instrum. Meas.* **66** 2056–2064

[9] Ibrahim M, Elgamri A, Babiker S, Mohamed A 2015 Internet of things based smart environmental monitoring using the Raspberry-Pi computer *5th Int. Conf. Digit. Inf. Process. Commun. ICDIPC* **2015** 159–164. https://doi.org/10.1109/ICDIPC.2015.7323023.

[10] Tai H, Celesti A, Fazio M, Villari M, Puliafito A 2015 An integrated system for advanced water risk management based on cloud computing and IoT *2nd World Symp. Web Appl. Networking, WSWAN* **2015** 1–7. https://doi.org/10.1109/WSWAN.2015.7210305.

[11] Ghosh N and Banerjee I 2019 IoT-based Seismic Hazard Detection in Coal Mines using Grey Systems Theory *15th Int. Wirel. Commun. Mob. Comput. Conf.* 871–876

[12] Yaseen M, Swathi D, Kumar T A 2018 IoT based condition monitoring of generators and predictive maintenance *Proc. 2nd Int. Conf. Commun. Electron. Syst. ICCES 2017* 725–729

[13] Ayaz M, Ammad-Uddin M, Sharif Z, Mansour A 2019 E.-H.M. Aggoune, Internet-of-Things (IoT)-Based Smart Agriculture: Toward Making the Fields Talk *IEEE Access* **7** (2019) 129551–129583. https://doi.org/10.1109/access.2019.2932609.

[14] Evita M, Djamal M, Zimanowski B, Schilling K 2015 Mobile Monitoring System for Indonesian volcano *Proc. - 2015 4th Int. Conf. Instrumentation, Commun. Inf. Technol. Biomed. Eng. ICICI-BME 2015* 278–281. https://doi.org/10.1109/ICICI-BME.2015.7401378.

[15] Supriyo B, Hidayat S, Suharjono A, Anif M and Koesuma S 2014 Design of real-time gas monitoring system based-on wireless sensor networks for Merapi volcano *1st Int. Conf. Inf. Technol. Comput. Electr. Eng. Green Technol. Its Appl. a Better Futur. ICITACEE 2014 - Proc.* 30–34. https://doi.org/10.1109/ICITACEE.2014.7065709.

[16] Zhang D, Fan H, Wang C and Zhou Y 2005 Low cost and high performance power system telemetry data transmission system based on embedded ethernet and ADSL *Proc. IEEE Power Eng. Soc. Transm. Distrib. Conf. 2005* p 1–5

[17] Ongo G and Kusuma G 2018 Hybrid Database System of MySQL and MongoDB in Web Application Development *Proc. 2018 Int. Conf. Inf. Manag. Technol. ICIMTech 2018* 256–260 https://doi.org/10.1109/ICIMTech.2018.8528120.

[18] Stepnov A, Konovalov A, Gavrilov A, Manaychev K 2017 Earthworm-based automatic system for real-time calculation of local earthquake source parameters *Seism. Instruments* **53** 267–279. https://doi.org/10.3103/s0747923917040107.

[19] Zhang X 2008 Design and realization of TCP-based nat traversal in P2P, Int. Conf. Apperceiving Comput *Intell. ICACIA 2008.* (2008) 284–286

[20] Sriyanto S and Ifantyana I 2018 Analisis Efek Tapak Lokal pada Lokasi Stasiun Seismik di Pulau Jawa *J. Fis. dan Apl.* **14** 68. https://doi.org/10.12962/j24604682.v14i3.3745.

[21] Battaglia J 2003 Location of long-period events below Kilauea Volcano using seismic amplitudes and accurate relative relocation *J. Geophys. Res.* **108**

[22] Lara-Cueva R, Benitez D, Carrera E, Ruiz M and Rojo-Alvarez J 2016 Automatic Recognition of Long Period Events from Volcano Tectonic Earthquakes at Cotopaxi Volcano, *IEEE Trans. Geosci. Remote Sens.* **54** 5247–5257

[23] Hidayati S, Ishihara K and Iguchi M 2007 Volcano-tectonic Earthquakes during the Stage of Magma Accumulation at the Aira Caldera, Southern Kyushu, Japan, Bull. *Volcanol. Soc. Japan*. 52 289–309

[24] Garcia L, Alguacil G, Titos M, Cocina O, De la Torre A and Benitez C 2019 Automatic S-Phase Picking for Volcano-Tectonic Earthquakes Using Spectral Dissimilarity Analysis, *IEEE Geosci. Remote Sens.* 1–5. https://doi.org/10.1109/lgrs.2019.2934220.

The influence of causalitic-learning model on problem-solving ability in terms of students' mastery of physics concepts

J Rokhmat[1,2], Hikmawati[2], Kosim[1,2], Wahyudi[2] and Y Sari[2]

[1] Department of Natural Science Education, Postgraduate Program, Universitas Mataram, Indonesia
[2] Physics Education, Faculty of Teacher Training and Education, Universitas Mataram, Mataram, Indonesia

E-mail: joni.fkip@unram.ac.id

Abstract. This study aims to investigate the effects of (a) the causalitic-learning Model (CLM) on (a) Problem-solving Abilities (PSA), (b) mastery of physics-concepts on PSA, and (c) the interaction between CLM and mastery physics-concepts on PSA. This research is a quasi-experimental design with an untreated control group design with a pre-test and post-test. The population is all students of class X Public Senior High School (PSHS) 1 Narmada, amounting to 396 people. Samples were taken using cluster random sampling technique and obtained students class X-1 as the experimental class (14 men and 16 women) and X-2 as the control class (9 men and 21 women). The data were collected using a PSA test and analyzed using parametric statistics two-way ANOVA test. With a significance of 0.05 and the criteria if the sig value <0.05 then Ha is accepted, it found that: (a) CLM affects PSA (sig. score = 0.02), (b) mastery physics-concepts affect PSA (sig. score = 0.007), but (c) there is no interaction effect between CLM and mastery physics-concepts on PSA (sig. score = 0.740).

1. Introduction

Physics is part of natural science and also a collection of knowledge, ways of thinking, and investigation. One of the challenges in learning physics is creating experiences by involving students in comprehensively constructing conceptual-knowledge. In the classroom the teacher helps students find concepts, principles, or facts for themselves, not giving lectures or controlling the class. The active involvement of students in learning is in line with causalitic learning that students are oriented to be able to create meaning from the material studied conceptually [1].

To the author's knowledge, in fact, in practice, there are still many educational problems, namely low learning outcomes. These are shown in part by the low physics scores of students. One of the reasons is that there are still quite a lot of teachers who organize teacher-oriented physics learning. In this case, the teacher only conveys information to students in an imposition (pouring) manner. Students only object to learning activities and do not have the opportunity to discover the physics-concepts they are learning on their own. Students directly receive ready-made knowledge from the teacher. The low mastery of this concept has an impact on the inability to solve physics problems. Mastery concept as a whole is an absolute requirement in achieving the success of learning physics and in problem-solving ability [2]. Therefore, the mastery concept as a whole needs to be developed to support problem-solving abilities.

Problem-solving ability (PSA) is one of the high-level cognitive abilities that allow students to acquire knowledge and skills [3]. Meanwhile, Rokhmat, Marzuki, Kosim, & Verawati [1] revealed that PSA is the ability of students to express their ability to choose or deductively predict various possible consequences of a phenomenon, which contains one or more causes given, and is able to identify how one or more of these causes can produce a selected or predicted result. PSA consists of six problem-solving indicators (IPSA), namely as follows: IPSA-1 understanding, IPSA-2 selecting, IPSA-3 differentiating, IPSA-4 determining, IPSA-5 applying, and IPSA-6 identifying.

The results of several studies show that learning physics with causalitic-thinking has a positive effect on problem-solving abilities [4], [5], [6], [7], [8]). In addition, the implementation of physics learning based on causalitic-thinking processes also received a very positive response from physics education students [9].

With regard to the title of this article, the discussion in this paper is oriented to answer the following two questions: first, what are the characteristics of the learning instrument with a causalitic-model, second, how do the empirical test results influence the implementation of the causalitic-learning model on (a) problem-solving abilities, (b) physics-concepts mastery, and (c) the interaction between CLM and mastery physics-concept on PSA.

The causalitic-learning is a new model that was developed in 2013. This learning model facilitates students to have an ability to think causally and think analytically. In causality thinking, students analyze the causal factors and predict the potential consequences for each problem or phenomenon. In a statistical sense, the causal factor is the independent variable, while the effect is the dependent variable. The main orientation of the development of this learning model is the improvement of problem-solving abilities. However, this learning model is also very suitable for improving the ability of conceptual understanding of physics material because, in learning, the conceptual discussion has a very large portion compared to mathematical discussion [10]

2. Methods

This type of research is quasi-experimental with an untreated control group design with a pre-test and post-test. In this research design, two classes were selected as samples, one class as the experimental class and the other as the control class. At first, a preliminary test was carried out for students in these two classes to determine their initial problem-solving abilities. Furthermore, in the research process, students in the control class were given conventional learning while students in the experimental class were given learning using a causalitic-model. At the end of the research activity, a final test was carried out for students to determine the level of final ability of students' problem-solving. The description of this research can be seen in Table 1.

Table 1. Research Design with Untreated Control Group Design with Pretest and Posttest.

Class	Pre-test	Treatment	Post-test
Experiment	O_{11}	X_1	O_{12}
Control	O_{21}	X_2	O_{22}

(Adobted from Crewell, [11])

O_{11} = experimental class before being given treatment,
O_{12} = experimental class after being given treatment,
O_{21} = control class before being given treatment,
O_{22} = control class after being given treatment,
X_1 = learning using a causalitic-model, and
X_2 = conventional learning.

2.1. Population

The population was 396 students of class X Mathematical & Natural Science (MNS) of Public Senior High School 1 (PSHS 1) Narmada in the 2018/2019 academic year. The sample was taken using cluster random sampling technique and obtained MNS-1 class students as the experimental group (14 male and 16 female) and MNS-2 class students as the control group (9 male and 21 female). The research treatment using causalitic-learning model was carried out for five weeks on momentum and impulse materials. The data for the Problem-solving Abilities (PSA) were taken by mean four descriptive questions, while the data for the concept-mastery were the results of physics-tests from the teacher at the school. In learning with this causalitic-model, uses instruments of the syllabus, lesson plans, and student worksheets. This Causalitic-learning Model (CLM) has four phases, namely (1) the orientation phase, (2) the exploration phase, and the development of the concept of causality, (3) the argumentation phase, and (4) the evaluation [1].

2.2. *Instruments*

The instruments in this study included the syllabus, lesson plans, and student worksheets, as well as test kits. The first three instruments are used for learning with a causalitic-model while the last instrument, a test tool, is used to retrieve data on students' problem-solving abilities.

There are three main components in the student worksheets. Those are the physics phenomenon, the causality table, and the identification space. First, the physics phenomenon is in the form of a simple-composite causal model (SCoCM). With this model, in the phenomenon, there are two or more elements of cause as well as elements of effect [4], [6], [1]). The second component is the causality-table for writing down all elements causes and effects resulted from the exploration. Finally, the identification-space is a place for students to compile arguments to explain how the conditions of each causal element so all of them produce each predicted effect element.

The problem-solving ability test kit has four main parts. The first part, namely the phenomena, is equipped with statements regarding the potential consequences. The second part is a space to write down the choice statements that students think potentially occur. The third is a place for writing students' opinions of other consequences outside of the provided choices. Finally, the fourth is a space for writing an explanation of how the conditions of each causal element together result in the effect that has been selected in the second section or predicted in the third section.

2.3. *Data analysis*

Data were analyzed using two types of statistics, descriptive and inferential. Descriptive statistics to analyze the position of group data, such as the mean and mean. Furthermore, inferential statistics are used to analyze the effect of using the causalitic-learning model in learning on problem-solving abilities in terms of mastery of the concept. The statistical test used is the Anova-test.

3. Result and Discussion

3.1. *Result*

The study aims to test the effect of (a) the Causalitic-learning Model (CLM) on students' Problem-solving Ability (PSA), (b) mastery of physics-concepts on students' PSA, and (c) of the interaction between CLM and mastery concept on students' PSA. The mastery concept, which is a moderating variable, is for classifying students who have high and low concept mastery. The score of concept-mastery is sorted from highest to lowest then divided into two groups. The first group (with the high score) is named the high-group while the rest (with the low score) the low-group. Given that the number of samples in the experiment and control groups is the same, namely 30 students, the number of students in the high and low groups in the experimental and control class is 15 students each. The data from the grouping results are presented in Table 2.

Table 2. Experiment and Control Class Student Grouping based on Concept-mastery

Class	Concept-mastery	N	Max	Min	Mean
Experiment	Low	15	98	24	58
	High	15			
Control	Low	15	94	20	59
	High	15			

Table 2 shows that for students in the experimental class, the highest, lowest, and mean scores for mastery of physics concepts were 98, 24, and 58, respectively. For students in the control class, the highest, lowest, and mean scores of mastery of physics concepts are 94, 20, and 59, respectively. Furthermore, the results of the analysis of Problem-solving Ability (PSA) before and after the implementation of learning with a causalitic model for the experimental class and conventional learning for the control class obtained results as in Table 3.

Table 3. Recapitulation of Problem-solving Ability Score

Problem-solving Ability		N	Min	Max	Mean	Std. Deviation
Experiment Class	Pre-test	30	12	50	30	
	Post-test	30	33	88	61	
Control Class	Pre-test	30	8	50	27	
	Post-test	30	33	72	52	

Table 3 shows a recapitulation of the score of physics Problem-solving Ability (PSA) in the experimental and control classes. In both, the minimum, maximum, and mean scores have all increased. The lowest score increased from 12 to 33, the highest score increased from 50 to 88, and the mean value increased from 30 to 61 (experiment class). In the control class, the lowest score from 8 to 33, the highest score increased from 50 to 72, and the mean value increased increased from 27 to 52.

Tabel 4. Average & N-Gain of Problem-solving Ability Score

Class	Pre-Test	Post-Test	N-Gain
Control	25.70	52.40	0.36
Esperiment	29.90	60.00	0.44

Table 4 shows the mean and N-gain scores of the students' Problem-solving Abilities (PSA) in the experimental and control classes.

Table 5. Research Hypothesis Test with IBM SPSS 24 Assistance

Dependent Variable: Problem-solving Ability (PSA) Significance Level: 0.05

Source	Type III Sum of Squares	Df	Mean Square	F	Sig.
Corrected Model	2504.183[a]	3	834.728	4.551	0.006
Intercept	190970.417	1	190970.417	1041.075	0.000
CLM	1033.350	1	1033.350	5.633	0.021
Mastery of Concept	1450.417	1	1450.417	7.907	0.007
CLM * Mastery of Concept	20.417	1	20.417	0.111	0.740
Error	10272.400	56	183.436		
Total	203747.000	60			
Corrected Total	12776.583	59			

The data on Problem-solving Ability (PSA) and concept mastery are then used as the basis for testing the effectiveness of the implementation of the Causalitic-learning Model (CLM) in improving PSA in terms of the students' mastery of physics concepts (Table 4).

Table 5 shows the results of the two-way ANOVA test using the SPSS 24 application. With a significance level of 0.05, the alternative hypothesis is accepted and the null hypothesis is rejected if the sig-score less than 0.05. In the table, it is found that the sig-score to test the effect of the implementation of the CLM on PSA is 0.021 (0.021 <0.05), to test the effect of mastery of physics concepts on PSA is 0.007 (0.007 <0.05). Meanwhile, to test the effect of the interaction between the implementation of the CLM and mastery of physics concepts on PSA, sig.-score is 0.740 (0.740> 0.05). With three sig.-scores, this means that the alternative hypothesis on the three tests is accepted, accepted, and rejected, respectively. Conversely, for the null hypothesis, respectively is rejected, rejected, and accepted. The effect of the interaction between the implementation of the learning model and the students' mastery of physics concepts on problem-solving abilities is presented in Figure 1.

Figure 1 shows no interaction effect between the Causalitic-learning Model (CLM) and mastery of physics concepts on Problem-solving Abilities (PSA). Graphically, the PSA for high group students is greater than that of the low group. These apply to students in the experimental class (doing learning with a causalitic-model) and students in the control class (doing conventional-learning). The two graphs (the top chart for students in the high group and bottom for students in the low group) appear roughly to be more likely to be parallel.

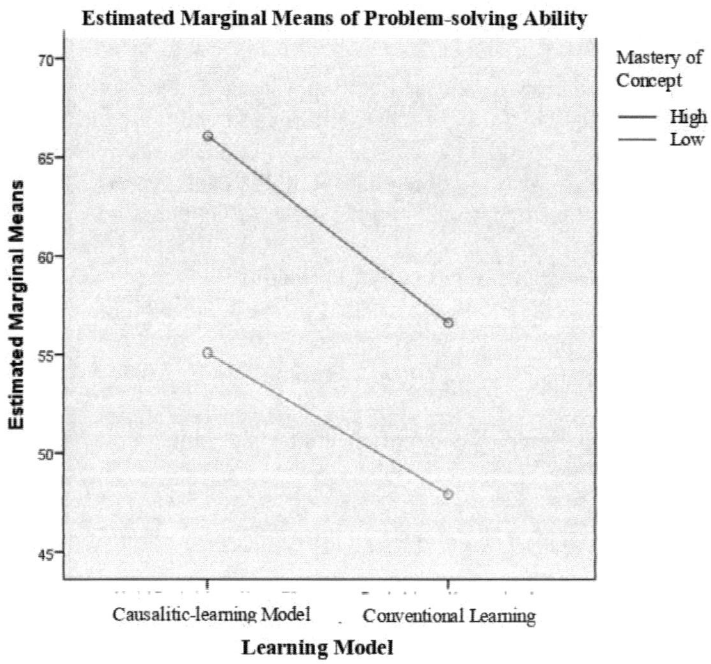

Figure 1. The Interaction Between the Causalitic-learning Model and the Students' Mastery of the Concept.

3.2. Discussion

Problem-solving Ability (PSA) must be explained as an estuary in every learning process. Practically, after completing the study, each student is expected to be able to solve problems related to everyday life. This underlies why in every lesson, PSA is placed as an essential orientation for every student.

As an estuary of all learning objectives, Problem-solving Ability (PSA) needs support by all higher-order thinking skills. In solving problems, it is necessary to carry out a critical analysis to obtain the key-problem. Next, it also needs to analyze the factors that cause the problem to occur. In this case, the problem is a result of these factors. Of course, PSA also needs the presence of knowledge related to these problems in students. The three sub-explanations above show that PSA is closely related to critical thinking skills, causality thinking, and conceptual mastery of knowledge. Of course, there are not only two thinking skills and the mastery of these concepts to optimize the PSA. These mean that other thinking skills are needed, such as synthesis thinking, systemic thinking, analytical thinking, and the other one.

The above paragraph can be interpreted that the process of developing PSA is complex. It is difficult to practice it in its entirety. Therefore, developing PSA needs to be done gradually and assistance at the early stage.

In developing critical thinking ability, one alternative way is to use multi-answer questions. This opinion is in line with Sarwanto, Fajari, & Chumdari [12] that stated the need to use open-ended questions for developing critical thinking. The use of these questions is relevant to the model of phenomena or problems used in causalitic-learning, namely phenomena or problems with a simple-composite causal model that have two or more elements of cause or effect [6], [10], [1].

Furthermore, the relationship of using knowledge as a support for developing PSA is also stated by Rudibyani, Perdana, & Elisanti [13]. They develop PSA based on knowledge assessment instruments. Meanwhile, related to the need for scaffolding in developing students' PSA in line with Rokhmat, et al ([6] & [1]), as well as Abugobar, Salheen, Yassin, Saed, & Yunus [14]. Rokhmat, et al [6] & [1] implement a scaffolding system in causalitic-learning, namely by providing information or exemplifying some of the work that must be done by students. Meanwhile, Abugobar, et al [14] uses scaffolding to help students learn English to be more fluent.

The word causalitic is an acronym for the words causality and analytic, so the term causalitic is related to the ability to think causality and analytics. Rokhmat [15] states that the process of causality and analytical thinking has five advantages, namely that with these two thinking, students are: 1) trained to analyze physical phenomena, 2) comprehend concepts comprehensively, 3) think critically and synthesize, 4) think comprehensively. divergent, and 5) solving problems based on the concept. Also, causal learning facilitates students to build their knowledge independently through group activities in solving physical phenomena and conducting classical discussions. This activity is by the principle of student-centered learning, that is, students are more active in learning. In line with this, Rokhmat [16] emphasizes that the causality thinking approach is centered on the activities of students. The student activity in question takes place when the student completes the tasks in the Student Worksheet and during discussion.

The results of hypothesis testing showed the influence of the causalitic-learning model on problem-solving abilities (sig. = 0.021 <0.05) (Table 5). Table 4 shows that the final achievement of students' PSA in the experimental class is higher than in the control class. The experimental class had an average score of 60.0 and an N-gain of 0.44, while the control class had an average score of 52.4 and an N-gain of 0.36. These two facts indicate that the implementation of physics learning with a causalitic model is more effective than conventional learning that is usually carried out by teachers in these schools. The results of this study are in line with research conducted by Tamami, Rokhmat, & Gunada [8], Nurmadiah, Rokhmat, & Ayub [17], Abdani, Rokhmat, & Rahayu [18], Anshori, Rokhmat, & Gunada [19], Faisal, et al [7], as well as by Nurjamilah, Rokhmat, Sahidu, & Harjono [20] all of which show that the causalitic-thinking approach and causalitic-learning have a positive effect on students' PSA.

Learning in the experimental class is focused on completing tasks and discussing them through causal thinking activities. This process is by the syntax of the Causalitic-learning Model (CLM), which consists of four phases, orientation, exploration, and development of the concept of causality, preparation of arguments, and the evaluation phase [1]. In learning, students are facilitated to be able to complete tasks in Student Worksheets (SWS) with the aim of training students in solving physics problems (momentum and impulse). Next, Nurjamilah, et al [20] stated that the implementation of causalitic learning improves students' reasoning abilities. Meanwhile, Anshori, et al [21] showed a positive effect of causalitic learning on creative thinking skills. The ability to reason does not stand alone but must be supported by other higher-order thinking skills, such as critical thinking and creative thinking. The ability to reason itself supports the development of Problem-solving Ability (PSA). Causalitic thinking can be divided into causality and analytic thinkings. The implementation of causality thinking requires other higher-order thinking skills, such as critical thinking, creative thinking, systems thinking [15]. Furthermore, analytic thinking in the context of causalitic learning requires the ability to reason. Some of these explanations show a strong relationship between causalitic-thinking ability and PSA. So through learning a causalitic model, the PSA of students can be improved.

Rokhmat, et al [4] stated that causality thinking includes understanding phenomena, determining causes, predicting effects, and distinguishing which causes as factors of every effect. Meanwhile, Rokhmat [15] stated when thinking analytically, students are required to be able to identify how the conditions of the causes that can result in each effect are based on the knowledge they already have which includes concepts, principles, theories, or physical laws.

The control class uses a conventional learning model. In this learning, students are still required to be active and given the opportunity to have discussions, presentations, and questions and answers. However, during the learning process, not all of the students were able to discuss well and understand the percentage that was delivered. During the question and answer process, students are less active and less enthusiastic. This is because most students only focus on how to solve the problems given, not on what concepts will be discussed. In addition, students cannot think openly in understanding a concept so that students' understanding of the concepts of momentum and impulse is not comprehensive and results in the problem-solving ability of the control class lower than the experimental class that uses the causalitic-learning model.

Comprehensive mastery of concepts is an absolute requirement in achieving the success of learning physics, and only by mastering the concept, all physics problems can be solved [2]. Therefore, by mastering concept as a whole, students have the potential to be able to develop problem-solving abilities more optimally.

The results of data analysis showed that the independent variable, namely the Causalitic-learning Model (CLM), had a positive impact on Problem-solving Ability (PSA). The average score of PSA in the experimental class is higher than the control class. The results also showed that the moderator variable, namely conceptual mastery, had a positive impact on students' PSA. This can be seen from the PSA of students who have high concept mastery, obtaining a better average score than students with low concept mastery both in CLM and conventional learning. In addition to mastery of physics concepts and problem-solving skills, the implementation of causalitic-learning has been shown to improve several other higher-order thinking skills, such as creative thinking skills, critical thinking, and reasoning abilities. Related to these three abilities, among others, as a result of research by Harwati & Rokhmat [22], Nurjamilah, Rokhmat, , Sahidu, Harjono, & Hikmawati [23], Rezeki, Rokhmat, Gunawan, Makhrus, & Wahyudi [24], and Warodiah, Rokhmat, Zuhdi, Kosim, & Ayub [25]

4. Conclusion

The results showed that the Problem-solving Ability (PSA) of students after learning with a causal learning model had increased. This means that there is an effect of the Causalitic-learning Model (CLM) on students' PSA. The CLM provides a better effect than conventional learning on the PSA of physics of students. Mastery of physics concepts also affects students' PSA. Students with high concept mastery have better PSA than students with low concept mastery. However, the interaction between the CLM and the mastery of physics concepts does not affect students' PSA.

Acknowledgement

We would like to thank the staff leaders of the Indonesian Ministry of Research, Technology, and Higher Education, Research Institutes, and all staff leaders at the University of Mataram who provided funding and facilitated this research. We also appreciate our family, and appreciate all the assistance given by students of the last year of the physics education study program who have contributed to helping carry out this research as well as by students of the Public Senior High School 1 (PSHS 1) of Narmada, Lombok, West Nusa Tenggara, Indonesia.

References

[1] Rokhmat J, Marzuki, Kosim and Verawati N N P 2020 The causalitic learning model to increase students' problem-solving ability *Journal of Physics: Conference Series,* **1572** (012068), doi:10.1088/1742-6596/1572/1/012068

[2] Hapsoro C A and Susanto H 2011 Penerapan Pembelajaran *Problem Based Intruction* Berbantuan Alat Peraga Pada Materi Cahaya di SMP *Jurnal Pendidikan Fisika Indonesia,* **7**(1), 28-32.

[3] Venisari R, Gunawan G and Sutrio S 2015 Penerapan Model Mind Mapping pada Model Direct Intruction untuk Meningkatkan Kemampuan Pemecahan masalah Fisika Siswa SMPN 16 Mataram *Jurnal pendidikan Fisika dan Teknologi*, **1**(3), 193-199.

[4] Rokhmat J, Marzuki, Hikmawati and Verawati N N 2017 Instrument development of causalitic thinking approach in physics learning to increase problem solving ability of pre-service teachers *American Institute of Physics* , 1-9.

[5] Rokhmat J, Marzuki, Hikmawati and Verawati N N 2017b The Causal Model in Physics Learning with a Causalitic thinking Approach to Increase the Problem-solving Ability of Pre-service Teachers *SOCIAL SCIENCES & HUMANITIES*, **25**(S), 153-168.

[6] Rokhmat J, Marzuki, Wahyudi and Putrie S D 2019 A Strategy of Scaffolding Development to Increase Students' Problem-Solving Abilities: The Case of Physics Learning with Causalitic-Thinking Approach *Journal of Turkish Science Education, **16**(4), p. 569-579. DOI:* 10.36681/tused.2020.8

[7] Faisal, Rokhmat J and Ardhuha J 2020 Kemampuan Pemecahan Masalah dan Kreativitas Peserta Didik dengan Pendekatan Berpikir Kausalitik Berscaffolding *Jurnal Pendidikan Fisika dan Teknologi, 6(1), p. 107-113, DOI:* http://dx.doi.org/10.29303/jpft.v6i1.1771

[8] Tamami F, Rokhmat J and Gunada I W 2017 Pengaruh Pendekatan Berpikir Kausalitik Scaffolding Tipe 2a Modifikasi Berbantuan LKS terhadap Kemampuan Pemecahan Masalah Optik Geometri dan Kreativitas Siswa Kelas XI *JurnalPendidikan Fisika dan Teknologi*, **3**(1), 76-83.

[9] Rokhmat J, Setiawan A and Rusdiana D 2012 Pembelajaran Fisika Berbasis Proses Berpikir Kausalitak dan Berpikir Analitik (PBK-BA), Suatu Pembiasaan Berpikir Secara Terbuka *Prosiding Seminar Biologi*, **9**(1), 391-397.

[10] Rokhmat J, Marzuki, Kosim and Verawati 2020 The Causalitic-learning Model to Increase Students' Problem-solving Ability. *Journal of Physics: Conference Series, **1572**(1)* 012068

[11] Creswell J W 2012 *Educational research: Planning, conducting and evaluating quantitative and qualitative research*. Boston: Pearson, p. 300-306

[12] Sarwanto, Fajari L E W and Chumdari 2021 Open-Ended Questions to Assess Critical-Thinking Skills in Indonesian Elementary School *International Journal of Instruction, **14**(1), p. 615-630.*

[13] Rudibyani R B, Perdana R and Elisanti E 2020 Development of Problem-Solving-Based Knowledge Assessment Instrument in Electrochemistry. *International Journal of Instruction, **13**(4), p. 957-974.*

[14] Abugohar M A, Salheen D A A, Yassin B, Saed H A and Yunus K 2020 Scaffolding Oral Fluency Mediating the Target Language in ELT to Tertiary-Level Students: A Follow-up Scheme *International Journal of Instruction, **13**(4), p. 331-346.*

[15] Rokhmat J 2013 Peningkatan Kemampuan Pemecahan Masalah Mahasiswa Calon Guru Fisika melalui Berpikir Kausalitas dan Analitik *Disertasi. Universitas Pendidikan Indonesia: tidak diterbitkan.*

[16] Rokhmat J 2015 Penerapan Pendekatan Berpikir Kausalitik Ber-scaffolding dalam

Meningkatkan Kemampuan Pemecahan Masalah Hukum Newton Tentang Gerak *Prosiding Seminar Nasional Fisika.*

[17] Nurmadiah N, Rokhmat J and Ayub S 2018 The Influence of Causal Thinking with Scaffolding Type 2a and 2b on Optics Problem-solving Ability *Jurnal Pendidikan Fisika Indonesia, **14**(2), p. 65-72, DOI:* 10.15294/jpfi.v14i2.12185.

[18] Abdani R, Rokhmat J and Rahayu S 2018 Pengaruh Pendekatan Berpikir Kausalitik Ber-Scaffolding dengan Pemberian Tugas Pendahuluan terhadap Kemampuan Pemecahan Masalah Siswa SMA *Jurnal Pendidikan Fisika dan Teknologi, **4**(2), p. 213-219*

[19] Anshori I A, Rokhmat J and Gunada I W 2019 Penerapan Model Pembelajaran Kausalitik dalam Meningkatkan Kreativitas Peserta Didik. *Jurnal Pendidikan Fisika dan Teknologi, **5**(2), p. 205-212, DOI:* http://dx.doi.org/10.29303/jpft.v5i2.1215

[20] Nurjamilah, Rokhmat J, Sahidu H and Harjono A 2020 Penerapan Model Pembelajaran Kausalitik untuk Meningkatkan Kemampuan Bernalar dalam Pembelajaran Fisika Masa Learning From Home Pandemi Covid-19 *Jurnal Pendidikan Fisika dan Teknologi, **6**(2), p. 183-192, DOI:* http://dx.doi.org/10.29303/jpft.v6i2.1960

[21] Anshori, Rokhmat, Gunada, Wahyudi, and Ayub 2021 The Effect of A Causalitic-Learning Model on Problem-Solving Abilities Reviewed From Creativity *Journal of Physics: Conference series,* **1816**(012041), p. 1-7, DOI: 10.1088/1742-6596/1816/1/012041

[22] Harwati K and Rokhmat J 2021 Development of Student Worksheet to Improve Creative and Critical Thinking Ability of Students in Causalitic-learning Model *Journal of Physics: Conference series,* **1816**(012038), p. 1-9, DOI: 10.1088/1742-6596/1816/1/012038

[23] Nurjamilah, Rokhmat J, Sahidu H, Harjono A, and Hikmawati 2021 Causalitic-learning Model to Improve Reasoning-Ability in Learning Physics in Terms of Student Creativity During The Covid-19 Pandemic *Journal of Physics: Conference series,* **1816**(012039), p. 1-9, DOI: 10.1088/1742-6596/1816/1/012039

[24] Rezeki H, Rokhmat J, Gunawan, Makhrus M, and Wahyudi 2021 Implementation of Causalitic-learning Devices to Improve Creative Thinking Ability and Problem-solving of Students In Physics *Journal of Physics: Conference series,* **1816**(012048), p. 1-7, DOI: 10.1088/1742-6596/1816/1/012048

[25] Warodiah Y N, Rokhmat J, Zuhdi M, Kosim and Ayub S 2021 The Strategy of Student-Worksheet Design with A Causalitic-learning Model to Improve Creative-Thinking Ability *Journal of Physics: Conference series,* **1816**(012047), p. 1-12, DOI: 10.1088/1742-6596/1816/1/012047

The use of causalitic learning model to encourage abilities of problem solving and creative thinking in momentum and impulse

J Rokhmat[1,2], I W Gunada[2], S Ayub[2], Hikmawati[2] and T Wulandari[2]

[1] Department of Natural Science Education, Postgraduate Program, University of Mataram, Mataram, Indonesia
[2] Physics Education, Faculty of Teacher Training and Education, University of Mataram, Mataram, Indonesia

E-mail: joni.fkip@unram.ac.id

Abstract. This study aims to test the use of a causalitic-learning model to encourage problem-solving and creative thinking skills. This study used a non-equivalent control group design. The population was students of class X Mathematics and Natural Sciences (MNS), Public Senior High School (PSHS) 3, the academic year 2018/2019, Mataram, Indonesia. Samples were taken using the purposive sampling technique and obtained by X-MNS-4 class students (33 students, 13 male, 20 female) as the experimental class and X-MNS-5 class students (28 students, 12 male, 16 female) as the control class. The data were taken using essay test questions and analysed using the MANOVA test but previously carried out the Lavene test and the Box's M test (with the help of the SPSS version 23 application). The results of statistical analysis, with a significance level of 0.05, the null hypothesis on both research hypotheses were rejected. The conclusion is that the implementation of the causalitic-model in physics learning on the subject of momentum and impulse affects the improvement of students' problem-solving and creative thinking abilities. Insights on innovation through technology integration, multi-answer phenomenon-based learning to promote creative thinking is also discussed. The implication is important for physics learning which emphasises conceptual understanding.

1. Introduction

Teaching and learning are two inseparable concepts in education. Teaching is defined as a change in human behavior that arises from the provision of knowledge. In line with what Fathurrohman [1] stated, teaching is a process of student-teacher interaction and learning resources in a learning environment. Learning has always been developing from time to time until now the term is known in learning, namely 21st-century skills.

According to Aji [2], in facing the challenges of the 21st century, teachers better prepare students to become investigators, problem solvers, thinkers critically and creatively. Competencies and survival skills needed by students in facing life, the world of work, and citizenship in the 21st century are emphasized on the following seven points: (1) critical thinking and problem solving, (2) collaborating and leading, (3) agile and adaptive, (4) initiative and entrepreneurial spirit, (5) communicating effectively both orally and in writing, (6) accessing and analyzing information, and (7) having curiosity and imagination [3].

In line with Zubaidah [4], new standards are needed in order students to have the necessary competencies in the 21st-century. Schools need to find ways to facilitate students to be successful in their work and life through the ability to think creatively, solve problems flexibly, collaborate and innovate. Therefore, it is necessary to create a learning environment that can facilitate the development of 21st-century skills.

One of the abilities above is solving problems. The ability to solve problems is a high-level cognitive ability. Problem-solving thinking is a level higher than the evaluation stage as part of Bloom's cognitive stage. As quoted by Suratno, Wahono, Chang, Retnowati, & Yushardi [5], Tok, Tok, & Dolapçıoğlu [6] states that problem-solving is related to higher-order thinking skills that require logical answers. Therefore, problem-solving strategies are also an important factor in facilitating students to reach a conceptual understanding of the material being studied.

Indicators of problem-solving abilities hereinafter referred to as IPSA, consist of six abilities: (1) understanding, (2) selecting, (3) differentiating, (4) determining, (5) applying, and (6) identifying ([7], [8], [9], [10]). This means that a student has problem-solving abilities when he can show these six performances. The combination of these six indicators allows students to solve problems that have more than one solution that is in line with the development of students' creative thinking abilities.

Three causal models, divergent models, simple combinations, and chain combinations [8] can facilitate the development of questions that have more than one correct answer. These models are in line with the development of creative thinking, especially for indicators of fluency, flexibility, and/or originality resulting from open-ended tasks [11]. Anwar, Aness, Khizatr, & Muhammad [12] stated that the effectiveness of these indicators also supports the development of elaboration skills. Rokhmat, et al [8] states that the fluency indicator shows how many answers are selected or predicted by students, flexibility indicators are related to how high the difficulty level of arguments has been made, originality shows that there are additional answers put forward by students, and the use of their sentences in the argument. Finally, the elaboration indicator shows how the arguments have been made in detail. With the rubric of the four indicators of creative thinking, the measurement of problem-solving and creative thinking abilities can be measured using the same problem model, namely problems that have more than one correct answer along with explanations or arguments.

In this 21st century era, teachers, students, and the educational environment must support the implementation of learning to develop 21st-century skills. The facts in the field show the opposite. The results of observations in class *X*-MNS (Mathematics and Natural Science) Public Senior High School (*PSHS*) 3 Mataram show that teachers and the learning environment have not been able to facilitate students to develop problem-solving abilities. The way the teacher teaches tends to be based on homogeneous questions, between examples of solving problems with questions for practice. When in practice, the questions are modified or slightly changed, students have difficulty in determining which equations are used. This indicates that students do not understand the concept of physics. Students only imitate what the teacher has given. This low understanding of physics concepts affects students' unable to solve problems independently. Silaban [13] states that the physics concepts that students already have cannot be mastered properly without practice solving problems related to the material being studied. That is, by improving problem-solving abilities, students will better master the concepts that exist in their cognitive structure, and train their thinking skills so that they can get better learning outcomes. This is in line with the opinion of Nisrina, Gunawan, & Harjono [14], namely the lower the ability of students to understand and master concepts, the more difficult it is for students to come up with new ideas.

In addition to seeing firsthand how the teacher taught, researchers also looked at the midterm questions used to evaluate the learning outcomes of class *X* students for the 2018/2019 academic year. The questions are in the form of multiple-choice totaling 20 numbers and 90% of the questions are based on mathematical calculations. This makes students focus solely on mathematical solutions and pay less attention to the basic concepts of the problems given. This was also revealed by Rokhmat [15] that teachers tend to introduce some equations (formulas) to related phenomena. In line with this, Helmi, Rokhmat, & Ardhuha [16] stated that the form of a physics written test that only relies on physics equations will hinder the development of students' problem-solving abilities. The existence of demands for answers in the form of numbers alone can impede creativity because it does not allow students to come up with new ideas that may arise due to the ability to think creatively.

The low ability to solve physics problems can be minimized by applying suitable learning scenarios. One of them is the selection of a learning model. Research conducted by Helmi et al [16], Tamami, Rokhmat, & Gunada [17], Yuliana, Rokhmat, & Gunada [18], and Faisal, Rokhmat, & Ardhuha [19] conclude that low problem-solving abilities can be minimized with an approach causalitic-thinking. During its development, this causal thinking approach was used as a reference in the preparation of a causal learning model by Rokhmat, et al [20]. So that the researcher chose an alternative causalitic-learning model.

This learning model is called the "causalitic-learning model" with the consideration that the main activity in learning is oriented to develop the potential of learners in causality thinking and analytic thinking. Thinking of causality, meaning that students are required to be able to determine the components of the cause, and based on these causal conditions, students are required to be able to predict all events (consequences) deductively that have the opportunity to occur. Whereas in analytic thinking, students are required to be able to identify how the conditions of the cause are, so that it can cause certain consequences based on the knowledge they have which includes concepts, principles, theories, and/or related physical laws [20].

Research on causalitic-thinking approaches has been conducted by Rokhmat [21], [22]. The result is that the application of the causalitic-thinking process can improve the physics problem-solving abilities of students and high school students. Meanwhile, Rokhmat, Marzuki, Wahyudi, & Putrie [9] examined scaffolding development strategies in physics learning with a causalitic-approach to improving the problem-solving abilities of high school students. The estuary of research on causality and analytical thinking approaches is the development of a causalitic-learning model. Rokhmat, et al [10] developed a causalitic-learning model to improve students' problem-solving abilities. One of the characteristics of this learning model is the activity of the teacher/lecturer/teacher in the form of guiding by showing some examples of problem-solving as a stage aid in the learning process.

The causalitic-learning model was developed in 2013. The idea of this learning model is to facilitate students to be able to identify the causal factors and predict, deductively, the potential consequences in each question or phenomenon. In statistical language, the causal factor is the independent variable, while the effect factor is the dependent variable. Furthermore, in the phase of arrangement of an argument, students are facilitated to be able to develop scientific explanations based on facts, concepts, principles, theories, and or physical laws to identify the condition of each causal factor so that with these conditions, certain consequences arise [20].

2. Methods

This research is a Quasi-Experimental type using untreated control group design with pretest and posttest. Quasi-Experimental Research is a research design that is carried out if there are external variables that cannot be fully controlled (Creswell, 2012)[23]. The form of the research design is shown in Table 1.

2.1. Population

The population in this study were all students of class *X-MNS* (Mathematics and Natural Sciences), Public Senior High School (*PSHS*) 3 of Mataram, in the 2018-2019 academic year. Sampling with purposive technique because the researcher has the consideration that the two classes of the sample taken have almost the same abilities, and obtained class *X-MNS*-4 (28 students with 12 male and 16 female) as the experimental class and *X-MNS*- 5 (33 students with 13 male and 20 female) as the control class.

Table 1. Research Design with Untreated Control Group Design with Pretest and Posttest

Class	Pre-test	Treatment	Post-test
Experiment	O_1	X_1	O_2
Control	O_3	X_2	O_4

(Adobted from Crewell, 2012: 300-306)

O_1 = experimental class before being given treatment,
O_2 = experimental class after being given treatment,
O_3 = control class before being given treatment,
O_4 = control class after being given treatment,
X_1 = learning using a causalitic-model, and
X_2 = conventional learning.

2.2. *Instruments*
The instruments in this study included the syllabus, lesson plans, and student worksheets, as well as test kits. The first three instruments are used for learning with a causalitic-model while the last instrument, a test tool, is used to retrieve data on students' problem-solving and creative-thinking abilities.

There are three main components in student worksheets, namely physical phenomena, causality tables, and identification spaces. Physical phenomena using a simple composite causal model (SCoCM). With this model, in a phenomenon, there are two or more elements of cause and two or more elements of effect (Rokhmat, et. Al, 2017b[8]; Rokhmat, et al, 2020[20]). The causality table is intended to list all the elements of cause and effect resulting from exploration. Finally, the identification space is a place for students to compile arguments as an explanation of how the conditions of each causal element are so that they all produce selected or predicted results.

The problem-solving and creative thinking abilities test kit has four main parts. The first part, namely the phenomenon is equipped with a statement regarding the effect that can occur. The second part is a space to write down the selected statements that students think have the potential to occur. The third is a place to write students' opinions about other consequences beyond the choices given. Finally, the fourth is space to write an explanation of how the conditions of each cause element together produce the effect that has been chosen in the second part or predicted in the third one.

The research data were analyzed using the multivariate analysis of variance test or the Manova test to test the effect of the independent variable of the causalitic-learning model on the dependent variable of problem-solving ability and creative thinking. Before the Manova test, the Lavene test and Box's M test were carried out using the help of the SPSS version 23 application. The Lavene and Box's M test was to determine whether the two population data had homogeneity of data distribution.

3. Result and Discussion

3.1. *Results*
The average score and standard deviation for problem-solving and creative-thinking abilities are The results of this study consisted of the results of the instrument test, as well as the analysis of the ability to think creatively and the ability to solve problems at the beginning and the end of the lesson. The data on problem-solving and creative thinking abilities were analyzed using the homogeneity test, normality test, and hypothesis testing using the Manova test.

3.1.1. Instrumen of Problem-solving Ability. The results of the validity, reliability, difficulty level, and difference power test results show that the four items of problem-solving ability are acceptable. The summary of the results of the feasibility test of the problem-solving ability test instrument is shown in Table 2.

Table 2. Results of the Problem-solving Ability Instrument Test

Number	Validity	Releability	Difficulty Level	Difference Power	Decision
1	Valid	Releabel	Moderate	Moderate	Accepted
2	Valid	Releabel	Moderate	Moderate	Accepted
3	Valid	Releabel	Moderate	Moderate	Accepted
4	Valid	Releabel	Difficult	Good	Accepted

3.1.2. Instrumen of Creative-thinking Ability. The results of the analysis of the difference in the ability to think creatively showed that there was one item that had sufficient difference in the category, while the other questions were categorized as good. A summary of the feasibility test results for the instrument is shown in Table 3.

Table 3. Results of the Creative-thinking Ability Instrument Test

Number	Validity	Releability	Difficulty Level	Difference Power	Decision
1	Valid	Releabel	Moderate	Good	Accepted
2	Valid	Releabel	Moderate	Good	Accepted
3	Valid	Releabel	Moderate	Good	Accepted
4	Valid	Releabel	Difficult	Sufficient	Accepted

3.1.3. Manova Test. The average score and standard deviation for problem-solving and creative-thinking abilities are shown in Table 4. In the table, the problem-solving ability of the experimental class students has a standard deviation (19.95614) greater than the control class students (14.98). The creative ability of the experimental class students had a standard deviation score (10.58) lower than the control class students (16.95). This fact shows that the score curve of the experimental class student's problem-solving ability is slower than the control class student. On the other hand, the score curve of the experimental class student's creative thinking ability is slimmer. In other words, the data on the problem-solving ability scores of the experimental class students were more varied than the control class students. On the other hand, the data on the score of the control class students' creative thinking abilities was more varied than the experimental class students.

Table 4. Average Score and Standard Deviation of Problem-solving and Creative-thinking Abilities

Dependent Variable	Class	Average Scor	Standard Deviation	*N*
Problem-solving Ability	Experiment	58.10	19.96	28
	Control	43.42	14.98	33
	Total	50.16	18.80	61
Creative-thinking Ability	Experiment	41.89	10.58	28
	Control	34.79	16.95	33
	Total	38.05	12.33	761

The homogeneity test of variance was carried out using the Levene test (Table 5). Table 5 is a recapitulation of Levene's results. The table shows that the F value for problem-solving ability is 2.72 with a sig. value 0.105, and the F value for creative thinking abilities is 2.13 with a sig. value 0.150. With a significance level of 0.05, the sig. value for both F values is greater than 0.05 so that H_0 is rejected and H_a is accepted. These results indicate that the problem-solving and creative-thinking abilities on momentum and impulse subject have homogeneous variants and so the Manova-test can be continued.

Table 5. Homogeneity Test Results Using Levene's Test

Dependent Variable	F	df_1	df_2	Sig.
Problem-solving Ability	2.72	1	59	0.105
Creative-thinking Ability	2.13	1	59	0.150

H_0 = The value of problem-solving and creative thinking abilities on the momentum and impulses of students has a non-homogeneous variant.

H_a = The value of problem solving and creative thinking abilities on the momentum and impulses of students has homogeneous variants.

Table 6. Box's *M* test results

Box's *M*	6.335
F	2.033
df_1	3
df_2	7040864.062
Sig.	.107

Table 7. Manova Test Results

Effect	Sig.
Pillai's Trace	0.008
Wilks' Lambda	0.008
Hotelling's Trace	0.008
Roy's Largest Root	0.008

H_0 = There is no effect of the causalitic-learning model on problem-solving and creative thinking abilities of students on momentum and impulse subject.

H_a = There is an effect of the causalitic-learning model on problem-solving and creative thinking abilities of students on momentum and impulse subject.

The Manova test has analytical prerequisites, namely the fulfillment of the Box's *M* test. Then the Box's *M* test is carried out as shown in Table 6. The results of the Box's *M* test, namely with a significance level of 0.05, the sig. (0.107) > 0.05 so that the conditions were met and the Manova test could be continued (Table 6). The Manova test is used to test whether there is an effect of the causalitic-learning model on several dependent variables, namely the problem-solving and creative thinking abilities. This sub-test of Manova includes the Pillae Trace test, Wilk Lambda, Hotelling Trace, and Roy's Largest Root test (Table 7). The table shows that the sig. each Manova sub-test was the same, ie 0.008 less than 0.05. This fact, with a significance level of 0.05, then in each sub-test, H_0 is rejected and H_a is accepted, and it means that there is an effect of the causalitic-learning model on problem-solving and creative thinking abilities on momentum and impulse subject.

3.2. Discussion
The problem-solving and creative thinking abilities are an ability that requires students to use higher order thinking skills. Both of these abilities have indicators. The first has six (IPSA = Indicator of Problem-solving Ability), while the second has four (ICT = Indicator of Creative Thinking). The two groups of indicators have a similarity. As an example, between IPSA-2 and ICT-1. The indicator IPSA-2 shows that students can choose at least one correct answer, while the indicator ICT-1 shows

how the students choose or predict the possible correct answer. So these two indicators coincide in terms of choosing or predicting the correct answers.

Another intersection also occurs between the IPSA-4 and IPSA-5 with the ICT-2. IPSA-4 and IPSA-5 are related to the selection and application of concepts, principles, theories, or laws of physics in developing arguments to explain why an effect occurs. Meanwhile, the ICT-2 indicates the level of difficulty of an argument that students have made, which also involves the use of physics concepts, principles, theories, or laws. So it is evident that the indicator of problem-solving and creative thinking abilities has some similarities in the student's work. This study used the MANOVA test to determine whether the causalitic-learning model has a simultaneous effect on problem-solving and creative thinking abilities.

The results of hypothesis testing using the MANOVA test showed that the causalitic learning model affected problem-solving and creative thinking abilities simultaneously (Table 7). The table shows that the sig. of all sub-tests are 0.008 or less than 0.05 as a condition that H_0 was rejected and H_a was accepted.

The existence of similarities in work about indicators of problem-solving and creative thinking abilities indicates that the development of these two abilities is mutually supporting one another. Increased problem-solving ability also increase the ability to think creatively and vice versa.

The indicator IPSA-2 is in line with ICT-1, which shows students can choose various answers. Furthermore, the indicator IPSA-4 is in line with the ICT-2 indicator, which indicates students can solve problems using concepts, theories, principles, laws of physics. Indicator IPSA-6 is in line with the ICT-4, in which students can compile a logical explanation that connects the conditions of each aspect of the causes with each answer.

The causalitic-learning model facilitates students to be able to understand physics conceptually and comprehensively. Rokhmat (2013b)[22] stated that causalitic-thinking trains students to understand the concept as a whole. This statement is in line with Pebriyanti, Sahidu, & Sutrio (2015)[24] that a comprehensive understanding of concepts can help students solve the physics problems faced. So, applying a causalitic-learning model can significantly improve students' problem-solving abilities.

The causalitic-learning model also has a positive influence on the creative thinking abilities of students. This results in line with Anshori, Rokhmat, Gunada, Wahyudi, & Ayub (2021)[25], Harwati & Rokhmat (2021)[26], Rezeki, Rokhmat, Gunawan, Makhrus, & Wahyudi (2021)[27], Warodiah, Rokhmat, Zuhdi, Kosim, & Ayub (2021)[28], and Nurjamilah, Rokhmat, Sahidu, Harjono, & Hikmawati (2021)[29]. From point of view of the causal model, creative thinking is in line with a divergent causal model (Rokhmat, et al, 2017b[8]) which is able to describe the various answers that may arise from a phenomenon. The use of questions with more than one correct answer is in line with the opinion of Meyer, et al (2015)[11] that the development of creative thinking can use semi-open questions. With this type of question, students have the flexibility to provide more than one answer. So the implementation of the causalitic-learning model facilitates students to find various solutions to one problem. The ability to think creatively can also be improved through formulating questions (Ramirez, 2021)[30].

3.3. Limitations

This research is not free from limitations. These are from the learning-model character, the readiness of students, the classroom situation, and the ability of the researcher himself. We note at least five limitations. Some are from other people's previous studies. First, the researcher finds it difficult to implement a causal learning model because it is still relatively new. Second, students feel unfamiliar with the researcher's strategy in building their knowledge, namely using multi-answer questions.

Related to the first limitation above, to overcome this, the researcher provides initial knowledge to students about how to learn with a causalitic-learning model. Researchers equip students with the knowledge of how to learn with this learning model at the beginning of learning, but a lot of time is needed so that the time allocation for learning is not enough.

Third, researchers experienced difficulties in developing students' abilities, especially related to problem-solving indicators of IPSA-4 and IPSA-5. Students are not used to answering physics problems using concepts, principles, theories, or laws of physics. The increase in the IPSA-4 and IPSA-5 indicator scores was the lowest, even though researchers repeatedly reminded them to use concepts, principles, theories, or laws in answering questions. For this reason, researchers continue to try to increase students' motivation to read conceptual explanation-oriented resource books instead of books that present more mathematical equations.

The causalitic-learning model facilitates students to read references before learning begins, namely by giving preliminary assignments. The assignments intended that students have initial knowledge of the material to be studied. They contain questions that can stimulate initial understanding. When applying this, researchers do not ask students to collect their work. Researchers only checked notebooks that can result in students did not work on the task. With this, the researcher also does not evaluate the results of the participants' preliminary work.

Fourth, students are not used to learning with conditions that require more than one answer in a problem. In this case, researchers must also get used to thinking openly, especially in developing instruments that can later stimulate the students' brains to think openly.

Fifth, many students find it difficult to express ideas verbally. They understand the meaning of the problem and have an idea as an answer, but they cannot state them in a sentence. Researchers noticed that some students give explanations to their classmates, but when asked to explain to other students in front of the class, they had difficulty arranging words. This problem can be overcome by accustoming students to take turns presenting the results of their group discussion and not pressing that the answers must be completely correct. There is also a need for assistance through individual or group teacher guidance. The stage of assistance itself already exists in the causalitic-learning model, namely the exploration and development stage of the concept of causality (Rokhmat, et al, 2020[10]). In this stage, the teacher guides by providing some examples of work that students must complete, including outlined in student worksheets or through direct explanations by the teacher.

Based on the explanation of the five limitations above, it is better if learning with this causal model is used as early as possible, starting from the pre-school level. Of course, this is a challenge to develop learning instruments with relevant causal models to be used at every level of education, from pre-school, elementary, junior high, high school, to tertiary levels.

From the perspective of scientific disciplines, laterally, this causalitic-learning model also potentially be implemented in all disciplines. So another challenge is developing instruments for causalitic-learning models for all disciplines. In general, the challenges that continue to be faced are developing the instruments and implementing causalitic-learning models for learning at every level of education and for all disciplines.

As a new learning model with an emphasis on conceptually and comprehensively discussion and requires the support of almost all types of higher-order thinking (such as analytical skills, synthesis, critical thinking, creative thinking, systems thinking, and evaluation skills), this learning model requires a long time of process to provide optimal benefits for the development of complete mastery of concepts in students. Efforts to optimize students to undergo learning with this causalitic model need process assistance with portions from large enough to a small, or even finally without this assistance. The stage assistance system is in line with the ideas of Rokhmat, et al (2019)[9] about the strategy of developing scaffolding in learning with a causal model.

4. Conclusion

The use of the causalitic-learning model is proven to be able to improve problem-solving and creative thinking abilities simultaneously. This learning model has been implemented in students of class X - MNS (Mathematics and Natural Science) Public Senior High School (PSHS) 3 Mataram in the 2018/2019 academic year. This learning model has the characteristics of being oriented towards a comprehensively conceptual discussion. To optimize the achievement of learning objectives, the following are recommended: introduction of learning strategies with this learning model, provide preliminary assignments, and using learning assistance (can be included in student worksheets).

Acknowledgement
We would like to express our gratitude to the Minister of Education and Culture and his staff, the leadership of the University of Mataram and his staff, as well as the Head of the Mataram University Research Institute and the staff who have helped finance the implementation of this research so that it can be carried out properly. Thanks are also conveyed to the leaders and staff of the Mataram Public Senior High School (PSHS) 3 who have provided a research place, as well as students of class X-MNS (Mathematics and Natural Science) of PSHS 3 Mataram who have been willing to be the subjects of this research. In particular, thanks are conveyed to the family and sister Tria Wulandari who have actively helped in the process of implementing and collecting data, as well as analyzing the data of this research.

References
[1] Fathurrohman M 2015 *Model-Model Pembelajaran Inovatif* (Jogjakarta: Ar- Ruzz Media)
[2] Aji D S 2017 Pengembangan Modul Pembelajaran Fisika Berbasis Problem Based Learning untuk Meningkatkan Kemampuan Pemecahan Masalah Fisika *Science Education Journal, **1** (1), 36-51*
[3] Wagner T 2010 *Overcoming The Global Achievement Gap (online)* (Cambridge, Mass., Harvard University)
[4] Zubaidah S 2016 Keterampilan Abad ke 21: Keterampilan yang Diajarkan melalui Pembelajaran *Seminar Nasional Pendidikan* dengan Tema "Isu-isu Strategis Pembelajaran MIPA Abad XXI **21**
[5] Suratno, Wahono B, Chang C Y, Retnowati A and Yushardi 2020 Exploring a Direct Relationship between Students' ProblemSolving Abilities and Academic Achievement: A STEM Education at a Coffee Plantation Area *Journal of Turkish Science Education, **17**(2), pp. 211-224, DOI:* 10.36681/tused.2020.22
[6] Toka T N, Tokb S, Dolapçıoğluc S D 2013 The Perception Levels of The Novice Teachers' Problem-Solving Skills *Procedia - Social and Behavioral Sciences* **116**, p. 415 – 420.
[7] Rokhmat J, Marzuki, Hikmawati and Verawati 2017a Instrument Development of Causalitic Thinking Approach in Physics Learning to Increase Problem Solving Ability of Pre-service Teacher *American Institute of Physics Conference Proceeding*, 03000-3 - 03000-5
[8] Rokhmat J, Marzuki, Hikmawati and Verawati 2017b The Causal Model in Physics Learning with A Causalitic-thinking Approach to Increase The Problem-solving Ability of Pre-service Teachers *Pertanika Journal of Social Science and Humanities (JSSH), **25**(S), 153-168.*
[9] Rokhmat J, Marzuki, Wahyudi and Putrie S D 2019 A Strategy of Scaffolding Development to Increase Students' Problem-solving Abilities: The Case of Physics Learning with Causalitic-thinking Approach *Journal of Turkish Science Education, **16**(4, pp. 569-579*
[10] Rokhmat J, Marzuki, Kosim and Verawati 2020 The Causalitic-learning Model to Increase Students' Problem-solving Ability *Journal of Physics: Conference Series, **1572**(1)* 012068
[11] Meyer A A and Lederman N G 2015 Creative cognition on secondary science: An exploration of divergent thinking in science among adolescents *International Journal of Science Education* http://dx.doi.org/10 .1080/09500693.2015.1043599
[12] Anwar M N, Aness M, Kizar A, Naseer M and Muhammad G 2012 Relationship of creative thinking with the academic achievements of secondary school students *International Interdisciplinary Journal of Education **1**(3), 44*
[13] Silaban B 2014 Hubungan antara Penguasaan Konsep Fisika dengan Kreativitas dan Kemampuan Pemecahan Masalah pada Materi Pokok Listrik Statik *Jurnal Penelitian Bidang Pendidikan, **20**(1), p. 65-75*
[14] Nisrina N, Gunawan G and Harjono A 2017 Pembelajaran Kooperatif dengan Media Virtual

untuk Peningkatan Penguasaan Konsep Fluida Statis Siswa *Jurnal Pendidikan Fisika dan Teknologi, 2(2), 66-72*

[15] Rokhmat J 2015 Penerapan Pendekatan Berpikir Kausalitik Ber-scaffolding dalam Meningkatkan KPM Hukum Newton tentang Gerak *Prosiding Seminar Nasional Fisika*

[16] Helmi F, Rokhmat J and 'Ardhuha, J 2017 Pengaruh Pendekatan Berpikir Kausalitik Ber-Scaffolding Tipe 2b Termodifikasi Berbantuan LKS Terhadap Kemampuan Pemecahan Masalah Fluida Dinamis Siswa *Jurnal Pendidikan Fisika dan Teknologi*, **3**(1), 68-75

[17] Tamami F, Rokhmat J and Gunada I W 2017 Pengaruh Pendekatan Berpikir Kausalitik Scaffolding Tipe 2a Modifikasi Berbantuan LKS terhadap Kemampuan Pemecahan Masalah Optik Geometri dan Kreativitas Siswa Kelas XI *Jurnal Pendidikan Fisika dan Teknologi, 3(1), 76-83*

[18] Yuliana I, Rokhmat J and Gunada I W 2017 Pengaruh Berpikir Kausalitik BerScaffolding Terhadap Kemampuan Pemecahan-Masalah Kalor pada Siswa SMA *Prosiding SNFA (Seminar Nasional Fisika dan Aplikasinya)* **2**, (85-92)

[19] Faisal, Rokhmat J and 'Ardhuha J 2018 Pengaruh Pendekatan Berpikir Kausalitik Scaffolding Tipe 2b Modifikasi Berbantuan LKS terhadap Kemampuan Pemecahan Masalah Optik Geometri dan Kreativitas Peserta Didik *Jurnal Pendidikan Fisika dan Teknologi, 3(1), 76-83*

[20] Rokhmat J, Marzuki, Kosim and Verawati N N P 2020 The causalitic learning model to increase students' problem-solving ability *Journal of Physics: Conference Series,* **1572** (012068)

[21] Rokhmat J 2013a Peningkatan Kemampuan Pemecahan Masalah Mahasiswa Calon Guru Fisika melalui Berpikir Kausalitas dan Analitik [The Increase of Problemsolving Ability of Physics Pre-service Student through Causality and Analytic Thinking] *Disertasi Doktor pada Pendidikan IPA* (Universitas Pendidikan Indonesia: tidak diterbitkan)

[22] Rokhmat J 2013b Kemampuan Proses Berpikir Kausalitas dan Berpikir Analitik Mahasiswa Calon Guru Fisika *Jurnal Pengajaran MIPA. 78-86*

[23] Creswell J W 2012 *Educational research: Planning, conducting and evaluating quantitative and qualitative research* (Boston: Pearson) p. 300-306.

[24] Pebriyanti D, Sahidu H and Sutrio S 2015 Efektivitas model pembelajaran Perubahan Konseptual untuk Mengatasi Miskonsesi Fisika pada siswa kelas X SMAN 1 Praya Barat Tahun Pelajaran 2012/2013 *Jurnal pendidikan Fisika dan Tekhnologi, **1**(2), 92-96*

[25] Anshori, Rokhmat, Gunada, Wahyudi and Ayub 2021 The Effect of A Causalitic-Learning Model on Problem-Solving Abilities Reviewed From Creativity *Journal of Physics: Conference series,* **1816**(012041), p. 1-7

[26] Harwati K and Rokhmat J 2021 Development of Student Worksheet to Improve Creative and Critical Thinking Ability of Students in Causalitic-learning Model *Journal of Physics: Conference series,* **1816**(012038), p. 1-9

[27] Rezeki H, Rokhmat J, Gunawan, Makhrus M and Wahyudi 2021 Implementation of Causalitic-learning Devices to Improve Creative Thinking Ability and Problem-solving of Students In Physics *Journal of Physics: Conference series,* **1816**(012048), p. 1-7

[28] Warodiah Y N, Rokhmat J, Zuhdi M, Kosim and Ayub S 2021 The Strategy of Student-Worksheet Design with A Causalitic-learning Model to Improve Creative-Thinking Ability *Journal of Physics: Conference series,* **1816**(012047), p. 1-12

[29] Nurjamilah, Rokhmat J, Sahidu H, Harjono A and Hikmawati 2021 Causalitic-learning Model to Improve Reasoning-Ability in Learning Physics in Terms of Student Creativity During The Covid-19 Pandemic *Journal of Physics: Conference series,* **1816**(012039), p. 1-9

[30] Ramirez, Hazel Joyce M 2021 Facilitating Computer-Supported Collaborative Learning with Question Asking Scripting Activity and its Effects on Students' Conceptual Understanding and Critical Thinking in Science *International Journal of Innovation in Science and Mathematics Education, Vol. 29(1), p. 31-45*

Journal of Physics: Conference Series **2165** (2022) 012053 doi:10.1088/1742-6596/2165/1/012053

The dielectric properties of skin damage and its correlation to free radical intensity caused by UVA/UVB radiation impact

M Faisal, U P Juswono and D R Santoso*

Physics Department, Brawijaya University, Jl. Veteran Malang, East Java 65145

E-mail: dieks@ub.ac.id

Abstract. The dielectric property in form of dielectric constant of material is a measure of material's response to a given electric field. Electric fields induce conductive and polarizing currents based on their effect on free charge attached to biological material. This phenomenon occurs at all levels of organization including cellular, molecular and ionic. When the skin is exposed to ultraviolet radiation there will be a series of chain processes that cause emergence of Reactive Oxygen Species (ROS) in the tissue. Reactive Oxygen Species (ROS) in tissue will be very dangerous because it can cause damaging free radicals. Excessive exposure to UVA / UVB radiation causes various kinds of damage which affect the low dipole moment and polarity of cells that make up skin tissue. Dipole moment and low cell polarity will decrease the dielectric constant value of tissue which is triggered by presence of free radicals in form of singlet oxygen ($^1O_2$) and superoxide anion (O_2^-*).

1. Introduction

The dielectric property in form of dielectric constant of material is a measure of material's response to a given electric field. Electric fields induce conductive and polarizing currents based on their effect on the free charge attached to biological material. This phenomenon occurs at all levels of organization including cellular, molecular and ionic [1]. The dielectric constant indicates polarization rate of molecule and density of electrostatic flux when given an electric potential in a biological or non-biological material [2].

Ultraviolet rays are a form of radiation. Ultraviolet light is classified as a type of electromagnetic wave which has a short wavelength and can ionize matter [3]. Skin is largest and widest organ covering entire the body [4]. Thick, thin and skin color of each person is different depending on various factors such as genetics, race, age, and gender [5]. Microscopically, the skin layer is divided into three parts starting from outermost is epidermis, underneath is dermis, and at bottom there is hypodermis or also called superficial fascia [6].

When skin is exposed to ultraviolet radiation there will be a series of chain processes that cause emergence of Reactive Oxygen Species (ROS) in tissue. Reactive Oxygen Species (ROS) in the body will be very dangerous because it can cause damaging free radicals [7]. This research will identify the correlation between dielectric properties and cell damage caused by free radicals triggered by UVA / UVB radiation.

2. Material and Methods

2.1. Preparation and treatments of mice

This research was started by trimming the back hair of mice and giving them acclimatization for 7 days so that the mice could adapt to environment. After acclimatization process was complete, mice were divided into six treatment groups, consisting of K0: mice that were not exposed to UV, K1: mice that were given UV exposure for 30 minutes, K2: mice that were given UV exposure for 60 minutes, K3: mice that were given UV exposure for 90 minutes, K4: mice given UV exposure for 120 minutes, and K5: mice given UV exposure for 150 minutes. UV radiation treatment is carried out for 21 days in a special black chamber as shown in figure 1 with a size of 40 cm x 20 cm x 20 cm which aims to get intense UV exposure. After exposure process was finished, the mice were dissected and skin tissue was taken.

Figure 1. Schematic of irradiation chamber

2.2. Measurement of free radicals and dielectric constant

Free radical measurements were carried out using Leybold-Heracus Electron Spin Resonance (ESR) instrument which worked at frequency of 13 MHz - 130 MHz with maximum current of 2 mA. The amount of free radical content will be indicated by the height of lissajous curve while type of free radical will be indicated by g-factor value. Furthermore, for measurement of dielectric constant ($\boldsymbol{\varepsilon_r}$) used Electrical Impedance Spectroscopy (EIS) unit which works at a frequency of 100 Hz - 100 kHz to get capacitance value (***C***) of material which is then put into the equation

$$\boldsymbol{\varepsilon_r} = \frac{C.d}{\varepsilon_0.A} \quad (1)$$

The cross-sectional area of plates (***A***) and the distance between plates (***d***) have been previously measured, while air permittivity value ($\boldsymbol{\varepsilon_0}$ = 8.85x10^{-12} F / m) refers to a reference.

3. Result And Discussion

The results measurement amount of free radical intensity contained in the mice skin samples due to exposure to ultraviolet radiation, it can be seen in Figure 2 which is shown a lissajous curve or a visualization graph that displays a combination of two sinusoidal waves with different phases in X and Y modes. ESR shows the intensity of free radicals contained in each treatment group. The more concave and steeper the curve shown, the more free radicals contained in the sample being measured. From the graph produced by ESR, it can be seen that control treatment that has been tracking at each frequency does not show any lissajous curve curvature that occurs, while for radiation exposure treatment it is found that the curve shape changes to be symmetrical like the letter V.

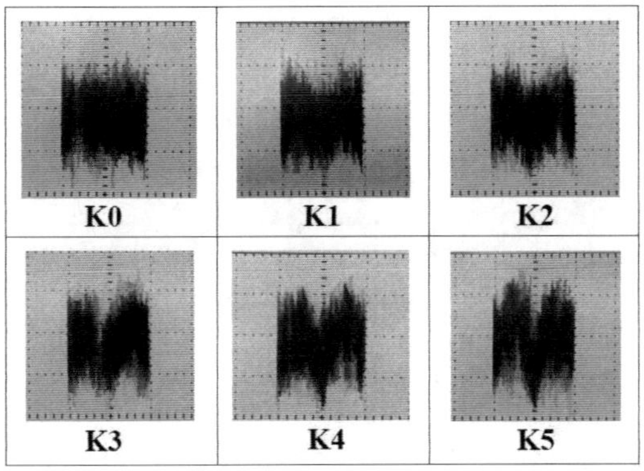

Figure 2. Graph of the results of measuring the intensity of free radicals in each treatment.

The curve shown becomes increasingly concave and steeper when measuring the skin samples of mice that receive longer radiation exposure. The comparison of the curvatures in each treatment can then be displayed in a graph to compare the number of free radical intensities contained in the skin tissue as shown in figure 3, where the highest concave is shown by the 150 minute exposure treatment of 90.64 ± 3.45 mm^2, whereas the lowest concavity is indicated by exposure to 30 minutes or as large 8.82 ± 1.8 mm^2. From the results of the comparison of observations of free radical intensity shown by the lissajous curve for each treatment, it can be seen that the longer the ultraviolet exposure given to the mice's skin will trigger greater oxidative stress and cause more free radicals in the skin tissue.

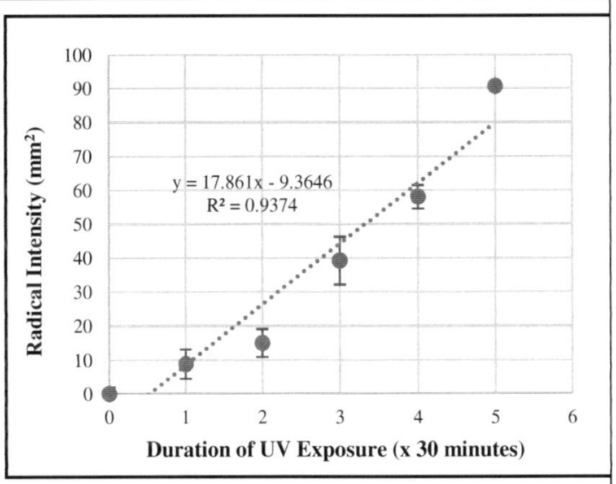

Figure 3. Graph comparison of free radical intensity in each treatment.

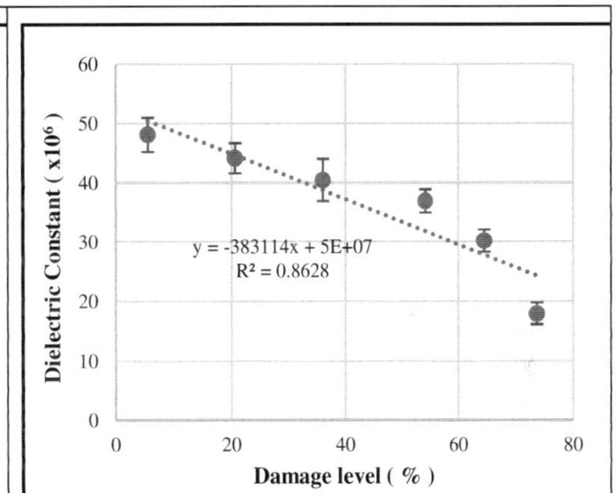

Figure 4. Graph correlation between the level of tissue damage and the dielectric constant value.

Determination types of free radicals contained in skin tissue due to exposure ultraviolet radiation can be known by looking at g-factor of each sample tested, as shown in Table 1.

Table 1. Results of identification of free radicals in mice skin samples

Treatment	Frequency (Hz)	I (A)	Lande factor	Lande factor (Alegre *et al*, 2006)	Radical types
K0	-	0.225	-	-	-
K1	19900000	0.225	1.5012	1.501	$^1O_2$
K2	19900000	0.225	1.5012	1.501	$^1O_2$
	19900000	0.224	1.5079	1.501-1.75	$O_2^{-\bullet}$
K3	19900000	0.224	1.5079	1.501-1.75	$O_2^{-\bullet}$
	20100000	0.225	1.5163	1.501-1.75	$O_2^{-\bullet}$
K4	20500000	0.225	1.5465	1.501-1.75	$O_2^{-\bullet}$
	20100000	0.225	1.5163	1.501-1.75	$O_2^{-\bullet}$
K5	19900000	0.224	1.5079	1.501-1.75	$O_2^{-\bullet}$
	19900000	0.225	1.5012	1.501	$^1O_2$
	20500000	0.225	1.5465	1.501-1.75	$O_2^{-\bullet}$
	20100000	0.225	1.5163	1.501-1.75	$O_2^{-\bullet}$

From the results of research that has been done, it was found there were two types of free radicals in the skin tissue of mice that were exposed to ultraviolet exposure. Two types of free radicals detected using ESR unit are singlet oxygen ($^1O_2$) and superoxide anion ($O_2^{-}*$). Both types of radicals are formed due to oxidative stress on skin tissue which gives rise to Reactive Oxygen Species (ROS) molecules. Reactive Oxygen Species (ROS) are short-lived molecules capable of causing oxidative damage to DNA, cellular proteins, and lipids. ROS molecules can cause permanent genetic changes and activate signal transduction pathways that promote skin carcinogenesis [8]. One of constituent components of skin which is very sensitive to oxidative stress is melanocytes because it has pro-oxidant properties of melanin [9]. The presence of ultraviolet exposure on skin means that energetic photons are transmitted through the skin layer and are absorbed through the chromophore cell molecules or photosensitizer, causing chemical-biological effects. Ultraviolet reacts with photosensitizers or chromophores on skin, such as cytochromes, riboflavin, heme and porphyrins. Chromophore absorbs energy from ultraviolet wavelengths [10].

Singlet oxygen ($^1O_2$) is oxygen which has one unpaired electron in its outer orbit and has a greater energy level, so it can form more reactive oxygen. Singlet oxygen ($^1O_2$) has two options, namely either transferring this energy to the surrounding organic material or continuing to form more reactive oxygen species. The mechanism for the formation of singlet oxygen can occur through enzyme reactions, chemical reactions and also sensitisers who are exposed to ultraviolet exposure, but the main thing is to use photosensitizers that are exposed to ultraviolet [11], as shown in figure 5.

Photosensitizer is a substance that can absorb light energy and transfer the excess energy to triplet oxygen to form singlet oxygen and then react directly with electron-rich compounds. When photosensitizer absorbs ultraviolet radiation energy and visible light rapidly becomes unstable and excited molecules. On the basis of the singlet sensitizer (^1XS) can be an excited singlet sensitizer ($^1XS^*$). In an excited singlet sensitiser ($^1XS^*$) can return to the baseline sensitiser (^1XS) through fluorescent emission or turning into a triplet sensitiser ($^3XS^*$) through a cross between systems [11].

Excited triplet sensitizer ($^3XS^*$) can lose a large amount of its energy through phosphorus energy or by reaction with triplet oxygen. The singlet oxygen is then formed through the energy transfer reaction from ($^3XS^*$) to triplet oxygen via triplet-triplet interaction mechanism. Life time of the excited triplet sensitizer ($^3XS^*$) longer than the excited singlet sensitizer ($^1XS^*$). Furthermore, to return to being a sensitiser in a basic state (^1XS) can start the cycle again to produce singlet oxygen. A sensitiser molecule can produce 10^3 until 10^5 singlet oxygen molecules before they become inactive or degraded by photobleaching by singlet oxygen or by other processes [12].

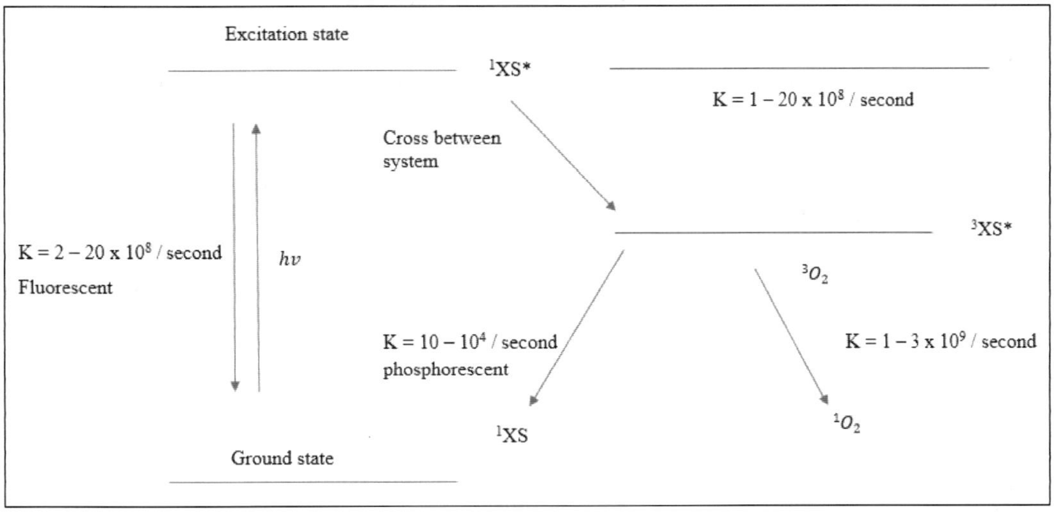

Figure 5. Mechanism of singlet oxygen ($^1O_2$) radical formation.

The second type of free radical is superoxide anion ($O_2^{-}*$). Superoxide anion radical ($O_2^{-}*$) formation because Reactive Oxygen Species (ROS) occurs in the tissue due to addition of one electron to the oxygen molecule. As we know oxygen is needed to produce ATP which is a source of energy through the oxidative phosphorylation process that occurs in mitochondria [7]. In this process there is a reduction O_2 become H_2O or simply can be written as equation 1

$$O_2 + 4H^+ + 4e^- \rightarrow 2H_2O \quad (2)$$

when the oxygen compound in the tissue has 2 unpaired electrons in different orbitals and has parallel spins, to form a covalent bond 2 electrons must be located in one orbital with opposite spins. This causes oxygen to gradually accept only one electron. Since the reduction of oxygen in the oxidative phosphorylation process requires the transfer of four electrons, this transfer cannot occur at once, but the electron transfer occurs gradually [13]. If there is oxidative stress, the transfer becomes imperfect which eventually forms superoxide anion ($O_2^{-}*$) compounds in the tissue as shown in figure 6

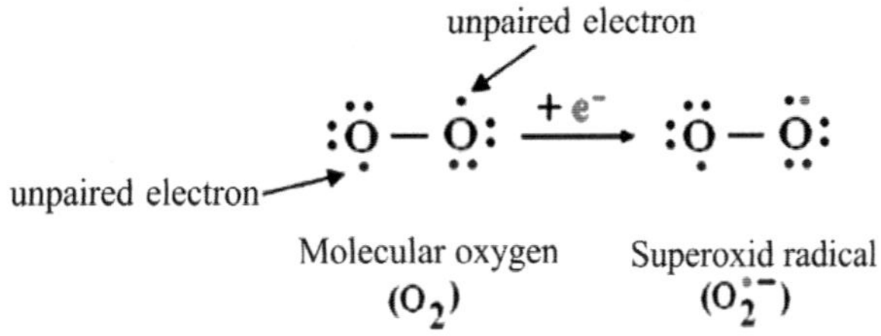

Figure 6. Mechanism of superoxide anion radical formation ($O_2^{-}*$).

When free radicals interact with compounds or molecules in the body, it will cause a series of chain processes that cause the emergence of new free radicals with an amount that cannot be tolerated by the body. Free radicals will attract electrons from surrounding biological macromolecules such as polyunsaturated fatty acids (PUFA) which contain at least three double bonds, protein, and deoxyribonucleic acid (DNA). Polyunsaturated fatty acids (PUFA) are important components of phospholipids in making up cell membranes [14]. DNA plays an important role as a genetic component of cells, while proteins have important functions such as enzymes, receptors, antibodies, and the cytoskeleton. The uncontrolled formation of ROS in the body will cause cellular damage to lipids, protein carbohydrates, and irreversible DNA structures. The body has the ability to defend itself from damage caused by free radicals through enzymes in the body to keep the number of free radicals at a low level, but when there is oxidative stress and cell dysfunction, free radicals will continue to increase rapidly and cause significant cell damage in body [15]. So that free radicals contribute in causing the pathogenesis of inflammatory diseases, erythema, premature aging, and genetic mutations that occur in skin tissue due to ultraviolet exposure. Environmental pollutants such as polycyclic aromatic hydrocarbons can be converted into quinone-mediated ROS. An in vitro and in vivo study shows polycyclic aromatic hydrocarbons and benzoapyrene work as photosensitizers together with UV-A exposure to synergistically produce singlet oxygen ($^1O_2$) and superoxide anion (O_2^-*) [9].

The dielectric constant is one of the specific properties possessed by each material. The dielectric constant describes the ability of a material to be polarized and store energy. Microscopically, the dielectric constant is also defined as the mobility or uniformity of the electric dipole due to an external electric field disturbance [2]. From the results of the research, the graph of relationship between the dielectric constant value of the skin tissue and the level of damage due to ultraviolet exposure is shown in figure 4, control treatment has the highest dielectric constant value, that is (48,05 ± 2,88) 10^6 with the lowest accumulated damage to the tissue of 5,54 %, Then for ultraviolet exposure treatment the dielectric constant value continues to decrease until the lowest is 150 minute exposure with the value (17,9 ± 1,83) 10^6 and has the highest accumulated damage of all treatments, which is equal to 73,70 %. The size of the dielectric constant value possessed by the skin tissue is influenced by three things, namely the number of healthy cells, the external electric field, and the dipole moment [16]. When the skin tissue is placed on two parallel plates with a charge, it will experience polarization. The cells that make up skin tissue are originally polar molecules and the ions in them are randomly positioned, which can then orient themselves toward the field. If more and more skin constituent cells such as melanocytes, keratinocytes, and mast cells are still healthy, the more dipole moments will be formed [2].

When the voltage source frequency is changed, the dielectric material in the capacitor will change the direction of the electric dipole to become more uniform. The response of the atoms making up biological materials will be easier to polar when conditions are normal and there is no cell damage. According to the Maxwell-Wagner theory, membranes are analogous to the parallel arrangement of capacitors and resistors in an electrical circuit [17]. The cell membrane has a very vital function in the cell work system. The cell membrane functions as a barrier and controller when a solution can enter and leave the cell. Membrane-separated solutions are a type of electrolyte that can conduct electric current based on the ability of the molecule to break down into ions. The ability of a molecule to break down into ions is determined by the degree of dissociation of a compound. A compound with a high degree of dissociation indicates that the compound is more conductor [1].

If the cells that make up the tissue suffer a lot of damage, the normal number of membranes will decrease, which causes the capacitance value in tissue to also decrease [18]. This explains that in the control treatment the condition of the skin cells is still healthy and the dipole moment and degree of polarity of the tissue are still very high. In contrast to the treatment of mice that received ultraviolet exposure, the number of healthy cells in the tissue continued to decrease as the length of time of exposure was given which made the dipole moment and polarity lower. The total dipole moment value of a material will be influenced by the dipole moment of each of the atoms making up the material, the external field strength, and the number of atoms [17]. Three electrical parameters that are important in the interaction between the electric field and the dipole moment of the dielectric are permittivity, conductivity and permeability. Permitivity ε (F/m) shows the number of polarizations and the partial rectification of a permanent electric dipole over a number of electric fields. Conductivity σ(S/m) shows the number of conduction current densities that occur due to a number of electric fields. The rectification of a permanent magnetic dipole is measured by permeability μ (H/m) [1].

4. Conclusion

From this research, UVA / UVB radiation has caused various kinds of damage which affect low dipole moment and polarity of cells that make up skin tissue. The low dipole moment and polarity of cell will affect the decrease in dielectric constant value of material. The damage that occurs can be triggered by the presence of Reactive Oxygen Species (ROS) which generates free radicals in form of singlet oxygen ($^1O_2$) and superoxide anion (O_2^{-*}). The longer the ultraviolet exposure given will increase free radical content and dielectric properties of material change.

Acknowledgments

This research is fully funded by the "Hibah Penelitian Guru Besar dan Doktor FMIPA-UB" with contract number 1585/UN10.F09/PN/2021.

References

[1] Tadapatri P Hiremath U S Yelamaggad C V and Krishnamurthy K S, 2010 Permittivity, conductivity, elasticity, and viscosity measurements in the nematic phase of a bent-core liquid crystal *J. Phys. Chem. B* **114**, 5 p. 1745–1750.
[2] Mayrovitz H N McClymont A and Pandya N, 2013 Skin tissue water assessed via tissue dielectric constant measurements in persons with and without diabetes mellitus *Diabetes Technol. Ther.* **15**, 1 p. 60–65.
[3] Lee J H Woo K Y Kim K H Kim H-D and Kim T G, 2013 ITO/Ag/ITO multilayer-based transparent conductive electrodes for ultraviolet light-emitting diodes *Opt. Lett.* **38**, 23 p. 5055–5058.
[4] Venus M Waterman J and McNab I, 2010 Basic physiology of the skin *Surg.* **28**, 10 p. 469–472.
[5] Khavkin J and Ellis D A F, 2011 Aging skin: histology, physiology, and pathology *Facial Plast. Surg. Clin.* **19**, 2 p. 229–234.
[6] Kolarsick P A J Kolarsick M A and Goodwin C, 2011 Anatomy and physiology of the skin *J. Dermatol. Nurses. Assoc.* **3**, 4 p. 203–213.
[7] Bolisetty S and Jaimes E A, 2013 Mitochondria and reactive oxygen species: physiology and pathophysiology *Int. J. Mol. Sci.* **14**, 3 p. 6306–6344.
[8] Ziech D *et al.*, 2010 The role of reactive oxygen species and oxidative stress in environmental carcinogenesis and biomarker development *Chem. Biol. Interact.* **188**, 2 p. 334–339.
[9] Bito T and Nishigori C, 2012 Impact of reactive oxygen species on keratinocyte signaling pathways *J. Dermatol. Sci.* **68**, 1 p. 3–8.
[10] Sherratt M J Bayley C P Reilly S M Gibbs N K Griffiths C E M and Watson R E B, 2010 Low-dose ultraviolet radiation selectively degrades chromophore-rich extracellular matrix components *J. Pathol.* **222**, 1 p. 32–40.
[11] Zheng Q Jockusch S Zhou Z and Blanchard S C, 2014 The contribution of reactive oxygen species to the photobleaching of organic fluorophores *Photochem. Photobiol.* **90**, 2 p. 448–454.
[12] Mosinger J *et al.*, 2010 Fluorescent polyurethane nanofabrics: a source of singlet oxygen and oxygen sensing *Langmuir* **26**, 12 p. 10050–10056.
[13] Brown G C and Borutaite V, 2012 There is no evidence that mitochondria are the main source of reactive oxygen species in mammalian cells *Mitochondrion* **12**, 1 p. 1–4.
[14] Schuchardt J P Huss M Stauss-Grabo M and Hahn A, 2010 Significance of long-chain polyunsaturated fatty acids (PUFAs) for the development and behaviour of children *Eur. J. Pediatr.* **169**, 2 p. 149–164.
[15] Azzam E I Jay-Gerin J-P and Pain D, 2012 Ionizing radiation-induced metabolic oxidative stress and prolonged cell injury *Cancer Lett.* **327**, 1–2 p. 48–60.

[16] Heileman K Daoud J and Tabrizian M, 2013 Dielectric spectroscopy as a viable biosensing tool for cell and tissue characterization and analysis *Biosens. Bioelectron.* **49** p. 348–359.

[17] Khan M F Rashid R B Rahman M M Al Faruk M RAHMAN M D M and RASHID M A, 2017 Effects of solvent polarity on solvation free energy, dipole moment, polarizability, hyperpolarizability and molecular reactivity of aspirin *Int. J. Pharm. Pharm. Sci* **9**, 2 p. 217–221.

[18] Shahzad A Khan S Jones M Dwyer R M and O'Halloran M, 2017 Investigation of the effect of dehydration on tissue dielectric properties in ex vivo measurements *Biomed. Phys. Eng. Express* **3**, 4 p. 45001.

AUTHOR INDEX